EARLIER VOLUMES IN THIS SERIES

Modifications of
Passive Films

*European Symposium on **Modifications of Passive Films**, Paris, France, 15–17 February 1993.*

European Federation of Corrosion
Publications

NUMBER 12

Modifications of
Passive Films

Papers presented at the European Symposium on
Modifications of Passive Films

Paris, France, 15–17 February 1993

Edited by

P. MARCUS, B. BAROUX AND M. KEDDAM

*Published for the European Federation of Corrosion
by The Institute of Materials*

THE INSTITUTE OF MATERIALS
1994

Book Number 577
Published in 1994 by The Institute of Materials
1 Carlton House Terrace, London SW1Y 5DB

British Library Cataloguing in Publication Data
Available on request

Library of Congress Cataloging in Publication Data
Available on application

ISBN 0-901716-52-9

Design and production by
PicA Publishing Services, Drayton, Nr Abingdon, Oxon

Made and printed in Great Britain

Contents

I *Chemical and Electronic Properties of Passive Films*

II Kinetics of Passivation and Electrochemical Properties of Passive Films

III Surface Treatments and Film Modifications

IV Passivity Breakdown

European Federation of Corrosion Publications
Series Introduction

The EFC, incorporated in Belgium, was founded in 1955 with the purpose of promoting European co-operation in the fields of research into corrosion and corrosion prevention.

Membership is based upon participation by corrosion societies and committees in technical Working Parties. Member societies appoint delegates to Working Parties, whose membership is expanded by personal corresponding membership.

The activities of the Working Parties cover corrosion topics associated with inhibition, education, reinforcement in concrete, microbial effects, hot gases and combustion products, environment sensitive fracture, marine environments, surface science, physico–chemical methods of measurement, the nuclear industry, computer based information systems, corrosion in the oil and gas industry, and coatings. Working Parties on other topics are established as required.

The Working Parties function in various ways, e.g. by preparing reports, organising symposia, conducting intensive courses and producing instructional material, including films. The activities of the Working Parties are co-ordinated, through a Science and Technology Advisory Committee, by the Scientific Secretary.

The administration of the EFC is handled by three Secretariats: DECHEMA e. V. in Germany, the Société de Chimie Industrielle in France, and The Institute of Materials in the United Kingdom. These three Secretariats meet at the Board of Administrators of the EFC. There is an annual General Assembly at which delegates from all member societies meet to determine and approve EFC policy. News of EFC activities, forthcoming conferences, courses etc. is published in a range of accredited corrosion and certain other journals throughout Europe. More detailed descriptions of activities are given in a Newsletter prepared by the Scientific Secretary.

The output of the EFC takes various forms. Papers on particular topics, for example, reviews or results of experimental work, may be published in scientific and technical journals in one or more countries in Europe. Conference proceedings are often published by the organisation responsible for the conference.

In 1987 the, then, Institute of Metals was appointed as the official EFC publisher. Although the arrangement is non-exclusive and other routes for publication are still available, it is expected that the Working Parties of the EFC will use The Institute of Materials for publication of reports, proceedings etc. wherever possible.

The name of The Institute of Metals was changed to The Institute of Materials with effect from 1 January 1992.

A. D. Mercer
EFC Scientific Secretary,
The Institute of Materials, London, UK

EFC Secretariats are located at:

Dr J A Catterall
European Federation of Corrosion, The Institute of Materials, 1 Carlton House Terrace, London, SW1Y 5DB, UK

Mr R Mas
Fédération Européene de la Corrosion, Société de Chimie Industrielle, 28 rue Saint-Dominique, F-75007 Paris, FRANCE

Professor Dr G Kreysa
Europäische Föderation Korrosion, DECHEMA e. V., Theodor-Heuss-Allee 25, P.O.B. 150104, D-6000 Frankfurt M 15, GERMANY

Preface

The publishing of this Proceedings Volume is the last event of the European Symposium on Modifications of Passive Films which was held in Paris, 15–17 February 1993*. Seventy-three communications (oral presentations and posters) were presented by participants from over fifteen countries. Fifty-eight manuscripts are published in this volume, providing a relevant and updated view of the advances and new trends in the field.

Improving the corrosion resistance of metallic materials by modifications of the passive films is a challenging objective which requires that substantial research efforts are made for a better knowledge of the chemical composition, the chemical states and the electronic properties of passive films. In addition to the set of surface analysis techniques, the promising emergence of STM and AFM probes in the study of passive films with atomic resolution was probably one of the highlights of the Symposium.

Kinetics of passivation and electrochemical behaviour remain very basic to our knowledge of passivity. These topics were illustrated by a more limited number of papers as the emphasis was on the modifications of the passive films.

A wide spectrum of film modifications and their effects on passivity and resistance to localised corrosion are described. Some of them are based on changes in the metal composition by traditional matrix alloying or are produced by more recent techniques using directed energy beams such as ion implantation. The focus of a large number of papers was on various film processings by chemical, electrochemical or physical means. They brought a large body of original data illustrating the role of chemical and structural features.

A precise assessment of the true potential of film modifications for further improvement of the corrosion resistance of passive materials is not yet possible. However there is no doubt that the data reported at this Symposium have contributed to identify the field as relevant in both fundamentals of passivity and development of more corrosion-resistant materials.

Paris, 26 October 1993
The Editors,
P. Marcus, B. Baroux and M. Keddam

*Financial support by CNRS and UGINE is acknowledged.

I

Chemical and Electronic Properties of Passive Films

Papers presented at the European Symposium on Modifications of Passive Films, organised by the Working Party on Surface Science and Mechanisms of Corrosion and Protection, held in Paris, France, 15–17 February 1993.

Influence of Ternary Elements on the Chloride Composition of Passive Films on Iron–Chromium Alloys

C. Hubschmid, H. J. Mathieu and D. Landolt

Laboratoire de métallurgie chimique, Département des matériaux, Ecole Polytechnique Fédérale de Lausanne (EPFL), 1015 Lausanne/Switzerland

Abstract

The presence of electrolyte anions in passive films formed under different conditions on Fe–25Cr and Fe–25Cr–X alloys (X = Mo, Si, V) was investigated by AES depth profiling and by XPS. The results indicate that chloride was incorporated into passive films provided this ion was present in the electrolyte during film formation. On the other hand, exposure of a preformed film to a chloride containing electrolyte, well below the pitting potential, did not lead to chloride incorporation. Addition of Mo, V or Si to the alloy enhanced the pitting resistance but did not significantly change the extent of chloride incorporation during film growth.

1. Introduction

It is generally accepted that anions from the electrolyte are incorporated into passive films during passivation [1]. Recent results obtained on an Fe25Cr alloy in acid solutions with varying chloride content showed that chloride incorporation during film formation depends on the chloride concentration of the passivation solution. Exposure of the passivated alloy to a chloride solution indicated that the pit initiation time decreases with increasing chloride content of the passive film [2].

In order to better understand the mechanism by which some alloying elements enhance the pitting potential of iron-chromium alloys, passive films on Fe–Cr–X alloys (X = Mo, V, Si) were formed in acid sulphate solutions in presence and absence of chloride. The films were analysed by AES and XPS in order to study the influence of ternary alloying elements on the chloride content of the passive films.

2. Experimental

The composition of the high purity alloys (Materials Research Toulouse) used in this study is given in Table 1. All electrochemical experiments were performed in deaerated solutions with a rotating disk anode at a rotation rate of 3000 rpm. This ensured good mixing of the electrolyte and minimised concentration gradients at the electrode. A platinum wire counter electrode and mercury sulphate reference electrode were employed. A Schlumberger model 1286 potentiostat with an Amel programmable function generator was used as current source.

Two types of disk electrodes were employed. For electrochemical experiments, the disks of 0.1 cm^2 were cast in epoxy. For samples involving surface analysis, no resin was employed. In this case, only the flat end of the electrode was in contact with the solution. Prior to polarisation, the surface was mechanically polished, ultrasonically cleaned, washed with absolute alcohol and dried in an Ar stream. Before polarisation, a cathodic prepolarisation of 60 s at –650 mV (SHE) was applied. After polarisation, the electrodes were removed from the electrolyte under anodic potential, rinsed with doubly distilled water and dried with Ar. They were then introduced through a fast entry lock into the ultra-high vacuum chamber of the surface analysis system.

Passivation was carried out at a constant potential of 500mV (SHE) for 60 min in solutions A and B (Table 2) respectively. In a third type of experiment passivation was carried out in absence of chloride (solution A, 1 h), then a concentrated chloride solution was added to bring the concentration to that of solution B and the anodic polarisation was continued for 30 min. This type of experiment, which allows to study possible chloride incorporation after passive film formation, is termed solution A+ in Table 2.

AES analysis was performed with a PHI 660 SAM system with a primary voltage of 5kV and a beam current of 100nA. For depth profiling, a 2 kV Ar$^+$ beam

Table 1 Composition of Fe–Cr and Fe–Cr–X alloys

ALLOY	Fe %w	Cr %w	X %w	Fe %at	Cr %at	X %at
Fe25Cr	76.6	23.1	-	75.5	24.5	-
Fe25Cr6V	69.2	24.8	6.0	67.6	26.0	6.4
Fe25Cr11Mo	63.9	25.1	11.0	65.7	27.7	6.6
Fe25Cr3Si	72.0	25.1	2.8	68.8	25.8	5.4
Fe31Cr	71.4	28.6	-	75.5	30.1	-

Table 2 Composition of electrolytes

SOLUTION	H_2SO_4 [M]	Na_2SO_4 [M]	NaCl [M]
A	0.1	0.4	0
B	0.1	0.4	0.12
C	0.1	0.4	0.5
A+	.1	0.4	0 (passivation) 0.12 (exposition)

was scanned over a 2×2 mm. Calibration of the sputter rate was done with respect to Ta_2O_5. The depth scale indicated in the figures of this paper is based on the assumption that the sputter rate of the passive films is 0.8 times that of tantalum oxide [3]. XPS analysis was carried out with a PHI model 5500 instrument, equipped with a 400W Al Kα source. More details on the experimental procedures can be found elsewhere [4].

3. Results and Discussion

Figure 1 shows chloride AES depth profiles measured on passive films on Fe25Cr polarised in solutions A, B and C respectively. As expected for solution A the chloride signal is below the noise level. From the data for solutions B and C it follows that the quantity of chloride incorporated into the film increases with the chloride concentration of the passivating solutions. Figure 2 represents the current evolution with time after immersion in a 1M NaCl solution at 25°C and 0 mV (SHE) of the prepassivated samples. The current

evolution and the induction time for pitting depend on the chloride content of the passivating solution. The data indicate that the chloride incorporation during film formation enhanced the pitting sensitivity of a passivated Fe25Cr alloy.

To compare the pitting resistance of different alloys polarisation curves were determined in a 0.1M HCl + 0.9M NaCl solution at 25°C. The results of Fig. 3 show that addition of 6 at.% of Mo, V, Si and Cr to Fe25Cr enhanced the pitting resistance. The behaviour is in agreement with previous observations [5].

AES depth profiles were measured on passive films prepared in solutions B and A+ respectively. Chloride was detected in significant amounts in the films only for those prepared in solution B. This is illustrated by the data of Fig. 4 obtained for the Fe–Cr–Si alloy. A similar behaviour was observed for all alloys studied. Figure 5(a) shows the maximum chloride concentration determined for films formed on different alloys in solution B. The problem was further studied using XPS. Figure 5 (b) shows results for the chloride concentration in the passive films formed on

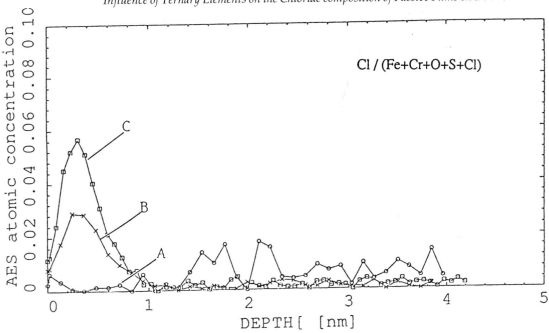

Fig. 1 AES chloride depth profiles measured on a Fe25Cr alloy after passivation of 60 min at 500mV (SHE) in solution A, B and C.

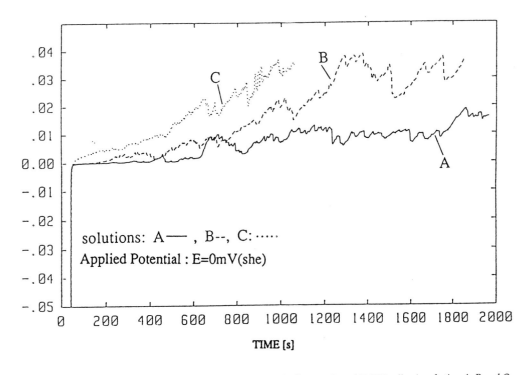

Fig. 2 Current evolution after immersion in a 1M NaCl solution at 0mV(SHE) of prepassivated Fe25Cr alloy in solution A, B and C.

different alloys. The data represent an average concentration derived from three measurements performed at three emission angles (because of an unfavourable signal to noise ration no attempt is made here to interpret the angle dependence of the data). The sequence of the relative chloride concentrations of the films for the different alloys is not the same as that determined by AES (Fig. 5(a)). This indicates that the observed differences are due to data scattering. Indeed, data scattering was relatively large because of the small chloride concentrations involved.

On the other hand, the data of Fig. 4 clearly show

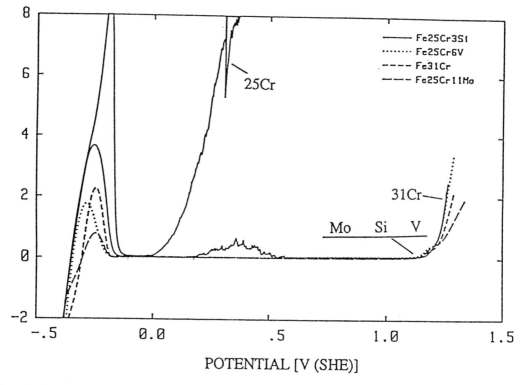

Fig. 3 Potentiodynamic polarisation curves of Fe–Cr–X alloys in a HCl 0.1M + NaCl 0.9M solution.

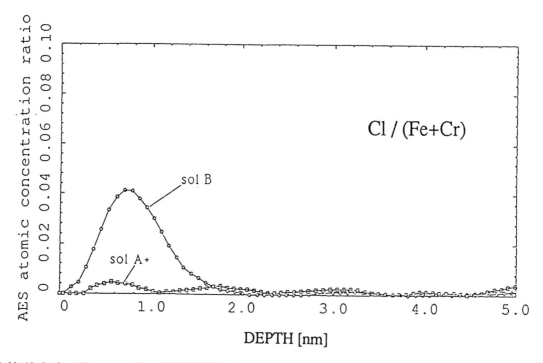

Fig. 4 AES chloride depth profiles measured on a Fe25Cr3Si alloy after passivation of 60 min at 500mV (SHE) in solution B and A+.

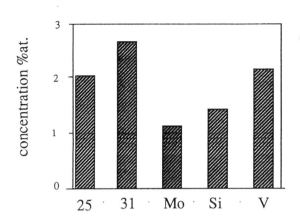

Fig. 5(a) Maximum of the AES chloride concentration depth profiles of Fe–Cr–X alloys passivated at 500 mV (SHE) in solution B.

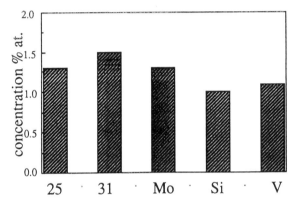

Fig. 5(b) XPS chloride concentration of passive films formed on Fe–Cr–X alloys at 500mV (SHE) in solution B.

that the chloride found in passive films formed in solution B is not a surface contamination or a precipitate salt film remaining after rinsing. This follows from a comparison of the results for solutions A+ and B, the

experimental conditions with respect to solution composition during exposure and to sample removal and rinsing being the same in the two cases. The data, therefore, conclusively show that during film formation in a chloride containing electrolyte these ions are incorporated. Although not shown here, sulphate was also present in the passive films in all cases [4]. The results of Figs 5(a) and (b) indicate that the addition of a ternary alloying element does not significantly influence the quantity of chloride incorporated into the passive film during its formation in solution B. The increased pitting resistance conferred to these alloys by the ternary alloying element is apparently due to other reasons, not yet well understood.

4. Conclusion

Chloride ions are incorporated into passive films formed on Fe–Cr alloys in chloride containing electrolytes. The breakdown resistance of such films was found to be lower than that of those not containing chloride.

Addition of Mo, Si, V enhanced the pitting resistance of Fe–5Cr, although the presence of the ternary alloying elements did not significantly modify the observed chloride incorporation during film formation.

References

1. Z. Szklarsksa-Smialowska, Pitting Corrosion of Metals, Chap. 2, NACE, Houston, 1986.
2. C. Hubschmid and D. Landolt, J. Electrochem. Soc., **140**, 1898, 1993.
3. R. L. Tapping, R. D. Davidson and T. E. Jackmann, Surf. Interf. Anal., **7**, 105, 1985.
4. C. Hubschmid, Ph.D. No. 1125 thesis EPFL, 1993.
5. R. Goetz, J. Laurent and D. Landolt, Corros. Sci., **25**, 1115, 1985.

Composition of Passive Films on Stainless Steels Formed in Neutral Sulphate Solutions

B. ELSENER, D. DEFILIPPO* AND A. ROSSI*

Institute of Materials Chemistry and Corrosion, Swiss Federal Institute of Technology, ETH Hönggerberg, CH-8093 Zurich, Switzerland
*Dipartimento di Chimica e Tecnologie Inorganiche e Metallorganiche, Università di Cagliari, Via Ospedale 72,1-09124 Cagliari, Italy

Abstract

The passivation of stainless steels (1.4301 with 18% Cr and 8% Ni and 1.4529 with 20% Cr, 25 % Ni and 6% Mo) was studied in neutral 1M Na_2SO_4 solutions by electrochemistry and XPS. The integral chromium Cr^{3+} content revealed in the passive films of the two stainless steels after 1 h of passivation is similar and it decreases with increasing passivation potential. The passive film thickness increases with the applied polarisation potential as well as the amount of oxygen bound to cations (MO). On the 1.4529 SS the film thickness is smaller and molybdenum in the Mo^{6+} and Mo^{4+} state are detected (mean concentration ca. 10% of the cations in the film). A low content of Ni^{2+} with respect to the bulk alloys is revealed in the passive films. A marked enrichment of Ni and Mo is observed in the metallic phase at the interface passive film / alloy independent of the passivation potential.

1. Introduction

Passive films of crystalline austenitic stainless steels have been studied mainly in acidic solutions. Generally a marked enrichment of chromium in the passive films and an increase of the film thickness with potential is found [1, 2]. The enrichment of molybdenum or the presence of nickel in the passive films is still under discussion. Only few papers refer to the behaviour in neutral solutions [3, 4]. Recent *in situ* studies using ellipsometry and potentialmodulated UV visible reflection spectroscopy on high purity Fe–Cr alloys and Fe–Cr–Ni alloys [5] show that in neutral sulphate solutions (pH 6) the cationic mass fraction of Cr^{3+} ions decreases gradually with increasing potential. The thickness of the films increased with increasing potential and at a given potential it decreased with increasing Cr content in the film and with decreasing pH.

The aim of this surface analytical study was to investigate the thickness and the chemical composition of the passive films formed on stainless steels 1.4301 and 1.4529 at different polarisation potentials in neutral solutions in order to establish a correlation with the very different corrosion resistance of the two materials [6].

2. Experimental

The composition of the two stainless steels is given in Table 1. The solutions were prepared from analytical grade $Na_2SO_4 \cdot 10H_2O$ and bidistilled water. They were stored in a glass reservoir of 2L volume and deaerated with argon gas for at least 4 h prior the experiment. After mechanical polishing to 1μm diamond paste the samples—already fixed on the XPS sample holder—were mounted on the electrochemical cell (150 mL) and the deaerated electrolyte was filled in under argon pressure. The samples were hold 15 min at the open circuit potential, then polarised in one step to the passivation potential and hold there for one hour, removed from the cell under applied potential, rinsed with bidistilled water, dried with nitrogen and transferred under a nitrogen stream to the spectrometer immediately.

XPS analyses were performed on an ESCALAB MkII spectrometer (Vacuum Generator Ltd., UK) using a MgKα (1253.6 eV) X-ray source run at 20 mA and 15 kV. The residual pressure in the spectrometer during data acquisition was always lower than 5.10^{-7} Pa. The unexposed surface of the sample was masked with a gold ring to allow the analysis of only the polarised surface. The spectra were obtained in the digital mode with a constant energy of 20 eV set on the electron analyser. The details of operation and calibration of the instrument are given in ref. [7]. To compensate for sample charging during the analysis all the binding energies were referred to the carbon 1s signal at 284.75 eV.

Spectra of pure iron, chromium, nickel and molybdenum after prolonged sputtering as well as iron oxide thermally grown in oven for 10 min at 400°C,

Table 1 Compositions of the stainless steels (mass%)

Material	Fe	Cr	Ni	Mo	Mn	S	P	other
1.4301	bal	18.1	8.7	0.06	1.3	0.003	0.029	Si 0.58
1.4529	bal	20.8	24.9	6.42	0.82	0.002	0.016	Si 0.34

chromium oxide thermally grown 1 h at 600°C, nickel oxide thermally grown at 800°C and molybdenum oxide were taken as XPS references. The model and the quantitative evaluation used in this study are given in the Appendix.

3. Results

The open circuit potential of all the samples tested decreased with time of exposure to the deaerated solutions. After 15 min a value of -400 ± 20 mV was reached for the 1.4301 steel at pH 5.8. For the 1.4529 steel values of -280 ± 20 mV were found. The current during one hour potentiostatic polarisation at all potentials from -0.2 to 0.6 V SCE decreased with time according to the expected power law. A steady state current density after one hour of polarisation is nearly reached only for the passivation potential $+0.6$ V SCE, while at all the other potentials the current continues to decrease. Prolonged potentiostatic polarisation may produce changes in the average composition of the passive films [4]. However the present results refer to one hour polarisation.

The calibration data for the metal elements and their oxides and for the alloys after ion etching are given in Table 2. Figure 1 shows the detailed XPS spectra of 1.4529 stainless steel after one hour polarisation at $+0.2$ V SCE in 1 M Na_2SO_4 solution (pH 5.8) after background subtraction [9]. The spectra of all the alloy constituents show contributions from the oxidised states in the passive film and the metallic state immediately beneath the film. The Mo3d spectra show contributions from two oxidised species, identified as Mo (VI) and a lower oxidation state, in addition to the metallic signal from the substrate. The Ni2p region shows the typical satellite structure for both the metallic and oxidised nickel. No attempt have been made at this stage to distinguish nickel and chromium oxide signals from their hydroxides. The O1s signal is not a single peak, the curve fitting always shows three components attributed to M-O-M (530 eV), M-OH and adsorbed H_2O.

With increasing passivation potential the binding energy (Fig. 2) of oxidised iron increases from 710.1 eV to 710.8 eV reflecting an increasing Fe^{3+} content [12]. The Cr^{3+} signal tends toward lower binding energies reaching 576.8 eV at $+0.6$ V SCE and this behaviour may be attributed to the presence of a higher chro-

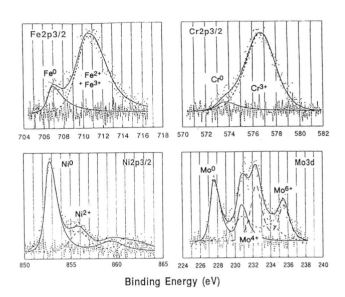

Fig. 1 Detailed XPS spectra of 1.4529 SS polarised for 1 h at $+0.2$ V SCE in 1 M Na_2SO_4 solution (pH 5.8) after background subtraction and curve fitting [7, 9].

mium (III) hydroxide content in the passive film at lower potentials as it can also be observed from the MO/O_{tot} ratio (Fig. 3). The binding energy of Ni^{2+} remains constant at 855.8 eV. In the 1.4529 SS with molybdenum, two oxidised species were detected: Mo^{6+} at a constant binding energy of 232.4 ± 0.2 eV and Mo^{4+} [11] with a binding energy increasing from 230.5 to 231.6 eV.

The *thickness of the passive film* after one hour of polarisation calculated according to the three layer model (ref. [7] and appendix) increased linearly with the passivation potential (Fig. 4). On the high alloyed stainless steel 1.4529 the passive film thickness is lower. The composition of the passive film vs the applied potential is shown in Fig. 5. The integral Cr^{3+} content in the film is nearly the same for the two stainless steels and decreases with higher potential values whereas the oxidised iron content increases. Only ca. 3–5% of Ni^{2+} is detected on the 1.4301 SS, increasing to about 10% on the 1.4529 stainless steel. About 10% of molybdenum is incorporated into the passive film on the 1.4529 regardless of the passivation potential.

In agreement with the results obtained in acidic solutions [8], the substrate immediately beneath the passive film is strongly enriched in metallic nickel for

Table 2 Binding energies (E_B) of the different chemical states

Sample	Chemical State	$E_B \pm 0.1$ (eV)
Iron (ion etched)	Fe° (2p3/2)	706.9
Chromium (ion etched)	Cr° (2p3/2)	574.0
Nickel (ion etched) sat	Ni° (2p3/2)	852.7 859.2
Molybdenum (ion etched)	Mo° (3d5/2)	227.5
1.4301 (ion etched)	Fe° (2p3/2) Cr° (2p3/2) Ni° (2p3/2)	706.7 574.0 852.8
1.4529 (ion etched)	Fe° (2p3/2) Cr° (2p3/2) Ni° (2p3/2) Mo° (3d5/2)	706.8 574.0 852.8 227.5
Oxidized Iron	Feox (2p3/2)	709.9
Oxidized Chromium (Cr_2O_3)	Cr^{3+} (2p3/2)	576.8
Oxidized Nickel (NiO) sat 1 sat 2	Ni^{2+} (2p3/2)	854.8 856.0 861.5
$Ni(OH)_2$ sat 1 sat 2	Ni^{2+} (2p3/2)	856.2 857.2 862.5
Oxidized Molybdenum (MoO_3)	Mo^{6+} (3d5/2)	232.5

both alloys (about twice the bulk content) and depleted in iron and to a minor extent in chromium. Molybdenum in the 1.4529 is present in the same concentration as in the passive film (*ca.*10%) but enriched compared to the bulk composition.

4. Discussion

Despite a very similar chromium content in the alloy the conventional 1.4301 and the high alloyed 1.4529 stainless steel show huge differences in electrochemical behaviour and pitting resistance in practical conditions [6]. The XPS investigations of this work have shown that oxidised nickel is depleted in the passive

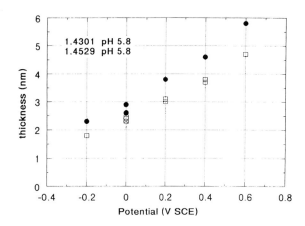

Fig. 4 Thickness of the passive films on the 1.4529 and 1.4301 SS formed after 1 h passivation in 1M Na$_2$SO$_4$ solution (pH 5.8) as a function of the potential.

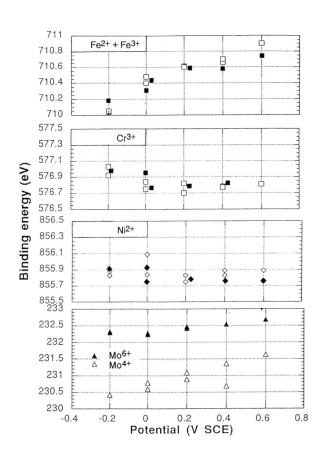

Fig. 2 Binding energies of the oxidised metals in the passive films of the 1.4529 (open symbols) and 1.4301 SS (closed symbols) as a function of the potential (1 h passivation in 1 M Na$_2$SO$_4$ solution at pH 5.8).

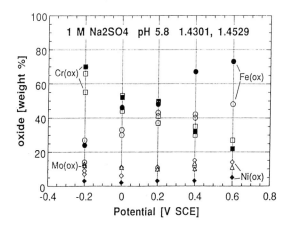

Fig. 5 Composition of the passive films on the 1.4529 and 1.4301 SS formed after 1 h passivation in 1M Na$_2$SO$_4$ solution (pH 5.8) as a function of the potential. Open symbols: 1.4529 SS, closed symbols 1.4301 SS.

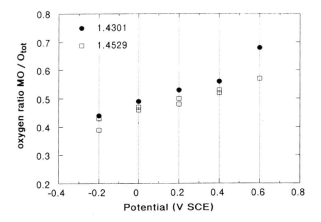

Fig. 3 Change in the ratio of oxygen bonded to metals (MO) to total oxygen in the passive films on 1.4529 and 1.4301 SS formed after 1 h passivation in 1M Na$_2$SO$_4$ solution (pH 5.8) with the potential.

films with respect to the bulk composition, oxidised molybdenum instead is incorporated at a level of *ca.* 10%. It is interesting to note that the integral Cr (III) oxy-hydroxide content is similar for both stainless steels, it decreases from 60 to 70% at –0.2 V SCE to only 25% at +0.6 V. These results are in good agreement with *in situ* results on the alloy Fe–20Cr–25Ni [5]. The major difference is thus the content of oxidised iron: on the 1.4301 the value of 50% Fe(ox) is reached at *ca.* 0.2 V SCE, at potentials > 0.4 V SCE *ca.*70% are found. On the high alloyed 1.4529, the content of Fe(ox) remains below 50% even at the most positive potentials of +0.6 V SCE. This lower oxidised iron content together with a smaller film thickness (Fig. 4) and a

9

higher hydroxide M-OH content (Fig. 3) in the film may be responsible for the more corrosion resistant passive film on the 1.4529 SS. In addition defects created by Fe^{2+} are compensated by Mo^{n+} present in the passive film.

The passive films in neutral media show concentration gradients that are time dependent [4]. The quantitative evaluation of this work considers the passive film to be homogeneous, thus integral values of the concentrations that comprise contributions from all the depths in the passive film are given. The presence of gradients in the film may imply that the 'true' concentration in the outermost layers are different from the integral values.

In agreement with other XPS studies on passive films of austenitic stainless steels a marked enrichment of nickel underneath the passive film is found in neutral solutions. This results in an iron content of only 30% in the outermost layers at the metal/film interface of the 1.4529 whereas *ca.* 60–65% Fe are found for the 1.4301 steel. This difference might be another reason for the elevated resistance to localised corrosion of the 1.4529 SS: the high content of nickel and the presence of molybdenum cause a strong shift of the anodic polarisation curve towards more noble potentials and reduces the (local) dissolution current when pits are initiated. Furthermore transition metals are known to be able to form intermetallic bonds with high stability, especially Ni-Mo [10]. This again inhibits the localised dissolution during film breakdown.

5. Conclusions

1. The integral Cr (III) oxy-hydroxide content as a function of the potential is similar on both steels, with increasing potential the integral Cr (III) content in the passive films diminishes.
2. The content of oxidised iron in the passive film of the 1.4529 SS is much lower due to the incorporation of oxidised nickel and molybdenum into the passive film. This might be a possible reason for the higher resistance to localised corrosion of this steel.
3. The thickness of the passive film on stainless steels in neutral solutions after one hour of passivation increases linearly with the applied potential, it is smaller on the 1.4529 SS than on 1.4301 SS.
4. A strong nickel enrichment (about a factor of two with respect to the bulk composition) is found in the alloy underneath the passive film of both stainless steels.

6. Acknowledgements

The work was financially supported by the Italian National Research Council (Comitato Tecnologico and Progetto Finalizzato di Chimica Fine e Secondaria).

References

1. H. Fischmeister and U. Roll, Fresenius Z. Anal. Chem., **319**, 639–645, 1984, and references therein.
2. K. Asami and K. Hashimoto, Langmuir, **3**, 897–904, 1987.
3. K. Sugimoto and S. Matsuda, Mater. Sci. Eng., **42**, 181, 1980.
4. W. P. Yang, D. Costa and P. Marcus, in 'Oxide Films on Metal and Alloys', B. R. MacDougall, R. S. Alwitt and T. A. Ramanarayanan (eds), Proc. Vol. **92–22** p. 516–529, The Electrochemical Society, Pennington NJ (1992) and references quoted therein.
5. N. Hara and K. Sugimoto, J. Electrochem. Soc., **138**, 1594, 1991.
6. H. Haselmair, R. Morach and H. Bohni, High alloyed Steels and Nickel Alloys for Fastenings in Road Tunnels, accepted by *Corrosion*, NACE.
7. A. Rossi and B. Elsener, Surf. Int. Anal., **18**, 499, 1992.
8. J. H. Scofield, J. Electron Spectroscopy Relat. Phenom., **8**, 129, 1976.
9. P. M. A. Sherwood, 'Practical Surface Analysis', ed. Briggs and M. P. Seah, Appendix 3, p.445, J. Wiley NY, 1983.
10. I. Olefjord and C. R. Clayton, ISIJ International, **31**, 134, 1991.
11. I. Olefjord and B. Brox, Passivity of Metals and Semiconductors, Ed. M. Froment, Elsevier Science Publishers, Amsterdam, 1983.
12. E. DeVito and P. Marcus, Surface Interface Analysis, **19**, 403, 1992.
13. M. P. Seah and W. A. Dench, Surface Interface Analysis, **1**, 1, 1979.
14. R. F. Reilman, A. Msezane and S. T. Manson, J. Electron Spectroscopy Relat. Phenom., **8**, 389, 1976.

Appendix—Quantitative Analysis

The quantitative evaluation of the XPS-data was based on the integrated intensities (I), i.e. peak areas in counts x $eV.s^{-1}$ obtained from the original spectra after non-linear background subtraction [9] and curve fitting, using the same parameters evaluated on XPS standards recorded in the same analysis conditions: the binding energies of pure metals, after ion etching and their oxides are given in Table 2.

The measured intensities of the components of the metal and of the oxide phase were used to evaluate the thickness of the surface layer and the composition of the interface alloy/passive film applying a three-layer model (contamination, oxide layer, substrate) under the assumption that each layer is homogeneous without concentration gradients. It follows that electrons originating from the metal substrate (taken as semi infinite) are attenuated exponentially by the oxide layer and the contamination layer, whereas the electrons originating from the oxide layer (with limited thickness t) are attenuated only by the contamination. The basic equations derived from this model are:

$$I_i^{OX} = [(g_i \cdot \sigma_i^{OX} \cdot C_i^{OX} \cdot \rho_i^{OX} \cdot \Lambda_i^{OX})/A_i] \cdot \{1 - \exp(-t/\Lambda_i^{OX})\} \cdot \exp(-1_C/\Lambda_i^{con}) \tag{1}$$

$$I_j^{m} = [(g_j \cdot \sigma_j^{m} \cdot C_j^{m} \cdot \rho^m \cdot \Lambda_j)/A_j] \cdot \exp(-t/\Lambda_j^{OX}) \exp(-1_C/\Lambda_j^{con}) \tag{2}$$

where

A atomic weight

g instrumental factor (proportional to $E_{kin}^{-0.5}$)

t passive layer thickness

l_C contamination layer thickness

ρ density (g cm^{-3})

$g_{Fe} = 0.0428$, $g_{Cr} = 0.0384$, $g_{Ni} = 0.0499$, $g_{Mo} = 0.031$, $g_O = 0.0372$, $g_C = 0.0321$,

$\sigma_{Fe} = \sigma_{Fe}^{OX} = 9.4249$, $\sigma_{Cr} = \sigma_{Cr}^{OX} = 6.8066$, $\sigma_{Ni} = \sigma_{Ni}^{OX} = 12.5664$, $\sigma_{Mo} = \sigma_{Mo}^{OX} = 8.90$, $\sigma_O = 2.4425$, $\sigma_C = 0.857$,

$\Lambda_{Fe}^{OX} = 2.24$, $\Lambda_{Cr}^{OX} = 2.50$, $\Lambda_{Ni}^{OX} = 1.92$, $\Lambda_{Mo}^{OX} = 3.07$, $\Lambda_{Fe}^{m} = 1.26$, $\Lambda_{Cr}m = 1.41$, $\Lambda_{Ni}^{m} = 1.08$, $\Lambda_{Mo}^{m} = 1.73$, $\Lambda_{Fe}^{Cont} = 2.03$, $\Lambda_{Cr}^{Cont} = 2.3$, $\Lambda_{Ni}^{Cont} = 1.74$, $\Lambda_{Mo}^{Cont} = 2.78$, $\Lambda_{C}^{Cont} = 2.71$, $\Lambda_{C}^{Cont} = 2.578$

The photoionization cross-section values $\{\sigma, \eta, \omega, \gamma, \iota\}$ were taken from [8] and corrected for the angular asymmetry function L(γ) with $\gamma = 49.1°$ for our instrument. It is assumed that the photoionization cross sections are the same in the different phase [metal, oxide, hydroxide]. For the calculation of the function L(γ), describing the intensity distribution of the photoelectrons ejected by unpolarized X-rays from atoms and molecules, the β values tabulated by Reilman and co-workers [13] were used. The mean free-path values $\Lambda(E_{kin})$ of the electrons were calculated as $\Lambda_i = B\sqrt{E_{kin}}$ with B = 0.054 for elements in their formal oxidation state zero; B = 0.096 for the oxidised forms; B = 0.087 for taking into account the attenuation due to the contamination layer [14]. The emission angle Θ of the detected electrons with respect to the normal to the sample surface was 0 degrees.

It has to be pointed out that the attenuation terms in the above equations contain the mean free path Λ_j for the oxide whereas the pre-exponential terms of Λ_j depend on the origin of the electrons (metal or oxide). The alloy density ρ_m was 7.93 g cm^{-3}. Considering the densities of anhydrous oxides (*ca.* 5.5 g cm^{-3}) and the presence of hydroxides and water in the passive films, a value for ρ^{ox} = 5 cm^{-3} seems to be reasonable.

The system of eqns (1) and (2), written for each species present in the surface film and in the substrate, was put in a parametric form for the unknowns t and l_C:

$$f_1(t\, l_C) = [(\rho^{OX}/\rho^m)\Sigma I_j^m k_j^m \exp(t/\Lambda_j^{OX}) \exp(l_C/\Lambda_j^{CON})] - \Sigma^{OX} \tag{3}$$

$$f_2(t, l_C) = [l^{CON} K_C \rho^{OX}/(1 - \exp(-l_c/\Lambda^{CON})] - \Sigma^{OX} \tag{4}$$

where $k_j^m = g_O\sigma_O\Lambda_O A_j/g_j\sigma_j\Lambda_j^m A_O$

$k_j^{OX} = g_O\sigma_O\Lambda_O A_j/g_j\sigma_C\Lambda_j^{OX} A_O$

$K_C = g_O\sigma_O\Lambda_O A_j/g_C\sigma_C\Lambda_C^{OX} A_O$

$\Sigma_{OX} = \Sigma Ij^{OX}K_j^{OX}\exp(-lC/\Lambda_C^{con})/(1 - \exp(-t/\Lambda_j^{OX}))$

(the subscript O indicating oxygen).

Values for t and Ic were obtained solving the parametric equations f_1 and f_2 using first 3D parametric plots. Numerical methods based on versions of Newton's method were used to find numerical approximations to the solutions of the equations. The composition of the oxide film and the metal layer under the film (substrate) were calculated simultaneously.

3

Characterisation of Surface Films on FeCrPC Alloys by XPS and X-ray Excited Auger Peaks

A. ROSSI AND B. ELSENER*

Dipartimento di Chimica e Tecnologie Inorganiche e Metallorganiche, Università di Cagliari, Via Ospedale 72, 1-09124 Cagliari, Italy
*Institute of Materials Chemistry and Corrosion, Swiss Federal Institute of Technology, ETH Hönggerberg, CH-8093 Zurich, Switzerland

Abstract

XPS spectra of phosphorus P2p and X-ray induced Auger transition PKLL have been measured on a series of reference compounds and on different P-bearing amorphous alloys in the as received state, after mechanical polishing and after anodic passivation. In both, the P2p and PKLL spectra, three chemical states could be detected after curve fitting based on standards. From both spectra the same quantitative informations are deduced. The data (P2p binding energy and X-ray induced Auger energy PKLL) are presented in a two dimensional chemical state plot (PKLL vs P2p). The three phosphorus species were identified as (i) P from the bulk alloy (129.4 eV and 1858.5 eV) showing a strong charge transfer from the metallic atoms, and (ii) phosphate P^{+5} in the passive film (133.4 eV and 1851.7 eV). The third 'intermediate' P signal (131.7 eV and 1855.6 eV) showed a similar Auger parameter as black and red P thus its chemical state might be attributed to elemental phosphorus. Angle resolved XPS measurements showed that this elemental phosphorus is located at the interface passive film/bulk alloy. The results are discussed with respect to the role of this elemental phosphorus in the high corrosion resistance of phosphorus bearing amorphous alloys.

1. Introduction

Previous studies of Fe–Cr amorphous alloys have shown that phosphorus is the most beneficial metalloid in improving their corrosion resistance [1–3]. Many investigations were carried out to determine the electrochemical and corrosion behaviour [4–7], clarifying the role of the alloy structure and composition and evaluating by means of XPS and/or AES the composition and the thickness of passive films as a function of the applied potential. The change in the alloy beneath the films has also been studied [8–17].

The P2p photoelectron spectra of different P-containing alloys exhibited a multicomponent character. Curve fitting routines allowed to reveal commonly three different P species: one at 133.3 eV attributed to phosphates (passive film), another at 129.4 eV assigned to phosphorus of the bulk alloy [18] and a third one at 131.7 eV which generally appears on mechanically polished and polarised samples. The identification of this component is still under discussion: some authors based on XPS results [11–14] and on AES studies [9,10] have suggested that it may be an intermediate oxidation state of phosphorus in the passive film.

The oxidation state is usually determined on the basis of chemical shifts of strong lines in XPS and AES.

However, a single line position might not give unambiguous informations on the oxidation state. It has therefore been suggested that the ambiguity in identification might be reduced considerably by using two dimensional 'chemical state plots' based on the Auger kinetic energy and the binding energy of the prominent photoelectron line [19–25].

In this work phosphorus P2p and PKLL spectra were measured on a series of reference compounds and on different P-bearing amosphous alloys before and after mechanical polishing and/or anodic passivation. Curves were fitted to determine the different P-species on XPS and X-ray excited Auger peaks. The data are presented in a two dimensional chemical state plot that allows to identify the chemical state of all the three P-species revealed.

2. Experimental

The phosphorus reference compounds analysed in this study were supplied by Aldrich. Most of the samples were powders and were dusted onto double-adhesive tape. During the analysis the sample probe was cooled with liquid nitrogen. The amorphous alloys studied were obtained by rapid quenching under helium. A detailed description of the production and characterisation is given in [15]. The following samples were analysed: (i) after sputtering and (ii) mechanically polished to 1μm and polarised at different passivation potentials in 1N HCl. After the electrochemical preparation [6,15] the samples were rinsed with bidistilled water, dried in a stream of nitrogen and transferred to the spectrometer.

The measurements were performed on an ESCALAB MkII spectrometer (Vacuum Generator Ltd., UK). The vacuum system consists of a turbomolecular pump, fitted with a liquid nitrogen trap, and a titanium sublimation pump. The residual pressure in the spectrometer during the data acquisition was always lower than $5 \cdot 10^{-7}$ Pa. The X-ray source was a non monochromatic Al Kα (1486.6 eV) twin anode run at 20 mA and 15 kV. This allows to measure the P-Auger lines with higher kinetic energies using the Bremsstrahlung. The spectra were obtained in the digital mode (VGS 1000 software on Apple IIe). The electron analyser was operated in Fixed Analyser Transmission (FAT) mode with a pass energy of 20 eV (FWHM Ag $3d_{5/2}$ = 1.1 eV), the PKLL lines were registered with a pass energy of 50 eV.

The instrument was calibrated according to [26]. The line position and the area of each peak were obtained by curve fitting the experimental spectra with gaussian/lorentzian functions after non-linear background subtraction according to [27, 28]. The P2p signal was fitted taking into account the spin orbit splitting ($2p_{3/2}, 2p_{1/2}$) of the single peaks. The gaussian/lorentzian ratio and the FWHM were hold constant and the energy of the peaks and their height were fitted using a least square algorithm.

3. Results

Figure 1 shows the P2p spectra and the PKLL Auger lines from one of the examined standards, Na3PO4, from a sputtered FeCrPC sample and from an anodic polarised amorphous alloy specimen after mechanical polishing. The original spectra (dotted line) after background subtraction, and the results of curve fitting (solid lines) are given in the same figure. The P2p and PKLL spectra of the amorphous alloys were fitted by using the parameters obtained from the standards. The sputtered alloy always exhibited a single peak at 129.4 ± 0.05 eV and a PKLL at 1858.5 eV. The spectra obtained on samples with airformed layer (as received) as well as on their surfaces after anodic polarisation show two peaks: one at 129.4 ± 0.03 eV and at 133.43 ± 0.18 eV with the corresponding PKLL signals at 1858.5 and at 1851.7 eV respectively. After mechanical polishing a third component appears which is characterised by the following values: 131.74 ± 0.06 eV and 1855.57 ± 0.25 eV for the P2p and PKLL signals respectively. The line widths for the individual peaks are

within the range obtained for pure compounds. The chemical state plot for phosphorus shown in Fig. 2. includes measurements from red and black phosphorus, phosphorus oxygen compounds and phosphorus from sputtered amorphous alloys. Lines with the same Auger parameter (AP) have +1 slope and the Auger parameter increases towards the upper left-hand corner. It is possible to identify three regions on the Wagner chemical state plot of P: one is located in the lower left-hand corner of the plot and corresponds to samples with highly electronegative ligands such as phosphates. The second region is in the upper right-hand corner where the P signals from the sputtered metallic glasses and from the bulk alloy occur. The third region is in the central part of the plot and the triangles represent data obtained from intermediate peak which appears on mechanically polished and polarised samples. This region has about the same AP as those found for P_{red} and P_{black}. The Auger parameter is calculated according to: α = KE (KLL) + BE (2p). Angle resolved depth profiles indicates that with respect to P5+ (133.4 eV) this species is located in the inner part of the film, at the interface with the bulk alloy.

4. Discussion

It is claimed that freshly mechanically polished amorphous alloys exhibit a P spectra which is composed of more than two P signals in different chemical states [9–11]. The results of curve fitting (Fig. 1) reveal three P peaks whose binding energies ($P2p_{3/2}$) and kinetic energies (PKLL) have been reproduced in several independent measurements. Based on the chemical shifts of the $P2p_{3/2}$ two of these peaks have been assigned to phosphates (133.4 ± 0.2 eV) and to phosphorus of the bulk alloy (129.4 eV) [18]. As far as the third species is concerned (131.7 eV) it was assumed to correspond to P bound in the +1 valence state in an unidentified compound located in the inner part of the passive film of the amorphous alloy [9, 10, 29]. This species actually exhibits a binding energy higher that of elemental P (130.4 eV) and lower than standards derived from P+1 compounds (NaH_2PO_2 at 132.2 eV) [13]. The analysis of the shape and position of the AES peaks allowed to assign the phosphate and the bulk alloy components whereas peak assignment on the basis of comparison with the AES spectrum of hypophosphite should be treated as preliminary [10].

In Fig. 2 the results obtained for the amorphous alloy (solid symbols) are distributed in three regions of the chemical state plot: the group having mean values at 133.4 eV ($P2p_{3/2}$) and 1851.7 eV (PKLL) corresponds to phosphates. The corresponding Auger parameter evaluated under the assumption that there is a fixed

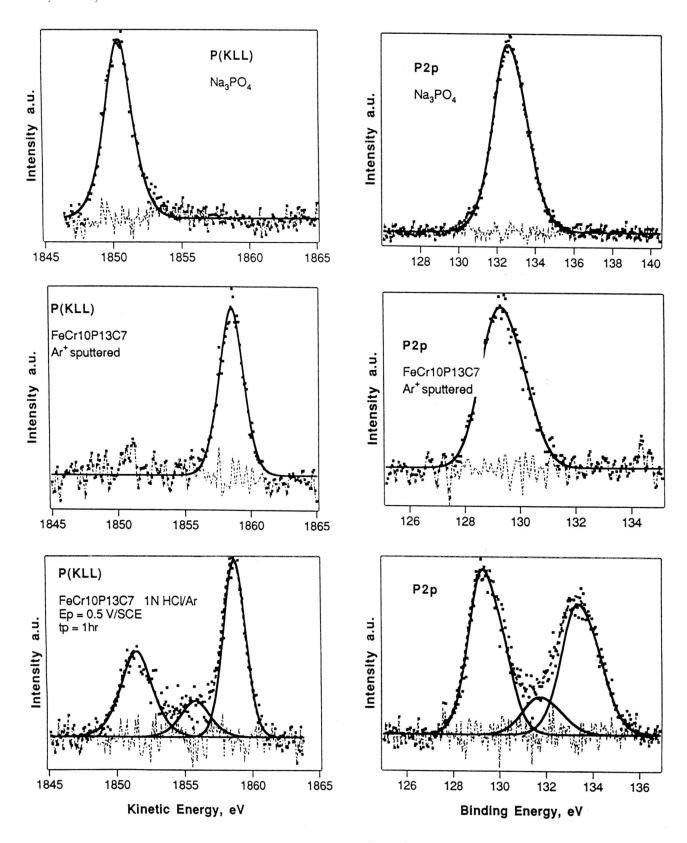

Fig. 1 Curve fitting of the detailed XPS and X-ray induced Auger spectra after background subtraction of phosphorus from Na₃PO₄ (top), sputtered FeCrPC alloy (centre) and the FeCrPC amorphous alloy, mech. polished and polarised for 1 h at +0.5 V SCE in 1 M HCl solution (bottom).

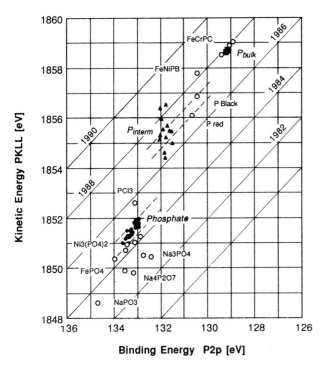

Fig. 2 Wagner chemical state plot for phosphorus showing P_{bulk} (■). P_{interm} (▲) and phosphates (●) from the passivated FeCrPC samples and standards (○) from this work and literature data.

difference between two energy lines of the same element in the same sample is 1985.1 ± 0.15, close to that evaluated for transition metal phosphates. The signals at 129.4 eV/1858.5 eV correspond to P in the bulk alloy. Phosphorus is apparently affected by the presence of iron and chromium. A series of sputtered samples is in fact found in the same region of the plot and has a similar Auger parameter. Measurements reported in the literature carried out with an Al/Al-Ag anode allowed to evaluate the atomic term of the static relaxation energy which was found to be very high for the FeNiPB alloy respect to P_{red} and P_{black} indicating the metallic properties of this sample [21]. The presence of negatively charged P similar to that of phosphides was first hypothesised by Asami *et al.* [18] for NiFeP amorphous alloys. They suggested a charge transfer from metallic atoms to P observing that the binding energies of the metallic components were both higher than those of iron and nickel metals. A binding energy of 129.3 eV for Cr_3P was also reported for reference [18]. The intermediate P (P2p$_{3/2}$ 131.74 eV, PKLL 1855.57 eV) occupies the region in the chemical state plot nearest to elemental phosphorus. This may suggest that elemental phosphorus is a potential candidate for the unknown compound. The possibility still exists that a chemical compound not yet recorded on the plot

may have these co-ordinates. The attribution to elemental P is consistent with the findings that its binding energy values remain constant with the applied passivation potential [15,16] and it is located in the inner part of the film.

5. Conclusions

The measurements of phosphorus P2p$_{3/2}$ and PKLL in the amorphous alloys after different pretreatments such as sputtering, mechanical polishing and anodic polarisation were compared with those of reference compounds using the Wagner chemical state plot and calculating the Auger parameters. These tools are a 'fingerprinting' but they may enable an empirical chemical state identification of P in different chemical environments. On the basis of the data here presented the following conclusions can be drawn:

- the curve fitting procedure of the X-ray induced PKLL spectrum recorded on mechanically polished samples gives three components as the P2p$_{3/2}$ signal;

- the X-ray induced Auger signal (PKLL) provides the same quantitative information as the P2p$_{3/2}$ spectrum;

- Phosphorus in the passive film is present as phosphate in the potential range examined;

- P in the bulk is strongly affected by the charge transfer from metallic atoms;

- The position of the third P signal in the chemical state plot and its Auger parameter suggest that it may be elemental phosphorus.

6. Acknowledgements

The Authors wish to thank the Institute of Physics at the University of Basel for the production of the amorphous alloys. The work was financially supported by the Italian National Research Council (Comitato Tecnologico and Progetto Finalizzato di Chimica Fine e Secondaria) and the Swiss National Research Foundation (NFP 19).

References

1. R. B. Diegle, N. R. Sorensen, T. Tsuru and R. M. Latanision, in Treatise on Materials Science and Technology, Vol. **23**, 1983, p. 63, Ed. J. Scully, Academic Press London.

2. K. Hashimoto, in Amorphous Metallic Alloys, T. E. Luborsky ed. Butterworth London, 1983, pp. 471–486.

3. P. C. Searson, P. V. Nagarkar and R. M. Latanision, in Modern Aspects of Electrochemistry, No. 21, Eds. R. E. White, J. O. bockris and B. E. Conway, Plenum Press New York, 1990, p. 121–160.

4. M. Naka, K. Hashimoto and T. Masumoto, Corrosion, **32**, 1976, 146.
5. B. Elsener, S. Virtanen and H. Bohni, Electrochimica Acta, **32**, 1987, 927.
6. S. Virtanen, B. Elsener and H. Bohni, J. Less-common Metals, **145**, 1988, 581.
7. B. Elsener, S. Virtanen and H. Bohni, Z. Nauk, AGH 'Metalurgia i Odlewnictwo', **16**, 63–72, 1990, and references therein.
8. D. R. Baer and M. T. Thomas, J. Vac. Sci. Technol., **18**, 722, 1981.
9. G. T. Burstein, Corrosion NACE, **37**, 549, 1981.
10. M. Janik-Czachor, ISIJ International, **31**, 149, 1991.
11. C. R. Clayton, M. A. Helfand, R. B. Diegle and N. R. Sorensen, Proc. Symp. on Corrosion, Electrochemistry and Catalysis of Metallic Glasses, R. B. Diegle and K. Hashimoto eds. Electrochem. Soc. Proc. Vol. **88-1**, p. 134, 1988.
12. T. P. Moffat, R. M. Latanision and R. R. Ruf, idem p. 25.
13. R. B. Diegle, N. R. Sorensen, C. R. Clayton, M. A. Helfand and Y. Lu, J. Electrochem. Soc., **135**, 1085, 1988.
14. D. De Filippo, A. Rossi, B. Elsener, S. Virtanen, Surf. Interface Analysis, **15**, 668, 1990.
15. A. Rossi and B. Elsener, Surface Interface Analysis, **18**, 499, 1992.
16. B. Elsener and A. Rossi, Electrochimica Acta, **37**, 2269, 1992.
17. B. Elsener and A. Rossi, Metallurgy and Foundry Engineering, **18**, 175, 1992.
18. K. Asami, H. M. Kimura, K. Hashimoto, T. Masumoto, A. Yokoyama, H. Komiyama and H. Inoue, J. non-cryst. Solids, **64**, 149, 1984.
19. C. D. Wagner and A. Joshi, J. of Electron Spectr. Rel. Phen., **47**, 1991, 283 and references quoted therein.
20. R. H. West and J. E. Castle, Surface Interface Analysis, **4**, 68, 1982.
21. R. Franke, Th. Chasse, P. Streubel and A. Meisel, J. of Electron Spectr. Rel. Phen., **56**, 381, 1991.
22. R. Franke, Th. Chasse, P. Streubel and A. Meisel, J. of Electron Spectr. Rel. Phen., **57**, 1, 1991.
23. L. S. Dake, D.R. Baer, K.F. Ferris and D.M. Friedrich, J. Vac. Sci. Technol., A **7**, 1634, 1989.
24. J. C. Riviere, J. A. A. Crossley and G. Moretti, Surface Interface Analysis, **14**, 257, 1989.
25. M. Schaerli and J. Brunner, Z. Phys. B, **42**, 285, 1981.
26. M. OP. Seah, Surface Interface Analysis, **14**, 488, 1989.
27. D. A. Shirley, Phys Rev. B5, 4709, 1972.
28. P. M. A. Sherwood, Practical Surface Analysis, ed. Briggs and Seah, Appendix 3, p. 445, J. Wiley Sons, 1983.
29. A. Rossi, D. De Filippo, S. Virtanen and B. Elsener, Proc. 11th Int Congress of Corrosion, Florence, 1990, Vol. **3**, p. 539–546.

4

Modifications of Passive Films formed on Ni–Cr–Fe Alloys with Chromium Content in the Alloy and Effects of Adsorbed or Segregated Sulphur

D. Costa and P. Marcus

Laboratoire de Physico-Chimie des Surfaces, CNRS-URA 425. Ecole Nationale Supérieure de Chimie de Paris, 11 rue Pierre et Marie Curie, 75005 Paris, France

Abstract

Combined electrochemistry and XPS analysis were used to investigate the passivation of three nickel-based alloys containing ≈10%Fe and 8, 19 and 34%Cr. The samples were passivated at 660mV/SHE in 0.05M sulphuric acid for 20 min and then transferred for XPS analysis without exposure to air. When the Cr content of the alloy increases, the kinetics of passivation is faster, the residual current in the passive state is lower and the chromium oxide concentration of the internal part of the passive layer increases, without significant change of the thickness of the film, showing that the stability of the passive layer is related to its chromium oxide concentration. The passive layers are ≈20Å thick. The influence of the presence of sulphur adsorbed or segregated on the surfaces of the 8%Cr and 19%Cr alloys was investigated. For both alloys, the passivation is slower in the presence of sulphur. The adsorbed sulphur is incorporated in the passive layer. The protective character of the passive layers decreases with increasing sulphur concentration on the surface and with decreasing chromium oxide concentration in the passive layer. These results are in agreement with a previous model of competition of chromium and sulphur during the growth of the passive layer.

1. Introduction

Nickel–chromium–iron alloys as alloy 600 are known for their high corrosion resistance. They are extensively used for steam generators in nuclear power systems. A limited number of surface studies of passivation and corrosion of nickel–chromium–iron alloys have been reported in the past [1–3] and more recently detailed works were performed in the laboratory on an alloy 600 in sulphuric acid [4–6]. A passivation mechanism emerging from the results of the XPS analyses was proposed [6]: (i) enrichment of the alloy surface with chromium by selective dissolution, and adsorption of OH$^-$ ions, (ii) growth of Cr_2O_3 islands within a layer of hydroxide and (iii) formation of a continuous layer of chromium oxide, covered by chromium hydroxide. The influence of sulphur on passivation has also been investigated [5]. In the presence of sulphur, the antagonistic roles of chromium and sulphur were pointed out: there is a competition between the growth of chromium oxide and the one of nickel sulphide islands; passivation occurs when the oxide is able to cover the sulphide.

In the present work, the study was extended to three nickel–chromium–iron alloys with different chromium concentrations (8, 19 and 34% chromium in the alloy). The electrochemical behaviour was characterised by potentiodynamic and potentiostatic measurements, and the passive layers formed at +660mV/SHE were analysed by XPS. The influence of the potential of polarisation on the composition of the passive layer on the 19%Cr alloy was investigated. Studies of the effects of adsorbed or segregated sulphur were also performed. The results are discussed in relation with the role of chromium and sulphur in the mechanisms of passivation.

2. Experimental Method

The samples studied were polycrystalline nickel-based alloys, containing 8, 19 and 34 at.% Cr and 9, 10 and 10 at.% Fe respectively. Discs of 8 mm dia. were cut by spark machining. The samples were mechanically polished with grinding paper and diamond paste (0.5μm), then washed with water and ultrasonically cleaned in acetone.

The electrochemical treatments were performed at room temperature in a cell mounted in a glove box under pure nitrogen atmosphere. The electrolyte (0.05M H_2SO_4) was deaerated with nitrogen before and during the electrochemical measurements. The potentials were measured with a sulphate reference electrode.

The area of the sample exposed to H_2SO_4 was 0.28 cm². The polarisation curves were recorded with a potential scan rate of 1 mV/s, after 10 min of cathodic reduction at –0.34 V. The potential range was –0.24 to +1.26 V/SHE. For the potentiostatic experiments, the samples were ion etched in the preparation chamber of the XPS spectrometer, then analysed by XPS. It was verified that the signals of oxygen and carbon were negligible. After that, the samples were transferred to the glove box in a special transfer vessel to avoid exposure of the sample to air. After less than one minute at the corrosion potential, the potential was stepped to +660 mV(SHE) (in the passive state). Polarisation at different potentials in the passive state (300 and 960 mV/SHE) were carried out on the 19% Cr alloy. After 20 min of polarisation, the sample was taken out of the cell under applied potential, rinsed with ultra-pure water and dried under nitrogen. Afterwards the sample was transported to the XPS spectrometer in the transfer vessel with an inert atmosphere of pure nitrogen in order to minimise the contamination which may alter the surface composition.

The XPS study was performed with a VG-ESCALAB MkII instrument. The use of the MgKα radiation (hv = 1253.6 eV) and of a triple detection system allowed us to obtain a good resolution with a high count rate. The binding energies were calibrated against the binding energies of Au 4f7/2 (84.0 eV) and Cu 2p3/2 (932.7 eV). The spectra were recorded in five regions corresponding to the core levels Ni 2p3/2, Cr 2p3/2, Fe 2p3/2, O1s and S 2p. A well-defined procedure for processing the spectra (satellite subtraction, background subtraction, peak synthesis), based on reference spectra recorded on reference materials, was used to obtain reliable information about the surface composition after the electrochemical treatments. The procedure for quantitative analysis and the calibration data of the metallic signals have been described elsewhere [6, 7]. The reference data for the oxides have been obtained on thermal oxides and are described in [8], as well as the spectral parameters used for hydroxides. The values of the ratios of the photoelectrons yield factors are Y_{Cr}/Y_{Ni} = 0.35 and Y_{Cr}/Y_{Fe} = 0.70 [8].

Angle-resolved XPS measurements were performed to obtain information about the distribution of the species in the surface films. The take-off angle θ is defined as the angle of the surface with the entrance of the hemispherical analyser.

The adsorption of sulphur was performed either by thermal segregation under vacuum (base pressure 10^{-10} mbar) or by chemisorption in $H_2S–H_2$. Sulphur segregation under ultra high vacuum was performed on all alloys. After ion etching in the preparation

chamber of the XPS spectrometer and verification by XPS measurement that no oxygen or carbon were detected on the surface, the samples were annealed at 600°C in the preparation chamber of the XPS spectrometer until the intensity of the sulphur signal (S2p, 162.3 eV), as measured by XPS analysis, had reached a constant value (typically 6 to 8 h). Sulphur chemisorption experiments were performed on the 8% and 19% Cr alloy. The samples were annealed at 970°C in pure hydrogen for 2 h, then treated at 970°C in gaseous $H_2S–H_2$ mixtures of total pressure 200 Torr, following a procedure already described [7]. The temperature Cu/Cu_2S mixture used to prepare H_2S by reaction with H_2 was 430°C, in order to maintain a partial pressure of H_2S/H_2 equal to 9.5×10^{-5}. This procedure allows the adsorption of one complete monolayer of sulphur on the surface [7]. The samples were then transferred under hydrogen into the XPS spectrometer for chemical analysis of the surface, and then transferred to the glove box in a special transfer vessel.

3. Results

3.1 Effect of chromium content on passive films

3.1.1 Surface composition of the samples before electrochemistry

After cleaning by ion etching, the surfaces were analysed by XPS. Oxygen, sulphur and carbon were not detected and the three alloying elements were present only in the metallic states. The ratios of intensities I_{Cr}/I_{Ni} and I_{Fe}/I_{Cr} do not change with the take-off angle (θ = 90° and 40°), indicating that the surface composition is identical to the bulk composition. The compositions of the alloys were thus calculated using the ratios of intensities I_{Cr}/I_{Ni} and I_{Fe}/I_{Cr} and the equations detailed in [7]. The values of the attenuation lengths λ_{Ni} = 8.7 Å, λ_{Cr} = 11.2 Å and λ_{Fe} = 10.1 Å, for the three alloys, calculated after [9]. With measured ratios of intensities of I_{Cr}/I_{Ni} = 0.012, 0.111 and 0.271 and I_{Fe}/I_{Cr} 2.067, 0.671 and 0.4°9, the surface and bulk compositions of the alloys were found to be Ni–8Cr–9Fe, Ni–19Cr–10Fe and Ni–34Cr–10Fe, respectively.

3.1.2 Electrochemical results

The potentiodynamic (i–E) curves, recorded at 1 mV.s⁻¹, all exhibit an activation peak, a passive and a transpassive region. The higher the chromium content of the alloy, the lower the current at the active peak, the more extended the passive domain and the lower the residual current in the passive state. This allowed us to choose the potentials for the potentiostatic experiments: +660 mV for the three alloys (in the middle of the passive region) and +300, +660 and +960 mV/SHE for the 19%Cr alloy (beginning, middle and end of the passive region).

The potentiostatic (i–t) curves recorded for the three alloys at +660mV/SHE are reported in Fig. 1. The kinetics of passivation becomes more rapid with increasing chromium concentration in the alloy. The value of the residual current measured after 20 min of passivation becomes lower with increasing chromium concentration (2.4 ± 0.1, 1.2 ± 0.1 and 0.5 ± 0.1 $\mu A.cm^{-2}$ for the 8%, 19% and 34%Cr, respectively), showing that the higher the chromium concentration in the alloy, the more protective the passive layer.

After polarisation of the Ni19Cr10Fe alloy, the different potentials, the residual current in the passive state is $1.2\pm0.1\mu A.cm^{-2}$ at +300 and +660 mV and 0.5 ± 0.1 $\mu A.cm^{-2}$ at +960 mV/SHE.

3.1.3 XPS analysis of the passive layers

Figure 2(a) shows the spectra recorded for the 8%Cr alloy after passivation at 660 mV(SHE), at two different take-off angles (90° and 40°). Figures 2(b) and 2(c) show the spectra of Cr2 p3/2 and O1s recorded for the 19%Cr (Fig. 2(b)) and 34%Cr (Fig. 2(c)) alloys at a take-off angle of 90°.

The nickel region exhibits a large peak at 852.8 eV which is characteristic of metallic nickel; in addition, for the 19%Cr alloy, a signal at 856.6eV, typical of nickel hydroxide $Ni(OH)_2$, is present, and for the 8%Cr-alloy, both oxidation states Ni^{2+} in NiO (854.6 and 856.2 eV) and in $Ni(OH)_2$ (856.6 eV) contribute to the

nickel spectrum. The chromium region exhibits a large peak of oxides, attributed to Cr^{3+} in chromium oxide (576.6 eV) and chromium hydroxide (577.2 eV). The iron region exhibits a very weak signal, in which contributions of metallic iron (706.9 eV), Fe^{2+} (709.4 eV) and Fe^{3+} (710.3 eV) in oxides are present. The oxygen region exhibit a signal attributed to O^{2-} in oxides (530.2 eV), hydroxyls OH^- (531.4 eV) and oxygen in water or SO_4^{2-} (532.2 eV). In the S 2p region, the signal is located at 169 eV and is due to sulphur in SO_4^{2-}. SO_4^{2-} is present only on the surface, as shown by the angle-dependent measurements (the signal intensity increases at the low take-off angle) and the fact that it is removed by a slight ion etching. Angle-dependent measurements allowed us to characterise the in-depth distribution of the different species. At low take-off angle, the signal from the species located in the external part of the film is enhanced. The ratios of the peak areas of OH^- and O^{2-} ($I_{OH^-}/I_{O^{2-}}$), of Cr^{3+} in hydroxide and oxide ($I_{Cr(OH)_3}/I_{Cr_2O_3}$) and of Ni^{2+} in hydroxide and oxide ($I_{Ni(OH)_2}/I_{NiO}$) increases with decreasing take-off angle, i.e. when the surface sensitivity increases. These results conclusively show that the hydroxides are located in the outer part of the passive film while the inner part is composed of oxides. The ratios ($I_{Fe^{2+}+Fe^{3+}}/I_{Cr_2O_3}$) do not change with take-off angle, which indicates that the iron cations are located in the inner part of the film. For the 8%Cr alloy, the variable angle experiments also revealed that the ratio ($I_{NiO}/I_{Cr_2O_3}$) and ($I_{NiO}/I_{Fe^{2+}+Fe^{3+}}$) increases with decreasing take-off angle, which proves that the

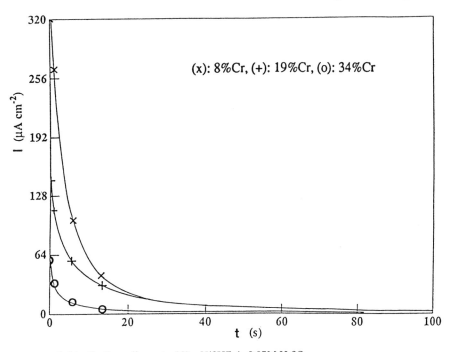

Fig. 1 Potentiostatic curves recorded for the three alloys, at +660 mV/SHE, in 0.05M H₂SO₄.

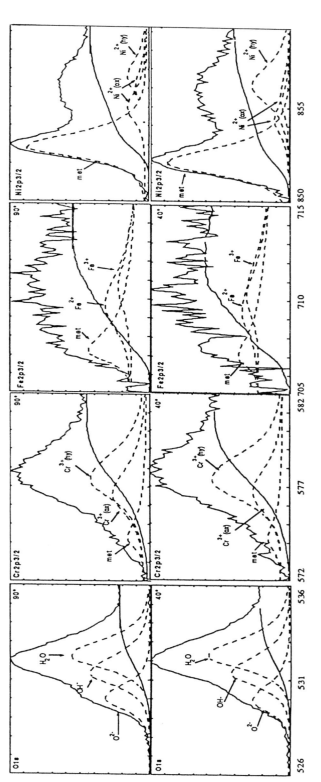

Fig. 2(a) XPS spectra recorded at two take off angles (90° and 40°) for the 8%Cr after passivation at +660 mV/SHE.

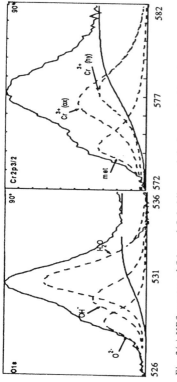

Fig. 2(c) XPS spectra of O1s and Cr2p3/2 recorded after passivation at +660 mV/SHE for the 34% Cr (take-off angle: 90°).

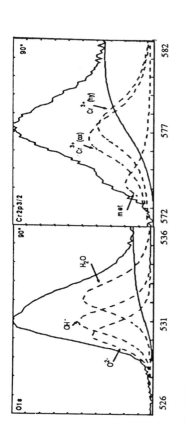

Fig. 2(b) XPS spectra of O1s and Cr2p3/2 recorded after passivation at +660 mV/SHE for the 19% Cr (take-off angle: 90°).

nickel oxide is located in the outer part of the oxide film.

The sputter-depth profiles were recorded for each alloy; they confirmed the results of the angle-resolved measurements. In all cases, the iron oxides are sputtered before the chromium oxides, showing an enrichment in chromium oxide in the internal part of the oxide film. The results of the angle-resolved measurements and sputtering-depth profiles support the conclusion that the passive films have a bilayer structure, composed of an external hydroxide layer and an inner oxide layer. Using this model, it was possible to calculate with the recorded intensities of the core-level signals, the compositions and the thicknesses of the two layers of the passive films (Table 1), and the composition of the metallic phase under the passive film, which often differs from the bulk composition of the alloy [5, 6)]. The thicknesses of the oxide and hydroxide layers are estimated with a precision of $\pm$ 1Å. The hydroxide layers consist mainly in chromium hydroxide $Cr(OH)_3$ for the three alloys and the oxide layer mainly in chromium oxide Cr_2O_3 for the 34% and 19%Cr alloys. For the 8%Cr alloy, the oxide layer is not homogeneous and may be considered as a nickel oxide NiO layer above a mixed (Cr_2O_3 + Fe_2O_3 + Fe_3O_4) inner layer. Thus, the composition of the oxide layer (considered as an homogeneous layer) given in Table 1 for this alloy is approximative. Therefore, in addition, we have calculated the chromium oxide content in the inner part of the oxide film, at the

alloy/film interface (Table 1). This content is the same as the average Cr^{3+} content in the oxide layer for the 19%Cr and the 34%Cr alloys, because the oxide layer is homogeneous, but different for the 8%Cr alloy.

The quantitative analysis of the oxygen spectra allowed us to confirm that the chemical species, are $Ni(OH)_2$, $Cr(OH)_3$ in the hydroxide layer and Cr_2O_3, Fe_3O_4, Fe_2O_3 and NiO in the oxide layer. The first metallic planes under the passive layer have either the same composition as the bulk (8%Cr), or are slightly enriched with nickel: we found that there was a layer of $\approx$15Å of composition $Ni_{65}Cr_{24}Fe_{10}$ on the 34%Cr alloy and a layer of $\approx$10Å of composition $Ni_{81}Cr_{15}Fe_4$ for the 19%Cr alloy.

For one of the alloys (with 19%Cr), the composition and thickness of the passive layers have been studied as a function of potential. The results for +300, +660 and +960 mV are reported in Table 2. When the potential increases, the nickel content of the passive layer increases, and at +960 mV, there is some NiO in the oxide layer. In this case, the variable angle experiments revealed that the ratio (I_{NiO} / $I_{Cr_2O_3}$) and ($I_{Fe^{2+}+Fe^{3+}}$ / $I_{Cr_2O_3}$) increases with decreasing take-off angle, which proves that the nickel and iron oxides are located in the outer part of the oxide film. So, the oxide layer is not homogeneous and consists in the outer (NiO +Fe_3O_4, Fe_2O_3 and an inner Cr_2O_3 layer. Both the oxide layer and the hydroxide layer thicken at high potential (+960 mV).

Table 1 Compositions and thicknesses of the passive layers formed on the three alloys after polarisation at +660 mV/SHE. (d) thickness of the hydroxide layer; (c) thickness of the oxide layer; [Cr^{OX}] inner: chromium oxide content (molar ratio of Cr_2O_3 in the oxide) in the inner part of the oxide layer, at the alloy/film interface; I_{res}: current density in the passive state

sample	hydroxide	d (Å)	oxide	c (Å)	[Cr^{OX}] inner	total (Å)	I_{res}
8%Cr	0.7 $Cr(OH)_3$ + 0.3 $Ni(OH)_2$	12	0.4 Cr_2O_3 + 0.4 NiO + 0.2(Fe^{3+},Fe^{2+})	9	0.66	21	2.4μA/cm^{-2}
19%Cr	0.8$Cr(OH)_3$ + 0.2 $Ni(OH)_2$	10	0.8 Cr_2O_3 + 0.2 (Fe^{3+},Fe^{2+})	10	0.8	20	1.2μA/cm^{-2}
34%Cr	$Cr(OH)_3$	10	0.9 Cr_2O_3 + 0.1 (Fe^{3+},Fe^{2+})	12	0.9	22	0.5μA/cm^{-2}

Table 2 Compositions and thicknesses of the passive layers grown after polarisation at +300, +600 and +960 mV on the 19% Cr alloy. (d): thickness of the hydroxide layer; (c): thickness of the oxide layer; [Cr^{OX}] inner : chromium oxide content (molar ratio of Cr_2O_3 in the oxide) in the inner part of the oxide layer, at the alloy/film interface; I_{res}: current density in the passive state

potential	hydroxide	d (Å)	oxide	c (Å)	[Cr^{OX}] inner	total (Å)	I_{res} (μA/cm^{-2})
+300	0.9 $Cr(OH)_3$ + 0.1 $Ni(OH)_2$	8.5	0.85 Cr_2O_3 + 0.15(Fe^{3+}+ Fe^{2+})	7.5	0.85	16	1
+660	0.8$Cr(OH)_3$ + 0.2 $Ni(OH)_2$	10	0.8 Cr_2O_3 + 0.2 (Fe^{3+}+ Fe^{2+})	10	0.8	20	1.2
+960	0.8$Cr(OH)_3$ + 0.2$Ni(OH)_2$	17	0.6 Cr_2O_3+ 0.2 NiO +0.2 (Fe^{3+}+ Fe^{2+})	10	1	27	0.5

3.2 Effects of sulphur on passivation

3.2.1 Surface composition of the samples before passivation

The thermal treatments involving sulphur adsorption or segregation are accompanied by chromium thermal segregation on the surface [7]. The quantitative treatments of the XPS spectra are thus complex because the composition of the surface plane differs from the composition of the bulk. Assuming that the total density of atoms in the surface plane is the same as in the bulk, the equations established by [7] give the sulphur coverage in fraction of monolayer Θ (by definition $\Theta = 1$ for a complete monolayer). The monolayer corresponds to an amount of sulphur of $\approx 42 ng.cm^{-2}$. This corresponds to 0.5 atom of sulphur per metal atom for the (100) crystallographic orientation. The composition of the first metallic plane was calculated using the procedure given [7]. The results of the XPS analysis of the surface of the 19% Cr and 8% Cr alloys after either sulphur segregation or sulphur chemisorption are reported in Table 3.

3.2.2 Electrochemical results

Figure 3 shows the kinetics of passivation of the sulphur-covered samples by comparison with the sulphur-free samples for the 8% Cr alloy (Fig.3(a)) and the 19% Cr alloy (Fig.3(b)). The presence of sulphur causes a slower passivation, even at a low sulphur coverage. The protective character of the passive layers, estimated by the residual currents in the passive state, is not changed at a low sulphur coverage (1.4 $\mu A.cm^{-2}$) for the 19% Cr with a sulphur coverage $\Theta = 0.2$, whereas it is significantly altered by a complete monolayer of sulphur (3.0 $\mu A.cm^{-2}$ for the 19% Cr and for the 8% Cr with a sulphur coverage $\Theta = 1$).

3.2.3 XPS analysis of the passive layers

The recorded spectra exhibit the same oxidation states of Ni, Cr and Fe as those of passive layers formed on the sulphur-free samples. The angle-dependent analysis and the sputter-depth profiles are also similar and the passive layers can be considered as bilayers of hydroxides and oxides. As for the sulphur-free alloys, the angle-resolved analyses showed that iron is present in the inner part of the oxide layer (the ratios $(I_{Fe^{2+}+Fe^{3+}} / I_{Cr_2O_3})$ do not change significantly with take-off angle) and the sputter-depth profiles revealed an enrichment in chromium in the inner part of the oxide layer.

The sulphur region exhibits two signals: a peak at 169.0 eV due to the adsorbed SO_4^{2-} ions and a peak at 162.3 eV, due to sulphur in a reduced state (adsorbed on the alloy surface or as sulphide) [5]. The intensity of the signal due to the reduced state varied, in the angle-dependent measurements and in the depth profiles, as the intensities of the signals of the oxides, showing that sulphur is incorporated in the internal part of the passive layer. This information allowed us to calculate the quantity of sulphur in the passive layer, considering that sulphur was present at the alloy/oxide interface and taking into account the attenuation of the S 2p signal due to the passive layer. In each case we have found a sulphur coverage after passivation close to the initial sulphur coverage of the alloy, showing that no sulphur was dissolved during passivation.

The thicknesses and compositions of the passive layers are reported in Table 4. The chromium oxide content in the inner part of the oxide layer was calculated taking into account the presence of sulphide in the oxide layer.

Table 3 Surface composition of the alloys in presence of sulphur. The sulphur coverage $\Theta = 0.2$ was obtained by segregation at 600°C in UHV, the sulphur coverages $\Theta = 1$ were obtained by adsorption in H_2S–H_2 at 970°C

SAMPLE	Cr surface concentration (at.%)	Θ = sulfur coverage (fraction of monolayer)
Ni8Cr9Fe $\Theta=1$	23	1
Ni19Cr10Fe $\Theta=0.2$	22	0.2
Ni19Cr10Fe $\Theta=1$	28	1

Table 4 Compositions and thicknesses of the passive layers formed after polarisation at +660 mV in the presence of adsorbed or segregated sulphur

sample	hydroxide	d (Å)	oxide	c (Å)	[Cr^{ox}] inner	total	I_{res}
8%Cr $\Theta=1$	0.7Ni(OH)$_2$+ 0.3Cr(OH)$_3$	20	0.7 Cr$_2$O$_3$+ 0.3 (Fe^{3+}+ Fe^{2+})	5	0.35	25	3μA/cm^{-2}
19%Cr $\Theta=0.2$	0.6Cr(OH)$_3$+ 0.4 Ni(OH)$_2$	12	0.8 Cr$_2$O$_3$ + 0.2 (Fe^{3+}+ Fe^{2+})	6	0.65	18	1.4μA/cm^{-2}
19%Cr $\Theta=1$	0.4Cr(OH)$_3$ + 0.6Ni(OH)$_2$	13	0.7 Cr$_2$O$_3$+ 0.3 (Fe^{3+}+ Fe^{2+})	6	0.35	19	3μA/cm^{-2}

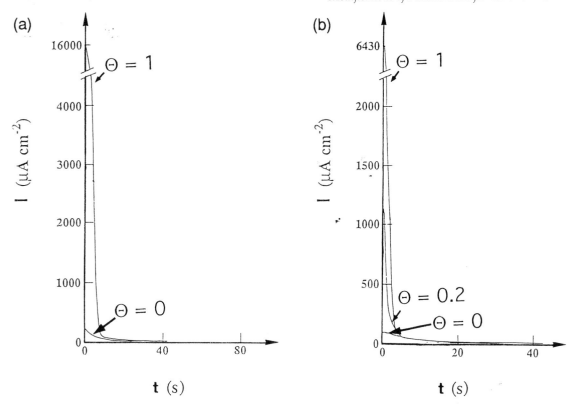

Fig. 3 Potentiostatic curves recorded for the sulphur-covered alloys, in comparison with the sulphur-free alloys, for the 8% Cr (3a) and 19% Cr (3b), at +660 mV/SHE, in 0.05M H_2SO_4.

When sulphur is present on the surface alloy, there is no significant variation of the total thickness of the passive layer, but there is a slight tendency of decrease of the thickness of oxide layer and increase of the hydroxide layer. For the 8%Cr covered by a complete monolayer of sulphur, the hydroxide layer is thicker, the oxide layer thinner and the total thickness increases.

The composition of the first metallic planes (10–15Å) under the passive layer was calculated in each case: $Ni_{85}Cr_8Fe_7$ for the 8%Cr with $\Theta=1$, $Ni_{70}Cr_{22}Fe_8$ for the 19%Cr with $\Theta=0.2$, and $Ni_{76}Cr_{15}Fe_9$ for the 19%Cr with $\Theta=1$. These results are not easily interpreted because the Cr surface concentration before passivation was higher than that of the bulk, due to the sulphur-induced surface segregation of chromium (see Table 3). It can be noted that for the 8%Cr, the composition is the same as that found under the passive layer without sulphur. This is likely related to the fact that on this alloy the dissolution during passivation was large in the presence of sulphur so that the first plane in chromium was dissolved.

4. Discussion

The passive layers formed on the alloys with different Cr contents have all a bilayer structure and are all mainly composed of chromium hydroxide and oxide, in agreement with previous results [6]. In all cases, even at high chromium content in the alloy, iron is present also in the inner art of the oxide. Nickel is present in the oxide, as NiO, only at very low chromium content in the alloy (8%). In this case the oxide layer is not homogeneous but is constituted of nickel oxide located above iron and chromium oxides. Applying such a model for the 8% Cr alloy, we have calculated the thicknesses of these two layers: 3Å for the NiO oxide and 6Å for the iron chromium oxides. The low value found for the thickness of the NiO layer indicates that NiO is not present as a continuous layer but rather as islands, as shown in Fig. 4. This schematic representation takes into account that the growth of the passive layer consists in the growth of oxide islands on the surface, as proposed in [6].

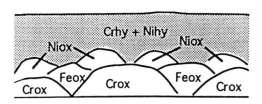

Fig. 4 Model of the passive layer formed on the 8% Cr alloy after polarisation at +660 mV/SHE.

The experiments performed at different potentials on the 19%Cr show that the passive layers contain more nickel oxides and thicken with increasing potential. The thickening of the passive layer with increasing potential was already observed on pure chromium [10] and on a Ni34Cr alloy [11]. At high potential (+960 mV), the oxide layer is constituted of a quasi continuous chromium oxide at the interface with the alloy, covered by nickel and iron oxides. This suggests that the enrichment in chromium oxide at the alloy/film interface is enhanced by increasing the potential.

The results did not reveal marked modifications of the composition of the alloy under the passive film, in contrast with the results obtained in [6] where the experiments were performed at +300 mV/SHE.

The presence of sulphur lowers the kinetics of passivation. The works performed in the past in the laboratory clearly showed that sulphur increases the dissolution of metals and remains on the surface [12], and it was shown for a Ni21Cr8Fe alloy that sulphur was incorporated in the layer as nickel sulphide [5]. The proposed mechanism was a competition on the surface between the growth of chromium oxide and nickel sulphide. Passivation may take place when the oxides are able to cover the sulphides. The present results are in agreement with this model. We also found that the adsorbed or segregated sulphur is not dissolved and remains as sulphide islands in the internal part of the passive layer, causing an enhanced dissolution in the passive state. The passive layer formed on the 8% Cr in presence of one sulphur monolayer contains no NiO (whereas there was some NiO in the passive layer formed on the sulphur-free sample), which is consistent with the idea that nickel is bonded to sulphur in the passive layer.

The general tendency is that in the presence of sulphur the inner oxide layers are less enriched in chromium oxide, they are thinner and the hydroxide layers are thicker.

The protective character of the passive layer increases with increasing chromium concentration in the alloy. As there is no marked change in the thicknesses (20Å), the increased protective character may be clearly attributed to the increase of chromium concentration in the passive layer.

The results of the analysis of the passive layers formed in the presence of sulphur are consistent with the roles attributed to the oxide and hydroxide layers in the protective character of the passive layer: the increase of the hydroxide layer thickness does not improve the protective character of the passive film. This shows that the oxide layer plays the main barrier role in the protection of the alloy.

Increasing the potential of passivation induces a better protective character, however the average chro-mium concentration in the passive layer is not higher. This result may be explained by the formation of a continuous Cr_2O_3 layer at the oxide/alloy interface .

Figure 5 reports the residual currents, measured in the passive state, as a function of the chromium oxide concentration in the internal part of the oxide layer (at the alloy/passive film interface). It appears that the protective character increases with increasing chromium oxide content at the alloy/passive film interface. This allows us to conclude that the major chemical aspect of the passive layer is the presence of a thin chromium oxide layer at the alloy/oxide interface.

5. Conclusion

Passive films formed on three nickel–chromium–iron alloys or different Cr-content have a bilayer structure composed of an outer hydroxide (mainly chromium hydroxide) and an inner oxide layer. The oxide layer contains mainly chromium oxide, with some iron oxides and at low Cr content in the alloy (8%), some NiO in the external part of the oxide. The protective character of this layer is related to its chromium oxide content in the inner part of the oxide layer (at the alloy/passive film interface). The most protective character is achieved when the inner part to the oxide layer is constituted of a continuous chromium oxide layer. At a given potential, the thicknesses of the passive layers are similar for the three alloys. For a given alloy, increasing the polarisation potential induces an incorporation of nickel oxide in the passive layer and a thickening of the layer.

The presence of adsorbed sulphur slows down the kinetics of passivation, enhances the dissolution and alters the protective character of the passive layer. This last effect is attributed to the presence of nickel sulphide is islands in the inner part of the passive layer. The adsorbed or segregated sulphur is not dissolved during passivation.

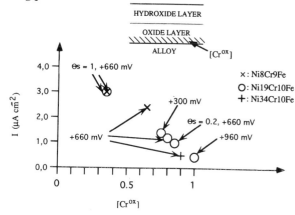

Fig. 5 Current density measured in the passive state vs the chromium oxide concentration in the inner part of the oxide layer (at the passive film/alloy interface, as shown in the upper part of the figure).

6. Acknowledgement

Partial support of this work by FRAMATOME is acknowledged.

References

1. K. Hashimoto and K. Asami, Corros. Sci., **19**, 427, 1979.

2. N. S. McIntyre, D. G. Zetaruk and D. Owen, J. Electrochem. Soc., **126**, 750, 1979.

3. P. Combrade, M. Foucault, D. Vançon, P. Marcus, J. M. Grimal and A. Gelpi, Proc. 4th Int. Symp. 'Environmental Degradation of Materials in Nuclear Power Systems-Water Reactors', ed. D. Cubiciotti, p.79, 1989.

4. J-M. Grimal and P. Marcus, Proc. UK Corrosion '88 and EUROCORR, Volume III, p.5, 1988.

5. P. Marcus and J. M. Grimal, Proc. 6th Int. Conf. on Passivity, Sapporo, Japan, September 1989; Corros. Sci., **31**, 377, 1990.

6. P. Marcus and J. M. Grimal, Corros. Sci., **33**, 805, 1992.

7. J-M. Grimal and P. Marcus, Surf. Sci., **249**, 171, 1991.

8. E. DeVito and P. Marcus, Surface Interface Anal., **19**, 403, 1992.

9. M. P. Seah and W. A. Dench, Surface Interface Anal., **1**, 2, 1979.

10. S. Haupt and H. H. Strehblow, Corros. Sci., **29**, 163, 1989.

11. T. Jabs, P. Borthen and H. H. Strehblow, Proceedings of the Symposium on 'Oxide Films on Metals and Alloys' (Toronto 1992), The Electrochemical Society, eds B. R. Macdougall, R. S. Alwitt, T. A. Ramanarayanan, P 294, 1992.

12. P. Marcus, A. Teissier and J. Oudar, Corros. Sci., **24**, 259, 1984.

5

Influence of Fe and Mo Alloying Elements on the Composition of Passive Films Formed on Ni–Cr Alloys in an Acidic Chloride Solution

N. JALLERAT, S. MISCHLER* AND D. LANDOLT*

Centre d' Etudes de Chimie Métallurgique, 15 rue Georges Urbaiu 94407 Vitry/Seine, France
* Ecole Polytechnique Fédérale de Lausanne, Département des Matériaux, 1015 Lausanne, Switzerland

Abstract

The pitting corrosion behaviour and the passive film composition of three nickel–chromium alloys (Ni15%Cr, Ni15%Cr10%Fe and Ni15%Cr10%Mo) in 0.1M H_2SO_4 + 0.4M Na_2SO_4 + 0.12M NaCl solution at 65°C were investigated with the aim to understand the respective effects of iron and molybdenum on the pitting resistance of NiCr alloys. The measured pitting potential varied in the increasing order NiCrFe < NiCr < NiCrMo. Surface analysis (carried out using AES depth profiling and XPS) shows that the passive films consist mainly of trivalent chromium oxide with incorporation of metallic cations and small quantities of anions. The passivation changes the composition of the metallic layer underneath the passive film. Possible relations between surface composition, pitting resistance and passivation kinetics are discussed.

1. Introduction

Ni–Cr based alloys are known for their high corrosion resistance associated with good mechanical properties. Among other things, they are used as steam generator tubing materials in many pressurised water nuclear reactor systems. The corrosion properties of Ni–Cr alloys in aqueous solutions are determined by the presence of a passive surface film. The study of the relation between corrosion behaviour and passive film composition is of importance for the understanding of the corrosion and protection mechanisms of these alloys. Surface analysis methods such as XPS (X-Ray Photoelectron Spectroscopy), AES (Auger Electron Spectroscopy) and SALI (Surface Analysis by Laser Ionisation) are well suited for the chemical characterisation of very thin oxide films and have been applied to the study of passive films formed on NiCr based alloy in different electrolytes by several authors [1–7]. The published results show that passive films consist predominantly of trivalent chromium oxide and have a thickness of approx. 1–2 nm. However, it is not clear to what extent alloyed elements and anions from the electrolyte are incorporated in the film. A comparison of data obtained by different groups with different alloys is difficult because of different experimental conditions employed and because of intrinsic limitations of the analysis methods used. Iron and molybdenum are the most important alloying elements added to NiCr. Mo improves the corrosion resistance in particular against pitting corrosion in chloride media.

Iron has a beneficial role on the mechanical properties. The present study was initiated with the aim to compare the influence of alloyed Fe and Mo on the electrochemical behaviour and on the composition of passive films formed on NiCr alloys in chloride environments. For this purpose three high purity alloys of nominal composition Ni15Cr, Ni15Cr10Fe and Ni15Cr10Mo were studied using electrochemical and surface analysis techniques at 65°C in a solution containing 0.1M H_2SO_4 + 0.4M Na_2SO_4 + 0.12M NaCl. The electrolyte was chosen in order to allow one to compare the results with those obtained previously with FeCr alloys [8]. The composition of the passive films was investigated using AES depth profiling as well as XPS in order to overcome shortcomings inherent in the use of single techniques.

2. Experimental

The studied alloys were prepared by melting pure metals (Johnson Matthey) in a levitation furnace under controlled helium atmosphere [9]. To homogenise the alloy concentration, the ingots weighing approx. 20 g were subjected to a heat treatment at 1150°C for 3 h under vacuum and then water quenched. The composition of the alloys measured by atomic absorption is reported in Table 1. Analysis by SEM (Scanning Electron Microscopy) and by EDAX indicated that the structure of the three alloys was homogeneous without detectable segregation or inclusions. The grains were of elongated shape (typical size 2 × 0.3 mm). The

Table 1 Alloy composition in wt%

Alloy	Ni	Cr	Fe	Mo
NiCrFe	76.4	15.4	9.6	0.021
NiCr	86.6	13.6	< 0.016	< 0.016
NiCrMo	76.2	14.8	< 0.016	9.94

electrodes were machined in the form of discs of 3.57 mm dia., used for polarisation studies, or 7.2 mm, used for surface analysis studies. For the measurement of polarisation curves, rotating electrodes were fabricated by embedding the metal disks (electrode area: 0.1 cm^2) in epoxy. The rotation rate was 3000 rpm. Electrode pretreatment consisted of mechanical polishing with SiC paper and with 1 μm diamond spray. The electrodes were washed with ethanol, doubly distilled water and dried with argon before being used. All experiments were carried out in de-aerated solutions of 0.1M H$_2$SO$_4$ + 0.4M Na$_2$SO$_4$ + 0.12M NaCl under argon. The temperature was 65°C. Polarisation curves were measured using a potentiostat Solartron 1286 controlled by a function generator AMEL 568. The potential was increased at a scan rate of 2 mV s^{-1}. All potentials are given with respect to the normal hydrogen electrode. The current was monitored using an HP 310 computer. Prior to each run, the samples were cathodically prepolarised –750 mV/she for 60 s. Electrodes used for surface analysis were polarised during 60 min at three anodic potentials, respectively: +200, +500 and +800 mV (except for NiCrFe samples submitted to a strong pitting corrosion at +800 mV). At the end of a polarisation experiment the electrodes were removed from the electrolyte without switching off the potentiostat. They were rinsed with doubly distilled water and stored in a dessicator until being

transferred into the UHV apparatus for surface analysis and immediately analysed. AES and XPS analysis were performed with a PHI 550/590 apparatus containing a CMA and differentially pumped ion guns. In AES, a constant primary beam current of 0.5 μA at 3 keV primary beam energy was employed and the electron beam was rastered over an area of 170 × 170 μm. Depth profile analysis was performed by rastering a 1 keV Kr$^+$ beam over an area of 2 × 1.5 mm. Each sample was analysed on two locations and each experiment was repeated twice. From the four measured AES profiles, a mean profile was calculated the reproducibility being of the order of 10%. Sputter time was converted into sputtered depth using a Ta$_2$O$_5$ standard and multiplying by 0.8, corresponding to the ratio between the sputtering rates of chromium oxide Cr$_2$O$_3$ and that of Ta$_2$O$_5$ films. The considered AES peaks, the electron escape depths and the sensitivity factors are reported in Table 2. In XPS, a X-Ray source was operated at 10 keV and 40 mA emission current using Mg Kα (1253.6 eV) radiation. The binding energies of XPS peaks are listed in Table 3 with area sensitivity factors and electron escape depths in the metal and in the oxide respectively. Different problems related to the quantitative analysis of passive films have been discussed previously [8, 10].

3. Results

3.1 Electrochemical behaviour

The potentiodynamic polarisation curves plotted in Fig. 1 show that Mo reduces the current density in the active potential region. A similar effect is found when Mo is added to ferritic stainless steels [11, 12]. It is not clear at present if this decrease in current density is related to a change in the reduction kinetics of H$^+$ ions

Table 2 AES peaks considered

Element	Transition	Kinetic energy [eV]	Escape depth [nm] *	Sensitivity Factor **
Cr	LMM	528	0.87	0.33
Ni	LMM	867	1.10	0.44
Fe	LMM	651	0.96	0.32
Mo	MVV	221	0.56	0.43
O	KLL	511	0.85	0.83
S	LMM	152	0.51	1.83
Cl	LMM	181	0.46	1.33
C	KLL	273	0.62	0.33

*   See Ref. 17, page 68

**   The AES sensitivity factors are taken from Ref. 17 except that for Ni, Fe and Cl taken from Ref. 20.

Table 3 XPS binding energies

Peak	Compound	Binding[1] energy [eV]	Sensitivity factor[2]	Escape depth[3] [nm]	
				Oxide	Metal
Cr 2p 3/2	Cr	574.1	0.87	1.22	0.74
Ni 2p 3/2	Ni	852.6	1.70	0.91	0.57
Fe 2p 3/2	Fe	706.9	1.20	1.09	0.67
Mo 5d	Mo	227.8	1.60	1.48	0.89
O 1s	Cr_2O_3	530.7	0.58	1.25	0.77
S 2p	Na_2SO_4	169.0	0.32	1.53	0.96
Cl 2p	NaCl	198.0	0.44	1.51	0.94

1) Data from Ref. 8 except for Ni, Cr, Fe and Mo determined on the studied alloys after sputter cleaning.

2) Data from Ref. 8 except for Ni and Fe determined on the NiCrFe alloy after sputter cleaning.

3) Escape depth $\Lambda = \lambda \cos\phi$, values of the electron mean free path λ calculated after Seah and Dench [21] for a Cr_2O_3 matrix or for a Ni matrix, respectively, $\phi = 49°$ for the present arrangement.

or to a change in anodic dissolution rate. The passivation potential does not change with alloy composition. The passivity breakdown at *ca.* 950 mV/she is due to the transpassive dissolution of Cr. Under the present conditions characterised by a rapid scanning rate (2 mV s⁻¹) the transpassive potential for chromium dissolution was attained before pitting occurred. Pitting required a finite incubation time and appeared only during prolonged potentiostatic polarisation.

Figure 2 shows current–time transients (plotted on a semi-logarithmic scale) measured at different applied potentials. The shape of the current decay gives an indication of the stability of the passive film. Pitting manifests itself by characteristic fluctuations due to unstable pit formation, followed by the irreversible increase of the current density corresponding to the growth of pit nuclei. Such effects are observed in Fig. 2 only for the alloy NiCrFe at 0.7 and 1 V. The peak observed on the transient of the NiCr alloy polarised at 1 V suggests that a pit has formed and subsequently repassivated. Thus the NiCr alloy seems to be sensitive to metastable pitting under these conditions. Pits were effectively observed only on the NiCrFe electrodes polarised at 0.7 and 1 V and on the NiCr alloy polarised at 1V. According to the data of Fig. 2 NiCrMo does not pit, even at high applied potential. From these experiment it is concluded that the stability against pitting of the different alloys increases in the order: NiCrFe < NiCr < NiCrMo. Film formation on the three alloys was examined using double logarithmic plots of the current–time curves (not shown). The slopes at times <100 s were found to have a value of 1 ± 0.3 independently of alloy composition. This indicates

that the film growth mechanism was the same in the three alloys. A change in the shape of the passivation transients was observed between 0.2 and 0.4 V for the NiCrMo alloy. This transition corresponds to the potential of transpassive dissolution of molybdenum [8]. The current density in the passive state was lower for the NiCrMo compared to the others alloys and never reached a steady state value. According to Vetter [13], the current density reaches a potential dependent steady state when the rate of formation of the passive films equals the rate of the film dissolution. The fact that a steady state current was observed for NiCr and NiCrFe but not for NiCrMo indicates that the rate of passive film dissolution was significantly lower in the case of the Mo bearing alloy.

3.2 Concentration and distribution of cations in the passive films

Figure 3 shows typical measured XPS spectra of Ni 2p3/2, Cr 2p3/2, Mo5d and Fe 2p 3/2 after background subtraction according to Sherwood [22]. Also indicated are the results of the applied deconvolution procedure. This procedure employs peak subtraction [10] for the separation of metallic and oxidised states. For this the measured curve is fitted by the measured peak of the element in the metallic state and then the metallic peak is subtracted yielding that for the oxide. Standard metallic peaks for the element Ni, Cr, Fe and Mo were obtained on sputter-cleaned NiCrFe and NiCrMo alloys. The Mo oxide peak was further deconvoluted by subtracting the Mo^{6+} standard peak obtained from p.a. MoO_3 powder supplied by Merck. The three peaks observed in Fig. 3(a) can be attributed

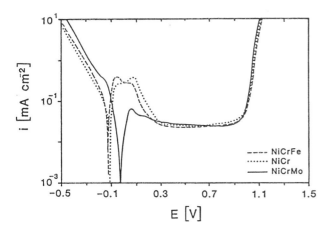

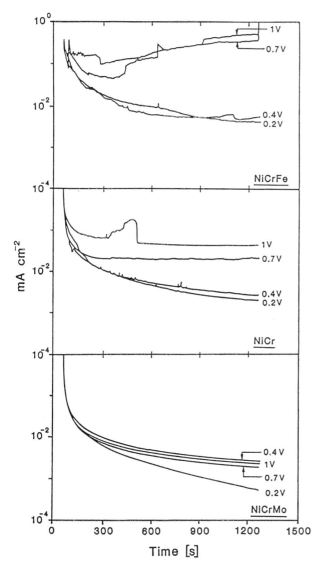

Fig. 1 Polarisation curves of NiCrFe, NiCr and NiCrMo alloys in 0.1M H_2SO_4 + 0.4M Na_2SO_4 + 0.12M NaCl solution at 65°C.

Fig. 2 Current–time transients at different passivation potentials of NiCrFe, NiCr and NiCrMo alloys.

to metallic Nickel (852.5 eV) and to an oxidised state of nickel (856.3 and 862 eV). The peak at high binding energy is supposed to be a satellite of the peak at 852.3 eV since the ratio of their intensities is a constant value. Nickel peaks of 856.3 eV binding energy have been observed on $Ni(OH)_2$ [7, 14, 15], on Ni_2O_3 [13] and on $NiFe_2O_4$ [7]. The peak energy of Cr2p3/2 corresponds to that of trivalent chromium oxide [8]. The deconvolution of the Mo5d peak indicates the presence of metallic molybdenum, hexavalent molybdenum oxide and molybdenum oxidised at a lower valence. Because of the superposition with the S2s peak (233 eV) it is difficult to determine the exact valence of oxidised molybdenum. The Fe 2p3/2 peak was deconvoluted into a metallic and oxidised part without subdividing the latter in different valence states.

The intensity of the metallic peaks shown in Fig. 3 decreases with increasing potential. From the signal intensity ratio of metallic and oxidised species of Ni, Cr, Mo and Fe the film thickness was estimated according the procedure described elsewhere [10] using the electron escape depths given in Table 3. The results are presented in Fig. 4 together with those obtained from AES profiles as described below. The data of Fig. 4 show that the passive film thickness does not depend significantly on alloy composition but is determined by the applied potential. The film thickness increases with the potential at an approximate rate of 1.3 nm/V. A similar value was observed on Fe–Cr alloys under similar conditions [8]. The oxide film observed on the cathodically treated samples are probably formed during rinsing with water or during the transfer to the surface analysis chamber. The fact that passive films are significantly thicker indicates that they are formed prevalently in the solution during polarisation in the passive range.

Average atomic fractions were calculated using the area sensitivity factors of Table 3 by assuming that metallic and oxidised signals correspond to the substrate and the film respectively. In order to correct for the signal attenuation due to the presence of the oxide film the intensities of the metallic peaks were multiplied by the factor exp (D/Λ) where D is the measured film thickness (Fig. 4) and Λ is the escape depth in the oxide given in Table 3. The results are shown in Figs 5 and 6. The oxide films formed at the cathodic potential contain significantly more Ni than the passive films (Fig. 5). The passive films formed on the NiCrFe alloy contains less Fe than the film formed on the cathodically polarised sample. The passivation of the NiCrMo alloy leads to an increased molybdenum content in the oxide film. The chromium content of the passive films varies between 80–85% for the NiCr and NiCrFe alloy and 75–80% of the NiCrMo

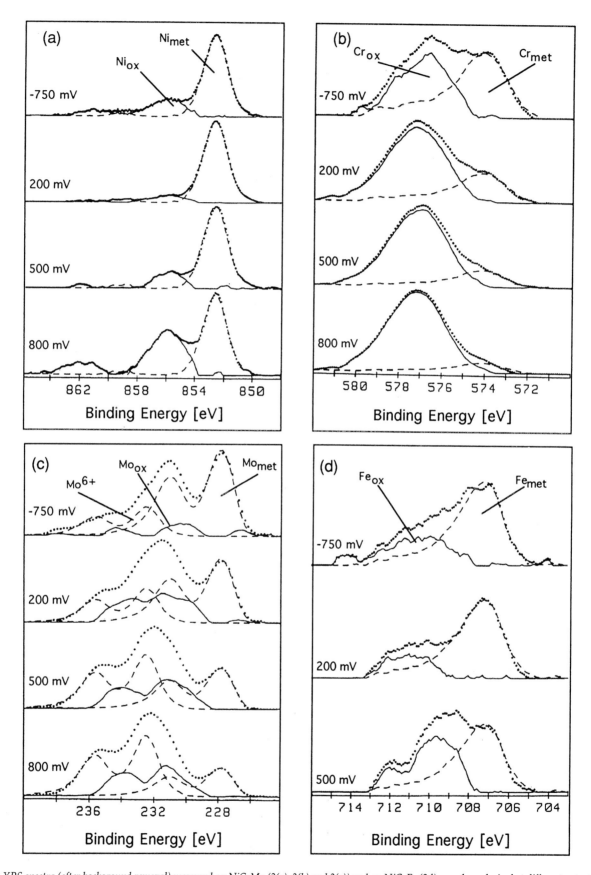

Fig. 3 XPS spectra (after background removal) measured on NiCrMo (3(a), 3(b) and 3(c)) and on NiCrFe (3d) samples polarised at different potentials.

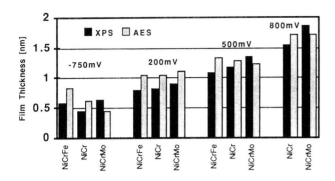

Fig. 4 Film thicknesses determined by AES and XPS on NiCrFe, NiCr and NiCrMo alloys polarised at different potentials.

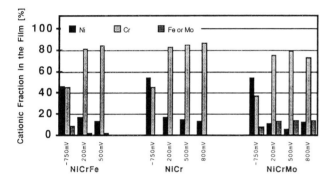

Fig. 5 XPS average cation fraction of Cr, Ni, Fe and Mo in oxide films formed on NiCrFe, NiCr and NiCrMo alloys polarised at different potentials.

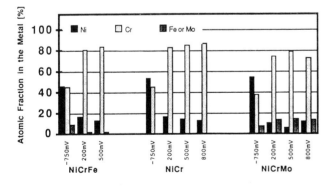

Fig. 6 XPS average atomic fraction of Cr, Ni, Fe and Mo in metallic layer underneath the oxide films formed on NiCrFe, NiCr and NiCrMo alloys polarised at different potentials.

alloy. The passive film, therefore, may be thought of being essentially made of trivalent chromium oxide into which relatively small amounts of ions of Ni and of the other alloying elements are incorporated.

In the case of cathodically polarised samples the atomic fraction calculated from XPS data in the metal underneath the oxide film (Fig. 6) corresponds well to the bulk concentration. Figure 6 indicates that during passivation chromium accumulates at the film/metal interface. This result is in agreement with the mechanistic model for the passivation of Ni–Cr–Fe alloys recently proposed by Marcus and Grimal [1]. According to this model an enrichment of the alloy surface with chromium by selective dissolution is a preliminary step in the formation of a Cr_2O_3 passivating layer.

Figure 7 shows the AES profiles of C, O, Cr, Ni and Mo measured on the NiCrMo alloy after passivation at 800 mV. The profiles shown are corrected for the electron escape depth. They indicate the presence of a chromium rich oxide film near the metal surface. The oxide film thickness was defined by the point of 50% intensity of the maximum oxygen amplitude. Carbon is present as surface contamination only. This is evidenced by the fact that the carbon signal goes to zero at depth of 0.5 nm. The Mo signal intensity is fairly constant in the film but increases steadily in the metal. A possible explanation is that preferential sputtering of alloy components with respect to molybdenum occurs in the NiCrMo system similarly to the FeCrMo system [16]. The Fe profile measured on the NiCrFe alloy exhibits a constant value in the film and in the alloy and the signal intensity measured in the alloy is about twice as much as in the film. The chlorine and sulphur profile will be discussed below. The measured oxygen depth profiles show a rather broad film/metal interface. This broadening may be due to several factors like initial surface roughness or sputtering induced topographical and compositional changes of the sample surface. For quantification purposes numerical methods have been developed [4] in order to correct the AES profiles for this broadening effects. The application of such models is however quite complex and requires the use of physical parameters known only approximately. For this reason only the correction for the escape depth was carried out in the present study. The observed profile broadening limits the interpretation of the depth profile of the metals. When the film thickness is smaller or of the same order than the observed profile broadening (approx. 1 nm) the Auger signal measured at average depths corresponding to the oxide contains significant contributions from the underlying metal. The passive films formed at 800mV are therefore expected to be less affected by the broadening effect since their thickness is well above 1 nm. The atomic concentration ratio profile Cr/Ni+Cr measured on the NiCr alloy at 800 mV, shown in Fig. 8, confirms the high Cr content of the film (*ca.* 80% atm) in agreement with the XPS results. Figure 8 suggests that no significant compositional gradients exists within

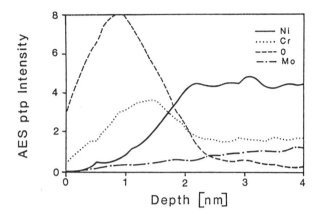

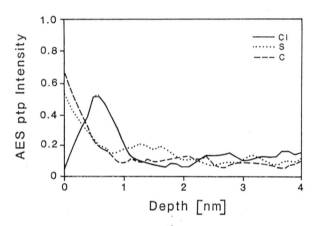

Fig. 7 AES amplitude–depth profiles after correction for the escape depth measured on the NiCrMo alloy polarised at 800 mV.

the film and in the underlying metal. Small gradients could, however, be masked by the broadening effect mentioned above. Similar results were found with the NiCrMo alloy passivated at 800 mV.

3.2 Concentration and distribution of anions

The binding energy of the measured XPS spectra of sulphur (169.4 eV) and chlorine (199.3 eV) shows that these elements are present on the surface as sulphate and chloride ion respectively. The average atomic fractions of the O, S and Cl anions were calculated using the sensitivity factors listed in Table 2 and plotted in Fig. 9. The concentration of sulphur is much higher in the passive film than on the cathodic samples thus indicating that passivation leads to the incorporation of sulphate into the passive layer. The presence of sulphate ions in passivated Ni–Cr–Fe alloy was also found by Kawashima *et al.* [5] and Seo and Sato [6] in sulphuric acid. The presence of Mo in the alloy reduces the amount of sulphate incorporated (Fig. 9). No significant variation of chlorine content with potential or alloy composition are found. This is probably due to the very poor sensitivity of XPS for chlorine which results in a very weak and therefore difficult to quantify signal.

AES depth profiling yields some valuable information concerning the depth distribution of anions. The sulphur profile (Fig. 7) indicates that sulphur is present at the surface as contamination, in the same way as carbon, and in the bulk of the passive film in the depth range 0.5 to 1.6 nm. The chlorine depth profiles shown in Fig. 10 are corrected for the escape depth. The apparent increase of the Cl LMM signal (181 eV) observed below the oxide formed on the NiCrMo alloy is due to the interference with the Mo MVV peak (186 eV). On the samples treated cathodically the chlorine is present only at the surface, but after passivation the chlorine is located deeper in the interior of the film as indicated by the second maximum appearing in the

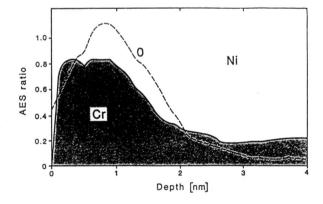

Fig. 8 AES atomic concentration ratio profile Cr/Cr+Ni measured on the NiCr alloy polarised at 800 mV. The O amplitude profile is also plotted to show the film/metal interface.

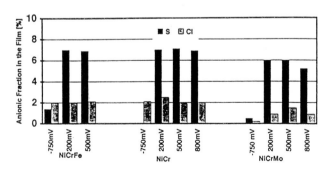

Fig. 9 XPS average anion fraction of S, Cl and O (not shown) in oxide films formed on NiCrFe, NiCr and NiCrMo alloys polarised at different potentials.

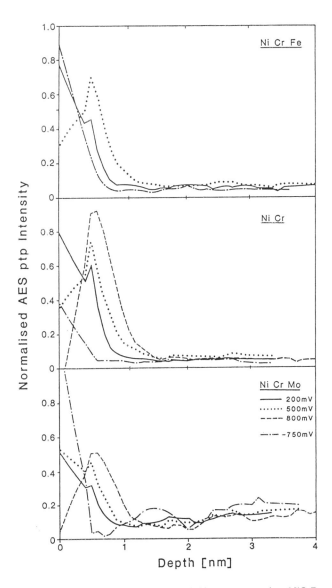

Fig. 10 AES amplitude–depth profiles of chlorine measured on NiCrFe, NiCr and NiCrMo alloys (amplitude normalised after correction for the escape depth with respect to the Ni signal in the bulk).

potential whilst molybdenum completely inhibits the pitting. It is generally accepted that pitting is initiated by the local breakdown of the passive film with subsequent exposition of the alloy to the electrolyte. This can be followed either by a repassivation of the metal or by the formation of a stable growing pit. The likelihood of passive film breakdown has been related to its composition and in particular to the presence of incorporated Cl⁻ ion [17,18]. In particular it has been shown that the pit initiation time is inversely proportional to the amount of chlorine present in the passive film formed by anodisation on a Fe24%Cr alloy [18]. Under the present conditions the chlorine content of the film cannot explain alone the observed differences in pitting behaviour. Mo effectively reduces the concentration of chlorine in the passive film but no difference with respect to the anions content is found between the NiCr and the NiCrFe alloy despite of the fact that the latter alloy is much more sensitive to pitting. The present results indicates that the presence of oxidised iron in the passive film may play a significant role in the chloride induced breakdown of the film. The good corrosion properties of austenitic stainless steels have been related to the accumulation of alloying elements like Ni, Cr and Mo in the metal layer underneath the passive film [19]. This enrichment of metals more noble than iron could facilitate the passivation process and the repassivation of initiated pits. Although the present results indicate an accumulation of Cr and, to a lesser extent, of Mo in the metal underneath the passive film, the composition of this layer does not differ fundamentally from the bulk alloy. Furthermore XPS measurements of alloy composition underneath passive films are subject to many uncertainties [10].

5. Conclusions

The polarisation behaviour of a Ni15Cr alloy is modified by alloying with iron or molybdenum. The alloy Ni15Cr10Fe is more sensitive to pitting than the binary alloy. Alloyed molybdenum on the other hand increases the resistance to pitting.

Passive films formed on Ni15Cr consist mostly of trivalent chromium oxide with incorporation of nickel cations and of chloride and sulphate anions. Alloyed iron and molybdenum do not affect the thickness of the passive film significantly but they alter its composition because of incorporation of iron or molybdenum ions.

Alloyed molybdenum reduced the amount of chloride and sulphate measured in the passive films.

The metallic layer underneath the passive film was found slightly enriched in chromium and molybdenum with respect to the bulk alloy.

profile at a depth of *ca.* 0.5 nm. The intensity of this maximum increases with the applied passive potential. At 800 mV chlorine is found only in the interior of the film but not at the surface. Similar chlorine profile were found by Lorang *et al.* [4] on passivated Inconel 600 and Hastelloy C4 alloys. Iron as alloying element does not influence the chlorine content in the film whilst molybdenum reduces it significantly.

4. Discussion

The results of the present work shows that additions of iron or molybdenum modify significantly the pitting behaviour of the Ni15Cr alloy in the studied electrolyte: iron promotes pitting by lowering the pitting

6. Acknowledgements

This work was financially supported by Fonds National Suisse, Bern and by the Centre d'Etudes de Chimie Métallurgique, CNRS Vitry/Seine. The authors thank J. Bigot and S. Peynot for the preparation of the alloys.

References

1. P. Marcus and J. M. Grimal, Corros. Sci., **33**, 805–814, 1992.
2. A. S. Lim and A. Atrens, J. Appl. Phys., **A54**, 343–349, 1992.
3. B. G. Pound and C. H. Becker, J. Electrochem. Soc., **138**, 696–700, 1991.
4. G. Lorang, N. Jallerat, K. Vu Quang and J. P. Langeron, Surf. Interface Anal., **16**, 325–330, 1990.
5. A. Kawashima, K. Asami and K. Hashimoto, Corros. Sci., **25**, 1103–1114, 1985.
6. M. Seo and N. Sato, Corrosion NACE, **36**, 334–339, 1980.
7. N. S. McIntyre, D. G. Zetaruk and D. Owen, J. Electrochem. Soc., **126**, 750, 1979.
8. S. Mischler, H. J. Mathieu, A. Vogel and D. Landolt, Corros. Sci., **32**, 925–944, 1991.
9. J. Bigot and A. I. M. Liège, Chauffage et Fusion par Induction 10, 1978.
10. S. Mischler, H. J. Mathieu and D. Landolt, Surf. Interface Anal., **11**, 182–188, 1988.
11 R. Kirchheim, S. Hofmann, D. Kuron and A. Schneider, Corros. Sci., **31**, 191, 1990.
12. R. Goetz and D. Landolt, Electrochim. Acta, **29**, 667, 1984.
13 K. J. Vetter, in Electrochemical Kinetics, p 754, Academic Press, New York, 1967.
14 W. D. Wagner, W. M. Riggs, L. E. Davis and G. E. Muilenberg, Handbook of X-Ray Photoelectron Spectroscopy, Perkin Elmer Corp., Minnesota, 1979.
15. P. Marcus, J. Oudar and I. Olefjord, J. Microscop. Spectroscop. Electron., **4**, 63, 1979.
16. H. J. Mathieu and D. Landolt, Appl. Surf. Sci., **3**, 348, 1979.
17. S. Mischler, Thesis No. 760, EPFL Lausanne, 1988.
18. C. Hubschmid and D. Landolt, to be published in J. Electrochem. Soc..
19. H. Fischmeister and U. Roll, Fresenius Z. für Anal. Chem., **319**, 639-645, 1984.
20. L. E. Davis, N. C. MacDonald, P. W. Palmberg, G. E. Riaeh and R. E. Weber, Handbook of Auger Eleetron Spectroscopy, Perkin Elmer Corp. Minnesota, 1976.
21. M. P. Seah and W. A. Dench, Surf. Interface Anal., **1**, 2, 1979.
22. P. M. A. Sherwood, in Practical Surface Analysis by Auger and Photoelectron Spectroscopy, D. S. Briggs and M. P. Seah (eds), App. 3, 455, Wiley, Chichester, 1983.

6

Depth Composition of Passive Films Grown on Fe–Cr Alloys in Borate Solutions: Influence of the Chromium Bulk Concentration

S. BOUDIN, C. BOMBART, G. LORANG AND M. DA CUNHA BELO

Centre d'Etudes de Chimie Métallurgique (CECM)/Centre National de la Recherche Scientifique (CNRS), 15, rue Georges Urbain-F 94407, Vitry/Seine Cedex, France

Abstract

The influence of the bulk chromium concentration $[Cr]_b$ on the depth composition of passive layers formed in de-aerated borate buffer solutions (+ 0.3 V/SCE at pH = 9.2) on the surface of Fe–Cr alloys (0–30 Cr%) was studied by means of Auger electron spectroscopy (AES) coupled with ion sputtering. Depth concentration profiles are derived from a previous quantitative approach based on the sequential layer sputtering (SLS) model. Oxidised chromium accumulates at every studied concentration against the internal film/alloy interface; simultaneously an extended metallic chromium depletion in the alloy is clearly pointed out under the passive film. Only above $[Cr]_b = 15 Cr\%$, the formation of some nearly complete chromium oxide ($\sim Cr_2O_3$) monolayers may occur at the interface, which we can relate to an effective protective barrier against corrosion of the alloy.

The electrochemical study mainly consisted in establishing cathodic reduction kinetics in diluted sulphuric acid (0.05 M) of the former passive films. Characteristic step-potentials were observed, which could be attributed to the reduction reactions of the successive external 'iron oxide' and internal 'chromium oxide' layers as revealed by the depth profiling analysis.

1. Introduction

It has widely been reported that passive films formed on the surface of Fe–Cr alloys were chromium 'enriched' [1–8]. When the bulk chromium concentration exceeds 12%, such enrichment was correlated with an improved corrosion resistance of the alloys in acidic and neutral solutions [5–8]. It was usually illustrated by the ratio of atomic fractions of oxidised Cr and Fe vs the bulk chromium concentration. In fact, sharp concentration gradients occur through the passive overlayers and this chromium oxide 'enrichment' remains located only in a very thin internal layer as shown in our latest studies [9–12] and cannot be merely quantified averaging the oxidised chromium amounts over the whole film thickness [2, 6–7]. So, in order to provide a better significant evolution of the oxidised chromium contents as a function of $[Cr]_b$, analytical results derived from AES and XPS sputter profiles should be preferably converted in 'true' depth concentration profiles, i.e. by absolute atomic densities (at.nm^{-2}), instead of atomic fractions, as a function of the eroded depth. In this direction, a suitable approach of AES depth profiling starting from the sequential layer sputtering (SLS) model [13] was developed in our laboratory for the analysis of thin oxidised films

[10–12]. In passive layers grown on stainless steels and thermal alumina films [12], its quantitative aspects were globally proved with the oxygen atomic balances deduced from AES depth concentration profiles which were found in accordance with the absolute oxygen film contents measured by nuclear reaction analysis.

In this work, AES depth profiling experiments were carried out on a series of Fe–xCr alloys, with x ranging from 0 to 31.5 at.%, passivated for 2 h in borate buffer solutions (pH = 9.2) at + 0.3V/SCE. The main purpose of this study was to point out that a critical $[Cr]_b$ concentration also exists in the borate medium for passivated Fe–Cr alloys, concentration around which large modifications of the oxidised atomic chromium densities are encountered at the film/alloy interface. This Cr concentration changes inside the film will be otherwise approached achieving a cathodic reduction of the previous passive layers. This method was employed as a fast and complementary technique for AES depth profile analysis because it allows to relate clearly the different potential plateau observed during the kinetic to the reduction of distinct oxide species as a function of their depth location while their lengths may be used to characterise qualitatively their stability or resistance to passivity breakdown [13–15].

2. Experimental

High purity metals containing less than 50 ppm impurities were melted into a plasma furnace (losses during melting are lower than 0.015%) to prepare the iron–chromium alloys. The atomic compositions of the alloys are reported in Table 1 as well as those determined by a typical quantitative AES analysis using pure standards [9]. Samples (2 cm^2 area) are cold-rolled, mechanically polished (up to alumina 0.3 μm) and then ultrasonically cleaned in demineralised water before their annealing 0.5 h at 900°C.

Electrodes were after that electrochemically polished in an aceto-perchloric solution (40 V, 1 mn), rinsed in water and dried in air; afterwards, they were cathodically reduced (10 mA.cm^{-2} at –1.8 V/SCE, potentials are referred to the saturated calomel electrode) during 5 mn in the passivating solution (0.075M Na tetraborate, 0.05M boric acid, pH = 9.2). Polarisation is applied directly to the passivation potential + 0.3 V/SCE for all studied alloys and iron during 2 hours (the residual current density is typically lower than 0.05 μA.cm^{-2}). The solution was evacuated and continuously replaced by water to rinse the samples which are finally covered in the bath with a small polypropylene (PP) sheet to keep a permanent thin liquid film on the passive layers during their transfer in the atmosphere and pumping in the fast entry storage chamber (10^{-6} Pa). At last, the PP sheet is removed in vacuum before the sample introduction into the Auger analysis chamber (10^{-8} Pa).

The same procedure was followed for the cathodic reduction in 0.05M H$_2$SO$_4$ solutions of the last passive films in order to protect them from the air contamination when the solutions were changed. Simultaneously with the sample immersion, a 1.25 μA.cm^{-2} current density was established between the platinum counter electrode and the working one.

Depth profiling was performed using high resolution Auger electron spectrometry in combination with ion sputtering. An external electron gun (30° incidence with the sample normal) operates with a defocused 9 keV primary electron beam rastering a sample area of *ca.* 250 × 250 μm^2 to minimise the possible damages of irradiation (1 μm dia. for the focused electron beam). A semi-dispersive kind of analyser (Riber Mac II) is employed, its axis making an angle of 40° with the surface normal to the sample (mean take-off angle: 55°). It works in the constant resolution ΔE mode over the whole energy range (0.75 eV energy resolution is presently achieved using a 11.5 eV pass energy). Digitised spectra are recorded in the direct mode N(E) between the sputtering sequences in order to avoid spectra distortions and energy shifts induced by ion bombardment. Sputter profiles are processed in the d[E.N(E)] dE derivative mode after the direct spectra have been smoothed and differentiated.

The scanning ion gun (Riber CI 50 RB), differentially pumped with a 1.3 10^{-4} Pa krypton base pressure, works at 3 keV onto a rastered sample area of 0.25 cm^{-2} (with a typical 1 μA ion target current and 50° incidence, the sputter rate lies around 1.0 nm mn^{-1} as

Table 1 Quantitative AES analysis of passivated Fe–Cr alloys in borate solutions 2 h at + 0.3V/SCE (nominal compositions in weight percents are in parentheses). I_{Cr} and I_{Fe}: normalised Auger intensities referred to pure iron; N_{Cr}, N_{Fe}: atomic densities in alloys. J: mean film thickness; ΣN_{Crox}, ΣN_{Crox}: atomic balances of oxidised Fe and Cr

[Cr]$_b$, at%	0	5.4	10.7	15.9	21.2	26.4	31.5	100
(nominal,wt%)		(5.2)	(10.2)	(15.0)	(20.05)	(25.0)	(31.3)	
AES [Cr]$_b$, at%		4.8±0.7	9.7±0.5	15.5±0.8	19.5±0.8	24.8±1.4	32.0±1.0	
I_{Cr}, a.u.		5.6±1.2	11.2±0.3	18.0±0.8	22.7±0.5	28.7±1.5	37.1±0.6	**114±2**
N_{Crb},at.nm^{-2}		0.9±0.2	1.9±0.1	3.0±0.1	3.8±0.1	4.8±0.2	6.2±0.1	**19.1**
I_{Fe}, a.u.	**100**	95.6±2.1	89.6±2.1	84.3±0.7	80.3±1.4	74.9±1.7	67.4±2.1	
N_{Feb}, at.nm^{-2}	**19.3**	18.4±0.4	17.4±0.4	16.3±0.2	15.5±0.3	14.5±0.3	13.1±0.4	
J thickness(ML)	16	21	20	20	16	20	18	
ΣN_{Crox}, at.nm^{-2}	-	22.0	31.4	59.2	69.5	80.1	62.2	
ΣN_{Feox}, at.nm^{-2}	103.3	116.8	73.8	59.7	56.3	54.1	43.0	

calculated for those passive layers). Sensitivity, β coefficients for elements in the film (ox) and in the alloy (met) are experimentally determined at 9 keV primary energy with pure standards (Fe, Cr, and *in situ* thin thermal Fe_2O_3 and Cr_2O_3 films [9]). For complementary information about our quantitative approach of depth profiling in passive layers and in alloys, the reader is invited to consult Ref. [9–12].

3. Results and Discussion

Depth analysis of the Fe–Cr alloys and iron passivated in the borate buffer medium are presented in Fig. 1 where only atomic densities of Fe and Cr elements as a function of the eroded depth are mentioned for clarity. Quantified results about Auger analysis of sputter clean alloys and passive overlayers are presented in Table 1.

The surface layers of the passive films, because of their prior contact with the electrolyte, contain mainly C, O, B and Cl atoms. Beneath this contamination layer, an iron 'oxi-hydroxide' phase (assimilated to Fe_2O_3 for the needs of the AES quantitation) predominates in the outerpart while the deeper layers near the alloy interface become more and more Cr-oxide (Crox) enriched with increasing Cr bulk content. The more important feature we can extract from this study deals with this Crox enrichment level.

All oxidised chromium gradients shown in Fig. 1 exhibit a maximum Crox atomic density higher than the corresponding $[Cr]_b$. Therefore a Crox enrichment exists at any Cr composition studied but remains always located in a narrow region immediately at the interface. With $[Cr]_b=15$ at.%, a steep increase of the maximum atomic density attests that, close to this Cr level, a critical concentration is attained in borate medium as already mentioned in acidic solutions [2,5,7]. Concurrently, the Crox amounts become the major oxidised component of the film (Table 1). For $[Cr]_b >15$ at.%, a nearly complete oxidised chromium barrier (4-6 ML thickness) containing more than 80% Cr_2O_3 (Crox in Cr_2O_3 = 8.8 at.nm^{-2}) is actually edificated, which is commonly thought to ensure a more effective protection of the metallic alloy.

Another important point should be underlined here, concerning the clear metallic chromium depletion (8–10 ML or *ca.* 2 nm) which occurs in the first layers of the underlaying alloy at any studied $[Cr]_b$. This is not an artefact of our computational treatment of profiles because this Cr loss is usually balanced by an equivalent Fe gain (while all profiles were separately processed). According to Landolt [8], such experimental evidence may justify the mechanism of a preferential oxidation of chromium. At the opposite, after passivation in acidic medium (Na_2SO_4, NaCl +

HCl solutions, pH = 2) [10], an oxidised chromium enrichment could be depicted in the film without any metallic Cr depletion in the underlayers of ferritic stainless steels: this speaks in this case for a simultaneous oxidation of the alloy constituents, followed by a selective dissolution of iron.

The film thickness, expressed by the number J of oxidised monolayers (1 ML = 0.215 nm [9]), is a parameter directly related to the sputter rate (assumed to remain constant in the film as in the alloy), which is deduced during the data profile processing [10–12]. In Table 1, the J parameter did not evolve widely (16 to 21 ML or 3.4 to 4.5 nm) in this medium, in opposition to what was noted at high bulk Cr contents in NaCl solutions (pH = 5.6) [6] and in borate solutions (pH = 8.98) [7] in which decreasing thicknesses were obtained. For the time being, the reported passive film thickness data at pH = 9.2 is not sufficiently substantial to justify whether a film growth according to the Cabrera-Mott model had occurred (as invoked by Kirchheim *et al.* [7]) or not.

Finally, cathodic reduction (CR) kinetic were carried out on the similar passive layers in order to perform a fast depth analysis which might be correlated with the surface analytical results [14–16]. The potential plateaux displayed in kinetics are attributed to the redox reaction of the alloy constituents being oxidised in the passive film. Then, with the help of standards taken with passivated chromium or iron corrosion products deposited on a Fe–Cr alloy, it was possible to estimate the potentials of anodic chromium and iron 'oxides' at respectively + 0.3 and – 0.4 V/SCE and for the oxide-free surface of alloys between – 0.5 and – 0.6 V/SCE [15].

Results for alloys and pure iron are reported in Fig. 2. Starting from a potential near + 0.4 V, the first reduced (outer layers of the film) species must be iron oxi-hydroxide mixtures at low bulk chromium contents ($[Cr]_b < 10\%$) as proved by the unique potential plateau observed before the reactivation of the surface alloy at *ca.* – 0.5 V. As reported by Seo *et al.* [17], the time required for this surface alloy reactivation allows to evaluate the corrosion resistance of some additions in stainless steels. In the same way, the plateau lengths could be used to evaluate the thicknesses of the respective iron and chromium oxide layers [15]. In fact, the different CR plateau lengths shown on Fig. 2 and attributed to iron 'oxide' at low $[Cr]_b$ cannot be clearly related with the small increase of the film thicknesses deduced from Auger analysis (Table 1); however, the CR results are globally in accordance with AES to point out that oxidised iron is the main component of those films. $[Cr]_b = 15 \%$ is confirmed as a critical concentration considering the change of kinetic lineshapes (Fig. 2). Thus the smooth potential decay

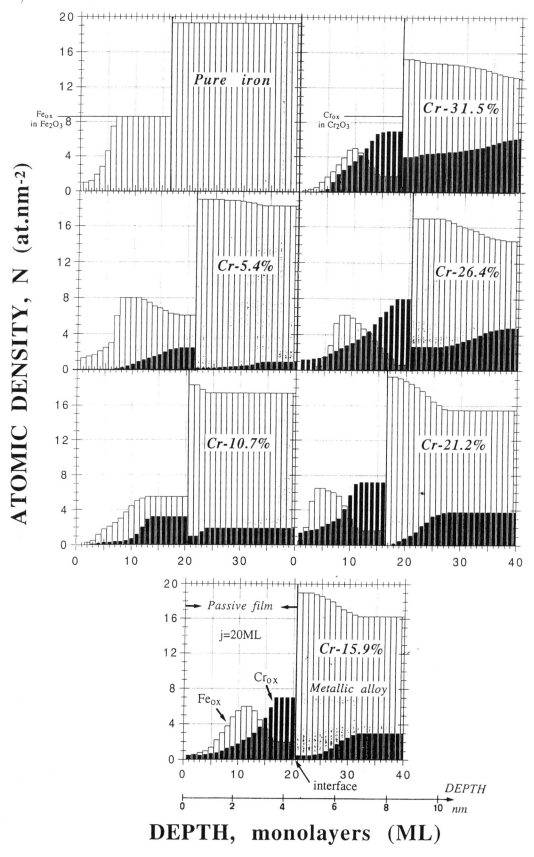

Fig. 1 Atomic concentration depth profiles of passive films formed during 2 h at +0.3V/SCE in borate solutions (pH = 9.2) at the surface of Fe–xCr alloys (0 ⩽ x ⩽ 31.5 at.%).

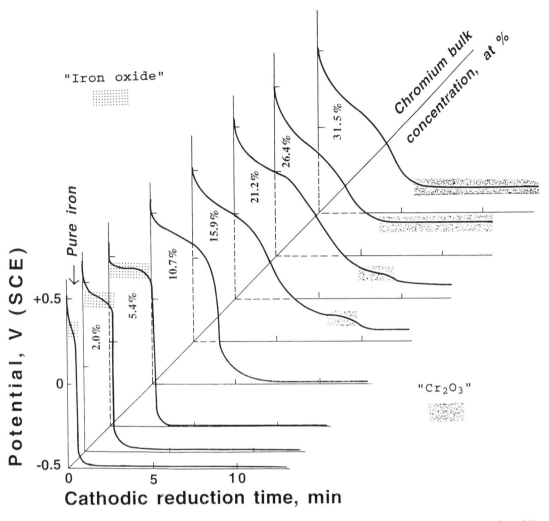

"Iron oxide"

Chromium bulk concentration, at %

31.5 %

26.4 %

21.2 %

15.9 %

Pure iron

10.7 %

5.4 %

2.0 %

Potential, V (SCE)

+0.5

0

-0.5

"Cr₂O₃"

Cathodic reduction time, min

0   5   10

Fig. 2 Cathodic reduction kinetics of passive films formed during 2 h at +0.3 V/SCE in borate solutions (pH = 9.2) at the surface of Fe–xCr alloys (0 ⩽ x ⩽ 31.5 at.%). Galvanostatic reduction is performed in a 0.05M H₂SO₄ solution with an applied current density of 2.5 mA cm⁻².

displayed between + 0.4 and – 0.4 V confirms that an oxidised Fe–Cr mixture constitutes now the external part of the film while the small plateau occurring at – 0.4 V, due to the reduction of the oxidised chromium [15], proves that Crox becomes the major component of the inner layers. With a plateau length increasing at higher [Cr]$_b$, the Crox barrier is assumed to be more completely formed as observed by AES in Fig. 1. Moreover, this reveals for those Crox species an enlarged resistance to reduction in agreement with the well-known beneficial influence of chromium in the films to ensure a good protection against corrosion in stainless steels.

4. Conclusions

The following conclusions concerning the influence of the chromium bulk concentration on the film composition of the Fe–Cr (0–30%) alloys passivated in borate

solutions were derived from AES depth profiling and cathodic reduction kinetics.

1. Elemental enrichment or depletion in depth profiles are quantified by means of absolute atomic densities.

2. An oxidised chromium enrichment is displayed in the inner part of the passive film against the film/alloy interface while a metallic Cr depletion simultaneously occurred beneath the film at any chromium bulk composition studied. Results speak for a mechanism involving a preferential oxidation of chromium at this pH. The passive film thickness does not significantly evolve with [Cr]$_b$ (3.5 to 5 nm).

3. A critical concentration [Cr]$_b$ around 15 % is confirmed in borate medium above which the building of an oxidised Cr barrier equivalent to complete Cr₂O₃ layers (4–6 monolayers) is practically achieved.

4. Cathodic reduction kinetics of the previously passive films exhibit two distinct potential plateau for the respective reduction reactions of the outer Fe oxide layers (+ 0.4 V) and the inner Cr oxide layers (– 0.4 V). In agreement with AES, rich 'iron oxide' layers are shown to occur at low $[Cr]_b < 10\%$, while a wide resistance to reduction of the nearly complete protective Cr_2O_3 layer is displayed when the critical $[Cr]_b$ passes beyond *ca.* 15 %.

References

1. R. P. Frankenthal and D. L. Malm, J. Electrochem. Soc., **123**, 186, 1976.

2. K. Asami, K. Hashimoto and S. Shimodaira, Corros. Sci., **18**, 151, 1978.

3. I. Olefjord, Mater. Sci. Eng., **42**, 161, 1980.

4. H. H. Strehblow, Surf. Interface Anal., **12**, 363, 1988.

5. H. Fishmeister and U. Roll, Fresenius Z. Anal. Chem., **319**, 619, 1984.

6. P. Brüech, K. Müller, A. Atrens and H. Neff, Appl. Phys., **A38**, 1, 1985.

7. R. Kirchheim, B. Heine, H. Fischmeister, S. Hofmann, H. Knote and U. Stolze, Corros. Sci., **29**, 899, 1989 and J. Häfele, B. Heine and R. Kirchheim, Z. Metallkd., **83**, 395, 1992.

8. D. Landolt, Surf. Interface Anal., **15**, 395, 1990.

9. G. Lorang, M. da Cunha Belo and J-P. Langeron, J. Vac. Sci. Technol., **A5(4)**, 1213, 1987.

10. G. Lorang, F. Basile, M. da Cunha Belo and J-P. Langeron, Surf. Interface Anal., **12**, 424, 1988.

11. G. Lorang, N. Jallerat, K. Vu Quang and J-P. Langeron, Surf. Interface Anal., **16**, 325, 1990.

12. G. Lorang, J. L. Xu and J-P. Langeron, Surf. Interface Anal., **19**, 60, 1992.

13. J. M. Sanz and S. Hofmann, Surf. Interface Anal., **8**, 147, 1986.

14. v. Mitrovic-Stepanovic, B. MacDougall and M. J. Graham, Corros. Sci., **24**, 479, 1984.

15. F. Basile and G. Lorthioir, Brit. Corros. J., **28**, 31, 1993.

16. M. da Cunha Belo, B. Rondot, F. Pons, J. Le Hericy and J-P. Langeron, J. Electrochem. Soc., **124**, 1317, 1977.

17. M. Seo, G. Hultquist, C. Leygraf and N. Sato, Corros. Sci., **26**, 949, 1986.

7

Chemical Composition of Naturally Formed Oxide on MgAl Alloys

J. H. NORDLIEN AND K. NISANCIOGLU

Department of Electrochemistry, Norwegian Institute of Technology N-7034 Trondheim, Norway

Abstract

Although magnesium oxide is not known for its corrosion resistance, certain magnesium alloys exhibit highly stable oxides in atmospheric exposure and in chloride solutions. The structure and chemical composition of naturally forming oxides on several magnesium–aluminium alloys are investigated by XPS with the purpose of understanding the factors influencing their corrosion properties. A significant enrichment of aluminium in the oxide relative to the aluminium concentration in the alloy is detected, and the significance of this result on the corrosion behaviour is discussed.

1. Introduction

Earlier work [1] has shown that alloying magnesium with aluminium leads to a significant improvement of the corrosion resistance in chloride media, especially as the aluminium concentration is increased from 2 to 4 wt % as shown in Fig. 1. Investigation of the various microstructural factors, such as the role of intermetallic phases, as possible causes of this improvement, did not provide satisfactory explanations [1]. It was decided, therefore, to investigate the effect of alloying with aluminium on the properties of the oxide film formed on magnesium and whether these properties can he related to the observed corrosion behaviour. The purpose of this paper is to measure the composition of the oxide forming naturally on clean metal surface as a function of thickness and aluminium content of the alloy and explore whether the information can be correlated with the corrosion behaviour in aqueous chloride media.

2. Experimental

Pure (99.99%) magnesium and commercial high-purity grade, die-cast MgAl alloys have been investigated. The compositions of the latter are given in Table 1. These alloys normally contain a small amount of manganese added to improve the corrosion resistance [2]. However, they can be regarded essentially as binary MgAl alloys since the manganese content does not exceed 0.5 %. The alloys were homogenised at 416°C for 16.5 h.

The oxide film growth and composition have been investigated by the XPS technique in a Vacuum Science Workshop (VSW) X-ray source using an Al anode

Table 1 Composition of the test materials obtained by spectrographic analysis (in wt%)

Alloy	Al	Mn
Mg-2%Al	2	0.5
Mg-5%Al	5	0.28
Mg-8%Al	5	0.26

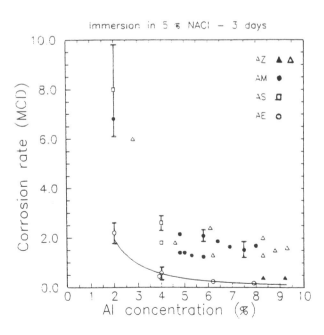

Fig. 1 Corrosion rate of various die cast alloys as a function of their Al content (wt %). MCD = $mgcm^{-2}day^{-1}$. From [1].

(Al Kα excitation hη = 1486.6 eV) in conjunction with a VSW HA-50 hemispherical electron analyser. The initial pressure in the chamber before introducing oxygen was *ca.* 10^{-10} torr.

'Clean' surfaces were obtained by mechanically scratching the surface *in situ* with a ruby stylus. The scratching method was selected instead of ion beam sputtering for removing the air-formed oxide because the latter enriches the surface with aluminium due to a higher rate of magnesium removal. Exposure to oxygen and chemical analysis were performed in an alternate sequence in the same chamber. Oxygen was introduced into the chamber using a controlled leak valve. All specimens were exposed to oxygen in the range 0 to 200 Ł (1 Ł = 1 Langmuir = 10^{-6} torr s).

3. Results and Discussion

The XPS Mg_{1s} peak data for the as scratched alloy surfaces are shown in Fig. 2. No chemical shift is discernable from these data with increasing aluminium content in the alloy within the resolution limits of the instrument (*ca.* 0.1 eV). This is probably because the Al atoms substitute for the Mg atoms in the lattice, causing small changes in the electric field of Mg_{1s} electrons. Figures 3 and 4 show, respectively, the shift in the Mg_{1s} peak of pure Mg and alloy Mg–5%Al as a function of oxygen exposure. The oxidation rates of all materials examined are quite similar. The shift in the Mg_{1s} peak as a result of oxide formation is *ca.* 1.3 eV for all alloys,

in good agreement with earlier data for pure magnesium [3,4].

Quantitative analysis of the Mg_{1s} peak suggests the presence of a small amount of MgO on the as scratched surfaces as shown in Fig. 5 in the case of alloy Mg–2%Al. Expressed in thickness, this oxide after about 1 h exposure to 10^{-10} torr is a fraction of an Ångstøm (*ca.* 0.6 Å on alloy Mg–2%Al), i.e. it does not yet provide a complete coverage of the surface.

The nature of the oxide on the as scratched surfaces was investigated further by analysing the Mg_{2s} and Al_{2p} peaks. The two peaks recorded simultaneously for a Mg–2%Al specimen are shown in Figs 6 and 7, respectively. Although the Mg_{2s} peak is more bulk dependent than the Mg_{1s} peak, the amount of oxide detected by the former is larger. Moreover, analysis of the Al_{2p} peak indicates that a significant fraction of the aluminium at the surface is oxidised.

In addition to oxygen leak into vacuum chamber, entrainment of the air-formed oxide and contamination from the ruby stylus, both during scratching, are under consideration as the possible causes of the initial oxide detected. Direct contamination by the stylus is not considered as very likely. Entrained oxide may be the cause of higher oxide concentration under the Mg_{2s} peak. However, neither of these two possibilities can account for the high Al/Mg ratio detected on the as scratched surface, as can he discerned from Fig. 8, in view of the small aluminium concentration in the metal. Whatever the source of this initial oxide, the

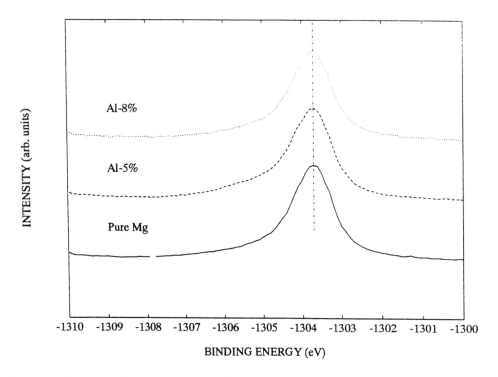

Fig. 2 Mg_{1s} peaks for three different materials measured on as scratched surfaces. No chemical shift is observed, as a result of alloying Mg with Al.

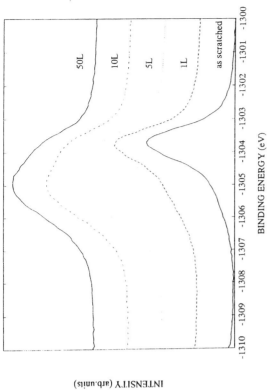

Fig. 3 The development of the Mg_{1s} peak for pure magnesium as a function of exposure to oxygen.

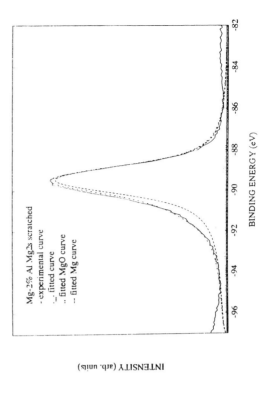

Fig. 4 Development of the Mg_{1s} peak for alloy Mg-5% Al as a function of exposure to oxygen.

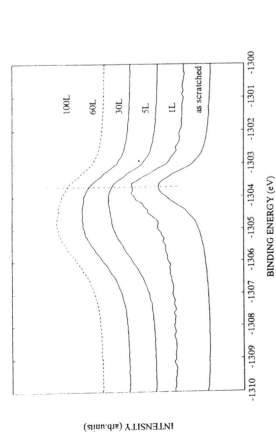

Mg-2% Al "clean" Mg_{1s} peak
o : experimental curve
-: fitted curve
..: fitted Mg curve
--: fitted MgO curve

Fig. 5 Mg_{1s} peak for as-scratched Mg-2% Al holding for 1 h at 10^{-10} torr.

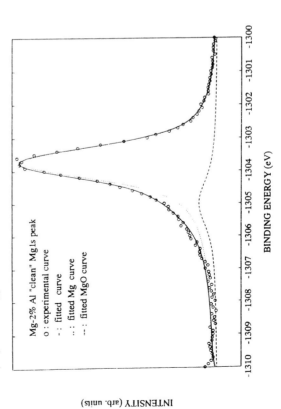

Mg-2% Al Mg_{2s} scratched
- experimental curve
-.: fitted curve
..: fitted MgO curve
--: fitted Mg curve

Fig. 6 Mg_{2s} peak for the as-scratched Mg-2% Al surface.

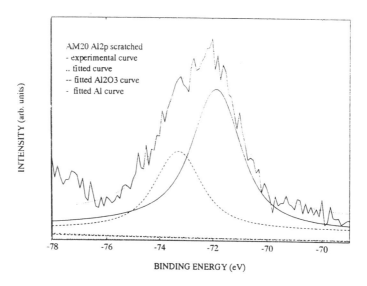

Fig. 7 Al$_{2p}$ peak for the as-scratched surface. Notice the size of the Al$_2$O$_3$ peak.

results indicate that the aluminium component of the alloy has a much stronger affinity to oxygen than does magnesium, such that the aluminium in the oxide is strongly enriched relative to the amount of aluminium in the alloy.

Figure 8 also shows, based on the Mg$_{2s}$ and Al$_{2p}$ peaks, the Al/Mg ratio profile as the oxide is allowed to form under exposure to oxygen. The aluminium concentration decreases relative to magnesium from the high value close to the metal surface in the subsequent oxide layers formed. With increasing oxide thickness, however, the Al/Mg ratio in the oxide still remains significantly higher than the Al/Mg ratio in the alloy for the three MgAl alloys tested.

The Al/Mg ratio in the oxide appears to attain a constant level at a certain oxide thickness measured from the metal surface. Figure 8 suggests that the aluminium concentration in this region is *ca.* 20 at.% for alloys Mg–5% Al and Mg–8% Al, i.e. relatively independent of the Al composition in the alloy, whereas it is *ca.* 7 at. % for alloy Mg–2% Al. These values agree well with the analyses of the air-formed oxides on the three alloys. It appears that the aluminium concentration in the oxide reaches some sort of a saturation level when the aluminium content of the alloy is 5 % or larger.

Based on these results, we believe that there is a certain correlation between the corrosion rate observed

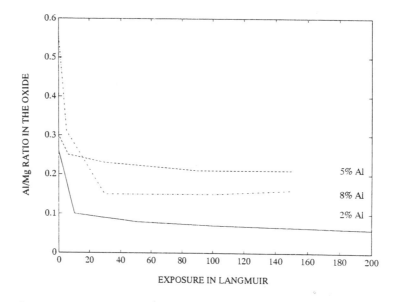

Fig. 8 Al/Mg ratio in the oxide as a function of exposure to oxygen. The steady compositions are reached already after ca. 60 Ł.

in the aqueous chloride media and the aluminium content of the oxide. First of all, the enrichment of the oxide with the aluminium component of the alloy is beneficial for corrosion resistance because aluminium stabilises the oxide in aqueous media of nearly neutral pH. Secondly, the increase in the aluminium oxide concentration with increase in the amount of alloyed aluminium up to *ca.* 4 % can he correlated with the significant improvement in the corrosion resistance in this composition range (Fig. 1). Further increases in the alloyed aluminium content does not cause further significant improvements in the corrosion resistance because the aluminium concentration of the oxide remains unchanged. We are in the process of obtaining XPS data for MgAl alloys with different aluminium compositions to give further support to this hypothesis. The nature and causes of the oxide on the as scratched surface are also under investigation.

4. Acknowledgements

This work was supported by Norsk Hydro a.s. and The Royal Norwegian Council for Scientific and Industrial Research.

References

1. O. Lunder, K. Nisancioglu and R. S. Hansen, SAE Technical Paper No. 930755, 1993.
2. O. Lunder, T. Kr. Aune and K. Nisancioglu, Corrosion, **43**, 291, 1987.
3. N. A. Braaten and S. Raaen, Physica Scripta, **43**, 430, 1991.
4. S. A. Flodström, C. W. B. Martinson, G. Kalkoffen and C. Kunz, Mat. Sci. Eng., **42**, 31, 1980.

8

Examination of Passive Layers on Fe/Cr, Fe/Ni and Cu/Ni Alloys with Ion Scattering Spectroscopy (ISS)

H.-H. Strehblow, C. Calinski, S. Simson, P. Druska, H.-W. Hoppe and A. Rossi*

Heinrich-Heine-Universitat Dusseldorf, Germany
*Universita di Cagliari, Sardegna, Italy

Abstract

Passive layers on Fe/Cr, Fe/Ni and Cu/Ni alloys have been studied with Ion Scattering Spectroscopy (ISS). The calibration of the method with standards permits the determination of depth profiles with a monolayer resolution. Layer models which are deduced from angular-resolved X-ray Photoelectron Spectroscopy (XPS) may be confirmed with ISS in more detail. The ISS method thus gives additional arguments for the multilayer models which are the basis for quantitative evaluation of XPS data. Strong arguments for XPS are its non-destructive nature and the detailed chemical information of this method provided by the chemical shift. Thus the combination of both methods provides a deeper insight to the chemical structure of these films especially when a systematic variation of the parameters of passivation on the basis of electrochemical studies is performed.

1. Introduction

Cr is one of the most important components of steels to achieve an optimum resistance to corrosion phenomena. It enlarges the potential domain of passive behaviour to negative potentials, reduces the current density of active dissolution and shifts the critical potential of localised corrosion in the presence of aggressive anions to more positive values in the passive range. The composition and structure of the passive layer and its electrochemical properties are responsible for its corrosion protection. Therefore a detailed knowledge about these properties is required. Many electrochemical and surface-analytical studies show the accumulation of Cr within the passive layer [1–8]. Among the surface methods X-ray Photoelectron Spectroscopy (XPS) [1,3], Auger Electron Spectroscopy (AES) [7] and Secondary Ion Mass Spectrometry (SIMS) have been applied. Most of these methods yield an average of the composition of the very thin passive layers of some few nm thickness according to their information depth. Angular-resolved measurements of XPS suggest the existence of a variation of the composition with depth. However, a quantitative evaluation encounters difficulties because of surface roughness and therefore not ideally flat multilayer structures. Despite these difficulties for rough surfaces passive layers have been examined successfully with XPS and the data have been evaluated quantitatively on the basis of a bilayer model [8,9]. In some few cases,

i.e. for Fe/Cr [4,5], Fe/Ni [10], Al/Cu [11] and Cu/Ni alloys, Ion Scattering Spectroscopy (ISS) has been applied to determine depth profiles with high resolution. ISS has a unique sensitivity to the topmost atomic layer of a solid surface. Thus combined with sputtering one may obtain a depth profile with atomic resolution. Careful interpretation is necessary as the results can be affected by possible sputter artefacts. This method provides elemental compositions and no additional chemical information like oxidation states. Therefore a close combination of ISS with XPS is rather promising. In addition a systematic variation of the relevant electrochemical parameters of preparation of the passive film is missing in many studies. It has been shown for several systems that the passive layer is submitted to well pronounced changes with time and potential of passivation. It changes its thickness and composition during its formation as well as during its reduction [6, 12, 20]. Similarly, the composition of the electrolyte and the metal phase, as well as the specimen pretreatment, usually have a pronounced influence on the structure of passive layers.

Polarisation curves for most systems provide a very instructive first insight which assists a proper choice for a subsequent surface-analytical examination and a related specimen preparation. Fig. 1(a) depicts a stationary polarisation curve of a Fe15Cr alloy in 0.5M H_2SO_4 with the indication of the potential domains of the soluble species which have been examined with the Rotating Ring Disc Electrode (RRDE).

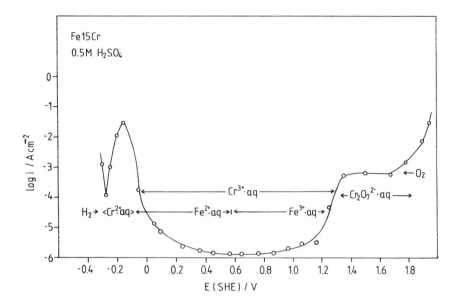

Fig. 1(a) Stationary polarisation curve of Fe15Cr in 0.5M H_2SO_4 with an indication of the soluble species and gases which are formed.

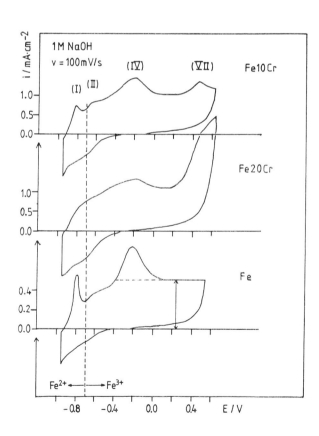

Fig. 1(b) Potentiodynamic polarisation curve of Ar sputtered Fe and Fe/ Cr alloys in 1M NaOH within closed system; — I: oxidation of adsorbed H_2; — II: formation of soluble Fe^{3+} and beginning Fe(III) oxide, IV Fe(II) to Fe(III) oxidation; — VII: transpassive formation of soluble CrO_4^{2-} indication of formation of soluble Fe(II) and Fe(III).

Similar indication of formation of soluble Fe(II) and Fe(III) to the behaviour of pure Cr, passivation occurs around 0 V when Cr^{3+} is formed first indicating the presence of a slowly dissolving Cr(III)oxide. The dissolution of Cr^{3+} from a passivated Cr electrode has been examined with the RRD technique with two concentric rings. Although the developed method worked well, the related corrosion current was too small even during passivation transients to permit a quantitative determination of the dissolution kinetics [21]. The dissolution of Cr^{3+} has been determined with radioactive tracer methods [22] and atomic absorption spectroscopy of the electrolyte [23].

XPS analysis of passive layers on Fe/Cr shows a change from Fe(II) to Fe(III) within the passive layer at the Flade potential of pure Fe, i.e. at E = –0.22 V in 1M NaOH, which is the extrapolated value from 0.58 V in 0.5M H_2SO_4 to pH 13.8 [9]. The same observation was made for pure Fe [17]. XPS studies of passive layers grown in 1M NaOH suggest an increase of the Fe content with potential [12]. Angular resolved measurements show a duplex structure with an Fe(III)-rich outer and a Cr(III)-rich inner part [17]. A confirmation of this idea will be obtained with ISS as a method with sufficient depth resolution. Similar questions arise for other alloys. Thus also Fe/Ni and Cu/Ni alloys have been studied by ISS in close combination with XPS and electrochemical investigations. In this publication our results for these systems are presented. This work shall demonstrate the value of the application of ISS in combination with XPS for a better understanding of the composition of passive layers.

47

2. Experimental

Fe/Cr, Fe/Ni and Cu/Ni alloys were obtained by melting 99.99% pure materials with appropriate mechanical and heat treatment to obtain homogeneous single phase metals. Circular specimens with a sharp edge were cut from the material with a 3 mm extension with a thread at the rear end to attach them to the specimen stub. The circular front plane was polished with diamond spray to a 1 μm final grading and cleaned with water and in ethanol with ultrasonic treatment. Finally the specimens were sputter cleaned with Ar (AG 21, VG Instruments, 5 min, 30 μA, A = 0.8 cm^2, 4 keV) and contacted with their front plane to the electrolyte under potentiostatic control at the electrode potential of interest. After passivation the samples were rinsed four times with pure water and transferred to the ultra high vacuum (UHV) of the analyser chamber within typically 5 min. For short time passivation transients of t < 10 s the specimens were contacted to the electrolyte at potentials of the start of hydrogen formation, i.e. E = – 0.96 V in 1M NaOH for 10 min and finally pulsed to the potential in the passive range. All specimen manipulations were performed without any contact to the laboratory atmosphere. Long-time passivation in the range of 1 h to 1 week was performed in an external electrochemical cell without previous Ar sputtering. After preparation the specimens were rinsed with deionised water and ethanol and immediately introduced into the UHV of the XP-spectrometer.

The electrochemical specimen preparations were performed with an electronic potentiostat. Fast potential changes were performed with pulse generators (Tektronix 26G3). The applied potential changes were fed into the potentiostat. The electrolytes were prepared with chemically pure substances (pro analysil) and deionised water (Millipore water purification system). Hg/Hg$_2$SO$_4$/ 0.5M H$_2$SO$_4$, E = 0.68 V and Hg/HgO/1M NaOH, E = 0.14 V served as reference electrodes. All potentials are given relative to the standard hydrogen electrode (SHE).

XPS and ISS measurements were performed in a commercial three chamber UHV system (ESCALAB 200 X, VG Instruments). The analyser chamber contains a twin X-ray source and a small spot ion source (EX 05 or AG61 VG Instruments) with scanning facilities for depth profiling. The preparation chamber contained the ion source for Ar sputter cleaning. The fast entry lock was connected to a fourth chamber which contained the electrochemical cell and the necessary feedthroughs for electrolyte, water and Ar supplies. All electrolytes and water were purged for at least 1 h with purified Ar (Oxisorb, Messer) before introduction into the small electrochemical cell with a volume of 2 cm^3. The quality of the specimen transfer and the transfer system with respect to traces of oxygen and other contaminants was routinely checked at least once a day. Seminoble metals like Cu did not form any oxide, reactive metals like Fe form only some few tenth of nm of oxide by water decomposition when exposed to pure water at open circuit conditions. The procedures of specimen preparation and the quality of the system as well as the its construction have been described in detail previously [12].

Depth profiling with the ion source (AG 61 or EX 05, VG Instruments) has been calibrated with passivated Ta specimens with an oxide thickness of 30 nm prepared according to the NPL procedure 1241. The specimens were sputtered with Ne$^+$ with an energy of 3 keV and 300 nA at 0.25 cm^2 surface area in the case of Fe/Ni and 0.36 and 0.75 cm^2 in the case of Fe/Cr and 0.75 cm^2 for Cu/Ni alloys. The sputter ion currents are given in the captions of the figures. For ISS spectra 3 keV 45 nA and 0.09 cm^2 and 30 nA and 0.09 cm^2 and 0.15 cm^2 were used for Fe/Ni, Fe/Cr and Cu/Ni respectively. The sputter rate during ISS measurements was included in the determination of depth profiles.

The structure of the ISS signals is determined by the energy distribution of the backscattered ions and the isotopic composition of the noble gas ions and the target materials. For a backscattering angle of 90 degrees the following simple relation will hold for the ratio of the energy E$_1$ of the backscattered Ne$^+$ ions and the primary energy E$_o$

$$E_1/E_o = (M_2 - M_1) / (M_2 + M_1) \qquad (1)$$

M$_1$ and M$_2$ are the masses of the incoming noble gas ions (Ne$^+$) and the target atoms respectively. The masses of Fe, Cr, Ni and Cu are very similar. A sufficient separation is possible, when a noble gas of a similar mass is used. Ar$^+$ induces a large sputtering background. Therefore Ne$^+$ is a good compromise. A primary energy of 3 keV instead of 1 keV also helps to get a sufficient energy resolution for the signals as well as reasonably high sputter rates. With the appropriate position of the specimen relative to the ion beam and the analyser the peaks of the signals coincide with the calculated energy position. The isotope composition of Ne has a pronounced influence to the shape of the signals. Figures 2(a) and (b) depict a measured ISS signal of a pure Cr and Fe standard respectively and their composition of the contributing signals of the isotopes according to their natural abundance if more than 1%. The signals of the individual isotopes may be described with pure Gaussians. The Gauss/Lorentz ratio was 0.995. For the computer-aided evaluation of

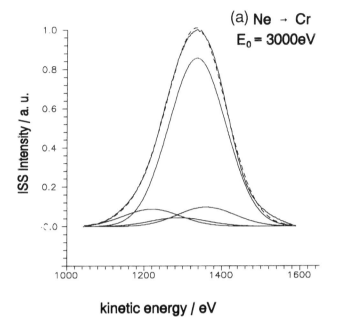

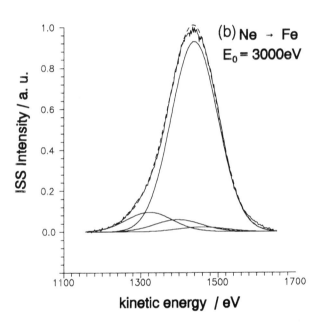

Fig. 2 ISS standard spectra and their deconvolution in relevant isotope combinations.

the ISS signals of standards and passivated alloys a linear background correction was experienced to be adequate. The spectra of actual specimen were composed with these standards by varying the height but fixing the relative height and energy position of the isotope signals among each other. A slight energy shift as a constant factor for E_0 of all partial signals and a small variation of the FWHM of the Gaussians was permitted to obtain an optimum fit. The energy shift should take care for small shifts of the primary energy

of the ion beam. The variation of the FWHM compensates for possible differences in roughness of the specimen.

The quantitative calibration of the method occurred with the background corrected and integrated signals of the pure sputter-cleaned metals. The relative sensitivity for Fe and Cr amounts to $k(Cr/Fe) = 1.95$. The sensitivity factor may be also obtained by evaluation of the spectra of sputter cleaned alloys i.e. of Fe5Cr, Fe10Cr, Fe15Cr, Fe20Cr, and Fe32Cr. The plot of the integrated intensities of Cr to Fe signals yields a line with a slope of –2.1 (Fig. 3(a)) which corresponds to the sensitivity factor. This method has been proposed previously by other authors [25]. As there is no preference to one of those values $k(Fe/Cr) = 2.05$ was taken as the average for further evaluation. Corresponding calibrations were performed for the other alloy systems.

Figure 3(b) depicts the bulk composition of different sputter-cleaned Fe/Cr alloys compared with their surface composition obtained from quantitative evaluation of ISS spectra. The straight 45° line demonstrates that the surface composition is not changed by preferential sputtering of one component which is expected because of their similar masses.

3. Results and Discussion

3.1 Passive layers on FeCr alloys

Figure 4 shows a selection of spectra during depth profiling an Fe15Cr specimen passivated for 15 min in $0.5M\ H_2SO4$ at $0.9\ V$. Apparently the maximum of the Cr enrichment is obtained after some min of sputtering and finally approaches the bulk composition. Figure 5 depicts the depth profile with the integrated ISS signals for polished FeCr specimen with different bulk composition with and without passivation in $0.5M$ H_2SO_4. The results for specimen with air-formed oxide films coincide with those of Frankenthal and Malm obtained with a completely different ISS spectrometer [4]. After 3 h passivation at 0.9 V in $0.5M\ H_2SO_4$ a pronounced enrichment of Cr in the centre of the passive layer is observed. It is assumed that the oxide/ metal interface during depth profiling is reached when the distribution approaches the bulk composition. XPS analysis of passivated specimens yields the same composition at the metal surface as the bulk value which justifies this assumption for the ISS profiles [9]. It should be mentioned that the results for the composition of the passive layer and the metal surface underneath may depend on the conditions of specimen preparation. Prolonged exposure of Fe/Cr alloys to strongly acidic electrolytes at potentials of active dissolution or eventually at open circuit conditions leads to a relatively uncontrolled accumulation of Cr at the

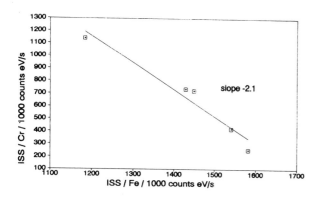

Fig. 3(a) Integrated ISS signals of Cr and Fe for different alloys allowing determination of k(Cr/Fe) for calibration purposes.

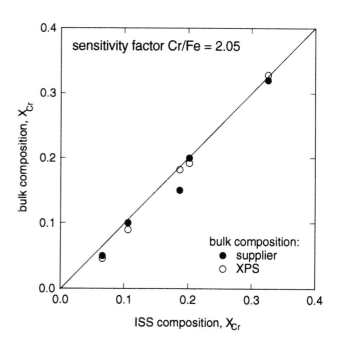

Fig. 3(b) Bulk composition of alloys according to supplier and XPS analysis with respect to ISS surface composition with k(Cr/Fe) = 2.05.

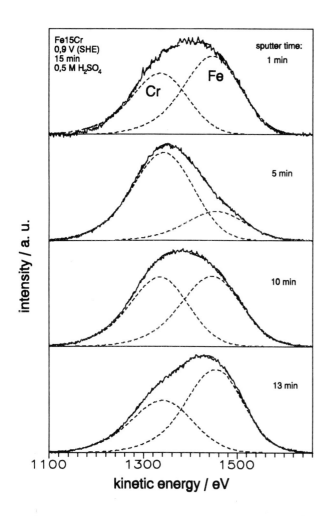

Fig. 4 ISS signal and its deconvolution in Cr and Fe contributions for a depth profile for Fe15Cr passivated in 0.5M H_2SO_4 at 0.9 V for 15 min.

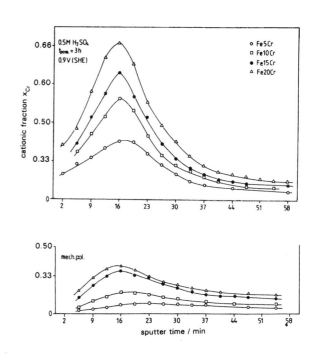

Fig. 5 ISS depth profile with 833 nA cm^{-2} sputter ion current density for different alloys externally passivated for 3 h at 0.9 V in 1M H_2SO_4 and mechanically polished specimen for comparison [5].

metal surface by preferential dissolution of Fe. This in turn causes a larger accumulation of Cr(III) within the oxide and of Cr at the metal surface underneath [9]. We therefore tried to maintain very carefully the conditions of specimen preparation as described above and started with a sputter-cleaned surface with known and well controlled composition. After this pretreatment and a specimen transfer within a closed system the samples were passivated without exposure to active dissolution. The contradicting results of different groups may result from the different pretreatent of the specimen. The influence of the passivation time on the thickness and composition of the passive layer is shown in Fig 6. Apparently the Cr enrichment is increasing with time up to more than 1 week. Similarly the growth of the layer with time is indicated by the width of the Cr distribution. The Cr enrichment also moves slowly to the oxide/electrolyte interface.

Depth profiles for samples passivated in 1M NaOH at different potentials are presented in Fig. 7. One clearly detects the increasing oxide thickness with potential by the widths of the profiles. The higher the potential the larger the thickness of an Fe-rich overlayer with a gradual increase of the Cr enrichment in the inner part of the film. At $E = -0.86$ V the Fe rich overlayer is missing. According to the Pourbaix diagrams for pH 13.8 for the pure metals at this potential the formation of Cr_2O_3 is expected whereas Fe should form soluble $HFeO_2^-$ and less protective $Fe(OH)_2$. Electrochemical and surface analytical investigations show just the beginning of the formation of oxide and soluble Fe(II). Thus the enrichment of Cr corresponds to a preferential dissolution of Fe and the formation of Cr_2O_3. This Cr(III) enrichment can be followed with time. After 20s only a slight enrichment is observed. In a simplified model at potentials well within the passive range a Cr rich layer is covered by Fe oxide, as one would deduce from XPS investigations. A detailed XPS study for identical specimen preparations of Fe10Cr and Fe20Cr alloys yields a decrease of the average Cr content within the passive layer with potential. Figure 8 depicts an example for Fe20Cr [9]. This result neglects however the elemental distribution across the film according to Fig. 7 and consequently leads to conclusions which may be improved by the details of an ISS analysis. The advantage of XPS investigations is however that sputter artefacts must not be considered as in the case of ISS depth profiles. The examination of the Cr and Fe metal signals yields the same composition of the metal surface underneath the passive layer as that of the bulk, independent of the potential of passivation. The chemical shift permits one to distinguish between Fe(II) and Fe(III) species. According to angular-resolved XPS measurements these species are in different sublayers, i.e. Fe(II) un-

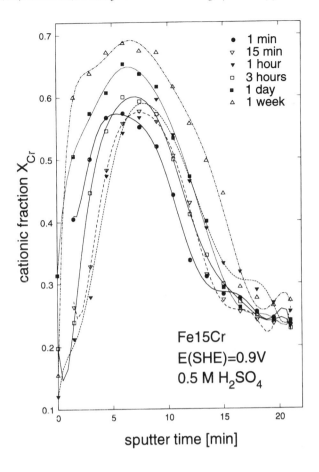

Fig. 6 ISS depth profiles of Fe15Cr with 400 nA cm⁻² sputter ion current density, passivated at 0.9 V in 0.5M H₂SO₄ for different times.

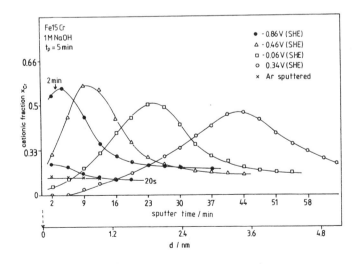

Fig. 7 ISS depth profile with 833 nA cm⁻² sputter ion current density for Fe15Cr passivated for 5 min in 1M NaOH at different potentials [5].

derneath Fe(III) [9]. These details are further valuable information for the chemical structure of the passive layer which cannot be obtained with ISS. Figure 8(a) shows that Fe(II) is oxidised to Fe(III) in the potential range of E =–0.6 to –0.2 V. In this range the Fe(II)-containing partial layer disappears whereas the thickness of the Fe(III) layer increases steeply. This process occurs additionally to the growth of the Fe(III)containing overlayer with potential which starts at E = –0.6 V. In conclusion this example demonstrates that both XPS and ISS yield different and supplementary results. Therefore studies with a close combination of both methods mutually confirm the results and lead to a more reliable model and interpretation of the data.

The characteristics of the depth profiles require some mechanistic explanation. Generally in all cases Cr(III) enters the passive layer simultaneously with Fe(II) and Fe(III) species. It will however accumulate because of its almost negligible dissolution rate. Fe^{3+} ions dissolve slowly from the surface of a passive layer but with a rate which is higher by at least one order of magnitude with respect to Cr^{3+} [23]. This situation consequently leads to the observed depth profile especially in acidic electrolytes. It is obvious that Fe(III) species formed in 1M NaOH cannot leave the metal surface as soluble corrosion products to a major extent. Furthermore, even at potentials more positive than –0.2 V Fe(II) is an intermediate product which forms most likely $Fe(OH)_2$ according to time resolved XPS studies of the passivation of pure Fe [17] and Fe/Cr alloys [9]. These Fe(II) species are oxidised in the course of the experiment to Fe(III) oxide. At potentials E ≤ -0.2 V this oxidation does not occur. Thus similar processes occur in the time as well as in the potential domain. Figure 8(b) gives an example for FeCr. The Fe-rich surface of sputter-cleaned Fe10Cr and Fe20Cr at the beginning of a passivation transient leads to the formation on large amounts of Fe(III) oxide which accumulate at the surface of the passive layer. In strongly acidic electrolytes these species will dissolve to a major extent as may be seen by the detection of Fe^{3+}

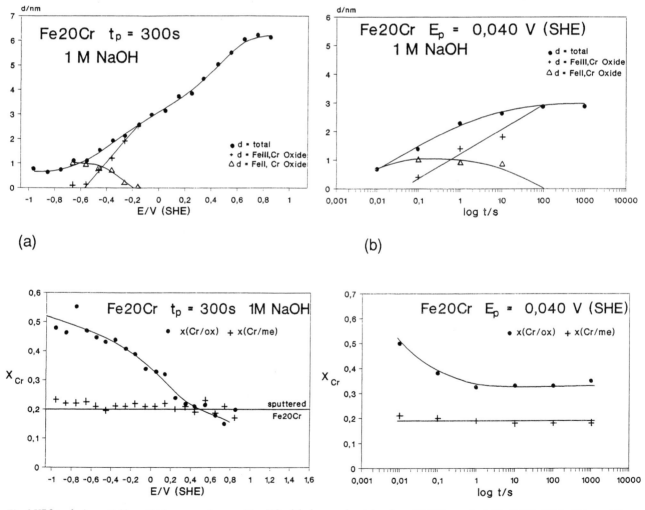

(a)

(b)

Fig. 8 XPS analysis, partial layer thicknesses and composition XC_r of the layer and metal surface of Fe20Cr passivated in 1M NaOH for 300 s at different potentials, (b) as (a) passivated at 0.04 V for different times [9].

ions with the rotating ring disc technique. Therefore for these conditions only a thin Fe(III)-rich oxide overlayer is found. The much faster dissolution of Fe^{3+} with respect to Cr^{3+} causes the loss of Fe(III) and a slow shift of the Cr(III)-rich zone to the oxide/electrolyte interface.

3.2 Passive Layers on Fe53Ni

Similar studies have been performed for FeNi alloys. For orientation the potentiodynamic polarisation curve is presented in Fig. 9. XPS investigations suggest a slight enrichment of Ni relative to the bulk within the passive layer [15] and especially at the metal surface underneath (Fig. 10). The main part of the layer is oxide with a thickness increasing linearly with the potential up to 5 nm. The thin hydroxide overlayer has a potential independent thickness of approx. 1 nm. The shift of the XPS binding energy of all oxide components towards higher values and of the work function by the opposite value at E = 0.7 V suggest the change of the $Ni(OH)_2$ layer to NiOOH which has been explained previously with a semiconducor model [14,15]. ISS depth profiles of identically prepared samples reveal many more details about the spatial distribution of the metal cations. Numerous ISS depth profiles have been recorded with passivation of Fe53Ni at different electrode potentials [10]. They all show the same shape. Figure 11 gives an example of a specimen passivated at E = 0.44 V well within the passive range. The enrichment in Fe for an outer layer of *ca*. 1.5 nm is followed by an inner part with Ni enrichment. It is interesting that the composition at the outmost part of the oxide film equals that of the bulk. Again as in the case of FeCr alloys the enrichment of one component occurs within the passive layer and not at the surface. These details

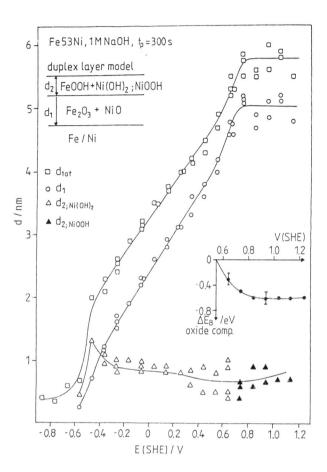

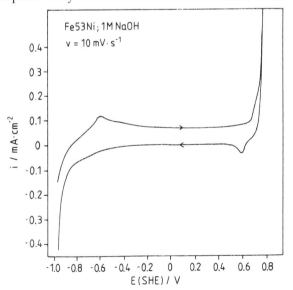

Fig. 9 Potentiodynamic polarisation curve of Ar sputtered Fe53Ni in 1M NaOH [15].

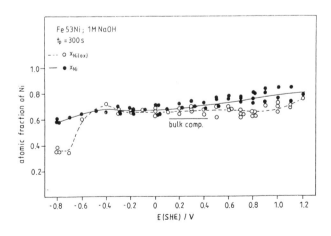

Fig. 10 Partial layer thicknesses and composition of oxide layer and metal surface for Fe53Ni passivated in 1M NaOH for 300 s at different potentials. Insertion shows shift of binding energy when approaching the transpassive range indicating the appearance of NiOOH [15].

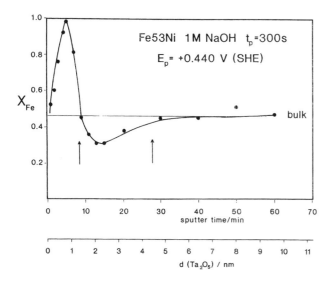

Fig. 11 ISS depth profile with 1200 nA cm^{-2} sputter ion current density of Fe53Ni, passivated in 1M NaOH for 300 s at E = 0.44 V [10].

cannot be found with XPS. This method is not sufficiently sensitive to detect changes within some few tenths of nm. It averages the film composition with an escape depth of the photoelectrons of some few nm. The depth resolution of angular dependent XPS measurements is also not sufficient to detect these profiles. An important question may be whether the Ni enrichment for a depth of more than 1.5 nm is located within the inner part of the oxide layer or at the metal surface. XPS studies with Fe50Ni suggest a pronounced enrichment of Ni at the metal surface and the presence of both cations within the passive layer [15, 16, 26]. Somewhat smaller oxide thicknesses up to 4 nm and no hydroxide were found for Fe53Ni in 0.005M H_2SO_4 + 0.495M K_2SO_4 (pH 2.9) in comparison with 1M NaOH [15,16]. Apparently the model of a hydroxide layer of approx. 1 nm followed by an oxide enriched in Ni(II) is a good approximation for this passive layer. The thicknesses of the hydroxide overlayer and the total layer deduced from the inflection points of Fig. 11 and related profiles for other potentials agree well with the results of XPS evaluations including the linear increase with the electrode potential. It seems however reasonable that at least part of the Ni enrichment has to be attributed to the metal surface. The depth resolution of XPS is too small to show all the details of ISS. It flattens out the features obtained by ISS depth profiles and suggests a small Ni enrichment within the oxide and apparently at the metal surface. Thus again XPS and ISS prove to be complementary methods to investigate the chemical composition and structure of passive layers.

The mechanistic explanation of the development

of the observed depth profile is similar to that of FeCr alloys. The difference in the dissolution rates of Fe and Ni is not as large as for Fe and Cr. Thus the accumulation of Ni in the passive layer is much less. Ni additionally is more noble than Cr and Fe. This may explain its accumulation at the metal surface. FeCr alloys on the contrary show the bulk composition at the metal surface.

3.3 Passive Layers on Cu20Ni

As a third example the investigation of ISS depth profiles of passive layers on CuNi is presented. Angular resolved measurements suggest an outer hydroxide and an inner oxide layer. Depending on the potential both sublayers may still be subdivided with the lower valent species as Cu_2O and $Ni(OH)_2$ being located inside and the higher valent species as CuO and NiOOH outside. Thus in the most complicated situation one detects a structure as presented schematically in Fig. 12. At least the duplex structure of the film is found with both methods, XPS and ISS. As the latter method cannot detect any chemical details such as oxidation states a further subdivision cannot be observed. However, the XPS composition and thickness coincides well with the results of the ISS depth profiles. Figure 13 depicts the results for three passivation potentials. The steps expected from the XPS evaluations are found almost quantitatively by ISS. Cu enters the outer hydroxide layer only when the potential is close to the transpassive range. For low potentials (E = –0.16 V) the inner oxide is very thin so that only a slight shoulder is found in the ISS profile.

The depth profiles for passivated CuNi alloys reflect the chemical difference of both metal components. Reactive Ni forms a pure $Ni(OH)_2$ layer during oxidation at negative potentials or during the first stages of the passivation process at more positive potentials. The more noble Cu enters the oxide layer which forms under the hydroxide at later stages when it is enriched at the metal surface. As in the case of Fe-containing alloys the lower valent Cu_2O is formed first

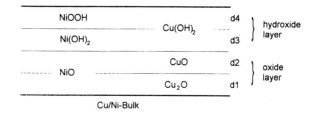

Fig. 12 Model of passive layer formed on Cu/Ni in 1M NaOH with the different sublayers existing at appropriate potentials [20].

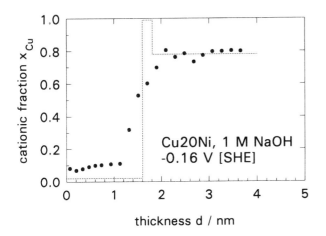

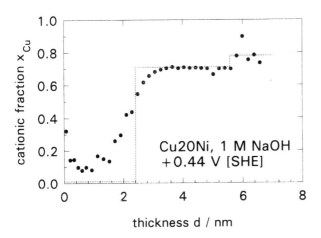

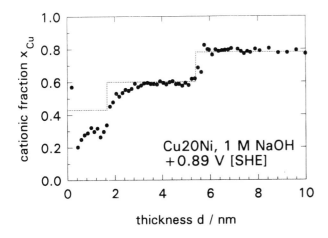

or at more negative potentials and will be oxidised to CuO at sufficiently positive potentials and during the later course of the passivation process. It finally may enter the hydroxide overlayer in the transpassive potential range.

4. Conclusion

Three examples of a combined investigation of passive layers by XPS and ISS have been presented. In all cases the bilayer or multilayer model suggested by angular-resolved XPS measurements has been justified at least as a good approximation. ISS measurements permit a more detailed insight into the depth profiles of these films. XPS and the evaluation of its data is non destructive and not affected by sputter artefacts. Thus the direct comparison of the results confirm the data of the two methods and rule out artefacts which might arise by the analytical method itself. XPS sputter profiles are affected by the restricted depth resolution of some nm and details of the profiles might easily be flattened. However, this method provides valuable chemical information as the oxidation state and the oxide or hydroxide nature of the passive layer or one of the sublayers. These details are not available from ISS. The discussed examples therefore demonstrate that a close combination of both methods is complementary. In addition, a specimen preparation under well-controlled electrochemical conditions and a systematic variation of the experimental parameters on the basis of electrochemical studies is one of the necessary requirements to obtain reliable results. Thus the examination of other systems forming surface films of only some few nm thickness with XPS and ISS seems promising.

5. Acknowledgement

The support of this work by the Deutsche Forschungsgemeinschaft with the necessary surface-analytical equipment in the project Str 200/5 is gratefully acknowledged.

References

1. H. Fischmeister and U. Roll, Fresenius Z. Anal. Chem., **319**, 639, 1984.

2. D.F. Mitchell and M.J. Graham, Surf. Interf. Anal., **10**, 259, 1987.

3. I. Olefjord and B. Brox, in Passivity of Metals, M. Froment and J. Kruger Eds., Electrochem. Soc. Princeton N.J., 5851, 1978.

4. R.P. Frankenthal and D.L. Malm, J. Electrochem. Soc., **123**, 186, 1976.

5. C. Calinski and H.-H. Strehblow, J. Electrochem. Soc., **136**, 1328, 1989.

6. H.-H. Strehblow, Surf. Interf. Anal., **12**, 363, 1988.

Fig. 13 ISS (v) sputter depth profiles with 400 nAcm⁻² sputter ion current density and XPS (..........) depth profiles of passive layers of Cu50Ni formed in 1M NaOH for 300 s at three potentials [20].

7. S. Mischler, A. Vogel, H. J. Mathieu and D. Landolt, Corros. Sci., **32**, 925, 1991.

8. P. Marcus and J. M. Grimal, Corros. Sci., **33**, 805, 1992.

9. H. W. Hoppe, S. Haupt and H.- H. Strehblow, submitted to Surf. Interf. Anal.

10. A. Rossi, C. Calinski, H. W. Hoppe and H.- H. Strehblow, Surf. Interf. Anal., **18**, 269, 1992.

11. H.-H. Strehblow and D. L. Malm, Corros. Sci., **19**, 469, 1979.

12. S. Haupt, C. Calinski, U. Collisi, H.W. Hoppe, H.-D. Speckmann and H.-H. Strehblow, Surf. Interf. Anal., **9**, 357, 1986.

13. S. Haupt and H.-H. Strehblow, J. Electroanal. Chem., **228**, 365, 1987.

14. H.-W. Hoppe and H.-H. Strehblow, Surf. Interf. Anal., **14**, 121, 1989.

15. H.-W. Hoppe and H.-H. Strehblow, Surf. Interf. Anal., **16**, 271, 1990.

16. H.-W. Hoppe and H.-H. Strehblow, Corros. Sci., **31**, 167, 1990.

17. S. Haupt and H.-H. Strehblow, Langmuir, **3**, 873, 1987

18. S. Haupt and H.-H. Strehblow, Corros. Sci., **29**, 163, 1989.

19. H.-H. Strehblow, Proc. NACE Corrosion '91, paper 76, Houston TX, 1991.

20. P. Druska and H.-H. Strehblow, NACE 12 Int. Corros. Congr., Houston, Sept. 1993.

21. S. Haupt and H.-H. Strehblow, J. Electroanal. Chem., **216**, 229, 1987.

22. Y. M. Kolotyrkin, Electrochim. Acta, **25**, 89, 1980.

23. R. Kirchheim, B. Heine, H. Fischmeister, S. Hofmann, H. Knote and U. Stolz, Corros. Sci., **29**, 899, 1989.

24. M.P. Seah, Community Bureau of Reference No. 261, 1983.

25. D. G. Swartzfager, Anal. Chem., **56**, 55, 1984.

26. P. Marcus and I. Olefjord, Corrosion, **42**, 91, 1986.

9

In Situ Xanes Studies of the Passive Film on Fe and Fe–26Cr

Alison J. Davenport, Jennifer A. Bardwell*, H. S. Isaacs
and B. MacDougall[†]

Department of Applied Science, Brookhaven National Laboratory, Upton NY 11973, USA
*Institute for Microstructural Sciences, National Research Council, Ottawa, Canada, K1A 0R9
†Institute for Environmental Chemistry, National Research Council, Ottawa, Canada, K1A 0R9

Abstract

In situ X-ray absorption near edge structure (XANES) has been used to study the passive film on iron and Fe–26Cr thin films under potential control in borate buffer. The technique enables information on both valence state and the amount of material lost through dissolution to be determined. The passive films on both Fe and Fe–Cr are formed with negligible dissolution. On reduction, iron is lost from both films due to reductive dissolution. In the case of Fe, it has been determined that some of the iron in the film is reduced back to the metallic state. Concurrent monitoring of the Fe and Cr edges has shown that in the case of Fe–Cr, loss of iron leads to enrichment of Cr in the film.

1. Introduction

The chemistry of passive films formed on pure metals and alloys in aqueous electrolytes is the subject of much experimental investigation. Surface analytical work has been carried out using e.g. Auger, X-ray Photoelectron Spectroscopy (XPS) and Secondary Ion Mass Spectroscopy (SIMS), leading to a greater understanding of the composition of passive films. However, these measurements must be made *ex situ* in UHV and it has been demonstrated [1] that the passive film on Fe–Cr alloys is not stable when removed from an aqueous environment for *ex situ* examination. Thus, an *in situ* technique is to be preferred, particularly if it can yield detailed information about the composition and valence states of elements in the film. These requirements are met by *in situ* X-ray Absorption Spectroscopy [2–4]. In the present work, X-ray Absorption Near Edge Structure (XANES) has been used to study the passive films on Fe and Fe–26Cr.

2. Experimental

The experimental setup for *in situ* XANES measurements has been described previously [3]. The design of the cell ensured good potential control of the sample at all times. Potentials are quoted with respect to a saturated mercurous sulphate reference electrode (~0.4V(SCE)). The electrolyte was a pH 8.4 borate buffer. The solution was de-aerated throughout the experiments by bubbling with nitrogen which also ensured that all dissolution products were removed

from the vicinity of the electrode. The samples consisted of a 6μm Mylar film onto which ~ 100Å of Nb or Ta were sputtered to provide electrical contact. 20 or 40Å of the alloy under investigation (Fe or Fe–26Cr) was then deposited on top of the conducting layer. Measurements were made at Beamline X19A at the National Synchrotron Light Source using a Cowan fixed exit height double crystal monochromator with a resolution ≤2eV (at the Cr(VI) pre-edge peak, 5993eV). Data were collected in fluorescence geometry using a 13-element solid state detector (Canberra). Dead time problems were avoided by ensuring that the total count rate in each detector element was less than 30 000 counts/s. A typical curve took ~25 min to collect counting at a rate of 2s/point above and below the edge and 5s/point around the edge region.

3. Results and Discussion

Figure 1 shows a polarisation curve of both bulk alloys and thin films of Fe and Fe–26Cr in the borate buffer at a sweep rate of 0.5mVs^{-1} in a deaerated solution. The displacement of the peaks for the thin films to higher potentials probably reflects the resistivity of the films (typically ~1kΩ) across a deposited disc ~2cm in diameter) which will introduce an ohmic potential drop so that the potential difference across the metal/solution interface is less than the applied potential. Apart from this problem, the polarisation curves of the thin films are very similar to those of the bulk alloys of the same composition implying that the properties of passive

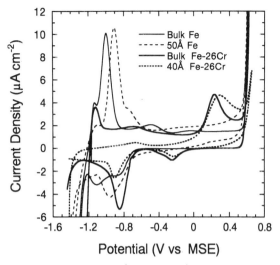

Fig. 1 Polarisation curves of 50Å of Fe and 40Å of Fe-26Cr on 100Å of Ta and bulk alloys of the same composition in a de-aerated borate buffer solution at a scan rate of 0.5mVs⁻¹.

and flat top to the edge is characteristic of metallic iron. On stepping the potential to –0.4V and then to +0.4V, the edge position shifts to higher energies and the top of the edge begins to sharpen into a peak (both are characteristic of oxides). The height of the edge at the right hand side of the figure does not change indicating that in the solution used (a borate buffer), oxidation of iron proceeds by a solid state reaction without any detectable dissolution. Reduction of the passive film on stepping the potential back to –1.5V results in a spectrum identical in shape to the first one, but the drop in the edge height indicates the dissolution of a significant amount of iron. Spectra from the next cycle show that this process is repeatable a second time. Subsequent cycles do not show such clear behaviour, probably due to a loss of integrity of the very thin film. The loss of material during reduction suggests that the 3-valent oxide is reduced to the soluble 2-valent state which dissolved away leaving a bare metal surface. If this reductive dissolution was the only reduction reaction, then it should be possible to reconstruct the XANES signal originating from the oxide alone by subtracting curve (d) from curve (c). The result of this subtraction is shown as [1]. Its shape is characteristic of 3-valent iron oxides [2] but the negative deviation in the curve around 7120 eV indicates that Fe in the metallic state has been gained upon cathodic reduction. Approximate fitting of this curve indicates that the ratio of reductive dissolution to solid state reduction is in the region of 2:1.

films formed on the thin films will reflect those on the bulk alloys.

Figure 2 shows a series of spectra collected at the iron *K* edge from a 40Å film of iron polarised cyclically in a borate buffer. It should be noted that the edges are not normalised so changes in the edge height represent quantitative changes in the amount of iron present. The initial curve was collected after cathodic reduction of the air-formed oxide film at –1.5V. The edge position

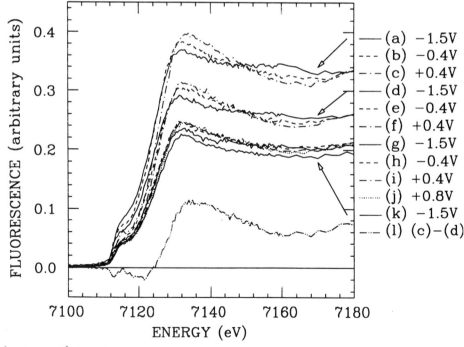

Fig. 2 Fe K edge of 40Å of Fe on 100Å of Ta. The spectra were recorded on stepping the potential in the sequence indicated. Curve (1) is the difference between (c) and (d).

The total amount of metal lost in each potential cycle is listed in Table 1. It is calculated on the following basis: it is assumed that the height of the Fe K edge for the as-deposited sample (which has a very thin air-formed oxide film), not shown, corresponds to 40Å of iron. If, for example, the edge height drops by 20%, this corresponds to a loss of iron equivalent to 8Å of metal. From Table 1, it appears that each potential cycle for pure iron involves the loss of approx. 7Å.

A similar set of experiments was carried out for thin films of Fe–26Cr. One of the advantages of using a multi-element solid state detector is that different detector elements can be used to collect fluorescence data from different chemical elements. Thus both iron and chromium K edges can be collected in a single spectrum by scanning over both edges whilst different detector elements collect Fe K_α and Cr K_α fluorescent photons. Figure 3(a) shows details of the Fe edge in a polarisation experiment on Fe–26Cr. The results are qualitatively similar to those found for pure iron: metallic Fe present at 1.5V undergoes oxidation at –0.4V and +0.4V without any loss of material, then undergoes significant dissolution as the sample is reduced back to –1.5V. This behaviour is repeated on a second potential cycle (not shown). However, the amount of iron lost in each cycle is ~2Å, considerably less than the ~7Å lost in the case of pure iron.

The chromium K edge, measured during the same two potential cycles, is shown in Fig. 3(b). Whilst the data are noisy due to the low concentration of chromium and the diminished number of detector elements used, it is clear that the data fall into two bands.

Table 1 Amount of metal lost during potential cycling measured from the edge height and converted into an equivalent thickness of pure metal

edge measured	sample	sample thickness /Å	cycle	metal lost /Å ±1
Fe	Fe	40	1	8
			2	6
			3	2
	Fe	20	1	7
			2	2
	Fe-26Cr	40	1	2
			2	2
	Fe-26Cr	20	1	1
			2	3
Cr	Fe-26Cr	40	1	0
			2	2
	Fe-26Cr	20	1	0
			2	>3†
	Fe-26Cr	40	1*	3
			2	2

*Sample was "activated" by stepping to +0.8V and back down to -1.5V.

†Sample sat in the transpassive region for several hours due to problems with the synchrotron.

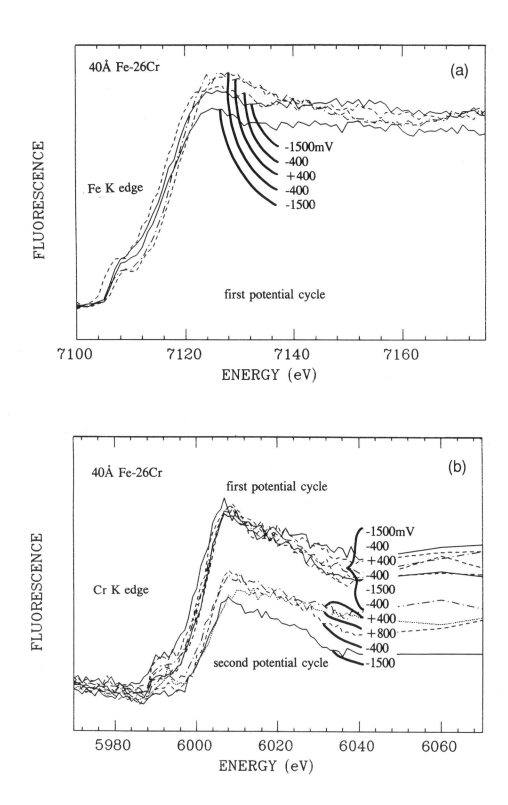

Fig. 3 40Å of Fe-26Cr stepped through two oxidation/reduction cycles with the potential sequence indicated: (a) Fe K edge (only the first potential cycle is shown); (b) Cr K edge.

All the spectra are roughly similar in shape, but there is a significant loss of chromium indicated by a drop in the edge height during the second potential cycle when the potential is raised into the transpassive region (increased from –0.4V to +0.4V). Concurrent monitoring of the iron and chromium edges demonstrates the interplay between the two elements in the film: on the first potential cycle, both elements are oxidised into the film. However, chromium is not present in the film in a sufficient concentration for significant transpassive dissolution to take place. During reduction of the oxide at the end of the first potential cycle, iron is lost by reductive dissolution leading to enrichment of chromium in the passive film. On the second potential cycle, the chromium is now no longer protected by the iron and is present in a sufficient concentration for transpassive dissolution to occur. This model is consistent with similar work carried out on AlCr alloys [5]. In the presence of a protective alumina layer, chromium is trapped in the film but if the alumina dissolves, then transpassive dissolution of chromium can take place.

In situ X-ray absorption spectroscopy is thus a powerful technique for studying alloy passivation. In addition to providing *in situ* valency information, the approach allows sensitive measurement of the quantity of material lost from the alloy through dissolution or passivation. In principle, it is possible to determine which components of an alloy undergo selective dissolution, which are enriched in the passive film, and which are enriched under the film in the metallic state.

4. Conclusions

1. Formation of the passive film on both iron and Fe–26Cr proceeds without any metal dissolution.

2. Cathodic reduction of the passive film on iron leads to the loss of the equivalent of ~7Å of metal in dissolution. The ratio of reductive dissolution to solid state reduction is approximately in the region of 2:1.

3. Cathodic reduction of the passive film on Fe–26Cr leads to the loss of only ~2Å of iron.

4. Transpassive dissolution of chromium from Fe–26Cr can only take place on the second potential cycle once iron in the film has undergone reductive dissolution leading to enrichment of chromium in the passive film.

5. *In situ* X-ray absorption spectroscopy of passive films on thin film electrodes is an excellent technique for obtaining quantitative data on valency and dissolution during alloy passivation.

5. Acknowledgement

This work was performed in part under the auspices of the U.S. Department of Energy, Division of Materials Sciences, Office of Basic Energy Science under Contract No. DE-AC02-76CH00016. Measurements were carried out at Beamline X19A at the National Synchrotron Light Source with help from Lars Furenlid. The authors would like to acknowledge the assistance of M. J. Graham and G. I. Sproule.

References

1. J. A. Bardwell, G. I. Sproule, D. F. Mitchell, B. MacDougall and M. J. Graham, J. Chem. Soc. Farad. Trans., **87**, 1011, 1991.
2. M. Kerkar, J. Robinson and A. J. Forty, Farad. Discuss. Chem. Soc., **89**, 31, 1990.
3. A. J. Davenport, H. S. Isaacs, G. S. Frankel, A. G. Schrott, C. V. Jahnes and M. A. Russak, J. Electrochem. Soc., **138**, 337, 1991.
4. J. A. Bardwell, G. I. Sproule, B. MacDougall, M. J. Graham, A. J. Davenport and H. S. Isaacs, J. Electrochem. Soc., **139**, 371, 1992.
5. A. J. Davenport, H. S. Isaacs, G. S. Frankel, A. G. Schrott, C. V. Jahnes, and M. A. Russak, J. Electrochem. Soc., in press.

10

Scanning Tunnelling Microscopy Study of the Atomic Structure of the Passive Film Formed on Nickel Single Crystal Surfaces

V. Maurice, H. Talah and P. Marcus

Laboratoire de Physico-Chimie des Surfaces, CNRS (URA 425), Université Paris VI, Ecole Nationale Supérieure de Chimie de Paris, 11 rue Pierre et Marie Curie, 75231 Paris Cedex 05, France

Abstract

Scanning tunnelling microscopy has been used to study Ni(111) electrodes passivated in 0.05M H_2SO_4. *Ex situ* atomic resolution imaging demonstrates the crystalline character of the oxide film in epitaxy with the substrate. Stepped surfaces are imaged which suggests a tilt of the surface of the film with respect to the interface between the film and the metal substrate. Local variations of the film thickness are likely to result from this tilt. The chemical nature of the atomic structure which is resolved is discussed.

1. Introduction

The chemical distribution of material in the passive films formed on pure metal and on alloy surfaces has been extensively studied along the direction normal to the surface thanks to depth sensitive spectroscopic techniques [1]. A bilayer model describes this distribution with the inner part of the film being an oxide component, and the outer part being an hydroxide or oxyhydroxide component. The atomic structure of passive films has been much less extensively investigated, although the amorphous character of passive films has been emphasised [2]. The relationships between, on one side, the extent of ordering within the film and the nature of the defects, and on the other side, the electrochemical behaviour and the corrosion resistance remain to be studied in details. In the case of pure nickel substrates, the thickness of the oxide inner part has been reported to range from 0.4 to 1.2 nm, and that of the hydroxide outer part have been reported to range from fractions of a monolayer to a complete monolayer (0.6 nm thick) [3–6]. Some controversy remains as to the extent of ordering within the passive film which is possibly related to the *ex situ* or *in situ* conditions of investigation of the structure [7,8].

We have undertaken a Scanning Tunnelling Microscopy (STM) investigation of the structure of the passive film formed on pure nickel substrate in acid electrolyte. In the previously reported *in situ* STM investigations related to passive films, the lateral resolution that would be requested to provide information on the atomic structure of the film was not achieved [9–

15]. Our choice of a (111) oriented Ni single crystal substrate was mainly dictated by the concern of selecting a system for which the passive film had previously been reported as being crystalline [7] in order to facilitate atomic resolution imaging. Preliminary experiments confirmed the possibility of achieving atomic resolution in *ex situ* investigations of passive films [16]. Another example of STM atomic resolution imaging on passivated FeCr surfaces is reported in these proceedings [17]. This paper reports further *ex situ* investigations on the atomic scale where the influence of the passivation potential was studied.

2. Experimental

Sample preparation from a Ni single crystal rod (purity 99.999%) involved successively: orientation within $\pm 1°$ by X-ray diffraction, spark machining, mechanical and electrochemical polishing and annealing at 1275 K for a few hours in a flow of purified hydrogen at atmospheric pressure. The sample was then transferred at room temperature and under the hydrogen atmosphere into a nitrogen-containing glove box where the electrochemical experiments were performed. Sulphuric acid (0.05M) was prepared from reagent grade H_2SO_4 and ultra pure water (resistivity of 18 MΩ.cm^{-1}). After immersion of the electrode in the electrochemical cell, the potential was stepped from the open circuit value to the passivation potential. Three values were selected: +550, +650 and +750 mV/SHE. After 12 min of anodic polarisation and completion of the formation of the passive film, the experiment was stopped by emersing the Ni(111) electrode at the applied potential. The sample was then rinsed with ultra pure water, dried in nitrogen gas and finally transferred to air for STM imaging.

STM imaging was performed with the Nanoscope II and III from Digital Instruments (Santa Barbara, CA) operating in atmospheric conditions. Maximum scan range both in X and Y axis of the scanner used for atomic resolution imaging was *ca.* 700 nm. Images were recorded for the most part in the constant current topographic mode of the STM. All reported topographic images have had a least square plane subtracted in order to remove any tilt of the scanning head relative to the sample surface. Sample bias voltages used for atomic resolution imaging were in the range from 30 to 200 mV, positive or negative, and setpoint currents were in the range from 0.5 to 1 nA. Tunnelling probe tips were made from W wire etched in 1M KOH.

3. Results and Discussion

STM observations of the metal substrate before the passivation treatment revealed the presence of the steps and terraces of the surface. Steps of various height could be imaged which were all aligned along the <–101> directions of the substrate, as determined from a crystal alignment performed with X-ray back diffraction. The width of the terraces was frequently found to exceed one hundred nanometers. Some roughness was measured on the terraces which is supposedly resulting from the formation of the native oxide. Atomic resolution was not achieved on these surfaces. After the passivation treatment, two major levels of structural modifications were observed. On a microscopic scale, a roughening dependent on the passivation potential was observed which likely results from metal dissolution and nucleation and growth of the passive film. These results are reported and discussed elsewhere [18]. On the atomic scale, a lattice forming stepped surfaces has been resolved. The parameters of this lattice and the characteristics of the stepped surfaces are found to be independent of the passivation potential.

Figure 1 shows a typical image recorded on the atomic scale after passivation. Stepped surfaces were imaged over large areas without any noticeable defects which indicates that the passive film is crystalline. A corrugated lattice is resolved on the terraces which has the following parameters: 0.32 ± 0.02 nm and 117° ± 5°. The agreement is excellent with the lattice parameters of a (111) oriented NiO structure: 0.295 nm and 120°. The orientation of the step edges was found to vary locally. The average orientation is within 5 to 10° from the main crystallographic direc-

tions (and their equivalent by threefold symmetry of the substrate): <–101> and <1–21>. Faceting of the step edges is evidenced by the presence of kinks along the close-packed directions as it can be seen on Fig. 1. The (111) oriented terraces have a width which ranges locally from 2 to 5 nm. The step height also varies locally but remain consistent with monolayer or double layer NiO spacing. This amounts to an average tilt of 8 ± 5° for these stepped surfaces with respect to the (111) orientation of the terraces. Taking into account the value of 8°, the average orientation of the surface of the passive film is (433) for steps along *[0–10]* and (765) for steps along *[1–21]*.

Figure 2 shows a higher magnification image of these stepped surfaces. The fine structure of a step locally oriented along the *[–101]* direction has been resolved. It shows a two stage transition between the upper and lower terraces with an intermediate terrace being resolved as indicated on the figure. This intermediate terrace is characterised by a higher corrugation.

The relation between the chemical nature of the passive film and the STM image is dependent on the tunnelling mechanism taking place for such thin film systems. A direct mechanism that would involve tunnelling from the metal substrate through the passive film is not likely. Such mechanism would probably require the penetration of the tip into the passive film because the overall thickness of the film is large with respect to the usual width of tunnelling barrier (*ca.* 0.5 nm [19]). In addition, the measured lattice

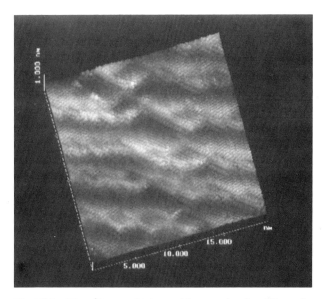

Fig. 1 (20 ×20) nm² topographic image of the stepped surface of the passive film formed on Ni(111) at + 750 mV/SHE in 0.05 M H₂SO₄, sample bias voltage Vₜ = +135 mV, and setpoint current Iₜ = 0.8 nA. The kinked structure of the step edges and the atomic lattice on the terraces are evidenced.

parameters are consistent with those of nickel oxide (0.417 nm) but not with those of metallic nickel (0.352 nm). Two other mechanisms can be invoked which differ depending on the role of the hydroxide outer layer. In the case of the mechanism involving electron transfer from the metal substrate to the conduction band of the oxide inner part followed by tunnelling through the hydroxide outer part, the structural information in the images would be relative to the oxide inner part of the passive film. In such a case, the thickness of the hydroxide outer layer should not exceed the width of the tunnelling barrier. The image shown in Fig. 2 where a two stage transition between upper and lower terraces has been resolved, could then be explained by a transition between two O planes (or Ni) via a Ni plane (or O) in the (111) oriented NiO structure which alternates O planes and Ni planes. In the case of a mechanism involving electron transfer through the oxide inner part to surface hydroxyls groups which would act as centres for tunnelling to the tip, the structural information of the images would be relevant to the hydroxide outer part of the passive film. The recorded images would then suggest that the surface hydroxyl groups form a complete ordered monolayer of (1×1) periodicity with respect to the underlying nickel oxide. The image shown in Fig. 2 could then be explained by a transition between two hydroxyl monolayers covering terraces of the underlying oxide. The intermediate area could correspond to a terrace not covered by hydroxyl groups in the vicinity of a step edge. The structural information contained in the recorded images does not allow us to favour any of these two possible mechanisms.

The average tilt of $8 \pm 5°$ with respect to the (111) oriented terraces is associated with the presence of steps at the surface of the passive film. The interface between the metal substrate and the passive film cannot be directly resolved in STM experiments. If we assume that the orientation of the substrate surface remains (111), the epitaxial relationships are: NiO(433) // Ni(111) with NiO[0–11] // Ni[0–11] and NiO(765) // Ni(111) with NiO[1–21] // Ni[1–21]. The expansion of the NiO lattice with respect to the Ni lattice is large, 16.5% based on the bulk values. It has been shown in the case of NiO(100) thin film grown on Ni(100) that this large lattice mismatch can be accommodated by a large coincidence unit cell corresponding to a (6×6)NiO // (7×7)Ni superlattice [20]. A similar large unit cell could also accommodate the interfacial stress for NiO(111) // Ni(111). However, the fact that a tilted epitaxy is measured indicates that a more favourable relaxation occurs at the interface in one direction.

Figure 3 shows the tilting of the surface of the passive film with respect to the orientation of the substrate. This tilting results in the presence of nano-facets, not shown on Fig. 3 for the sake of clarity, corresponding to O planes and Ni planes on the oxide side of the interface if a bulk termination is considered. It is quite likely that these nano-facets are strongly attenuated by the atomic relaxation effects. A major consequence of the tilt of the terraces of the passive film with respect to the interface, as illustrated on Fig. 3, would be local variations of the thickness of the film. Taking into account an average thickness of 1 nm, the presence of steps at the surface of the film would induce variations of thickness of *ca.* 25%. The thickness could even decrease by *ca.* 45% in the case of the presence of monolayer substrate steps at the interface. The bottom of the steps could constitute preferential sites for the breakdown of the passive film where the barrier effect of the film on cation diffusion would be drastically decreased. Considering the average width of 3.5 nm for the terraces and one of such sites per step, the density of these preferential sites of breakdown would be *ca.* 3×10^6 cm^{-1}. It is suggested that the most critical breakdown sites are the sites where a step of the passive oxide coincide with a step of the substrate giving a film thickness estimated to 0.56 nm in Fig. 3. The density of such sites is evidently much lower than the density of steps on the oxide surface.

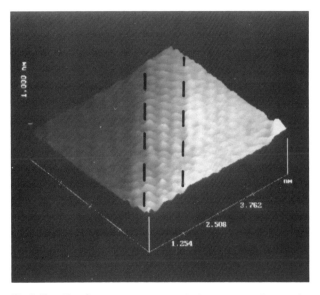

Fig. 2 (5 × 5) nm^2 *topographic image of the fine structure of a step edge along the [-101] direction of the surface passivated at + 750 mV/SHE, V_t = +113 mV, I_t = 0.5 nA. The lines indicate a two stage transition between the upper terrace (right) and the lower terrace (left).*

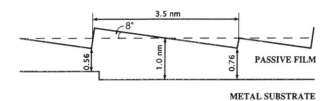

Fig. 3 Schematic representation of the orientation of the surface of the passive film with respect to the orientation of the metal substrate. A thickness of 1 nm and a tilt of 8° have been considered. Monolayer oxide steps (0.24 nm) and double layer oxide steps (0.48 nm) are represented at the surface as well as a monolayer substrate step (0.20 nm) at the interface.

4. Conclusion

After passivation of Ni(111) single crystal surfaces in 0.05M H_2SO_4, *ex situ* STM observation shows an epitaxial crystalline film. On the atomic scale, stepped surfaces are imaged. Their characteristics are independent of the passivation potential. The measured lattice parameters of 0.30 ± 0.02 nm and $117 \pm 5°$ fit those of NiO(111), the inner component of the passive film. The stepped surface of the passive film is attributed to a tilt of $8 \pm 5°$ of the surface of the film with respect to the orientation of the metal substrate. This tilt may result from the accommodation of the lattice misfit at the interface. A major consequence of this tilt would be local variations of the thickness of the film which may constitute preferential sites of breakdown.

References

1. see e.g. P. Marcus, in Electrochemistry at Well-Defined Surfaces, eds. J. Oudar, P. Marcus and J. Clavilier, Special Volume of J. Chimie Physique, **88**, 1687, 1991.

2. J. Kruger, in Advances in Localized Corrosion, eds. H. Isaacs, U. Bertocci, J. Kruger and S. Smialowska, NACE, 1987, 1, and references therein.

3. P. Marcus, J. Oudar and I. Olefjord, J. Microsc. Spectrosc. Electron., **4**, 63, 1979.

4. B. P. Lochel and H.-H. Strehblow, J. Electrochem. Soc., **131**, 713, 1984.

5. F. T. Wagner and T. E. Moylan, J. Electrochem. Soc., **136**, 2498, 1989.

6. D. F. Mitchell, G. I. Sproule and M. J. Graham, Appl. Surf. Sci., **21**, 199, 1985.

7. J. Oudar and P. Marcus, Appl. Surf. Sci., **3** , 48, 1979.

8. R. Cortes, M. Froment, A. Hugot-Legoff and S. Joiret, Corros. Sci., **31**,121, 1990.

9. O. Lev, F. R. Fan and A. J. Bard, J. Electrochem. Soc., **135**, 783, 1988.

10. F. R. Fan and A. J. Bard, J. Electrochem. Soc., **136**, 166, 1989.

11. R. C. Bhardwaj, A. Gonzalez-Martin and J. O'M. Bockris, J. Electrochem. Soc., **138**, 1901, 1991.

12. R. C. Bhardwaj, A. Gonzalez-Martin and J. O'M. Bockris, J. Electrochem. Soc., **139**, 1050, 1992.

13. K. Sashikata, N. Furuya and K. Itaya, J. Vac. Sci. Technol., **B9. 2**, 457, 1991.

14. M. Szklarczyk and J. O'M. Bockris, Surf. Sci., **241**, 54, 1991.

15. J. S. Chen, T. M. Devine, D. F. Ogletree and M. Salmeron, Surf. Sci., **25B**, 346, 1991.

16. V. Maurice, H. Talah and P. Marcus, Surface Sci. Lett., **284**, L431, 1993.

17. M. P. Ryan, R. C. Newman, S. Fujimoto, G. E. Thompson, S. G. Corcoran and K. Sieradzki, these proceedings, pp.66–69.

18. V. Maurice, H. Talah and P. Marcus, Proc. 12th Int. Corros. Congr., Vol.3B, p.2105. Houston, TX, USA, NACE International, September, 1993.

19. G. Binnig, H. Rohrer, Ch. Gerber and E. Weibel, Phys. Rev. Lett., **49**, 57, 1982.

20. R. S. Saiki, A. P. Kaduwela, M. Sagurton, J. Osterwalder, D. J. Friedman, C. S. Fadley and C. R. Brundle, Surface Sci., **282**, 33, 1993.

11

Atomic-Resolution STM of Passive Films on Fe–Cr Alloys

M. P. Ryan, R. C. Newman, S. Fujimoto, G. E. Thompson, S. G. Corcoran* and K. Sieradzki*

UMIST, Corrosion and Protection Centre, PO Box 88, Manchester, M60 1QD, UK
*The Johns Hopkins University, Department of Materials Science, Maryland Hall, Baltimore, MD 21218, USA

Abstract

Atomic resolution of primary passive films formed on sputter-deposited microcrystalline Fe–25Cr has been achieved *ex situ* in air, with promising results *in situ*. The *ex situ* surface of several aged passive films showed a long-range triangular lattice structure with a spacing of 3.1 Å. This is consistent with a projection of the rhombohedral Cr_2O_3 structure or, more likely, with a 2D lattice of OH groups on top of this structure. The latter interpretation is supported by local observations of a typical ring-like oxide image underneath the triangular lattice. Fresh films, or films examined *in situ*, showed less evidence of crystallinity.

1. Introduction

There have been only a few atomic-resolution STM studies of passive oxide films on metal surfaces. Maurice, Talah and Marcus [1] showed *ex situ* that the passive film formed on Ni(111) is crystalline; the lattice observed corresponded to NiO(111). *In situ* measurements on the same system may yield similar results. Moffat, Fan and Bard [2] have observed atomically ordered domains on passivated chromium surfaces; however the exact nature of this lattice has yet to be established. In the same study the active-passive transition of Cr in sulphuric acid solution was examined *in situ* at reduced resolution.

It is not surprising that passive films on pure metals should be crystalline; one can readily devise a geometrical scheme for film formation on low-index planes. However, *alloy* passivation, where one element is highly enriched in the passive film, is a different matter, and it has long been known that Fe-Cr alloys can form disordered films, especially at higher Cr contents [3].

In this paper we report *ex situ* and preliminary *in situ* measurements on Fe–25Cr surfaces prepared by sputter deposition. The passivation was carried out in the potential range where only Cr is likely to participate in film formation.

2. Experimental Procedure

Fe-Cr alloys were prepared at Salford University as thin films on chemically-polished silicon wafer substrates [4]. The average film thickness was *ca.* 150 nm, prepared by ion-assisted ion beam co-deposition using adjacent pure-metal targets. This technique uses a stationary substrate and produces a linear variation of the alloy composition across the wafer. In this study a composition of 25 at.% Cr was selected, and small samples were cleaved from the wafer at the required composition. Low-resolution STM of these specimens showed that the surface relief due to the crystallite structure was less than 30 Å (Fig. 1). The grain size was *ca.* 30 nm. Electron diffraction of similar films deposited on polymer substrates showed sharp rings in the bcc diffraction pattern, i.e. no obvious evidence of texture.

High-resolution STM examination was carried out using a NANOSCOPE 2 (Digital Instruments Inc) with *in situ* electrochemical capability. Freshly-etched tungsten tips were used throughout, with a wax coating for *in situ* work. Bias voltages and tunnelling currents were normally *ca.* 600 mV and 2 nA respectively. For *ex situ* examination, the Fe–25Cr specimen was cathodically treated at –1 V [SCE] in 0.1M H_2SO_4, then the potential was stepped to –400 mV for 5 min. This potential is in the primary passivation region as defined by Frankenthal [5] (Fig. 2). The amount of charge passed during passivation was *ca.* 4 mC cm^{-2}; in conjunction with the cathodic treatment, this is enough to destroy any memory of the original surface with its air-formed oxide (at least at the resolution of conventional electrochemical measurements [4]), though obviously no reactive alloy surface can be treated like a gold monocrystal and some future work on the role of cathodic treatment will be necessary. The specimen

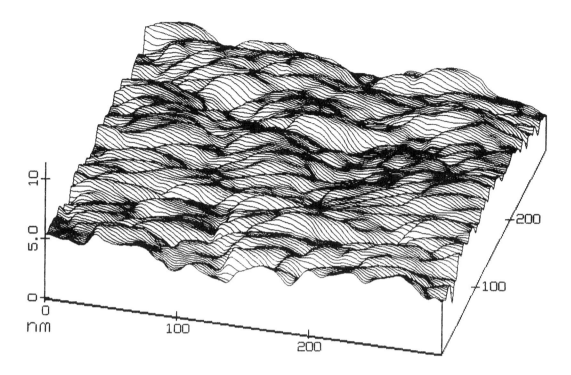

Fig. 1 Low-resolution STM image (height mode, ex situ) of Fe–25Cr in air, showing typical surface relief and faceting.

was rapidly removed from the acid (open-circuit potential measurements showed that no reactivation occurred over the relevant time scale), rinsed with deionized water, and dried in a stream of nitrogen. STM examination was either performed within minutes or, in one case, after prolonged (several weeks) storage in nitrogen. For *in situ* examination, the passivation conditions were those given previously except that a platinum pseudo-reference electrode was used, possibly introducing an error of up to 25 mV in the set potential. The tip was retracted during the first stages of passivation and lowered after 5 min to begin imaging of the passive film under potential control.

The passivation conditions are such that only Cr is expected to take part in the primary passive film formation; Fe will dissolve relatively freely until the surface is blocked by the Cr_2O_3 film. According to the percolation model of alloy passivation [6–9], the film would not be crystalline at this stage, but structural relaxation leading to crystallisation cannot be ruled out during slower long-term growth.

3. Results and Discussion

Ex situ STM examination of passivated surfaces of Fe–25Cr immediately showed evidence of atomic resolution and, on aged samples, it was possible to observe a long-range triangular lattice structure with a = 3.1 Å (Figs 3, 4). Many 2D Fourier transforms of the quality

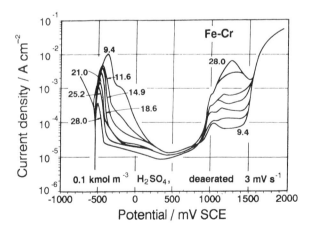

Fig. 2 Polarisation curves for Fe–Cr thin film alloys in 0.1M H_2SO_4.

shown in Fig. 5 could be obtained on different regions of such surfaces. In some areas, the triangular structure appeared to be on top of a more typical ring-like oxide structure similar to that observed on oxidised silicon in UHV [10]; a detailed study of these two apparent structures will be presented elsewhere [11]. Freshly passivated samples and *in situ* images (Figs 6,

67

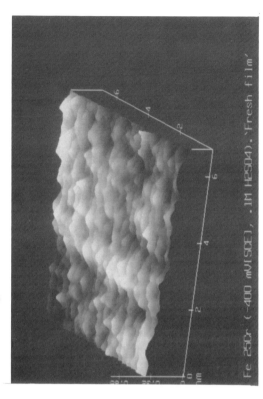

Fe 25Cr. (-400mV, .1M H2SO4), stored N2.

Fig. 4. Filtered 3D representation of the triangular lattice on passivated and aged Fe–25Cr.

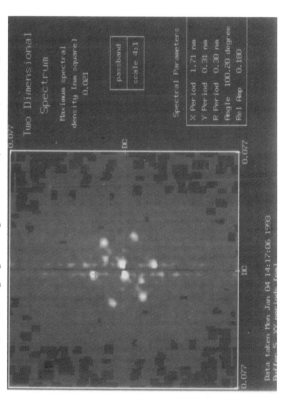

Fe 25Cr (-400 mV[SCE], .1M H2SO4). 'Fresh film'

Fig. 6 Disordered but stable atomic corrugations observed on freshly passivated Fe–25Cr ex situ. The corresponding 2D Fourier transform showed no evidence of order.

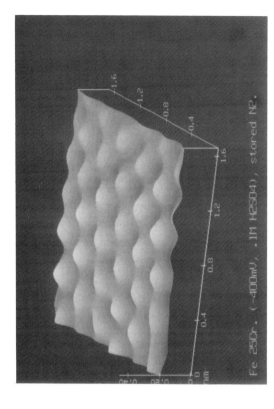

Horizontal distance [nm]	0.62	0.92	1.86	
Vertical distance [nm]	0.01	0.03	0.04	
Angle [deg]	1.09	1.81	1.15	

Spectral period [nm] 0.31

fixed distance

Zoom 4:1

Fig. 3 STM image (current mode, ex situ) of surface structures observed on Fe–25Cr after passivation at –400 mV [SCE] and ageing in nitrogen. The lattice parameter of the triangular lattice is 3.1 Å.

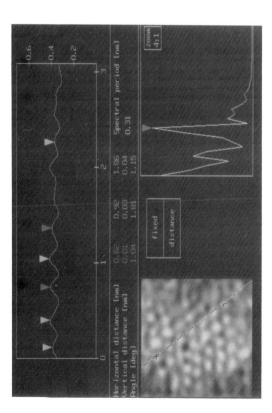

Two Dimensional
Spectrum

Maximum spectral
density [nm square]
0.021

passband
scale 4:1

Spectral Parameters

X Period	1.71 nm
Y Period	0.31 nm
R Period	0.30 nm
Angle	100.20 degree
Rel Amp	0.180

Data taken Mon Jan 04 14:17:06 1993

Fig. 5 2D Fourier transform of the triangular lattice shown in Figs 3, 4.

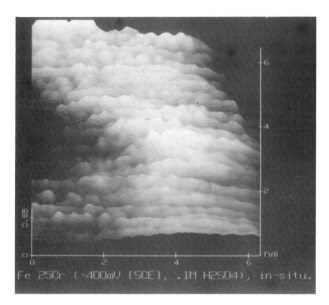

Fig. 7 Preliminary in situ *image of Fe–25Cr showing atomic resolution but more noise than in the* ex situ *studies.*

7) showed more disordered structures with atomic corrugations, but only local evidence of a lattice. Obviously much more work is required before we can say conclusively that passive films are disordered, but this is now a reasonable possibility.

The triangular lattice observed on aged samples (and on fresh films formed on Fe–15Cr [11]) could be a projection of the unit cell of the rhombohedral chromium oxide structure (a = 5.38 Å) viewed along the trial axis. The arrangement and spacing of the oxygen ions in this structure is roughly consistent with the observed images; the axial angle is usually quoted as 55 rather than 60 degrees, but whether this would be maintained near a surface is questionable. Additionally, one must consider the likelihood that the surface is in fact covered with a 2D lattice of OH groups, or even that there is a hydroxide phase in the outer part of the film. The literature does not show any precedents for imaging OH in this fashion, but at present

this is the most likely explanation, and it is consistent with the local observation of a conventional thin-oxide structure, similar to oxidised silicon, in areas where the 'OH' species are not detected. The widespread observation of the triangular lattice suggests that, while the thin alloy films may not have much texture, they probably have low-index surface facets which favour the observed orientation of the oxide.

4. Acknowledgements

This work would not have been possible without the deposition of special materials by our colleagues at Salford University—J. S. Colligon, H. Kheyrandish and S. P. Kaye.

References

1. V. Maurice, M. Talah and P. Marcus, Surf. Sci. Lett., **284**, L431, 1993.
2. T. P. Moffat, Fan Fu-Ren and A. J. Bard, extended abstracts of the NACE Corrosion Research Symposium, Nashville, 1992.
3. C. L. McBee and J. Kruger, Electrochim. Acta, **17**, 1337, 1972.
4. S. Fujimoto, G. S. Smith, R. C. Newman, H. Kheyrandish, J. S. Colligon and S. P. Kaye, Corros. Sci., in press (proceedings of Advances in Corrosion and Protection, Manchester, 1992).
5. R. P. Frankenthal, J. Electrochem. Soc., **114**, 542, 1967.
6. K. Sieradzki and R. C. Newman, J. Electrochem. Soc., **133**, 1979, 1986.
7. Song Qian, R. C. Newman, R. A. Cottis and K. Sieradzki, J. Electrochem. Soc., **137**, 435, 1990.
8. Song Qian, R. C. Newman, R. A. Cottis and K Sieradzki, Corros. Sci., **31**, 621, 1990.
9. D. E. Williams, R. C. Newman, Q. Song and R. G. Kelly, Nature, **350**, 216, 1991.
10. J. P. Pelz and R. H. Koch, J. Vac. Sci. Technol. B, **9**, 775, 1991.
11. M. P. Ryan, R. C. Newman and G. E. Thompson, in preparation.

12

AFM Studies of EFC Round Robin Alloy 516 Relating to the Role of Mo in Passivation

J. E. Castle, X. F. Yang, J. H. Qiu* and P. A. Zhdan

Department of Materials Science and Engineering, University of Surrey, Guildford Surrey GU2 5XH, UK
*Nanyang Technological University, Singapore

Abstract

In previous studies, using XPS and solution analytical techniques to study passivation, we have raised the possibility that the passivating film on FeCrMo alloy grows from discrete nucleation centres. In this paper we describe the application of Atomic Force Microscopy (AFM) to investigate the growth of the passivating film on a typical alloy (Fe15Cr4Mo) which we believe likely to show this phenomenon. Surface images have been obtained on samples exposed in water and subsequently, after replacement of the water by electrolyte, at potentials controlled in the passivating range. Changes in surface topography are consistent with the growth of nuclei which do appear to grow and merge over increasing periods of elapsed time. Measurements of surface roughness have been made and related to the formation of the surface deposit.

1. Introduction

In a series of papers, concerning electrochemical passivation of Fe17Cr and Fe15Cr4Mo alloys [1–3], we have shown that Mo greatly enhances the efficiency of film formation. Analyses of the electrolyte and the film led us to the conclusion that the film formed during the initial stages of passivation was incomplete, allowing direct dissolution of the alloying elements. The decrease in passivation current with increasing time corresponded, in our model, with the blocking of pores or open areas in the film. One possible explanation for the action of molybdenum is that the precipitation of molybdenum compounds (probably the oxide) provides the nuclei on which the solid phase of chromium oxide grows at the onset of passivation. Such a process of inoculation would account for the prime feature of molybdenum addition to the alloys— that it reduces the magnitude of the current peak at passivation and decreases by a factor of two the amount of charge appearing in the flux of ions to the electrolyte.

In this work we have examined the enrichment of molybdenum in the surface film on Fe15Cr4Mo, as a function of potential. We thus identify a potential range in which the enrichment of molybdenum exceeds that of chromium, and then seek evidence of precipitation phenomena in the film using *in situ* AFM.

2. Experimental

Fe15Cr4Mo alloy was provided (as alloy 516 used in the EFC Round Robin [4]) by the Max Planck Institute in Düsseldorf, Germany. It was heat-treated at 800°C for 20 min, followed by furnace cooling, rolled to 0.5 mm sheet and cut to test-pieces of 1.3 cm^2. The surface of the sample was ground and polished to a 1 μm finish. Electrochemical exposure in 0.1M H_2SO_4 for 1 h at different potentials was followed by immediate analysis by XPS using the procedure described in several previous publications [1, 3].

2.1 AFM Studies

The *in situ* measurements reported in this paper follow on extensive ex-situ measurements which gave invaluable experience in obtaining the correct imaging parameters. The AFM system used was the Nanoscope II made by Digital Instruments Incorporated, USA. The head D was used with a scan range of 1 to 0.3 μm. In the electrochemical passivation we have followed as closely as possible the conditions used in the EFC Round Robin, i.e. cathodic reduction of 5 mAcm^{-2} for 5 min, open circuit for 5 min, followed by a jump to the passivation potential (in this case –104 mV SHE). This procedure had to be modified for the *in situ* experiments as the rapid production of bubbles damaged the cantilever at cathodic reduction. The sample was therefore held at –104mV (SHE) directly. Bubble formation was a problem and the bubbles were carefully removed from time to time during the passivation process. The cell used was supplied as standard by Nanoscope and is shown in Fig. 1. The exposed area of the working electrode was 0.41 cm^2. The reference electrode used was lead/lead sulphate and was calibrated against a calomel electrode in 0.1M H_2SO_4. The surface, in water, is shown in Fig. 2. This field of view

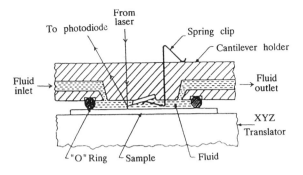

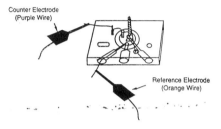

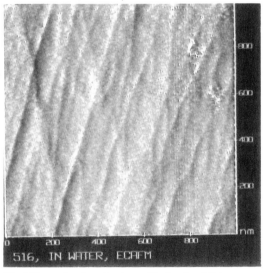

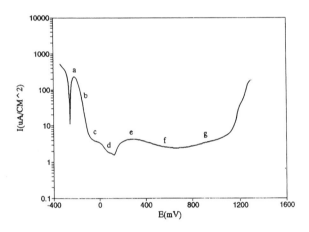

Fig. 1 AFM electrochemical cell. The working electrode forms the horizontal base to the cell which is fitted with a platinum counter electrode and a Pb/PbSO₄ reference electrode.

Fig. 2 The surface in water, imaged by in situ AFM.

is maintained throughout the subsequent exposures shown below.

3. Results

3.1 Surface Analysis

The XPS data were obtained at the representative potentials shown in the polarisation areas, Fig. 3, points a–g. The surface composition was transformed to an enrichment factor:

$$F(Mo) = \frac{Mo / (Fe + Cr + Mo)_{surface}}{Mo / (Fe + Cr + Mo)_{alloy}}$$

or

$$F(Cr) = \frac{Cr / (Fe + Cr + Mo)_{surface}}{Cr / (Fe + Cr + Mo)_{alloy}}$$

and the values obtained are plotted as a function of potential in Fig. 4. We can see immediately that F(Mo) is greater than F (Cr) for potentials through the active range and the passivation peak. The Mo 3d spectra are given in Fig. 5 and show that the metallic component Mo is least intense for the potentials at which surface enrichment is the greatest. We thus conclude that an oxidised form of molybdenum ((Mo^{+6}) is involved in the formation of the initial solid product on the metal surface. The potential, c; –104mV (SHE), is approximately the maximum for molybdenum enrichment and this was therefore used for the AFM studies.

Fig. 3 Polarisation curve for Fe15Cr4Mo alloy in 0.1 M H₂SO₄.

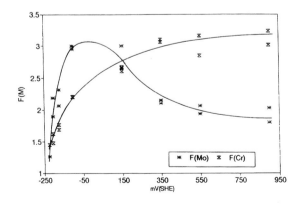

Fig. 4 The enrichment factors for Mo and Cr measured after polarisation of Fe15Cr4Mo alloy in H₂SO₄ for 1 h. There is a selective enrichment of molybdenum relative to Cr and Fe over the interval –250 mV to 900 mV (SHE). The AFM studies were carried out at –104 mV (SHE) in the low potential of the passive region.

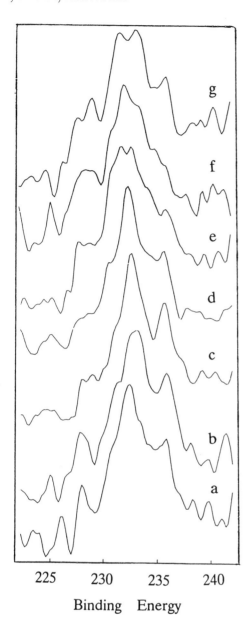

Fig. 5 XPS spectra for Mo. Both metallic and oxidised forms are present at most potentials. The proportion of oxidised molybdenum is greatest for the potentials which show the enhanced enrichment. The spectra are lettered according to the potentials indicated in Fig. 3.

3.2 Atomic Force Microscopy (AFM)

In situ scans were made in water, as a function of time at –104mV. The time series are shown in Fig. 7. The current decreased with exposure time as shown in Fig. 6. In each case a scan was made in height mode followed by a scan in force mode.

The time series given in Fig. 7 show very dramatically the growth of a surface deposit. Careful comparison of images shows that the individual platelets of the deposit form and then grow in thickness to reach

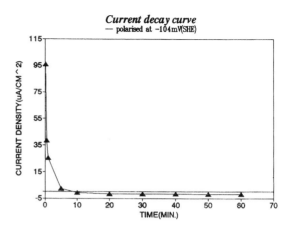

Fig. 6 Current decay curve of Fe15Cr4Mo alloy polarised at –104 mV (SHE).

maturity at an overall size of ≈ 60 nm × 200 nm. The comparison shows:

(a) inter-growth (e.g. at 'a' on the 5 min and 20 min images);

(b) loss (e.g. at 'b' on the 20 min and 30 min images);

(c) Breakage (e.g. at 'c' on the 20 min and 30 min images).

The broken plate at 'c' is shown in a zoom enlargement of the data in the inset micrograph on Fig. 7. This reveals the sharp fracture which has occurred. The final thickness of individual platelets was measured using the AFM in height mode, on the fields of view shown in Figs 8(a) and 8(b): it was of the order of 10 nm. The force mode and height mode scans for a given time of exposure were made sequentially. Careful comparison showed significant loss and breakage between the scans at a given time indicating that this damage was probably caused by the scanning tip itself. In particular it can be seen that the broken platelet featured in the zoom shot was in fact whole in the micrograph accompanying the height measurement in Fig. 8(b). The actual platelet across which the height cursor has been placed has been completely lost as shown at point 'b' in Fig. 7 (30 min scan).

The fact that such disturbance can be caused confirms that the platelets are truly a surface deposit. The three-dimensional nature of the platelets is illustrated by the presence of further plates beneath those which are broken– as at 'd' in the 30 min scan.

The generation of this platelet deposit, whilst dramatic, is a transient phenomenon. A more mature surface is eventually created. Platelets are lost or dissolved from the surface more rapidly than they are created from *ca.* 20 min and by the one hour scan they

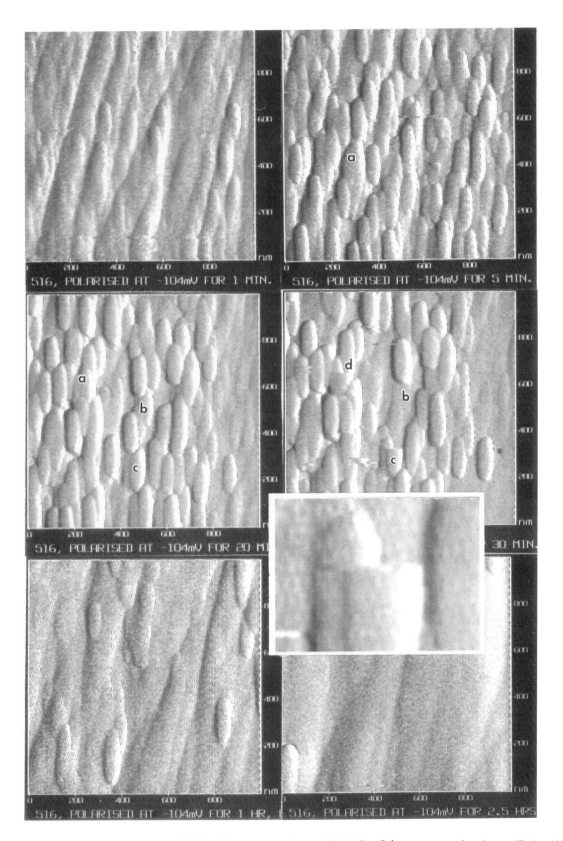

Fig. 7 AFM scans of a time series from 1 min to 2.5 h. The field of view is as shown in Fig. 2. Detailed comments are given in text. The inset image is a zoom reconstruction of the data at the point 'c' in the 30 min scan.

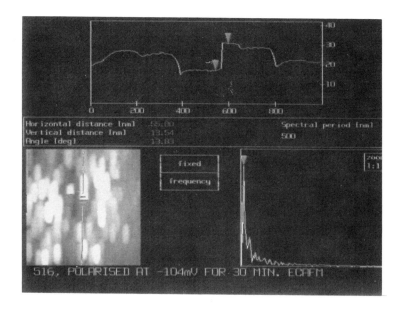

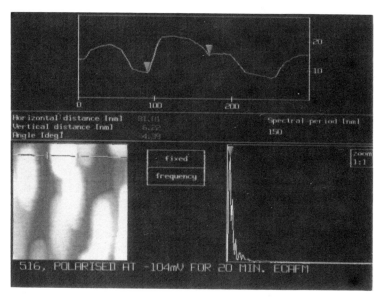

Fig.8 (a) and (b): Examples of the micrographs obtained in height mode used for measurement of the thickness of the platelets at broken edges.

have almost disappeared. The residual featureless background is presumably the true passive film and it is this which will be studied in more detail in further work.

4. Discussion

The platelets observed in this work were an unexpected and unusual phenomenon. They have been separately observed in many *ex situ* studies and are not an artefact of a single run under *in situ* conditions. We find their area to be approximately the same in the different runs but they are sometimes square, rather than oblong, in shape. We have not been able to see

them in the SEM, presumably because of lack of contrast in such a thin deposit. Repeated AFM measurements suggest that they reduce in thickness on vacuum exposure, suggesting that they might be hydrated.

The character of the platelets, in particular the fact that they can be lost electrochemically or dislodged by the scanning tip, suggests that they do not arise from partial protection of the passivating surface. In our first sight of the closely packed plates we did consider that the surface could be a mosaic of passivated steps on the metal surface. However we believe that images of the type shown in this paper show the correct interpretation to be that of a deposit.

We cannot yet be clear on the way in which these

platelets relate to the XPS analyses: the test pieces for XPS analysis have, in the past, been held vertically in the cell and our exposure time for XPS analysis is generally 1 h. Nevertheless, *ex situ* measurement show that the platelets still remain on the surface at this time. However, the volume of deposit seems to make it unlikely that this is a chromium-rich deposit. During the time at which they appear there is a very large exflux of iron from the surface and common-sense indicates that this might be the source of the precipitate. We have already undertaken scans on the 17Cr alloy and can find no similar deposit. It is possible therefore that the platelets owe their precipitation to molybdenum enrichment at the surface and that it is they which account for the greater current efficiency for the formation of the film. We are actively seeking to obtain the composition of platelets at the present time. We do not necessarily see them as part of the long-term protective film.

5. Acknowledgement

We thank British Gas plc. for their support of work on scanning probe microscopies by their provision of a Fellowship (PAZ) and the AFM/STM system.

References

1. J. E. Castle and J. H. Qiu, 'A co-ordinated study of the passivation of alloy steels by plasma source mass spectrometry and X-ray photoelectron spectroscopy: Part I — Characterisation of the passive film', Corros. Sci., **29**, 5, 591–603, 1989.
2. J. E. Castle and J. H. Qiu, 'A co-ordinated study of the passivation of alloy steels by plasma source mass spectrometry and X-ray photoelectron spectroscopy: Part II — Growth kinetics of the passive film', Corros. Sci., **29**, 5, 605–616, 1989.
3. J. E. Castle and J. H. Qiu 'The application of ICP-MS and XPS to studies of ion selectivity during passivation of stainless steels', J. Electrochem. Soc., **137**, 7, 2031-2038, 1990.
4. P. Marcus and I. Olefjord, Corros. Sci., **28**, 589, 1988.

13

Photoelectrochemical Study of Passive Films

F. Di Quarto, S. Piazza and C. Sunseri

Dipartimento di Ingegneria Chimica dei Processi e dei Materiali, Università di Palermo, Viale delle Scienze, 90128 Palermo, Italy

Abstract

The application of photoelectrochemical techniques to the *in situ* characterisation of passive layers on metals and alloys is reviewed. Complications arise when dealing with thin films having strongly disordered structure. Some selected examples are presented concerning model systems, which form stable surface layers with constant composition in large ranges of electrode potential and film thickness, as well as corrosion product layers on metals and alloys of commercial interest.

1. Introduction

In the last years different powerful spectroscopic techniques have been developed for the analysis of very thin passive layers on metals and alloys. Although useful information can be gathered both from *in situ* and *ex situ* techniques, there is a large agreement on the advantages offered by the former specially in the case of studies on very thin passive films, where a risk exists to change structure and composition of passive films on going from the electrochemical cell, under potentiostatic control, to the vacuum [1].

Owing to the semiconducting or insulating nature of passive films grown on many metals and alloys of technological interest, a growing interest has been devoted toward photoelectrochemical techniques. Photocurrent Spectroscopy (PCS) can give *in situ* information on:

(a) Electronic structure and (indirectly, through the optical gap value) chemical composition of passive films.
(b) Energetics at the passive film/electrolyte interface (flat band potential determination, conduction and valence band edge location).
(c) Mechanism of generation and transport of photocarriers.
(d) Kinetics of growth of photoconducting films.

The first two aspects are important for a deeper understanding of the corrosion behaviour of the metallic substrate. Moreover, the advantages of this technique can be appreciated by considering that PCS is:

(i) poor demanding for the surface finishing;
(ii) non-destructive and it does not need high light intensity;
(iii) a low cost technique.

As for the sensitivity limits, in several cases very thin passive films (1–2 nm thick) can be scrutinised by improving the signal detection technique by means of a lock-in/mechanical chopper system.

In this work we will present the results of some older and very recent investigations carried out in our laboratory which help to stress the importance of PCS in passivity studies.

2. Theoretical Background

As previously mentioned, it is possible to use the PCS technique to scrutinise semiconducting or insulating passive layers. In such a case by irradiating the electrode with light of suitable energy (the energy of incident photons, hv, must be higher than the optical gap of the film, E_g^{opt}), electrons can be excited from occupied electronic states (usually states of the valence band, V.B.) into vacant ones (usually states of the conduction band, C.B.), according to the reaction:

$$film + hv = e^- (C.B.) + h^+ (V.B.)$$

where e^- (C.B.) represents an electron in empty states of the conduction band of the solid and h^+ (V.B.) is the hole (a vacant electronic state) created in the valence band. Photoeffects (photocurrent and photopotential) can be detected as a consequence of the modification of the charge distribution into the electrode under illumination. Under application of an external bias the generated photocarriers drive an electrochemical reaction at the film/electrolyte interface. The possible photoreactions depend on the kind of photocarrier arriving at the interface. Anodic or cathodic photoreactions will occur respectively for n-type and p-type semiconductors, whilst in the case of insulators both types of reaction can occur depending on the direction of electric field. The study of the photocurrent as a function of electrode potential and wavelength is the basis of PCS.

Although photoeffects have been reported long time ago, only in the last decades they have been investigated quantitatively for the single crystal semiconductor/electrolyte [2, 3] and, more recently, for the passive films/electrolyte [4, 5] junctions. In the last case difficulties arise because passive films are usually very thin and amorphous or strongly disordered.

2.1 Optical band gap and electronic structure

Different authors [3–7] have suggested that it is necessary to take into account the differences in the electronic structure and transport properties between crystalline and amorphous semiconductors (a-SC) in order to explain the photoelectrochemical response of passive films. A simplified scheme of the density of electronic states (DOS) as a function of energy is reported in Fig.1, showing the existence of a finite density of states within the mobility gap (E_C–E_V) of the a-SC. In Fig.1 all the possible optical transitions from occupied to empty states are reported too. According to this picture, the main contribution to the measured photocurrent in amorphous passive films can be attributed to optical transitions between extended states of the V.B. (initial states) and those of the C.B. (final states).

For optical transitions in the vicinity of the mobility edge of an a-SC (or band edge in a crystalline SC) it has been shown theoretically [8] that it is possible to get the following relationship between the absorption coefficient and the energy of the incident photons:

$$\alpha h\nu = \text{const}\left(h\nu - E_g^{opt}\right)^n \tag{1}$$

where n=2 is often experimentally found in a-SC, whilst for crystalline materials different n values are reported for indirect (n=2), direct allowed (n=0.5) or direct forbidden (n=1.5) optical transitions [8]. From this equation it comes out that the extrapolation to zero of the $(\alpha h\nu)^{1/n}$ vs hν plot allows to get the E_g^{opt} value in a quite straightforward way.

Although eqn (1) has been obtained assuming a parabolic DOS distribution both in the V.B. and the C.B. (Tauc approximation), it has been shown by Mott and Davis [8] that a similar dependence for α (n = 2) can be obtained for a-SC by assuming a linear distribution of DOS in the band tails and by neglecting optical transitions between localised states. In this last case the optical gap is a measure of the smaller energy difference between localised states of the valence band and extended states of the conduction band or vice versa ((E_C–E_B) or (E_A–E_V) in Fig. 1, whichever is the smaller). According to the theoretical expectations, the Mott-Davis model should be valid in a narrow range of energy below the mobility gap, usually of the order of *ca.* 0.3 eV. This fact could help to discriminate between the optical gap in the sense of Mott-Davis and the mobility gap of amorphous materials. This is particularly important in the case of amorphous passive films, where the value of the optical gap is usually employed for the identification of the nature of the film. In absence of large deviations from stoichiometry this difference is in the order of 0.2–0.3 eV, strongly supporting the Mott-Davis suggestion that the long range disorder affects only slightly the band gap of materials. Larger differences between the E_g^{opt} value of the film and that relative to the crystalline counterpart must be interpreted as an indication that significant deviation from stoichiometry is present in the passive film. Large non-stoichiometry can imply also changes in the density of the passive film, which is known also to affect the value of optical gap in amorphous materials [8].

In some cases at energies below the mobility gap an exponential dependence of the absorption coefficient on the energy of the incident photons has been reported and usually referred to as Urbach tail:

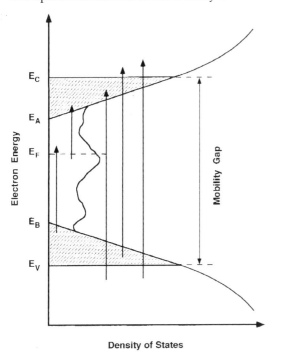

Density of States

Fig. 1 Schematic density of states distribution in an amorphous semiconductor with localised electronic states (dashed regions) and defect states within the mobility gap. Vertical arrows show possible optical transitions.

$$\alpha \propto \exp\left[\frac{h\nu - E_g^{opt}}{E_0}\right] \tag{2}$$

The presence of this tail has been associated with a region of transition from the strong band to band absorption to deep tail, and E_0 has been assumed as a measure of the disorder or tail width. According to such an interpretation, the upper limit of validity of eqn (2) should coincide with the mobility gap determined according to eqn (1). The use of such a method for the determination of the mobility gap of passive films has given quite contradictory results in the photoelectrochemical literature. For this reason in the following we will make use of the Tauc approximation (eqn (1)) for the evaluation of the mobility gap of passive films.

2.2 Photocurrent modelling

The most easily accessible quantity in PCS is the photocurrent generated at the film/electrolyte interface under illumination with light of suitable wavelength. In order to get the optical band gap of a passive film by means of eqn (1) we need to correlate the light absorption coefficient with the photocurrent intensity at different wavelengths.

In a series of papers [7,9–11] we have shown that the most important differences in modelling the photocurrent behaviour of passive films with respect to the case of crystalline SCs are essentially two.

(i) The possibility to neglect, in many cases, the contribution to the measured photocurrent of the diffusive term with respect to the migration term in the transport mechanisms of photocarriers. This can occur owing to the very low mobility of the photocarriers or because the film is so thin that the electric field extends across the entire thickness.
(ii) The low mobility of photocarriers implies also a short thermalisation length of the photogenerated electron-hole pair and then the need to take into account geminate recombination effects. These are strongly dependent both on the energy of incident photons and on the electric field intensity.

According to these considerations, we have shown that the photocurrent behaviour of amorphous passive films grown on different metals can be modelled, in a large range of thickness, electrode potential and photon energy, by assuming for the photocurrent in quasi steady-state conditions the following expression [11]:

$$i_{ph} = |e|\varphi\overline{\eta}[1-\exp(-\alpha x_{sc})]\frac{\mu\tau\overline{F}}{x_{sc}}\left[1-\exp\left(-\frac{x_{sc}}{\mu\tau\overline{F}}\right)\right] \quad (3)$$

where φ is the photon flux corrected for the reflection losses which takes into account possible light interference effects [10], $\overline{\eta}$ is the average efficiency of free carrier generation in the average field approximation which takes into account possible geminate recombi-

nation effects, α the absorption coefficient of the light and $\mu\tau$ (μ the mobility and τ the lifetime of the photocarriers) the drift range before deep trapping. From eqn (3) it is possible to show that, by neglecting eventual variations of $\overline{\eta}$ and of the reflection losses with the photon energy, a direct proportionality holds between α and i_{ph} when $\alpha x_{sc} << 1$, so that eqn (1) can be rewritten as:

$$(i_{ph}h\nu)^{1/n} \cong const(h\nu - E_g^{opt}) \quad (4)$$

where the i_{ph} values must be corrected for the spectral emission of the lamp/monochromator system. For insulating films $x_{sc} = D_{ox}$ and the total oxide thickness, D_{ox} replaces the space-charge region width in eqn (3).

In previous papers [9,11] we reported on the influence of the electric field and the photon energy on η; expressions both for the efficiency of free carriers generation and for the width of space charge region in a-SC were derived. We have shown that an increase of the photon energy (longer thermalisation lengths) and of the electric field will favour the escape of the electron-hole pair from their mutual attracting coulombic barrier, leading to increased photocurrent yield.

When geminate recombination effects are operating, an increase of energy of the incident photons should be reflected in a change in the shape of the photocharacteristics from supralinear to sublinear [11,12]. This finding can help to ascertain whether a supralinear photocharacteristic is related to geminate recombination or to strong surface recombination effects. The former are important with supraband gap weakly absorbed light, whilst the latter must be considered in presence of strongly absorbed light at low band bendings and become negligible at not too high band bendings (> 0.5–1 V).

Finally we stress that for incident light having energy lower than the optical gap a larger dependence of the photocurrent on the electric field values is expected owing both to the geminate recombination effects and to the onset of a mechanism of transport of photocarriers by hopping between deep localised states. In such cases an exponential behaviour of the photocurent can be expected as a function of the electric field [13].

3. Indentification and Characterisation of Passive Films

In this section we will present some selected examples of the experimental work performed in our laboratory in order to illustrate the capability of PCS in providing *in situ* information as well as its ability to monitor changes in composition or structure of passive films due to changes in the experimental conditions.

3.1 Model systems

This class of systems includes metals (mainly the so-called valve metals) which anodised in suitable solutions are able to form stable passive films having well defined composition and spanning a large range of thickness, from very few up to hundreds nanometers. In Tables 1(a), (b) we report the optical band gap values of anodic oxide films grown on valve-metals showing different electrical and photoelectrochemical behaviour. In Table 1(a) we have grouped three oxide films which exhibit typical n-type semiconducting behaviour [7,9–11,14]. Moreover, they present two common aspects:

(i) an optical gap larger in the amorphous state than in the crystalline counter-part;
(ii) a clear dependence of the photocharacteristics upon the energy of the incident light, under illumination with photons having energy higher than the optical band gap.

For all these oxides a tail in the photocurrent spectra is present at energies lower than the optical gap value, derived according to eqn (4). These results have been interpreted as a clear indication of the formation, in the experimental conditions reported in the original papers, of amorphous oxide films having a short range order very similar to the crystalline phase. The long-range disorder in such oxides, originates for each oxide different defective structures and DOS distributions, as evidenced by the impedance study [14].

An interpretation of these results can be put forward, in agreement with the scheme of band structure in a-SC formerly depicted, by assuming that the reported optical gap is a measure of the mobility gap of such amorphous oxides. Moreover, the shrinkage of optical gap after crystallisation (0.2–0.3 eV) is in agreement with the estimate made by Mott [8] for the effect of lattice disorder on the extent of localisation of DOS close to the mobility edges. The disappearance of the influence of the light wavelength on the shape of the photocharacteristics after crystallisation is in agreement with the hypothesis that geminate recombination effects are now negligible owing to the increased thermalisation length of the photocarriers.

Table 1(b) includes three large band gap oxide films showing impedance and photoelectrochemical behaviour typical of insulating films. Differences and analogies were observed in their photoelectrochemical behaviour. In the case of passive films on tantalum [12] and aluminium [13,15] we have recently reported on the presence of cathodic photocurrent at electrode potentials negative with respect to the their flat band potential and under illumination with sub-band gap

Table 1(a) Optical band gap values of semiconducting oxide films grown anodically on valve metals. Band gap values of the crystalline counterparts are also reported

Anodic Film	E_g^{opt} , eV	E_g^{cryst} , eV	Influence of λ on photocurrent characteristics
a-Nb$_2$O$_5$	3.35	3.15*	yes
a-TiO$_2$	3.2 - 3.4	3.0 - 3.2	yes
a-WO$_3$	3.05	2.75*	yes

* Value obtained after crystallization under argon of amorphous films.

Table 1(b) Optical band gap values of insulating oxide films grown anodically on valve metals. Band gap values of the crystalline counterparts are also reported

Anodic Film	E_g^{opt} , eV	E_g^{cryst} , eV	Influence of λ on photocurrent characteristics
a-Al$_2$O$_3$	6.0 - 6.3	6.3**	no
a-ZrO$_2$	4.7	?	?
a-Ta$_2$O$_5$	3.9 - 4.2	?	yes

** Calculated theoretically (Xu and Ching, Phys. Rev. B, **43**, 4461 (1991)).

light. An internal electron photoemission process from the underlying metal into the oxide C.B. was suggested for explaining cathodic photocurrent energy thresholds lower than the anodic ones for both oxides. In the case of zirconium oxide it has not yet ascertained whether an internal photoemission process is operating too at cathodic potentials. The optical gap has been determined by means of eqn (4) in the case of tantalum and zirconium oxide which display a band gap less than 5 eV (see Table 1(b)), whilst in the case of aluminium oxide it has been indirectly inferred on the basis of a detailed study of anodic photocurrent spectra in different solutions [13,15]. At variance with tantalum oxide thin films [12], zirconium (see Fig. 2) and aluminium oxide [13,15] films show appreciable anodic photocurrent under illumination with photons having energy lower than the band gap of the oxides since very low thicknesses. It is noteworthy that in the case of aluminium sub-band gap anodic photocurrent was observed only with films formed in electrolytic solutions containing organic anions (tartrate or citrate) [13].

In the case of zirconium oxide the anodic photocurrent spectrum is modified changing the film thickness (Fig. 2). On the other hand the two anodic photocurrent thresholds, obtained according to eqn (4) and reported in Fig. 3, remain practically constant by changing the nature of the electrolytic solution (H_2SO_4, K_2SO_4, NaOH) or the film thickness. A possible explanation for the presence of two threshold energies can be put forward by assuming that oxide films grown on zirconium metal present a duplex structure with the external layer, having a nearly constant thickness, consisting of hydrated oxide and the internal one consisting of anhydrous oxide. This suggestion is supported by the following arguments:

(i) The extrapolations at lower (~ 3.0 eV) and higher (~ 4.7 eV) photon energy remain unchanged for films formed both in sulphate and sodium hydroxide solutions. This would exclude sub-band gap transitions involving electronic states within the gap generated by the incorporation of sulphate ions into the oxide film.
(ii) The formation of a hydrated $ZrO_2 (H_2O)_n$ thin surface layer is in agreement with XPS results reported in the literature [16].
(iii) The constancy of the thickness of the hydrated layer (or zirconium hydroxide) is suggested by the sharp decrease of the photoresponse at long wavelengths with increasing the film thickness (see Fig. 2).
(iv) The formation of a ZrO_2 film, whose thickness is a function of the applied voltage, is well established in the literature and the anodic photocurrent threshold of ~ 4.7 eV reported in Fig. 3 is in agreement with the value of band gap expected for such insulating oxide [17].

We like to stress that the photoelectrochemical results could help to explain the large electronic current usually observed during the anodization of this metal in spite of the large band gap reported for crystalline ZrO_2.

Regarding the high band gap insulating oxides of Table 1(b), only for tantalum oxide it has been established a clear dependence of the shape of the photocharacteristics on the energy of incident photons under supra-band gap irradiation, similar to that observed for oxide films having lower optical gap (Table 1(a)). In the case of zirconium the limited energy range above the band gap accessible for exploring such an

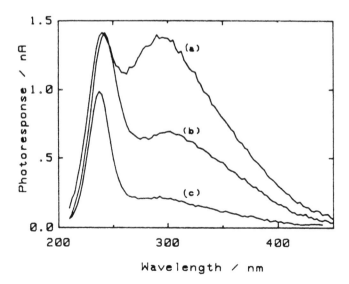

Fig. 2 *Raw photoresponse (without correction for the photon efficiency) as a function of wavelength for ZrO_2 anodic films grown at 8 mA cm⁻² in different sulphate solutions. (a) 1N K_2SO_4, D_{ox} = 25 nm; (b) 1N H_2SO_4, D_{ox} = 25 nm; (c) 1N K_2SO_4, D_{ox} = 80 nm.*

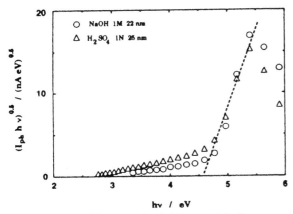

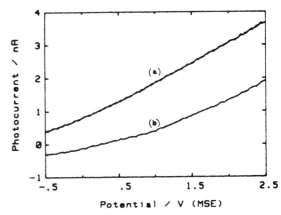

Fig. 3 Determination of the energy threshold for two ZrO_2 films grown in 1M NaOH solution up to 22 nm (circles) and in 1N H_2SO_4 solution up to 25 nm (triangles), respectively.

Fig. 4 Photocurrent vs potential curves at different wavelengths for a ZrO_2 film grown in 1N H_2SO_4 solution up to 13 nm. (a) $\lambda = 230$ nm; (b) $\lambda = 300$ nm.

effect, did not allow to give a definite answer and more investigation is necessary in order to interpret the supralinear photocharacteristics observed under both supra and sub-band gap monochromatic irradiation (see Fig. 4).

As for the aluminium oxide, it has been shown in previous papers [13,15] that the anodic photocurrent changes in exponential way at high fields and almost linearly at low electric fields. A reasonable explanation for this behaviour could be put forward assuming a trap-limited mechanism of transport of the photocarriers in localised electronic states. In this last hypothesis the use of a field-dependent mobility term in eqn (3) would be able to fit the experimental data.

3.2 Base metals and alloys

The importance of PCS in the characterisation of passive films on metals like copper, iron, chromium, nickel and their alloys is now well established [4,5,18–22]. Owing to their extreme thinness and/or to the possibility of formation of different phases (even changing the electrode potential within the passivity range) with different oxidation states, hydration degree and electronic properties, the photoelectrochemical study of these systems presents more difficulties.

Still in these more complex systems the PCS technique can be usefully employed to monitor the changes of corrosion films grown on the metallic substrate as a function of the experimental parameters. With this aim we present some results regarding corrosion layers grown in free corrosion conditions on copper and on a commercial Al-brass, whose composition is re-

ported in Table 2. Several photoelectrochemical studies carried out on pure copper have demonstrated conclusively the critical role of the solution composition and ageing time in determining the semiconducting properties (n or p-type) of Cu_2O films formed in a wide range of pH, from weakly acidic to basic [4, 5,18].

In a previous study on the corrosion of copper in sodium sulphate solution [18], we have shown the existence of an intermediate range of pH (3.5 < pH < 6) where the formation of a protective n-Cu_2O oxide films occurs at the free corrosion potential, as well as the favourable effect of the addition of Cu^{++} ions to the solution for stabilising the formation of n-type oxide. These results are summarised in Fig. 5. In the same figure we report also the results relative to Al-brass showing that the protective layer formed on this alloys in the reported conditions is substantially Cu_2O. This conclusion was reached by comparison of the photocurrent spectra for copper and Al-brass showing both the same features and a practically coincident band gap (see Fig. 6 and Ref.[18]). This finding confirms that a strong dezincification process occurs during the surface layer formation and suggests a negligible effect of Al addition on the properties of the corrosion layer. The only appreciable effect of the alloying elements can be traced out to the small widening of the pH range where p-type Cu_2O films are formed. This last effect is independent of the presence of Cu^{++} ions in the solution and could be attributed to small local changes in the pH values, due to the dissolution of alloying elements. The influence of the ageing time on

Table 2 Chemical composition of the investigated Al–brass alloy

Element	Cu	Zn	Al	Pb	Fe	As	P
weight %	76.70	21.21	2.03	0.002	0.005	0.027	0.001

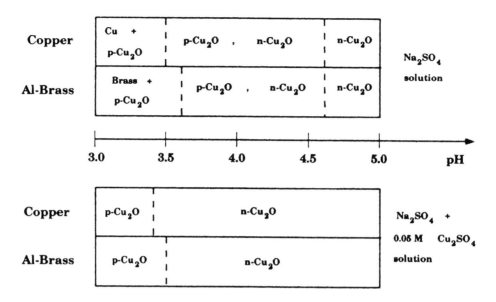

Fig. 5 *Nature and conductivity type of surface layers identified by PCS at different pH values on copper and Al-brass in aerated 0.1M Na$_2$SO$_4$ solution with or without addition of Cu^{2+} ions after 24 h immersion at open circuit conditions.*

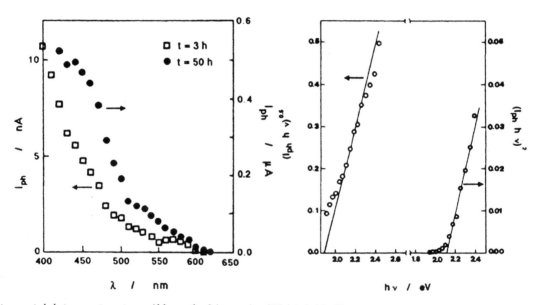

Fig. 6 *Left part: corrected photocurrent spectra on Al-brass after 3 (squares) and 50 (circles) h of free corrosion in aerated 0.1M Na$_2$SO$_4$ solution (pH ≈ 5). U$_{corr}$ ≈ −20 mV(SCE). Right part: determination of the optical gap of the layer.*

the conductivity type of Cu$_2$O layers formed on Al-brass is evidenced in Fig. 7.

A further demonstration of the usefulness of PCS as *in situ* analytical tool comes out from the result of Fig. 8 showing the effect of a change in the electrolyte composition on the nature of the corrosion layer on copper. In the figure the photocurrent spectrum of the corrosion layer formed on copper in presence of a large amount of chloride ions is displayed. The optical gap has been obtained by assuming direct optical transitions for the crystalline corrosion layer. A value of 3.15

eV has been derived in very good agreement with the theoretical value [23] for crystalline CuCl. At the measuring potential the photocurrent was anodic and very low in agreement with the insulating nature of CuCl. The photoelectrochemical identification was further supported by the X-ray patterns of the electrode.

4. Conclusions

Although recent and powerful surface analytical techniques (XPS, XANES, EXAFS) are still the best candi-

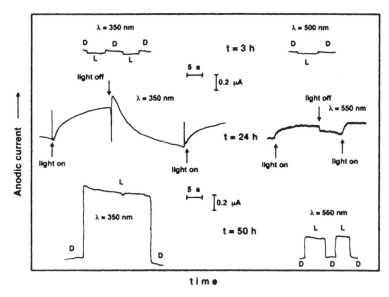

Fig. 7 Total current in the dark and under monochromatic irradiation of different wavelengths for the sample of Fig. 6 after different immersion times (electrode potential = U_{corr} +10 mV).

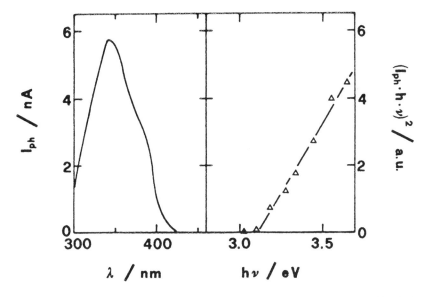

Fig. 8 Photocurrent action spectrum and determination of the optical gap for a CuCl film grown on copper in 0.1M Na_2SO_4 + 0.05M NaCl + 0.05M $CuSO_4$ solution (pH $\approx$ 3.0) after 24h free corrosion ($U_{corr} \cong 0$ mV(SCE)).

dates for getting decisive information on the chemical composition and structure of thin passive films, it is in our opinion that supplementary information is necessary to reach a deeper knowledge of the different electrochemical aspects involved in the formation of a stable passivating layer. The importance of the role played by the solid state properties of the films in determining the kinetic of electronic and ionic exchange at the passive film/electrolyte interface makes unavoidable the use of *in situ* electrochemical techniques, capable of monitoring changes in the electronic properties of the passive film.

In this frame the PCS technique can play a key role in providing valuable information on the electronic properties of passive films and on their changes as a function of different experimental parameters. Of course, owing to the complex nature of the passive films on alloys and metals, PCS alone may be not sufficient to reach ultimate conclusions on such systems. Further improvement of the technique can be devised in the use of tunable lasers and very small light spots, which allow to characterise small surface areas on corroding alloys. These improvements will probably expand our knowledge on passive films as well as on the role of inhomogeneities in determining local breakdown of passivity.

5. Acknowledgements

The financial support by Italian C.N.R Technological Committee (contract no. 9004208.MZ78) and M.U.R.S.T. (40% and 60% funds) as well as the thesis work performed by our students are gratefully acknowledged.

References

1. B. R. MacDougall, T. A. Ramanarayanan, R. S. Alwitt Eds., Oxide Films on Metals and Alloys, Vol. **92–22**, The Electrochem. Soc. Inc., Pennington, N.J. , 1992.
2. Yu. V. Pleskov and Yu. Ya. Gurevich, Semiconductor Photoelectrochemistry, Consultants Bureau, N.Y., 1986.
3. H. Gerischer, Electrochim. Acta, **35**, 1677, 1990.
4. U. Stimming, Electrochim. Acta, **31**, 415, 1986, and references therein.
5. L. Peter, in Comprehensive Chemical Kinetics, vol. **29**, p.353, Elsevier, Amsterdam ,1989.
6. S. R. Morrison, Electrochemistry of Semiconductor and Oxidized Metal Electrodes, Plenum Press, N.Y., 1980.
7. F. Di Quarto, G. Russo, C. Sunseri and A. Di Paola, J. Chem. Soc. Faraday Trans., 1 **78**, 3433 (1982).
8. N. F. Mott and E. A. Davis, Electronic Processes, in Non-Crystalline Materials, 2nd edn., Clarendon Press, Oxford, 1979.
9. F. Di Quarto, S. Piazza, R. D'Agostino and C. Sunseri, J. Electroanal.Chem., **228**, 119, 1987.
10. F. Di Quarto, S. Piazza and C. Sunseri, J. Chem. Soc. Faraday Trans., 1, **85**, 3309, 1989.
11. F. Di Quarto, S. Piazza and C. Sunseri, Electrochim. Acta, **38**, 29, 1993.
12. F. Di Quarto, C. Gentile, S. Piazza and C. Sunseri, Corros. Sci., in press.
13. F. Di Quarto, S. Piazza, A. Splendore and C. Sunseri, in Ref.1, p.311.
14. S. Piazza, C. Sunseri and F. Di Quarto, AIChE J., **38**, 219, 1992.
15. F. Di Quarto, C. Gentile, S. Piazza and C. Sunseri, J. Electrochem. Soc., **138**, 1856, 1991.
16. P. Meisterjahn, H. W. Hoppe and J. W. Schultze, J. Electroanal. Chem., **217**, 159, 1987.
17. A. K. Vijh, in Oxides and Oxide Films, vol.2, J. W. Diggle Ed., M. Dekker Inc., N.Y., 1973.
18. F. Di Quarto, S. Piazza and C. Sunseri, Electrochim. Acta, **30**, 315, 1985.
19. C. Sunseri, S. Piazza and F. Di Quarto, J. Electrochem. Soc., **137**, 2411, 1990.
20. C. Sunseri, S. Piazza, A. Di Paola and F. Di Quarto, J. Electrochem. Soc., **134**, 2410, 1987, and references therein.
21. G. Dagan, W. M. Shen and M. Tomkiewicz, J. Electrochem. Soc., **139**, 1855, 1992.
22. P. Schmuki and H. Bohni, J. Electrochem. Soc., **139**, 1909, 1992.
23. S. Lewonczuk, J. G. Gross, J. Riengeissen, M. A. Khan and R. Riedinger, Phys. Rev. B **27**, 1259, 1983.

14

Modifications of the Structural and Electronic Properties of Thermochemically Formed Passive Films on Fe–Cr Alloys during Long Term Ageing: Influence of Silicon Addition

*D. LIU, D. GORSE AND B. BAROUX**

CECM-CNRS, 15 rue Georges Urbain, 94407, Vitry/Seine, France
*Centre de Recherche d'Ugine, 73400, Ugine, France

Abstract

This work is concerned with the influence of silicon segregation on the electronic, structural and electrochemical properties of passive films thermochemically formed on Fe–17%Cr alloys with various silicon contents. We show that long term exposure to pure water changes drastically the above mentioned properties, depending on the Si content in the alloy. Ageing may cause a progressive charging in silicon cations, at a rate increasing with the amount of silicon in the alloy. The film modifications produced by ageing are evidenced at the same time by a peak of photoresponse, a discontinuity in rest potential and a change in the film structure (visible by Grazing Incidence X-ray diffraction), in their representation as function of the ageing time.

1. Introduction

The importance of ageing in modifying the electrochemical properties of passive films stems from a few recent works [1–3]. They were motivated by the observation that some passivated stainless steels (AISI 434...) exposed for various periods of time in air or, more efficiently, to pure water exhibit an improved resistance either to localised corrosion [1, 4] or to uniform depassivation in acid medium [1].

However, this beneficial influence of ageing is still unexplained and concerns passive films formed either by polarisation in aqueous chloride containing solution (prepared by wet mechanical polishing, followed by 24 h air ageing prior to immersion in the test solution), or by thermochemical passivation.

We concentrate in this work on passive films formed by thermochemical passivation (also called 'bright annealing'). This treatment consists in annealing the specimens under hydrogen atmosphere containing a residual water partial pressure: under these conditions, no bulk oxide but only a thin (few nanometers thick) protective and adherent (passivating) film is formed at the steel surface. The passive films are distinguished as a function of both the annealing temperature (T) and dew point (σ).

As already stated, these thermochemically formed passive films, aged at open circuit in pure water, obey very slow evolution kinetics:

- one month of ageing has been found insufficient to detect noticeable modifications in the film composition [1];

- ageing for long periods, up to few months, are required to modify the film composition [2, 3, 5].

On the other hand, changes in the photoelectrochemical (PEC) behaviour occurring after various ageing periods in water were reported in a previous paper. It has been shown that these PEC changes occurring after long term ageing in water can be 'classified' as a function of the formation dew point.

In parallel, the chemical composition of the so formed films has been investigated at various stages of ageing [5]. It was observed that thermochemical passivation favours the segregation in the minor oxidizable elements in the passive films, mainly silicon, with enrichment factors depending also on the annealing dew point, in otherwise identical formation conditions. The composition aspects of the problem are discussed elsewhere [5].

In the present work, we focus on the modifications of the electronic and structural properties occurring during long term water ageing, up to about one year, and try to correlate them with the evolution of the electrochemical behaviour. Special attention will be paid to the role of silicon segregation on the above mentioned properties. To this aim, we shall compare the properties of passive films thermochemically formed on either a pure Fe–17%Cr–1.2%Si alloy, or an industrial stainless steel containing 0.4%Si, in various annealing conditions.

We shall see that long term exposure to pure water influences significantly the silicon segregation, and that, in turn, this factor apparently controls largely the evolution of both the electrochemical and physico–chemical properties of the so formed passive film.

2. Experimental

The alloys studied are either an industrial ferritic steel containing 0.4%Si as minor alloying element (AISI434), or a synthesised Fe–17%Cr–1.2%Si alloy. The films are formed by annealing in H_2O containing H_2 atmosphere for 15 min at 820°C, so called 'dry' and 'wet' annealing for $P_{H_2O}/P_{H_2} = 4 \cdot 10^{-6}$, and $1.3 \cdot 10^{-5}$ respectively. The corresponding dew points are $\sigma = -67°C$, $-58°C$ for 'dry' and 'wet' annealing respectively [1]. Then the specimens are aged in water, used as a means of accelerating air ageing, the ageing time being noted t_A in the following. The electrochemical and photoelectrochemical (PEC) tests are performed in phosphate solution (KH_2PO_4, pH 4.4) degassed by N_2 bubbling prior to the test.

In this paper, we discuss (i) the rest potential values measured at immersion in phosphate solution (noted $V_{rest(t=0)}$), and (ii) the photocurrents (I_{ph}) detected at 0.5V/SCE under white light illumination (by the lock-in technique), as functions of the ageing time in water (t_A).

The shape of the V_{rest} vs t_A curve does not depend on the time at which the rest potential is measured, over the whole test duration (~ few hours). Only a slight decrease of the slope dV_{rest}/dt_A was observed, as a result of few hours of immersion in the phosphate solution, indicating that neither significant chemical dissolution nor repassivation processes take place at open circuit in the test solution under conditions of short exposure time. Accordingly, the potential for PEC measurements (0.5V/SCE) is chosen close to $V_{rest(t=0)}$ below the range of oxygen evolution.

Moreover, films formed on the industrial 0.4%Si containing alloy under similar annealing conditions were tested for the first time by using the grazing incidence X-ray diffraction technique. The tests were performed on a standard Philips diffractometer (PW1830 generator: 40-kW, I = 35mA, wavelength: λ(Co K_α) = 1.79 Å). The operating conditions (1/30° divergence slit, sample, collimator, monochromator,

proportional detector, step angle: 0.025°, sample time: 55 sec) do not allow a precise determination of the film structure, since no peak indexation is made possible. A graze angle of 0.6° (in 2θ) is chosen, based on a rough estimate of the critical angle deduced from reflectivity profiles. The spectra recorded during a 2θ scan are shown in Figs 3 and 4. In all cases, two texture peaks are visible located at 2θ = 52.47° and 77.37° respectively, corresponding to the (110) and (001) peaks of the b.c.c structure pertaining to the underlying ferritic stainless steel are visible, with reproducible intensity on all spectra (mainly for the more intense peak at 52.47°, the other one decreases or increases depending on either the film thickening or thinning).

Photocurrent measurements were performed by the lock-in technique under white light illumination, by using a 150 W xenon lamp, a mechanical chopper (EG&G Brookdeal 9479) allowing to vary the light modulation frequency between 5Hz and 3KHz and a PAR 5208 two phase lock-in amplifier coupled to a PAR 273 potentiostat. The I_{ph} values reported in Fig. 2 were measured at the frequency of maximum yield, and corrected for the variations of the lamp emissivity (using a Bentham DH-Si detector).

3. Rest Potential Evolution in Time

An increase in rest potential with water ageing time t_A was observed for the two alloys whatever the formation conditions, as seen in Fig. 1. However, noticeable differences between the films were noted, for example regarding the slope dV_{rest}/dt_A, as a function of the formation conditions and silicon content of the alloy (see Table 1, p.88). Particularly, the following points are noticeable:

(i) the slope dV_{rest}/dt_A increases with oxidising power (dew point) of the annealing atmosphere (compare Figs 1(a) and (b)) and silicon content of the alloy (C_{Si}). It changes from 0.15mV/day for $\sigma = -67°C$ and $c_{Si} = 0.4\%$ to 0.84mV/day for $\sigma = -58°C$ and $C_{Si} = 0.4\%$ and to 1.68mV/day for $\sigma = -58°C$ and $C_{Si} = 1.2\%$.

(ii) for the 'wet' annealed 0.4%Si containing alloy, a clear tendency in the V_{rest} vs t_A curve emerges after an ageing time $t_A \sim 3$ months,

(iii) for the 1.2%Si containing alloy, $V_{rest(t=0)}$ exhibits a steep drop, once reached the apparently 'critical' ageing of 4 months for which the rest potential attained its anodic limit of order 0.4V/SCE,

(iv) then the slope falls off to a value equal to 0.71 mV/day approaching that found on the 'wet' annealed alloy with a lower Si content.

4. Photoelectrochemical Evolution in Time

To facilitate the comparison with the electrochemical evolution of the specimens with ageing described

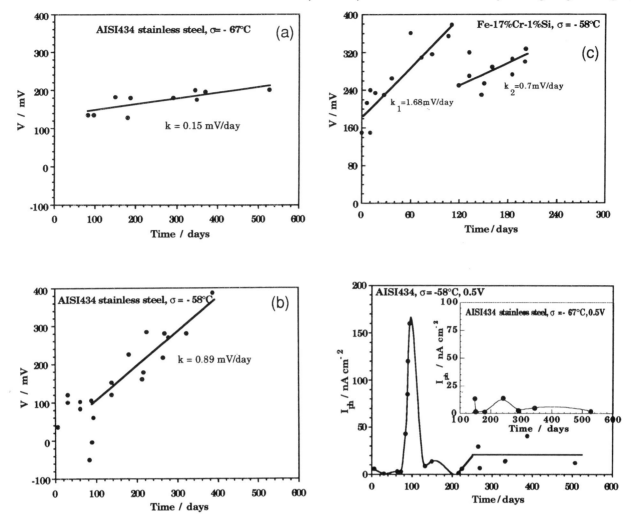

Fig. 1 *Plot of the rest potential at immersion in the phosphate solution vs time of exposure to pure water, t$_A$ (in days):*
(a) for a 'dry' annealed ferritic stainless steel (σ = –67°C and C$_{Si}$ = 0.4%);
(b) for a 'wet' annealed ferritic stainless steel (σ = –58°C and C$_{Si}$ = 0.4%);
(c) for a 'wet' annealed synthetized alloy (σ = –58°C and C$_{Si}$ = 1.2%).

Fig. 2 *Plot of the photocurrent measured at 0.5V/SCE applied voltage in phosphate solution (pH 4.4) under white light illumination vs time of exposure to pure water, t$_A$ (in days), for a 'wet' annealed ferritic stainless steel (σ = –58°C and C$_{Si}$ = 0.4%). In inset: same as previously for a 'dry' annealed ferritic stainless steel (σ = –67°C and C$_{Si}$ = 0.4%).*

above, the photocurrent measured at 0.5V/SCE (see above) is reported vs ageing time t$_A$ in Fig. 2. A peak of photocurrent response occurs, concomitant with the abrupt change in rest potential behaviour visible on Fig. 1(b), once reached a 'critical' ageing, in between 2 and 3 months for the 'wet' annealed 0.4%Si containing alloy. In inset, are represented the photocurrent values obtained under identical conditions on a 'dry' annealed specimen. These I$_{ph}$ values are all < 10 nA cm^{-2}. Note that the lower slope dV$_{rest}$/dt$_A$ was also observed in these formation conditions, suggesting that 'dry' annealing retards the film evolution.

As for the 'wet' annealed 1.2%Si containing alloy, no photocurrent was measured, with some exception (i) occurring at the very time of the potential drop (t$_A$ ~ 4 months), (ii) of unexpected amplitude for such system (I$_{ph}$ ~ 800 nAcm^{-2} in otherwise unchanged

illumination conditions), (iii) largely potential independent over the 0.3–0.6 V/SCE range and decreasing abruptly after 0.6V/SCE, below range of O$_2$ evolution. This result is not consistent with the 'classical' photoexcitation of carriers through the band gap of an n-type semiconducting film (and collection according to Gärtner [6]). We suggest that it could be possibly explained by an internal photoemission process taking place in the metal through the energy barrier existing at the interface between the alloy and the film made insulating in this range of ageing times for some 1.2%Si containing specimens.

The more noticeable point is that only a peak of photoresponse is detected in its representation as a function of ageing time, the photosignal remaining close to the background noise most of the time. No clear understanding of this behaviour can be given at

Table 1 slope of the $V_{rest(t=0)}$ vs ageing time t_A curve for various alloys and annealing conditions, in mV /day

$dV_{rest(t=0)}/dt_A$	$\sigma = -67°C$	$\sigma = -58°C$	
ageing time in water	t_A ($\leq$ 15 months)	$t_A \leq 4$ months	$t_A \geq 4$ months
AISI 434 (0.4%Si)	0.15	0.84	
Fe-Cr-1.2%Si		1.68	0.71

present. However it pertains to non-metallic and disordered solids, rather than to a semiconductor.

For the 'wet' annealed industrial alloy, another feature must be noted in Fig. 2: after 7 months ageing, I_{ph} increases again, the photocurrent versus ageing time curve presenting a 'staircase shape' ($I_{ph} \sim 30nA$ cm^{-2}). This behaviour could be consistent with a beginning of film ordering.

It is worth noting that, in the present experiment where one measures the photocurrent produced under white light, and not in spectrometric mode as usual in semiconductors photoelectrochemistry, two time constants are evidenced at low ($\omega_1 \sim 100Hz$) and high ($\omega_2 \sim 1000Hz$) frequency respectively for the Fe–17%Cr–1.2%Si containing alloy, while for the 0.4%Si containing alloy only one time constant is visible over the same frequency range.

5. GIXD Results

A comparison of the GIXD spectra for the industrial alloys annealed in 'wet' atmosphere for 15 min (Fig. 3. (a)) and 4 h (Fig. 4) shows that the intensity of the peak corresponding to a silicon oxide increases with increasing the annealing time. This result corroborates previous AES observations [1, 3]. In addition, this technique shows that once reached $t_A = 20$ days, the peak at 41.3° (in 2θ, corresponding to d = 2.54Å) disappeared in the background noise (Fig. 3 (b)). Further ageing in water leads to an increase in intensity over a large angular region in between 20° and 66° in 2θ, occurring suddenly once reached almost the same ageing time considered as 'critical' for both the electrochemical and photoelectrochemical behaviour (Fig. 3 (c)–(e)). Likewise, the GIXD intensity falls off to a level comparable with that measured for the unaged specimen after about four months (Fig. 3 (f)), no structured spectra being recorded in the following for exposure times to water between 4 and 6 months. However for ageing times attaining *ca.* 10 months (Fig. 3 (g)), some features appear on the spectra, corresponding to large d-spacing ranging from 4 to 5 Å (from $2\theta \sim 25°$ down to 21°).

Other peaks, located at $2\theta \sim 44.3°$ and 48.8° respec-

tively (d-spacing 2.37Å, 2.16Å) are visible on Fig. 3 (a), which are kept during further ageing up to 57 days (Fig. 3 (c)), even after the disappearance of the silicon oxide signal, which could be assigned to iron oxides or hydroxides. These peaks are no more visible on Fig. 3 (d) (70 days ageing) and following where the spectrum changes drastically, and are not recovered after 4 months (Fig. 3 (f)).

6. Discussion

We have compared the results obtained by electrochemistry, Photoelectrochemistry and Grazing Incidence X-ray diffraction techniques. It was proved that, for fixed formation conditions, changes in the rest potential behaviour, photocurrent response and GIXD spectra occur once reached the same 'critical' ageing time in water, t_A.

By making a parallel between the ageing effects evidenced on thermochemically passivated alloys with two respective Si contents, 0.4% and 1.2% Si for an industrial stainless steel and a synthetized alloy respectively, we have shown that the film evolution during ageing is controlled by both the silicon content in the alloy and the dew point of the annealing atmosphere (for same annealing times), that is to say by a 'unique' parameter, the enrichment factor in silicon, allowing at the present time to rationalise the results. Further work is still necessary to understand the role of silicon in modifying the electrochemical, electronic and structural properties of the film.

However, the possible hydration of the film caused by exposure to water for long periods might also play a non negligible role in the ageing process and should be now carefully examined at first, for example, on pure Fe–Cr alloys.

7. Conclusion

In recent works, it was shown that long term exposure to pure water of stainless steels, passivated either electrochemically or thermochemically, contributes to reinforce the corrosion resistance of the passive film.

This paper enlightens the role of the minor oxidiz-

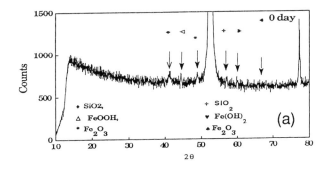

Figure 3.a

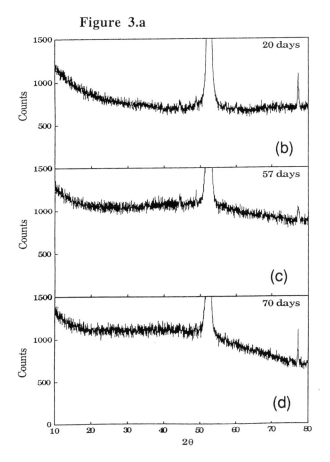

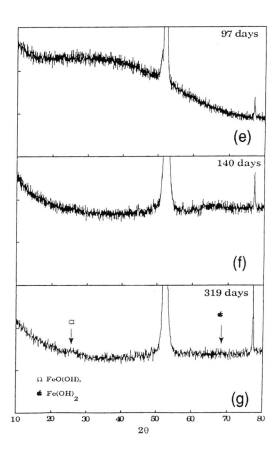

Fig. 3 Diffraction spectra in 2θ obtained at 0.6° graze angle on a 'wet' annealed industrial stainless steel (15 min annealing time) at various stages of ageing, as indicated in the figure: (a) unaged, (b) t_A = 20 days, (c) t_A = 57 days, (d) t_A = 70 days, (e) t_A = 97 days, (f) t_A = 140 days, (g) t_A = 319 days.

able elements in the alloy, particularly silicon, on the electrochemical and physico–chemical properties of the thermochemically formed passive films.

It is shown that the silicon segregation in the passive film occurring during the film formation annealing apparently determines its evolution, due to long term exposure to pure water for times up to one year.

At last, there is apparently a time correlation between the observed changes in the rest potential values, photocurrent response and grazing incidence X-ray diffraction spectra recorded at various stages of film ageing in water.

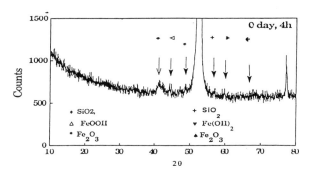

Fig. 4 Same as Fig. 3 for a 'wet' annealed industrial stainless steel, the annealing time being equal to 4 h instead of 15 min.

References

1. D. Gorse, J. C. Joud and B. Baroux, Corrosion Sci., **33**, 1455, 1992.
2. C. Goux, D. Liu, D. Gorse, J. C. Joud and B. Baroux, Materials Science Forum Vol.11–112, pp.139–146, 1992.
3. D. Liu, D. Gorse and B. Baroux, to be published in Proc. 11th European Corrosion Congress, Barcelona, Spain, 5–8 July, 1993.
4. B. Baroux, in Proc. 11th International Corrosion Congress, Florence 2-6 April, 5, 469, 1990.
5. B. Legrand, C. Coux, C. Senillou and J. C. Joud, this Symposium, pp.188–194.
6. W. Gartner, Phys. Rev., **116**, 84, 1959.

15

Polarisation and pH Effects on the Semiconducting Behaviour of Passive Films Formed in Chloride Containing Aqueous Solutions

J.-P. Petit, L. Antoni and B. Baroux*

LIES-G, ENSEEG. INP Grenoble, BP 75, F- 38402 Saint-Martin d'Hères Cedex, France
*LTPCM, ENSEEG, INP Grenoble, and Centre de Recherches d'Ugine, F- 73403 Ugine, France

Abstract

The present work deals with the semiconducting properties of passive layers formed on a ferritic stainless steel, and more precisely with their modifications induced by a prior polarisation in acidic or neutral chloride electrolytes, for times ranging from a few minutes to about one day. Impedance spectroscopy allowed to derive linear Mott-Schottky plots and therefore to deduce the apparent flatband potential, E_o, of the passive film/electrolyte interfaces. The dependence of E_o on the film formation conditions was investigated. An unexpected dependence of E_o on the measurement pH value, $-180\,mV/pH$ unit, was found and discussed. The photoelectrochemical study showed bell shaped photocurrent vs potential curves; the potential E_{peak} of maximum of photocurrent depends on the film formation and film measurement conditions.

1. Introduction

Passive layers on metals and alloys often exhibit semiconducting properties which may control, at least partially, the electrochemical processes at the Passive Film/Electrolyte (PF/El) interface. Therefore, impedance spectroscopy and photoelectrochemistry, which have been in the past two decades widely and successfully used to characterise semiconducting materials and to study their behaviour in contact with an electrolyte, have attracted well-deserved attention as valuable *in situ* techniques for the study of passive films and their modifications in corrosive media [1–4].

The latter techniques were used in this work to study passive films on stainless steels in chloride containing aqueous electrolytes. More precisely, the aim of this study is to assess the modifications of the semiconducting properties of the passive layer due to a prior polarisation of the latter in the corrosive medium itself, and, as far as possible, to connect these modifications with the induced pitting resistance improvements that were otherwise observed [5, 6].

2. Experiment

2.1 Formation of the passive film

All experiments were made with the same electrode material, i.e. an industrial titanium stabilised Fe17Cr alloy (17.4%Cr, 0.025%C, 0.4%Ti). Unless otherwise stated, the electrolyte was a 0.02M aqueous sodium chloride solution, whose temperature was kept at 23°C and which was deaerated with either argon or an Ar10%H$_2$ mixture.

The passive layers were formed on mechanically polished (SiC P1200 under water) samples, which were first stored in air for 24 h. The native films were then modified by applying to the stainless steel electrode a well-defined prepolarisation potential. E_p = +0.2 V vs SCE, the prepolarisation times, t_p, ranging from 0.5 to 15.5 h. The pH value of the electrolyte during the film formation was either pH$_f$ = 3.0 or pH$_f$ = 6.6; in the next paragraphs, these films will be called respectively acidic and neutral films.

2.2 Impedance and photoelectrochemical measurements

The measurements were mostly performed in the film formation medium itself; in some experiments, the films were characterised first keeping the pH of the electrolyte at pH$_f$, then deliberately changing it to other values, called pH$_m$ in the following.

Special attention has been paid to the impedance measurement procedure. In a typical experiment, the impedance of the PF/El interface under dark conditions is measured as a function of frequency, f, (10 kHz $\geqslant$ f $\geqslant$ 10 Hz), at successive potential values (-0.1 V/ SCE $\leqslant$ E $\leqslant$ 0.6 V/SCE). Before measurement, the electrode is allowed to relax at rest potential before applying the required potential; the impedance measurements are started up when a quasi-stationary dc current is reached. Further details are available in reference [7].

The photoelectrochemical measurements have been performed by means of a typical photoelectrochemical set-up, using the white light of a xenon lamp ($\lambda > 230$ nm) or the monochromatic lines of an argon ion laser. Due to the low amplitude of the monochromatic photocurrents, the latter were generally measured under modulated light using a mechanical chopper and the lock-in technique; last, some additional experiments have been made under stationnary conditions using both monochromatic and white light.

3. Results and Discussion

3.1 Impedance spectroscopy
When studying a Semiconductor/Electrolyte (SC/El) junction, and by extension a PF/El interface, it is important to determine its so-called flatband potential, E_{fb}, i.e. the potential where there is no potential gradient between the surface and the bulk of the semiconductor. The knowledge of the E_{fb} value indeed is essential to built the energetic diagram of the semiconductor vs a reference potential in the electrolyte, and therefore to account for the (photo) electrochemical processes at the interface. Flatband potential determination can be achieved by means of various methods; among them the impedance spectroscopy method, do more or less directly measure the capacitance of the space charge region of the semiconductor, C_{SC}, as a function of the applied potential, E, and then make use of the linear relationship between C_{SC}^{-2} and E which is expressed by the Mott-Schottky relation:

$$C_{SC}^{-2} = 2/\,(\varepsilon\varepsilon_o N_D q S)\,(E - E_{fb} - kT/q),$$

where N_D is the carrier concentration and the other parameters have their usual meaning. Passive layers, in particular those formed on stainless steels, are often so thin that the applicability of the Mott-Schottky theory to such films can with good reason be brought into question. Nevertheless, works have been reported in the literature showing that linear Mott-Schottky plots could be obtained at PF/El interfaces, even in the case of few nanometers thick films. This is also true in the case of all the passive layers that were investigated in the present work.

Figure 1 illustrates an example of our results, which were deduced from complex impedance diagrams. As already shown in reference [7], the latter are closely described, at least in the 1 kHz–10 Hz frequency range, by an equivalent circuit where a constant resistance, R_s, is associated in series with a con-

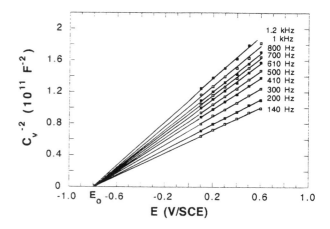

Fig. 1 *Mott-Schottky plots obtained at a Fe17Cr/0.02M NaCl, H_2O, Ar10%H_2 interface (t_p = 15h30, pH_f = 6.6),(Surface S = 0.785cm^2).*

stant phase element, Z, i.e. two frequency dependent components, a resistance $R_v(f)$, and a capacitance, $C_v(f)$, which verify the following relations:

$$R_v = (a\omega^n)^{-1}, C_v = b\omega^{n-1}, \quad 0 \leq n \leq 1$$

$$R_v C_v\,\omega = b/a = \tan(n\pi/2) = \text{constant}, \omega = 2\pi f$$

The values of the constants R_s, a, b and n can easily be determined from the impedance spectra. This model, or very similar ones, have already been used by numerous authors [8]. It should be pointed out that it does just represent a practical formulation of a complex admittance usually written as: $Y = 1/Z = j\,C\,\omega\,/(j\,\omega\tau)^{1-n}$, where τ is a time constant. As can be seen in Fig. 1 and in Ref. [7], C_v obeys a Mott-Schottky relation (for an n type semiconductor). Whatever the frequency value, the $C_v^{-2}(E)$ diagrams are linear over more than 0.5 V, and do intercept the potential axis at the same potential value, E_o, which we will call here the apparent flatband potential. The C_v^{-2} vs E slopes are clearly frequency dependent, and, as shown by Fig. 2, this frequency dependence can here be expressed as:

$$P(\omega) = dCV^{-2}/dE = K\omega^{2-2n}$$

where K is a constant, so one can write: $C_v^{-2}(E) = K\,\omega^{2-2n}\,(E-E_o)$. Frequency dependent Mott-Schottky plots have often been reported in the literature, even in the case of bulk semiconductors, and were attributed to various causes, such as a continuous distribution of relaxation times originating for instance from deep traps, surface states, surface roughness [8], non uniform chemisorption [9], an amorphous character of the electrode [10], or dielectric relaxation phenomena [11], for instance connected with an hydrogen content of the semiconducting material [12]. It is not the purpose of

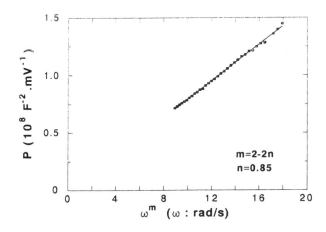

Fig. 2 Frequency dependence of the slopes $P = dCv^{-2}/dE$ of the Mott-Schottky graphs presented in Fig. 1.

this paper to discuss in details the origin of this frequency dependence, which appears here to be located in the passive layer because frequency variations do not affect the E_o value [13].

This frequency dependence does not allow to directly deduce N_D from the Mott-Schottky slopes. However, some informations can be derived from the K values when using reasonable assumptions and approximations, particularly for the value of time constant τ: for instance, when testing films formed in chloride medium, it appears that the N_D value of acidic films is a few times that of neutral films; furthermore, the space charge region width estimated at the most anodic used potential is of the same order of magnitude than the film thickness which was estimated from AES measurements ($d \approx 3$–$4\,nm$). More quantitative results will be given in a forthcoming paper [6].

But it has to be emphasised that, in this study, the parameter n, which practically quantifies the amplitude of the frequency dispersion, is nearly independent of the applied potential, and is characteristic of the investigated passive film, as suggested in Table 1. The latter indeed shows that the n value obtained for a film formed and measured in chloride medium (Run 1) is noticeably different than the n value relative to films formed and measured in sulphate electrolyte (Run 2); moreover, if the passive layer is formed in chloride medium and then measured successively in sulphate (Run 3a) and in chloride (Run 3b) media, the same n value is obtained in both experiments, and this value is close to that measured in Run 1. It is also noteworthy that the n value is greater for films formed at $pH_f = 3.0$ than for those formed at $pH_f = 6.6$, and that, in both cases, n appears to decrease with the prepolarisation time, t_p, due possibly to an higher hydrogen content in acidic films than in neutral ones, proton deinsertion decreasing this hydrogen content in the course of time. This idea will be discussed in details in a short coming paper.

Figure 3 illustrates the effects of t_p and pH_f on the value of the apparent flatband potential, E_o. It can be observed that E_o does not depend on t_p for films formed in acidic medium: on the contrary, the E_o value of neutral films varies strongly and quickly during the first prepolarization hours and then reaches a limit after *ca.* 10–12 h. Another interesting point is evidenced on Fig. 4: whatever the pH_f value and the electrolyte that have been used for the formation of the passive layer, E_o depends linearly on pH_m, the pH value of the electrolyte during the measurements, but the slope value measured in this work, $dE_o/dpH_m = -180\,mV/pH$ unit, is quite different from its generally reported value, i.e. more or less $-60\,mV/pH$ unit. For explaining this unexpected behaviour, a simple model is proposed here, which is presented in the case of the chloride electrolyte, but can be easily adapted to that of the sulphate electrolyte. Assuming that (H_2O) and (Cl^-) adsorb on cations acidic sites $[M^{3+}]$ on the outer part of the passive film, the following equilibria are expected:

$$[M^{3+}]_f + 2(H_2O)_{sol} <=> [MOOH]_f + 3(H^+)_{sol} \quad K_1 \quad (1)$$

$$[M^{3+}] + 2Cl^- <=> [MCl_2]^+ \; [MCl_2]^+ \quad\quad K_2 \quad (2)$$

Table 1 Dependence of parameter n on the conditions of film formation and measurement

	Run 1	Run 2	Run 3a	Run 3b
Film formation in aqueous	NaCl 0.02M	Na$_2$SO$_4$ 0.02M	NaCl 0.02M	
Film measurement in aqueous	NaCl 0.02M	Na$_2$SO$_4$ 0.02M	Na$_2$SO$_4$ 0.02M	NaCl 0.02M
n	0.86	0.91	0.87	0.87

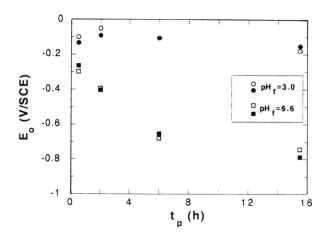

Fig. 3 *Dependence of the apparent flatband potential, E_o, on t_p and pH_f at Fe17Cr/0.02M NaCl, H_2O, Ar($\square$, $\bigcirc$) or Ar10%H_2 ($\blacksquare$, $\bullet$) interfaces.*

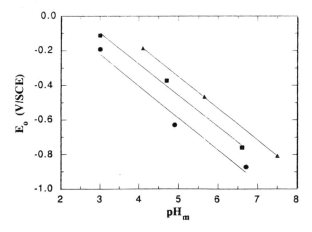

Fig. 4 *Dependence of E_o on the measurement pH value for passive films formed on Fe17Cr (t_p = 15h30) in 0.02M NaCl, H_2O, Ar10%H_2, pH_f = 3.0($\bullet$) or pH_f = 6.6($\blacksquare$) and in 0.02M Na_2SO_4, H_2O, Ar10%H_2 pH_f = 6.6 ($\blacktriangle$). Measurements were performed in the film formation medium.*

Then, provided K_2 and/or the chloride concentration, c, are large enough, the potential drop, V_H, in the double layer can be expressed as:

$$V_H = 3\frac{kT}{q}\ln(H^+) + \frac{kT}{q}\ln\frac{K_1 K_2 c^2}{A}$$

where parameters k, T and q have their usual meaning and A is the total amount of M^{3+} in the film, and the –180 mV/pH unit dependence is found. At the opposite, for low K_2 and/or c values, it can be shown that the usual slope of –60 mV/pH unit is obtained. It should also be pointed out that the presence of the second term in the expression of V_H can account for the fact that, at a given pH_m, the measured E_o value depends on the conditions of the formation of the passive layer.

3.2 Photoelectrochemistry

We present here some first results dealing with the above mentioned systems. First of all the tested passive films, even those obtained using the shortest polarisation times (0h30), exhibit anodic photocurrents. This confirms the n-type semiconducting character of the passive layers which was expected from the above presented Mott-Schottky plots. These results are in agreement with the literature data [3]. The bandgap value, E_G, is nearly constant from one sample to another, i.e. $E_G = 2.15\,eV$, which is the bandgap energy of Fe_2O_3.

More interesting is the fact, that, in the whole 1 Hz–3 kHz light modulation frequency range, the photocurrent versus potential curves, $I_{ph}(E)$, show the same bell-shaped response illustrated in Fig. 5. Before focusing on this particular shape of the $I_{ph}(E)$ curves, it should be noted that, looking at both stationary and transient $I_{ph}(E)$ curves, the photocurrents measured with films formed at $pH_f = 3.0$ are much lower compared to those obtained with neutral films; this cannot be explained only by a difference in the potential gradient or in the carrier concentration in the passive layer. Neutral films indeed give lower photocurrents when being tested at $pH_m = 3.0$ but these photocurrents remain much higher than those obtained, at the same pH value, with acidic films. It is more likely that acidic layers contain more traps than neutral ones and that some of these traps could be hydrogen species.

Figure 6 shows that the potential, E_{peak}, where I_{ph} is at its maximum, depends on the conditions of the formation of the passive layer, which is not surprising. E_{peak} is also a function of pH_m, the pH value of the electrolyte during the measurements: whatever the pH_f value, E_{peak} depends linearly on pH_m, with a –60 mV/pH unit slope. The latter value could be explained by a shift of the passive layer band edges of –60 mV/pH unit, which would mean that the passive layer is modified under illumination conditions so that the above mentioned conditions for a –180 mV/pH unit dependence are no more fulfilled under light irradiation. This explanation cannot be ruled out, but another interpretation seem reasonable, which is connected with the existence of the photocurrent peak. Figure 7 shows that the beginning of the photocurrent decrease coincides with an abrupt increase of the anodic dark current, which cannot be attributed to an irreversible modification of the passive layers, since, when scanning the potential in a cyclic way, the photocurrent and dark current exhibit reversible behaviour. This suggests that the decrease of the photocurrent could be connected with the creation or the filling or emptying of surface states located within the bandgap, involving oxygen evolution processes possibly associated with

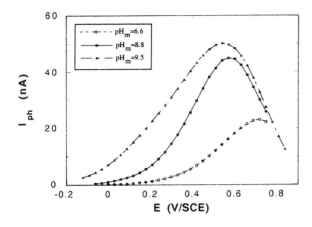

Fig. 5 Photocurrent vs potential curves for Fe17Cr/0.02M NaCl, H₂O, Ar10%H₂ interface (t$_p$ = 15h30, pH$_f$ = 6.6). Photoelectrochemical conditions: λ = 501.7 nm, P # 60mW.cm⁻², light modulation frequency: 45Hz.

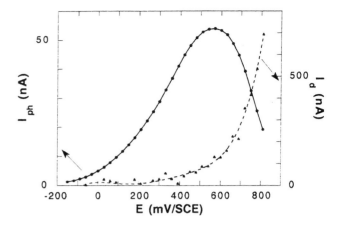

Fig. 7 Simultaneous recording of the photocurrent I$_{ph}$ and of the dark current I$_d$ as a function of applied potential, E. Film formed and measured (pH$_m$ = 9.4) in 0.02M NaCl, H₂O, Ar (t$_p$ = 15h30, pH$_f$ = 6.6). Photoelectrochemical conditions: λ = 501.7nm, P#60mW/cm², light modulation frequency: 45Hz. (S = 0.785cm²).

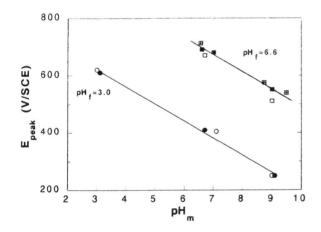

Fig. 6 Dependence of E$_{peak}$ on pH$_m$ for passive films formed and measured in 0.02M NaCl, H₂O, Ar10%H₂ (t$_p$ = 0h30 (O) or 15h30 (other symbols), pH$_f$= 3.0 or 6.6). Photoelectrochemical conditions: monochromatic (λ = 501.7nm) (⊞ 1.5) or white light (other symbols) illumination, light modulation frequency: 45Hz (⊞ 1.6) or 205 Hz (other symbols).

proton deintercalation. This idea will also be discussed in a forth coming paper.

4. Concluding remarks

Passive films formed on ferritic stainless steels were prepolarised in chloride aqueous electrolytes and the induced modifications of their semiconducting properties were investigated by means of impedance spectroscopy and photoelectrochemical techniques. Interactions between the passive layer and protons could at least partially explain the reported results. The relation with the prepolarisation induced pitting resistance improvement is currently being investigated.

References

1. S. R. Morrison, Electrochemistry at Semiconductor and Oxidized Metal Electrodes, Plenum Press, New York, 1982.
2. Yu. V. Pleskov and Yu. Ya. Gurevich, Semiconductor Photoelectrochemistry, Consultants, Bureau, New York, 1986.
3. U. Stimming, Electrochim. Acta, **31**, 415, 1986.
4. F. Di Quarto, this symposium.
5. B. Baroux, Proc. 11th Int. Congr. of Corrosion, Florence. April 1990, 5, 469.
6. B. Baroux, J. P. Petit and L. Antoni, to be published in the proceedings of the Electrochemical Society Symposium on Chemistry, structure and stochastic process in the breakdown of passivity, Honolulu, USA, 16–21 May, 1993.
7. J. P. Petit, L. Antoni, B. Baroux, 6e Forum sur les impedances electrochimiques. Montrouge, France, 26 Nov., 1992.
8. J. F. McCann and J. S. P. Badwal, J. Electrochem. Soc., **129**, 551, 1982.
9. C. K. Braun, A. Fujishima, K. Honda and L. Nadjo, Surface Sci., **176**, 367, 1986.
10. F. Di Quarto, C. Sunseri and S. Piazza, Ber. Bunsenges. Phys. Chem., **90**, 549, 1986.
11. F. Fransen, M. J. Madou, W. H. Laflere, F. Cardon and W. P. Gomes, J. Phys. D: Appl. Phys., **16**, 879, 1983.
12. D. M. Tench and E. J. Yeager, J. Electrochem. Soc., **120**, 164, 1973.
13. W. P. Gomes, Proc. of the Ecole d'hiver du CNRS L'Interface semiconducteur/electrolyte, Aussois, France, 1984.

16

Modification of the Passivating Film on Copper in the Presence of Benzotriazole: Photoelectrochemical Characterisation

E. Sutter, D. Lincot* and C. Fiaud

Laboratoire d'Etudes de la Corrosion, Ecole Nationale Supérieure de Chimie de Paris, 11 rue Pierre et Marie Curie, 75231 Paris Cedex 05, France
*Laboratoire d'Electrochimie Analytique et Appliquée URA 216. Ecole Nationale Supérieure de Chimie de Paris, 11 rue Pierre et Marie Curie, 75231 PARIS Cedex 05, France

Abstract

As-grown oxide layers on copper in aerated 0.1M sodium acetate and 0.5M sodium chloride solutions, with and without benzotriazole, are studied using photoelectrochemical techniques.

In the sodium acetate solution, BTA does not basically modify the semi-conducting properties of the oxide layers. Moreover in this electrolyte, the oxide film shows simultaneously, at the rest potential, p-and n-type behaviours associated with different energy gaps. In the sodium chloride solution, a p-type behaviour is observed for short immersion times, but for longer immersion times, an additional anodic component appears in the higher wavelength range. In presence of BTA, the oxide layer shows the same semi-conducting behaviour in sodium chloride than in sodium acetate solutions.

1. Introduction

Naturally grown oxide layers on copper consist of simple cuprous oxide or have a duplex structure with an inner Cu_2O layer followed by a CuO or a $Cu(OH)_2$ layer depending on the pH of the aqueous solution. The semiconducting properties of the surface layers grown on a metal have often been taken into account [1] to explain their different electrochemical stabilities when they are grown in different experimental conditions.

In this work photoelectrochemical techniques have been used for a better understanding of the mechanism of protection of copper in presence of benzotriazole which is a well known inhibitor for copper corrosion. The influence of the surface treatment of the electrode prior to immersion is also discussed. The electronic properties of the copper oxide layers formed in a weakly basic (0.1M sodium acetate) solution are compared to those of the surface layers formed in presence of chloride ions (0.5M NaCl).

2. Experimental

The electrochemical set up consisted of a potentiostat PAR273, computerised with a IBM-PC. For photoelectrochemical experiments, a Jobin Yvon H20 monochromator equipped with a Xenon lamp (75W) was used, and a mechanical chopper from Optilas

modulated the incident light at a frequency of 16 Hz for which a good signal-to-noise ratio was obtained. The modulated photocurrent was detected by a lock-in amplifier (PAR5208) through the potentiostat. For spectral response measurements, the incident spectrum, $\Phi_o(\lambda)$, of the Xenon lamp was measured with a calibrated Si photocell. The external quantum efficiency is then calculated using the relation ($\Phi E(\lambda) = Iph(\lambda)\Phi_o(\lambda)$) uncorrected from the reflexion coefficient. Contrary to the classical use of the module of the photocurrent in the previous relation, we have found that using the in-phase component, adjusting the phase to zero, was a very convenient way to characterise also the sign of the photocurrent [2].

Before immersion, the copper electrodes were mechanically polished (grade 1000), rinsed with deionized water and air dried.

3. Results and Discussion

3.1 Behaviour in a sodium acetate solution with and without BTA (0.1M)

The spectral responses of a copper electrode near the rest potential are reported in Fig. 1. They show a main optical transition at 0.35 µm and a hump near 0.50 µm in both cases. The hump at 0.50 µm seems to be more pronounced in presence of BTA and corresponds to a positive photocurrent, whereas the short wavelength response is negative. Figure 2 shows the influence of the applied potential on the spectral response in the BTA containing solution, the same evolution being obtained without BTA. Large changes of the response are occurring within *ca.* 200 mV around the rest potential: negative polarisations lead to a large increase of the short wavelength response, together with the almost complete extinction of the hump at higher wavelengths. For anodic polarisations the behaviour is reversed (Fig. 2(b)).

It is remarkable that at the rest potential both positive and negative photocurrents are present together. The anodic photocurrents of the oxide films on copper are noteworthy since the bulk oxides are known to show only p-type conductivity due to cation vacancies in the oxide lattice: p- and n-type behaviour of oxide films on copper have been observed by several authors [3–5] as a function of the applied potential, but the new point, here, is that these two behaviours, anodic and cathodic currents, are present at the same potential near the rest value, and further associated with different wavelength ranges.

In order to determine the band gap values of the film, $(\Phi E.h\nu)^2$ and $(\Phi E.h\nu)^{1/2}$ versus $h\nu$ plots have been calculated and are reported in Fig. 3 for a spectral response obtained in the cathodic range. The best fit is obtained for $(\Phi E.h\nu)^{1/2}$ with a good linearization over almost the whole wavelength range between 2.7 and 3.5 eV. The 2.7 eV value obtained for the band gap is higher than the results found in the literature where a value between 1.9 and 2.35 eV is generally reported [5-8] for copper oxide films formed by anodic polarisation in alkaline solutions or by immersion in acidic solutions (review of the literature values are reported in [6]). The higher value found in our case could be attributed to a different structure of the cuprous oxide (even amorphous) or to the effect of lower thickness of the film formed in the sodium acetate solution, a decrease of the thickness leading to an increase of the band gap value [8]. The optical transitions calculated from Fig. 2 as a function of the applied potential show that the band gap determined in the anodic and the cathodic range leads to the same values in the short wavelength range, indicating that they correspond to the response of the same material but with opposite directions of the electric field (Fig. 4). The additional transition corresponding to the anodic hump obtained at the rest potential, leads to a gap value of about 1.9 eV which is also reported for Cu_2O [5]. However at the present time, care must be taken to associate definitively the observed transitions to Cu_2O for lack of complementary experiments. The formation of a n-type Cu_2O has sometimes been correlated with the presence of cupric ions in the electrolyte [3,5]. Siripala proposes the simultaneous existence of spatially sepa-

rated n- and p-type regions on the same Cu_2O film, the n-type Cu_2O being due to an excess of copper ions in a part of the oxide layer. Whatever the origin of the phenomenon, it corresponds to a change of the field direction in the film perpendicular to the surface, directed towards the solution in the lower band gap material (to explain the anodic photocurrent) and directed towards the back contact in the larger gap material (to explain the cathodic photocurrent).

3.2 Influence of the surface preparation on the photoelectrochemical behaviour

The electronic properties of an oxide film on copper are known to vary greatly depending on its method of preparation. Indeed when a copper electrode has been submitted to a cathodic pretreatment before the formation of its oxide layer by immersion in the sodium acetate solution, an enhancement of the anodic component of the photocurrent can be observed in the following conditions: in Fig. 5, after a 4 h immersion time, an anodic potential (E_{corr} + 170 mV) has been applied for 15 min followed by a cathodic polarisation (E_{corr} –230 mV) for 15 min. The photocurrent obtained at this cathodic potential is anodic, the quantum efficiency is greatly enhanced and the gap value determined in such conditions is 1.9 eV. According to Siripala *et al.* [3] ,the origin of this n-type photosensitive layer can be attributed to the reduction of Cu^{2+} ions mainly obtained during the anodic polarisation, followed by the electrodeposition of a Cu_2O layer from the solution during the cathodic polarisation. It is noticeable that such a behaviour is not observed on a just polished copper sample which always shows a cathodic photocurrent in the cathodic potential range, as shown in Fig, 2(a).

3.3 Behaviour in a 0.5M NaCl solution without BTA

Figure 6(a) shows the spectral response of a copper sample after a 7 h immersion time in a NaCl solution at the rest potential (–220 mV/ECS). Compared to the response in sodium acetate solutions (Fig. 1), the quantum efficiency shows a higher value (about two orders of magnitude) and its maximum value occurs in a larger wavelength range (between 0.3 and 0.5 µm). $(\Phi E.h\nu)^2$ and $(\Phi E.h\nu)^{1/2}$ vs $h\nu$ plots show that the band gap value is *ca.* 1.9 eV for the indirect transition which corresponds to the second transition observed in the sodium acetate solution at 0.50 µm. A possible effect of the thickness of the film could be involved to explain this evolution. It is noticeable that for a few hours immersion time in the NaCl solution, in the whole wavelength range considered, the oxide film never shows an anodic photocurrent and behaves like a p-type semiconductor, contrary to the behaviour in a sodium acetate solution.

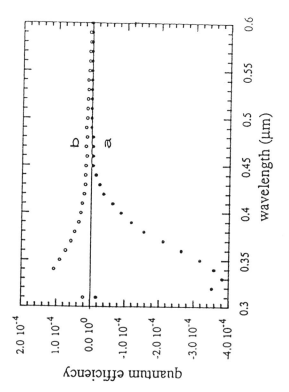

Fig. 2 Spectral responses of copper in a 0.1M sodium acetate solution after 18 h immersion (a) in the cathodic and (b) the anodic potential range.

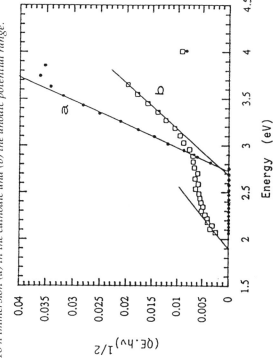

Fig. 4 Optical transitions in a sodium acetate solution. (a) in the anodic (400 mV) or the cathodic (−400 mV) range; (b) near the rest potential (30 mV).

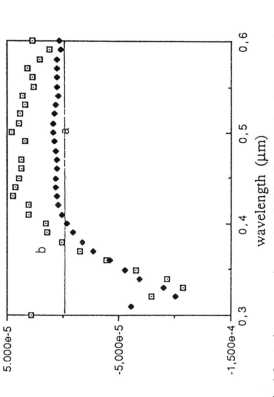

Fig. 1 Spectral responses of copper in a 0.1M sodium acetate solution near the rest potential (30 mV/SCE) after 18 h immersion (a) without, and (b) with BTA.

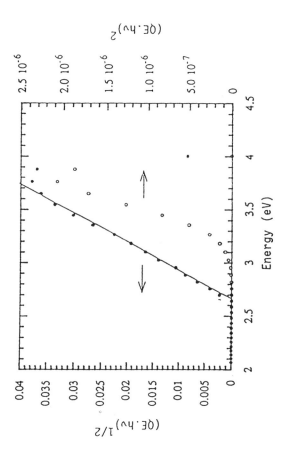

Fig. 3 $(\Phi E.h\nu)^2$ and $(\Phi E.h\nu)^{1/2}$ vs $h\nu$ for a spectral response obtained in the cathodic range (at −400 mV).

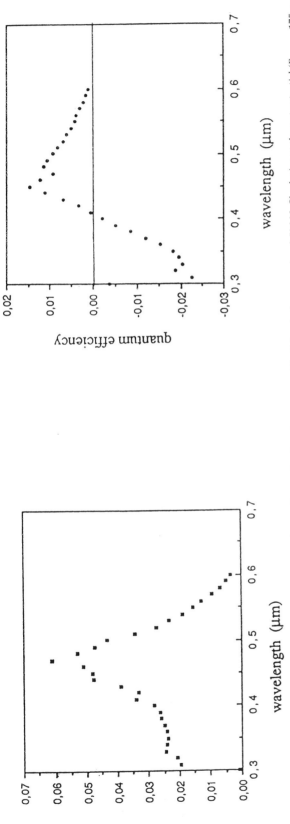

Fig. 5 Spectral response at (E_{corr} -230 mV) in a 0.1M sodium acetate solution after application of an anodic potential (E_{corr} +170 mV); E_{corr} = 30 mV.

Fig. 7 Spectral response in a 0.5M NaCl solution at the rest potential (E_{corr} = -175 mV) after a 65 h immersion time.

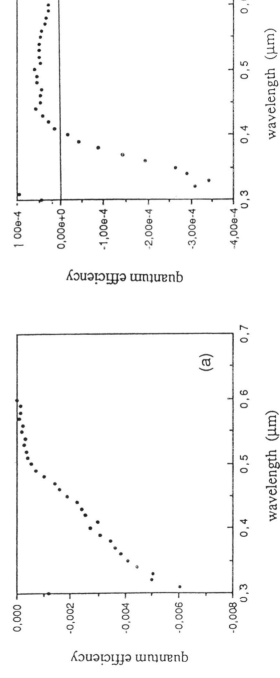

Fig. 6 Spectral responses in a 0.5M NaCl solution at the rest potential after 7 h immersion (a) without BTA (E_{corr} = -220 mV) (b) with BTA (E_{corr} = -60 mV).

For longer immersion times in the same electrolyte, (65 h immersion time in Fig. 7) the photocurrent exhibits an additional anodic component in the higher wavelength range (0.4–0.6 μm). Such a p → n transition with increasing immersion times in chloride containing solutions has been observed by Di Quarto et al. [5] and attributed to the formation of a new phase containing copper in the highest oxidation state. A $Cu_2(OH)_3Cl$ layer is known to form on the cuprous oxide surface in NaCl solutions after a few days immersion time [9], but no data are available concerning the semiconducting properties of this hydroxychloride.

3.4 Behaviour in a 0.5M NaCl solution with 0.1 BTA

When BTA is added to the sodium chloride solution, the spectral response is very similar to the response obtained in the sodium acetate solutions with or without BTA and shows at the rest potential a main cathodic contribution near 0.35 μm followed by an anodic photocurrent in the higher wavelength range (Fig. 6b). It seems that BTA hinders the modification of the oxide film introduced by Cl⁻ ions and allows the formation of oxide films with electronic properties similar to those obtained in sodium acetate solutions.

4. Conclusion

In a sodium acetate solution where the corrosion process of copper is assumed to be very low, both anodic and cathodic photocurrents have been observed near the rest potential, whenever BTA is present or not in the electrolyte during the immersion time. Nevertheless BTA seems to enhance the anodic contribution. The transition energies found for the film are almost independent upon the presence of BTA, indicating that no specific photoactivity of the film occurs in this wavelength range due to the presence of the inhibitor in the acetate solution. The simultaneous existence of spatially separated n- and p-type regions on the same Cu_2O film could be responsible for such a behaviour but the mechanism of formation of these two layers is not yet clear.

In a NaCl solution, in which the oxide film on copper is known to be poorly resistant, the quantum efficiency is higher, possibly due to an increase of the oxide layer thickness. At short immersion times only a p-type behaviour is observed which turns into a n-type behaviour for longer immersion times, probably due to the formation of some new layer of Cu(II). When the film is grown in a BTA containing NaCl solution, the spectral response is similar to that observed in an acetate solution corresponding to an oxide film with similar electronic properties and with a high electrochemical stability.

References

1. N. Sato, J.Electrochem.Soc., **129**, 255, 1982.
2. E. M. M. Sutter, C. Fiaud and D. Lincot, Electrochim. Acta, **38**, 1471, 1993.
3. W. Siripala and K. Premasiri Kumara, Semicond. Sci. Technol., **4**, 465, 1989.
4. A. Aruchamy and A. Fujishima, J. Electroanal. Chem., **266**, 397, 1989.
5. F. Di Quarto, S. Piazza and C.Sunseri, Electrochim. Acta, **30**, 315, 1985.
6. U. Collisi and H. H. Strehblow, J. Electroanal.Chem., **210**, 213, 1986.
7. W. Paatsch, Ber. Bunsenges. Phys. Chem., 1977, **81**, 645.
8. U. Collisi and H. H. Strehblow, J. Electroanal. Chem., **284**, 385, 1990.
9. M. Drogowska, L. Brossard and H. Menard, Corrosion, **43**, 549, 1987.

II KINETICS OF PASSIVATION AND ELECTROCHEMICAL PROPERTIES OF PASSIVE FILMS

17

Kinetics of Film Formation on Fe–Cr Alloys

R. KIRCHHEIM

Max-Planck-Institut für Metallforschung, Institut für Werkstoffwissenschaft, Seestr. 92, 7000 Stuttgart-10, Germany

Abstract

Ellipsometric, surface analytical and electrochemical studies were conducted on the kinetics of passive film formation on iron and iron–chromium alloys in various electrolytes at room temperature. It will be shown that with increasing pH of the solution and increasing Cr content of the alloy the thickness of the passive film growths by a logarithmic time law in agreement with the Cabrera–Mott model. However, at low pH and low chromium contents the passive film is formed in a very short time and its thickness remains constant with time or even decreases in some cases. This experimental finding and the time dependence of galvanostatic transients are indicative of film formation by precipitation from the solution. The experimental results are compared with a simple model on solid film formation. The concept of precipitation provides an explanation for the abrupt change of the critical current density around a Cr content of 10 wt%. The chromium enrichment in the passive film is determined by the conditions during precipitation and the magnitude of the passive current in the passive state.

1. Introduction

Among the common models on film formation on iron there is one proposed by Hoar [1] and Evans [2], where the oxide is formed by releasing water from an adsorbed layer of hydroxyl ions. For iron in sulphuric acid a visible layer is sometimes formed before passivation. Frank [3] suggested that this layer is composed of $FeSO_4$ and that it growth to a certain thickness, where the current densities in the pores becomes so high that passivation underneath the film will be possible. After passivation the sulphate film will be dissolved in the solution.

For Fe–Cr alloys Sieradzki and Newman [4] proposed that a percolation of oxygen bridges among Cr atoms occurs around 10 at.% Cr, which can be used to interpret experimental results of Rocha and Lennartz [5] with respect to the activation potential changing abruptly at about the same concentration. But the passivation potential as obtained from stationary current potential diagrams [6] increases only smoothly as the Cr-concentration is raised from 1 to 18%. The critical current density necessary for passivation, however, exhibits a sudden drop between 6 and 10%, which besides the lower passivation potential is one of the major reasons for the corrosion resistance of stainless steels.

It is the major goal of this study to present some evidence for a new concept of passive film formation, where precipitation from a supersaturated solution is responsible for the initial film growth. This film will change later on with respect to thickness and compo-sition approaching a steady state. Before the processes are discussed in more detail the steady state is characterised first.

2. Steady State of Passive Fe–Cr Alloys

There are numerous surface analytical studies showing that the passive film on Fe–Cr alloys, namely the one formed in acidic solutions, is enriched in chromium. This is explained by a preferential dissolution of Fe^{3+} ions from the film. However, in the framework of a high field mechanism [6] the cations jump in the forward (from the alloy towards the solution) direction only and, therefore, the Cr enrichment will be restricted to the first cation layer. Assuming further that the dissolution rates per Fe and Cr ion are the same as the ones of the pure metals the passive current densities have been calculated for 0.5M H_2SO_4 as a function of Cr-concentration [6]. In addition the contributions of the two elements, i.e. the partial current densities, can be obtained in good agreement with experimental results as shown in Fig. 1. No adjustable parameter has been used to obtain the curves in Fig. 1.

As the current densities of passive Fe and Cr differ by a factor of ca. 500 in 0.5M H_2SO_4 at 298 K the first layer of the passive film will contain mostly Cr cations, even for very low alloy contents. Therefore, the dissolution rate of Cr from a passive film on Fe-1 at.% Cr is about the same as for pure chromium (cf. Fig. 1). The enrichment of the following layers of the passive film is not a consequence of the preferential Fe dissolution, as outlined before, but has to be explained by the

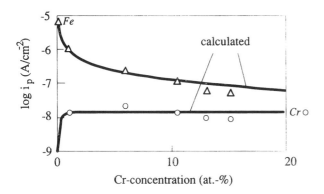

Fig. 1 Partial dissolution current densities of passive Fe–Cr alloys as a function of the Cr-content. The lines are calculated as described in Ref. [6] and the circles (Cr) and triangles (Fe) are determined by chemical analysis of the electrolyte. The value for pure Cr is shown at the right hand side of the frame.

additional assumption that Fe-ions migrate faster in the high field compared to Cr-ions. For an eight times larger mobility the Cr-enrichment in passive films can be calculated in good agreement with experimental findings [6].

The attainment of the stationary composition of the passive film requires a very long time of passivation in sulphuric acid (of the order of one week). Because of the low current densities in neutral and alkaline solutions the steady state cannot be reached. Under these conditions the higher mobility of Fe-ions gives rise to a more or less pronounced separation of a chromium rich passive film adjacent to the alloy and an iron rich film on top of it [7]. This is shown schematically in Fig. 2.

The concept developed for the passivity of Fe-Cr alloys can be applied to other alloys as well and it is interesting to note that for times as long as 3000 years a steady state appeared to be attained for Cu-Sn bronzes, where the distribution of Sn between alloy and outermost patina follows the rules developed before [8].

Assuming that the Cr-enrichment occurs during film formation or in the active state, too, this allows us to describe the concentration dependence of the passivation potential for Fe-Cr alloys [6].

3. Passive Film Formation by Precipitation

The preceding section has shown that we can describe the steady state of Fe–Cr passivity quite well, whereas kinetic aspects like the abrupt change of the critical current density around 10 at.% Cr remains unexplained. In order to do so, we propose that the passive film is formed by precipitation. This concept is applied to iron first and extended to the Fe–Cr alloys later.

For iron in 0.5M H_2SO_4 we have a pronounced current plateau in the active region (cf. Fig. 3) which is considered to be a diffusion limited current densities due to the formation of a prepassive film. It is black and visible, if the corresponding potential is imposed long enough. The prepassive film becomes partially or totally passive if the potential is increased above the passivation potential.

The kinetics of prepassive and passive film formation as evaluated from potentiostatic step experiments [9] yields the same product of the initial current densities i_o and the square root of a characteristic time tp, where the current drops due to film formation. In standard text books it is shown that dissolution of a species and concomitant diffusion into the electrolyte

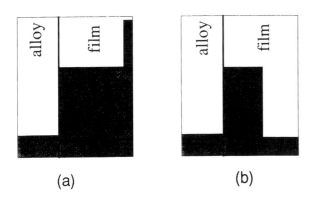

Fig. 2 Schematic presentation of the Cr-distribution (cross hatched area) in passive films on Fe–Cr alloys as caused by a different mobility in the high electric field of the film and (a) preferential dissolution of Fe (acidic solutions) or (b) no dissolution (neutral and alkali solutions).

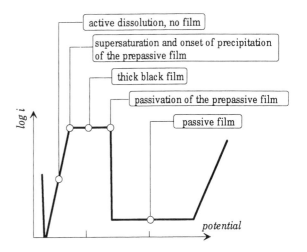

Fig. 3 Schematic polarisation curve for iron in 0.5M H_2SO_4 and characteristic potentials for anodic dissolution and film formation.

increase the concentration in front of the electrode according to a square root time law. If the concentration reaches the solubility limit CS, a solid film is precipitated more or less rapidly depending on the nucleation and growth conditions. Then the product of $i_o \div t_p$ is given by:

$$i_o \sqrt{t_p} = \frac{z}{2}(c_s - c_b)\sqrt{D\pi} \qquad (1)$$

where c_b is the initial concentration in the electrolyte and D is the diffusion coefficient and z the valence of the dissolved species. Thus passive and prepassive films form the same way as solid films occurring during electropolishing [10]. Because precipitation requires nucleation barriers to be overcome, it appears to be more realistic and in agreement with experimental findings for electropolishing that a supersaturation of the electrolyte occurs [10].

The same square root dependence as described by eqn (1) can be extracted from galvanostatic transients, where the onset of passivation is accompanied by a steep potential increase. The relation between imposed current densities and passivation times as measured in Ref. [9] is shown in Fig. 4. For iron the data can be described by eqn (1) yielding a slope of –0.5 in the double logarithmic plot of Fig. 4. Below the critical current densities of 0.1 A.cm^{-2} no passivation occurs due to convective transport in the solution, which was neglected during the derivation of eqn (1) but which can be included by assuming the presence of a Nernst's diffusion layer in front of the anode [9,10].

For low Cr contents corresponding values of i_o and t_p are the same as for pure Fe. Deviations occur for the first time at 6%Cr, but for low current densities the same critical value of i_o is approached. For 18%Cr the values are close to the ones for pure chromium and they can be described above the critical current densities by a straight line of slope 1. The same dependency is observed for pure iron in neutral solutions [11]. If the corresponding electrical charges $i_o t_p$ are converted to film thicknesses assuming oxides to be formed, values of *ca.* 1–5 nm are obtained. It is believed that under these circumstances the solubility and/or diffusivity in the solution is very low, precipitation of the film occurs immediately, i.e. as isolated nuclei which were fed by the continuing dissolution during time t_p until the whole surface is covered by a solid film. Nevertheless, diffusion cannot be totally neglected because still a critical current densities exists for these systems.

Possible precipitation reactions for pure iron are the following:

$$Fe^{2+} + SO_4^{2-} \rightarrow FeSO_4 \qquad (2)$$

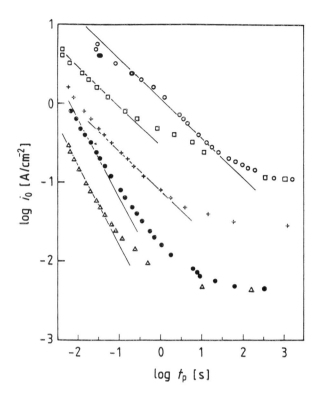

Fig. 4 Current density i_o vs passivation time t_p as obtained from galvanostatic transients in 0.5M H_2SO_4 for various Fe–Cr alloys. The lines drawn through the data points have either slope –0.5 or –1. Open circles: Fe, squares: Fe–6%Cr, crosses: Fe–10%Cr, full circles: Fe–18%Cr, and triangles: Cr. The data points for 0.5, 1 and 3%Cr were omitted for clarity because they fall next to the points for pure iron.

$$Fe^{2+} + H_2O \rightarrow FeO + 2H^+ \qquad (3)$$

$$2Fe^{3+} + 3H_2O \rightarrow Fe_2O_3 + 6H^+ \qquad (4)$$

or a combination of eqns (3) and (4) yielding magnetite. Among these choices eqn (2) can be eliminated because an increase of the sulphate concentration by a factor of 3 did not change $i_o t_p$ significantly. It is interesting to note that Engell and Herbsleb [12] found a strong change of the kinetic behaviour by adding sodium sulphate to sulphuric acid. But the observed effect is due to a concomitant change of the pH. The latter has a strong effect on the reactions described by eqns (3) and (4). In addition Auger analysis of the prepassive film shows [9] that it contains a small amount of sulphate only, the major component being iron oxide (or hydroxide).

It is usually assumed that films formed by precipitation are thick and porous. However, they may very well be thin and compact, if they are formed quickly like some of the films produced during electropolishing.

The prepassive film once formed in the active plateau (*cf.* Fig. 3) is easily transformed into a passive film by a further potential increase, even if the prepassive film had an appreciable thickness [9]. During the corresponding passivation process the measurement of the dissolution rate of iron in the electrolyte by atomic absorption spectroscopy [9] yielded values which are the same as the current density and, therefore, the passive film must grow on top of the prepassive film. Otherwise the currentless dissolution of the prepassive film would have supplied large amounts of iron to the electrolyte. With the assumption that the prepassive film has the composition of magnetite the passivation of it would correspond to an extraction of Fe^{2+}ions yielding γ-Fe_2O_3. At the same time the ionic conductivity of the film will drop drastically giving rise to the small passive currents.

4. Ellipsometric Studies

In order to provide more evidence for the proposed model, ellipsometric measurements were conducted. The thickness of the film was obtained in the usual way from the parameters Δ and ψ [9]. For low pH and low Cr-content, data evaluation was not possible due to high dissolution rates. Thickness and current densities were monitored after a potential step from the free corrosion potential to 300 and 900 mV(SCE). In good agreement with the ellipsometric data, film thickness was also determined by photoelectron spectroscopy (XPS) at the end of an experiment. More details on data evaluation and results are presented elsewhere [9].

Ellipsometric measurements have been conducted before for iron and Fe–Cr alloys in neutral and alkali solutions demonstrating the validity of the high field mechanism by Cabrera and Mott [13], which predicts that the current densities decrease proportional with the reciprocal passivation time. If the corresponding charge is totally consumed for film formation, the thickness is described by a logarithmic time law. In agreement with previous measurements and the high field mechanism, our measurements had the predicted dependencies for film thickness and current density in borate buffer (pH = 9). However, in acidic solutions, the thickness is already very large after the first second of passivation and hardly increases in the following time as shown in Fig. 5 (in a few cases a decrease has been observed). This is considered to be an additional piece of evidence for a precipitation process as a precursor of passivation.

In the light of the new concept the results would be interpreted qualitatively as follows (some aspects can be treated quantitatively as shown in Refs. [9] and [10]). A concentration profile of corrosion products piles up in front of the electrode and precipitation of

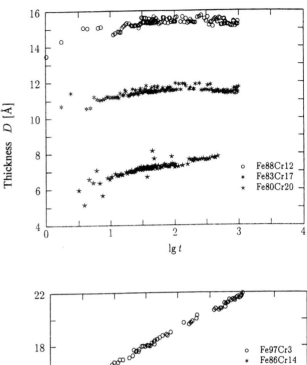

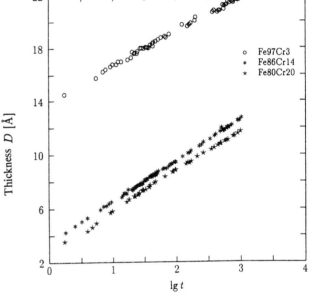

Fig. 5 Thickness of passive films formed as a function of passivation time for the alloys shown in the legend and for (a) $0.5M\,H_2SO_4$ (pH = 0.3) and (b) borate buffer (pH = 9).

the prepassive film occurs after a critical supersaturation of the solution is reached. With the onset of nucleation and growth of the film the concentration is decreased to its equilibrium value and back diffusion from the supersaturated solution will take place which feeds the growing film. For low pH and/or low Cr content, the solubilities and related values of supersaturation are large yielding pronounced back diffusion and, therefore, thick films form at the very beginning of passivation. Then the electric fields, as the driving force for additional film growth, will be small and might be even lower than the corresponding steady state value causing a thinning of the passive film. But for high pH and/or high Cr-content, the terminal solubilities and presumably the critical value for supersaturation, too, will be very small and, therefore, a very thin passive film (or a few islands of it) are formed leading to large electric fields and a further growth by the high field mechanism. In the extreme case of very low solubilities we may not be able to distinguish between a precipitation from solution on the one hand and direct film growth onto the metal surface on the other hand because the short times required for reaching the necessary concentration for precipitation will be so small that the ions can migrate only a very short distance from the electrode.

5. Consequences for Fe–Cr Alloys

Within the framework of the concept of the precipitation of a prepassive film and its following passivation an explanation can be provided for the sudden decrease of the critical current density for passivation of Fe–Cr alloys from a value typical for pure iron to a value near that of pure chromium occurring around an alloy composition of 10%Cr (*cf.* Fig. 4). In Fig. 6 the concentration profiles of Fe and Cr are shown schematically for the pure elements and two alloys (one low and one high in its Cr content). The lower critical current densities of Cr corresponds to a lower solubility and/or diffusivity (which determines the

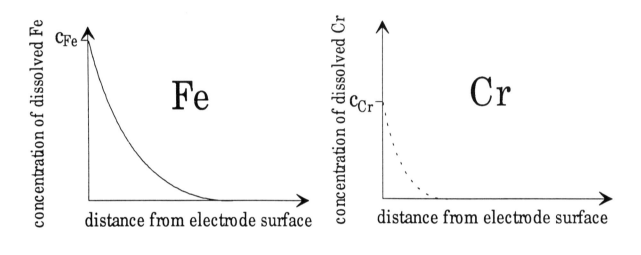

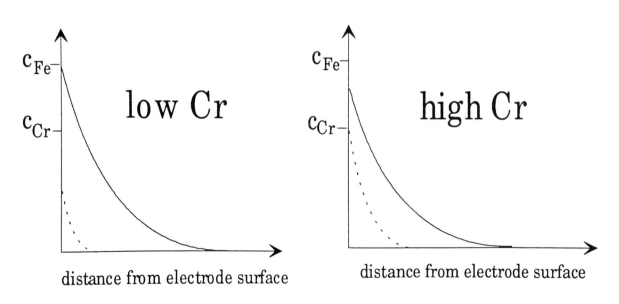

Fig. 6 Schematic concentration profiles for Fe and Cr in front of iron, chromium and Fe–Cr alloys for the time where one of the two species reaches its supersaturation value (c_{Fe} or c_{cr}).

extension of the profiles into the solution). Having an alloy, both components are dissolved and the two species compete about which will reach the solubility limit first. For low Cr contents it will be the dissolved Fe and the precipitated film will be Fe-oxide (some Cr may coprecipitate), whereas for high Cr contents the Cr ions will reach the solubility limit first and Cr-rich oxides will be precipitated. As the critical current density is related to the dissolution rate of the film [9] the sudden change from an Fe-rich to an Cr-rich film causes a corresponding change of the critical current densities, too.

Despite the preferential dissolution of iron from the passive film, it will take a very long time for the iron rich film formed on an alloy with low Cr content to reach the stationary composition with its high Cr concentration, because the current densities are so low in the passive state. This has been observed experimentally [14] for alloys with less than 10 at.% Cr. Above this concentration the Cr-rich films are precipitated and further changes of the Cr content in the film are small. This behaviour of the Cr concentration becomes also obvious by comparing the results of Asami *et al.* [15] for passivation times of 1 hour and the steady state concentrations determined in Ref. [6]. Below 10 at.% Cr the Cr concentration in the film is much smaller than the steady state values, whereas above 10 at.% Cr the difference between the film composition for 1 h and for more than 48 h is small. The latter case corresponds to the precipitation of a Cr-rich film yielding a Cr concentration which is close to the one of the steady state.

References

1. T. P. Hoar, J. Electrochem Soc., **117**, 17C, 1970.
2. U. R. Evans, Electrochim. Acta, **16**, 1825, 1971.
3. U. F. Frank, Z. Naturforsch., **4a**, 378, 1949; and Z. f. Elektrochem., **62**, 649, 1958.
4. K. Sieradzki and R. C. Newman, J. Electrochem. Soc., **133**, 1979, 1986.
5. H.-J. Rocha and G. Lennartz, Arch. Eisenhuttenwes., **26**, 117, 1955.
6. R. Kirchheim, B. Heine, H. Fischmeister, S. Hofman, H. Knote and U. Stolz, Corros. Sci., **29**, 899, 1989.
7. S. Boudin, C. Bombart, G. Lorang and M. Da Cunha Belo, this volume.
8. L. Robbiola, C. Fiaud and A. Harch, this volume.
9. J. Hafele, B. Heine and R. Kirchheim, Z. Metallkde., **83**, 395, 1992.
10. R. Kirchheim, K. Maier and G. Tolg, J. Electrochem. Soc., **128**, 1027, 1981.
11. B. Heine, PhD Thesis, University of Stuttgart, Germany, 1988.
12. G. Herbsleb and H. J. Engell, Z. f. Elektrochem., **65**, 881, 1961.
13. N. Cabrera and N. F. Mott, Rep. Prog. Phys., **12**, 267, 1948.
14. R. Kirchheim, unpublished results.
15. K. Asami, K. Hashimoto and S. Shimodaira, Corros. Sci., **18**, 151, 1978.

18

Influence of Anions on the Electrochemical Behaviour of Passive Iron

K. E. Heusler, L. Jaeckel and F. J. Nagies

Abteilung Korrosion und Korrosionsschutz, Institut für Metallkunde und Metallphysik
Technische Universität Clausthal, D-38678 Clausthal-Z., Germany

Abstract

Adsorption isotherms of chloride and sulphate on passive iron, and current efficiencies for film growth and for dissolution of iron(III) from the film at different current densities, sulphate concentrations and pH-values in borate and acetate buffers were measured using a quartz frequency balance and a rotating ring-disc electrode. The influences of electrolyte composition on the steady state current densities and the charges stored in the film were determined.

The results are interpreted by the theory of parallel transfer of iron ions and oxygen ions at the oxide/electrolyte interface. Mechanisms of the transfer reactions of oxygen ions and iron ions are proposed. Sulphate ions accelerate the dissolution by formation of a complex with iron(III).

1. Introduction

The oxide film on any passive metal in an electrolyte solution grows at potentials positive to the equilibrium potential of formation of the oxide by transfer of metal ions from the metal and of oxygen ions from the electrolyte into the film[1–3]. Usually, most of the Gibbs Energy is consumed by the ion transport through the film. However, the steady state corrosion rate and the current efficiencies for film growth and metal ion dissolution are determined by the electrochemical reactions at the interface between the film and the electrolyte.

The transfer reactions of metal ions and of oxygen ions at this interface proceed in parallel and are coupled by the common interfacial potential difference. In the steady state at constant electrode potential, the film thickness remains constant, oxygen ion exchange is in equilibrium, and the current density is due to irreversible metal ion dissolution at the interfacial equilibrium potential difference for oxygen ion exchange. The interfacial potential becomes more positive during film growth. Thus, the rate of metal ion dissolution grows with the incorporation rate of oxygen ions. During film dissolution, the overpotential with respect to oxygen ion exchange becomes negative. Oxygen ions flow from the oxide into the electrolyte and the rate of metal ion dissolution is slower than in the steady state. In the different steady states at constant film thickness and constant rates of growth or dissolution, the same total current densities also flow through the inner interface between the metal and the film and through the film.

The rates of the reactions at the outer interface between the film and the electrolyte depend on the compositions of both the oxide and the electrolyte. The composition of the oxide essentially is given by the overpotential vs the equilibrium potential of formation, or the equivalent partial pressure of oxygen. For passive iron, the composition of the oxide at the interface towards the electrolyte is practically constant between the Flade potential and the potential of oxygen evolution, i.e. the activities of oxygen ions and of iron (III) ions are independent of the electrode potential. In this potential region, the rates of the ion transfer reactions at passive iron are only determined by the composition of the electrolyte and the interfacial potential difference which changes with pH and total current density, but only indirectly with the electrode potential. In particular, one expects the rates of the ion transfer reactions to depend on the nature and the activity of anions participating in the ion transfer reactions. The influence of several anions at different pH-values is reported.

2. Experimental

The dissolution rates of iron (III) were measured at a rotating ring-disc electrode [4] as described in earlier papers [5, 6]. The working electrode was made from coarse-grained zone refined iron with > 99.98% Fe, the ring electrode from pure gold. The counter electrode was a platinum disc parallel to the rotating electrodes. Two separate rigid reference electrodes and two Luggin capillaries were adjusted for minimal electrolyte resistance towards the disc or ring electrodes. The resolution of the ring current was < 1 nA. The collection efficiency was determined at $N_0 = 0.32$.

A quartz frequency balance [7] was used to obtain the rates of film formation. AT-Quartz blancs of 20 mm dia. and a resonance frequency of 10 MHz were sputtered on both sides with gold and, in addition, on one side with < 10 µm pure iron. The iron side was exposed to the electrolyte maintaining equal pressures at both sides of the quartz disc in order to avoid the influence of surface stress [7, 8] on the frequency measurements. Frequencies were measured with a 8-digit counter (Racal Dana Mod. 1991). A mercury sulphate electrode usually was the reference.

From the changes dm/dt of mass and from the total current density j the current densities j_c of iron (III) dissolution and j_a of incorporation of oxygen ions during steady state film growth were calculated [9] with the molar masses M_O and M_{Fe} of oxygen and iron, respectively, from

$$j_a = [(dm/dt) + (M_{Fe}/3F)j]/[(M_O/2F) + (M_{Fe}/3F)] \quad (1)$$

The electrolytes were borate and acetate buffers, 4.5 s pH < 7.5, at 25.0°C, with different concentrations of Na_2SO_4. Oxygen was removed by a stream of pure nitrogen.

3. Results

A collection [1] of steady state current densities of passive iron measured by several authors in different acid electrolytes is shown in Fig. 1. Although the authors obtained somewhat different results for the same electrolyte, there is a clear trend towards an increase of the steady state current densities from sulphate via phosphate to phthalate solutions. Differences among the authors are often due to different waiting times for the steady state.

According to the present experiments, steady state current densities of the order nA cm^{-2} were established in borate buffer, pH 7.4, at any constant electrode potential during times of at least one week after passivation. The steady state current densities were one order of magnitude larger in acetate buffer than in borate buffer at the same pH. The addition of sulphate or chloride to a borate buffer increased the steady state current density in absence of pitting. Qualitatively, the acceleration of the dissolution of iron(III) from the oxide is explained by complex formation which may be preceded by chemisorption of the anions. According to measurements with the quartz frequency balance, chloride and sulphate ions are in fact adsorbed on passive iron. Figure 2 shows that adsorption can be described by Langmuir isotherms. While the mass Γ_{max} for full coverage with chloride is of the expected magnitude, Γ_{max} for sulphate is much larger indicating coadsorption, e.g. of water.

The influence of sulphate on the kinetics of passive iron in neutral solutions was investigated in detail. Figure 3 shows that the current density grew upon addition of sulphate. Prior to the addition of sulphate a quasi-steady state current density in the range 15–45 nA cm^{-2} was established after one day at constant potential in borate buffer. It is also shown in Fig. 3 that after addition of sulphate the dissolution current of iron (III) jumped to a relatively high value and approached the current through the iron electrode after *ca.* 600 s. The difference between the dissolution current of iron (III) and the current through the iron electrode is used for thinning the oxide film. Integration of the difference of the currents yields the charge consumed for oxide dissolution between the two quasi-steady states. This charge q grew with the activity of

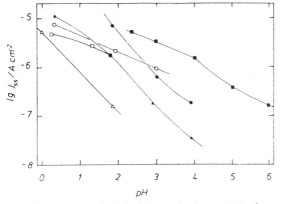

Fig. 1 Steady state current densities j_{ss} for passive iron vs pH in ($\triangle$, $\blacktriangle$, o) 0.5 M sulphate, ($\bullet$) 0.15M phosphate and ($\blacksquare$) 0.05 phthalate solutions at 20–25°C.

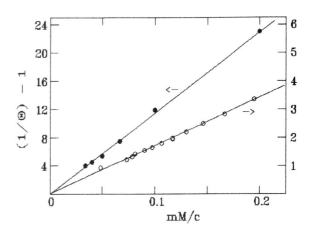

Fig. 2 Optimal fit to Langmuir isotherms for adsorption of chloride (o) and sulphate ($\bullet$) on passive iron with $\Gamma_{max}(Cl) = 285$ ng cm^{-2}, $1/K(Cl) = 17mM$; $\Gamma_{max}(SO_4) = 3060$ ng cm^{-2}; $1/K(SO_4) = 116$ mM.

sulphate according to Fig. 4. The differences of the quasi-steady state current densities with sulphate towards those without sulphate were proportional to the concentration of sulphate.

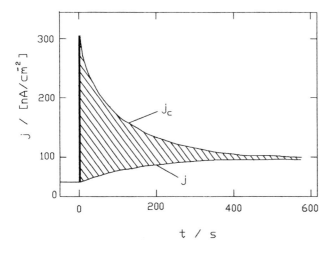

Fig. 3 Current density j at passive iron in 0.4M borate buffer, pH 7.4, 25 °C, E_{HSS} = 1.05V vs the hydrogen electrode in the same solution, after addition of 56 mM sodium sulphate at the time t = 0, and current density j_c of iron (III) dissolution.

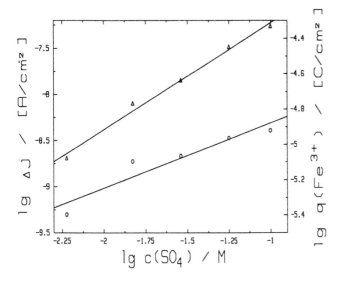

Fig. 4 Increase Δj of the steady state corrosion current density (Δ) due to the addition of sulphate at passive iron in 0.4 borate buffer, pH 7.4, E_{HSS} = 1.05 V, and charge q (o) for dissolution of iron (III) consumed during 600 s in the transition from the steady state in absence of sulphate to the steady state in its presence vs the concentration c(SO$_4$) of sodium sulphate at 25° C.

Some time after applying constant anodic current densities exceeding the steady state corrosion current density, steady state rates of film growth and iron(III) dissolution were established. In this steady state, the rates dE/dt of change of the electrode potential E with time t are constant as shown in Fig. 5 indicating constant field strength in the film and constant interfacial potential differences. Double logarithmic plots of the rates j_c of iron (III) dissolution vs the rates j_a of incorporation of oxygen ions during steady state film growth were linear. Figure 6 shows measurements at different pH-values with iron in acetate buffer. There was no systematic dependence of the slope d ln j_a/d ln j_c = 1.1 (± 0.1) on the pH-value. The slope was the same in acetate and borate buffers.

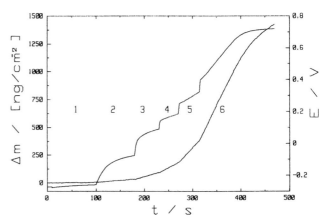

Fig. 5 Electrode potential E (upper curve) vs the mercury sulphate electrode (E_H = 0.64 V) and mass changes Δm of passive iron in 0.3 M acetate buffer, pH 5.5, vs time t at different anodic current densitites j/(mA cm^{-2}): (1) 0.48; (2) 4.8; (3) 24; (4) 47.9; (5) 96.5; (6) 193.5.

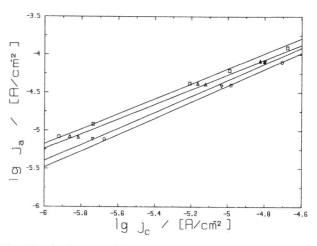

Fig. 6 Current density j_a of film growth vs current density j_c of iron(III) dissolution in 0.5 M acetate buffer at (o) pH 5; (□) pH6; (Δ) pH 7; (∇) pH 7.5.

Upon addition of sulphate the current efficiency of film growth, or j_a at any constant j_c, decreased as shown for measurements with the rrde in Fig. 7. The effect was larger at small current densities than at high ones, because the slope $d \ln j_a/d \ln j_c$ rose with the concentration of free SO_4^{2-} ions to *ca.* 1.7 at $c = 0.46$ M as shown in Fig. 8.

The concentration $C(SO_4)$ of free sulphate ions was calculated from the total concentration c of sulphate and the concentration $c(H)$ of hydrogen ions with $K = 0.012M$:

$$c(SO_4) = c/[1 + \{c(H)/K\}] \tag{2}$$

The current densities j_c for dissolution of iron(III) at constant rate of film growth $j_a = 10$ µA cm^{-2} are plotted in Fig. 9 vs the concentration of sodium sulphate. The apparent reaction order in this particular case was $d \ln j_c/d \ln c = 0.36$, but it decreased with j_a. Figure 10 shows that the current efficiency for film growth grew with pH in 0.3 M acetate buffer for pH $\leq$ 6 and decreased for pH $\leq$ 6.5. At smaller concentrations of the acetate buffer or at higher concentrations of sulphate the current efficiency of film growth became lower, in particular at high pH.

3. Discussion

The experiments are interpreted in terms of the theory of oxide electrodes [1–3]. Oxygen ions and metal ions are exchanged between the electrolyte, 3, and the oxide, 2, on a metal, 1. The rate j_a of oxygen ion transfer at the oxide/electrolyte interface is

$$j_a = k_a^+ (_3a_{OH^-})^q \exp \gamma FE_{2,3} / RT - k_a^{-2}a_{O^{2-}}$$
$$(_3a_{OH^-})^{q-2} \exp -(2-\gamma)E_{2,3} / RT \tag{3}$$

with the activity of OH$^-$ in the electrolyte and oxygen ions in the oxide. If the reaction order is q in the anodic reaction, it must be q–2 in the cathodic reaction for thermodynamic reasons.

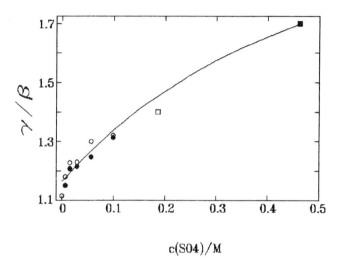

Fig. 8 Slope $d \ln j_a/d \ln j_c = \gamma/\beta$ vs concentration $c(SO_4)$ of sulphate; (o) 0.4 M borate buffer, (•) 0.3 M acetate buffer, (□) 1M H_2SO_4 + 0.05 M Na_2SO_4 + 0.05 M H_2SO_4 [10].

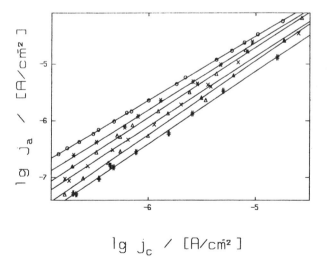

Fig. 7 Current density j_a of film growth vs current density j_c of iron (III) dissolution in 0.3M acetate buffer, pH 7.3, for different concentration c/mM of Na_2SO_4:
(o) 0; (*)6; (Δ) 15; (×) 29; (▲) 56;(#)99.

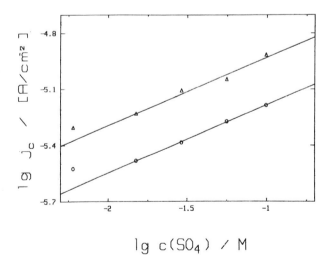

Fig. 9 Current density j_c of iron(III) dissolution at the current density j_c = 10 mA cm^{-2} vs concentration of sulphate (Δ) in 0.3 M acetate buffer, pH 7.3, and (o) in 0.4M borate buffer pH 7.4.

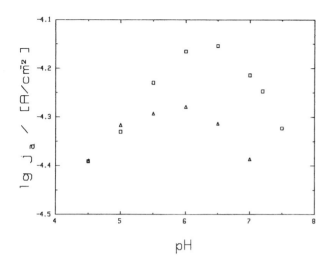

The rate j_c of iron(III) ion transfer is

$$j_c = k_c^+ \, 2^a Fe^{3+} \, 3^a L^P \exp \beta FE_{2,3}/RT \qquad (4)$$

for cations $c = Fe^{3+}$ assuming the activity $3a(Fe^{3+})$ of iron (III) ions in the electrolyte being small compared to the solubility of iron (III) from the film substance. In eqn (4), the participation in the iron ion transfer of some ligand L with the reaction order p is considered.

The steady state interfacial potential difference $E_{2,3}$ at the interface between γ-Fe_2O_3 and the electrolyte corresponds to equilibrium with respect to O^{2-}/OH^- or O^{2-}/H_2O. The Nernst equation for this case yields

$$E^e_{2,3} = E^o_{2,3} + (RT/2F) \ln 2aO^{2-}/(3^aOH^-)^2 \qquad (5)$$

After elimination of $E_{2,3} = E^e_{2,3}$ from eqns (4) and (5) one obtains the pH-dependence of the steady state corrosion rate jst for L= OH^-:

$$(d \ln j_{st}/d \, pH)E^e_{2,3} = p - \beta \qquad (6)$$

For some other ligand L like sulphate one expects

$$(d \ln j_{st/d} \ln a_L)pH = P_L \qquad (7)$$

The rates j_a of film growth and j_c of dissolution of iron (III) in a given electrolyte determine the ratio of the transfer coefficients. For $j^+_a >> j_a^o$ one finds

$$(d \ln j^+_a/d \ln j_c)_{a(oH-)a(L)} = \gamma/B \qquad (8)$$

The pH-dependence of the current density of film growth at constant rate jc of the dissolution of iron(III) is given by

$$(d \ln j^+_a/d \ln 3 a_{OH^-})_{jc} = q - p\gamma/\beta \qquad (9)$$

Similarly, the dependence on the activity of any other ligand is

$$(d \ln j^+_a/d \ln 3 a_L)_{jc} = -p_L\gamma/\beta \qquad (10)$$

Other ligands L establish new reaction paths with their particular kinetic parameters in parallel to the reaction paths proceeding in electrolytes in which only hydroxyl ions are available. It follows from Fig. 4 and eqn [7] that there is an additional dissolution rate $j_{c,d}$ which is first order with respect to sulphate.

Immediately after addition of the sulphate, the increased anodic rate of iron(III) dissolution is compensated by the cathodic rate of oxygen ion dissolution, i.e. there is a 'currentless' dissolution of the oxide. The charge q involved in thinning the oxide to the new steady state is in agreement with the experimental data in Figs. 3 and 4 [9]

The influence of sulphate on the kinetics of film growth is understood with two parallel iron ion transfer reactions, i.e.

$$j_c = k_c \exp \beta_1 FE_{2,3}/RT + kc' \, a(SO_4) \exp \beta_2 FE_{2,3}/RT \qquad (11)$$

For the anodic oxygen ion current density $j_c^+ >> j_c^-$ in eqn (3) from eqn (11) one expects the slope

$$(d \ln j^+_a/d \ln j_c)_{a(OH^-),a(L)} = \qquad (12)$$

$$\frac{\gamma[1+B \, (SO_4)]}{\beta_1[1+(\beta_2/\beta_1)B \, a(SO_4)]}$$

with $B = (k_c'/k_c)(k_a/j_a)^{(\beta_1-\beta_2)/\gamma}$

The dependence of the effective slope $(g/b)_{eff}$ on the activity of sulphate in Fig. 8 was calculated from eqn (12) as an optimal fit to the experiments with $\gamma = 1.43$ (1: p.350; 10), $\beta_1 = 1.23$ and $\beta_2 = 0.6$. The parameter B follows from the effective slopes at $a(SO_4) = O$ and $a(SO_4) = 0.46$ M. According to Fig. 7 the effective reaction order decreases with j_a as predicted from

$$\{d \ln j_c\}/d \ln a(SO_4)_{pH,j_a} = 1 + B^{-1} \qquad (13)$$

The kinetic parameters correspond to the following mechanisms of ion transfer: The rate determining step of oxygen ion transfer [1: p.350; 10] is

$$_{ad}OH^- \rightarrow {}_2O^{2-} + {}_3H^+ \qquad (14)$$

Adsorbed hydroxyl ions are formed in the equilibrium

$$H_2O = {}_{ad}OH^- + {}_3H^+ \qquad (15)$$

The effective transfer coefficient [11] then is $\gamma = 1 + \alpha$ with $\alpha = 0.43$ for step (14).

In presence of sulphate, iron ions are transferred according to

$$_2Fe^{3+} + {}_3SO_4^{2-} \rightarrow {}_3FeSO_4^+ \qquad (16)$$

with the transfer coefficient $\beta_2 = \alpha_2 = 0.6$ and the reaction order $y = 1$ with respect to sulphate.

In absence of sulphate, the effective transfer coefficient $\beta_1 = 3\alpha_1 = 1.23$ suggests the simple mechanism

$$_2Fe^{3+} \rightarrow {}_3Fe^{3+} \qquad (17)$$

References

1. K. E. Heusler, The Electrochemistry of Iron, in Encyclopaedia of the Electrochemistry of the Elements, Vol. **IXA**, p. 230, ed. by A. J. Bard, Dekker Inc., New York and Basel, 1982.
2. K. E. Heusler, Electrochim. Acta, **28**, 439, 1983.
3. K. E. Heusler, Corros. Sci., **29**, 131, 1989.
4. K. Nachstedt, Ph.D. thesis, Clausthal, 1987.
5. K. E. Heusler, Ber. Bunsenges. physik. Chem., **72**, 1197, 1968.
6. K. E. Heusler and L. Fischer, Werkst. Korros., **27**, 697, 1976.
7. K. E. Heusler, A. Grzegorzewski, L. Jaeckel and J. Pietrucha, Ber. Bunsenges. physik. Chem., **92**, 1218, 1988.
8. K. E. Heusler and J. Pietrucha, J. Electroanal. Chem., **329**, 1992, 339
9. K. E. Heusler. L. Jaeckel and F. J. Nagies, ECS Proceedings Vol. **92–22**, 1992, 505.
10. K. J. Vetter and F. Gorn, Werkst. Korros., **21**, 703, 1970; Electrochim. Acta, **18**, 321, 1973.
11. K. D. Allard and K. E. Heusler, J. Electroanal. Chem., **77**, 35, 1977.

19

Modifications of Passive Films on 304 Stainless Steel Investigated by Capacitance Measurements

R. Devaux, C. Masson, D. Vouagner, A. M. de Becdelievre, C. Duret-Thual* and M. Keddam**

Laboratoire de Chimie Appliquee et Génie Chimique (URA 417 du CNRS), Université C. Bernard, Lyon I, 43 Bd du 11 novembre 1918, 69622 Villeurbanne, Cedex, France
* IRSID-Unieux, BP 50 42702 Firminy, France
¨Laboratoire de Physique des liquides et Electrochimie (UPR 15 du CNRS), Université P. et M. Curie, Tour 22, 4 place Jussieu, 75252 Paris, Cedex 05, France

Abstract

A.c. impedance measurements were carried out in order to study frequency dispersion of capacitance of passive 304 stainless steel in various solutions. Complex capacitance diagrams were plotted in the Cole-Cole plane in order to study the relaxation processes and establish their relationship with passive film properties.

1. Introduction

The semiconducting behaviour of passive film is often deduced from photoelectrochemical or capacitance measurements. By light illumination, the modification of charge distribution in the film induces a current at constant potential, a potential at constant current and a variation of the electrode capacitance [1–6]. However Mott-Schottky plots used to characterise the semiconductor properties of passive film on stainless steel cannot be derived from impedance data measured over a wide frequency range. In this work, combined a.c. impedance measurements extended over a large frequency domain (10^3–5.10^{-2} Hz) with surface analyses (XPS, SIMS) are used to investigate the behaviour of passive film in various experimental conditions. Complex capacitance C* were calculated from experimental impedance data and represented in Cole-Cole diagrams. The frequency dependence of the capacitance was then tentatively related to some properties of the passive layer.

2. Experimental

The electrodes of AISI 304 stainless steel were disks of 15 mm dia., they were first mechanically polished on wet emery papers (grit 240–1200) then with diamond pastes (3–1 μm). They were finally degreased in acetone and ethanol and rinsed with distilled water, then dried.

Some electrodes were implanted with molybdenum ions (2.5×10^{16} ions.cm^{-2}, 100 keV). Implantation was performed in an industrial facility†.

The electrochemical measurements were performed in a three-electrode cell described elsewhere [7]. The contact surface with the solution was 1 cm². The electrolytes were NaCl and Na_2SO_4 aqueous solutions (RP PROLABO) de-aerated in a separate vessel by pure nitrogen bubbling then transferred to the experimental cell under controlled atmosphere. Both the 0.02M and 0.5M NaCl solutions have respectively the same ionic strength as the 0.007M and 0.2M Na_2SO_4 ones. Anhydrous ethanolic solutions of $Zn(NO_3)_2$ (80 g.L^{-1}) were also used.

The ac impedance diagrams were recorded in potentiostatic mode at the rest potential, in the frequency range between 10^3 and 5.10^{-2} Hz with a Tacussel Z computer. The transformations and simulation curves were obtained from a simulation and fitting program developed specifically for this study.

3. Results

The Nyquist plots of impedances measured at the rest potential in neutral aqueous media containing either chloride or sulphate anions reveal only the HF trend of a large capacitive loop. In order to get more information, the impedance Z is replotted as a complex capacitance C* after elimination of R_{HF} (series resistance) and R_{LF} (parallel resistance) respectively the HF and the LF limits.

†NITRUVID, Usine du Parc, rue de l'Ondaine, 42490 Fraisses, France.

$$C^* = C' - jC'' = \frac{1}{j\omega Z_c}$$

where $\dfrac{1}{Z_c} = \dfrac{1}{Z - R_{HF}} - \dfrac{1}{R_{LF}}$

C^* exhibits a frequency dispersion extending within two clearly separated domains (Fig. 1(a)) obeying the characteristic depressed circular shape of the Cole-Cole distribution law:

$$C = C_\infty + \frac{C_0 - C_\infty}{1 + (j\omega\tau)^{(1-\alpha)}}$$

where the α exponent accounts for the deviation from the Debye dispersion [8].

Formally the capacitance behaviour of the film can be represented by the electrical network shown in Fig. 1(b) where the $C_{s1}R_{s1}$ and C_2R_2 time constants are involved in the Cole-Cole law according to:

$$C^* = C_{\infty 1} + \frac{C_{01} - C_{\infty 1}}{1 + (j\omega C_{s1}R_{s1})^{(1-\alpha_1)}} + \frac{C_2}{1 + (j\omega C_2 R_2)^{(1-\alpha_2)}}$$

The evolution of these two loops was studied as a function of experimental conditions.

The HF domain seems to arise from the dielectric behaviour of the passive film as the C_{01} and $C_{\infty 1}$ values are consistent with the thickness estimated by XPS and SIMS [7]. After ageing in chloride solution, the evolu-

tion of the XPS Fe signal for unimplanted samples showing an enhancement of the metallic iron component (Fig. 2) indicates a decrease of the oxide film thickness. This correlates with the observed increase of $C_{\infty 1}$ and C_{01} (Table 1). Moreover the superimposition of the oxygen profiles (Fig. 3) as recorded for Mo-implanted steel (curve 2) and unimplanted steel (curve 1) samples reveals that oxide layers are thicker on implanted surfaces. The fitted parameters (C_{01} and $C_{\infty 1}$) deduced from the simulation are also found to be lower (Table 1).

The increase of C_{01} with the electrolyte concentration in chloride and sulphate solutions (Table 2) can be associated to either a decrease of the passive film thickness or an increase of the dielectric constant or both. This is not inconsistent with the interpretation of C_2 changes proposed below.

It can be observed that the C_2 values estimated in a chloride medium are in any case higher than in a sulphate medium (Table 2). By comparing these two media, it is found that increasing the concentration (Table 2) or the exposure period (Table 1) gives rise to changes of C_2 in opposite directions. C_2 goes up in the presence of Cl^- and goes down in a sulphate solution.

The increase of the C_2 values through ageing in chloride medium could be ascribed to changes in the film properties related to dehydration and partial recrystallisation [9,10] allowing a higher rate of ion-exchange with the electrolyte than the hydrated amor-

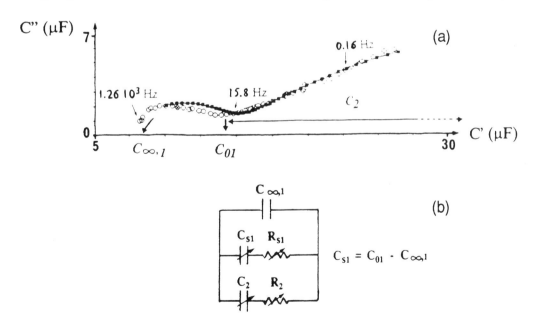

Fig. 1(a) Cole-Cole capacitance curves obtained with a stainless steel sample immersed 100 h in aerated NaCl (0.02M) solution.
Experimental curve ○ ○ ○ ○
Fitting curve ━■━ ━■━
$C_\infty = 7.1\ mF$; $C_{01} = 13.9\ mF$; $a_1 = 0.33$; $C_2 = 45.2\ mF$; $R_2 = 7.10^5$; $\Omega\ a_2 = 0.62$.
(b) Electrical network.

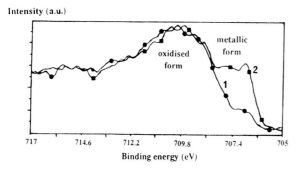

Intensity (a.u.)

oxidised form

metallic form

2

1

717   714.6   712.2   709.8   707.4   705

Binding energy (eV)

Fig. 2 XPS spectra of the Fe $2p^{3/2}$ peak.
Unimplanted steel samples immersed in de-aerated 0.02M NaCl solutions.

1 - 1 h of immersion

2 - 100 h of immersion.

Counts

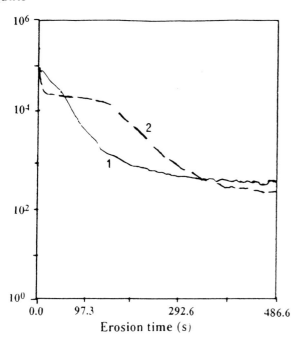

10^6

10^4

2

1

10^2

10^0

0.0   97.3   292.6   486.6

Erosion time (s)

Fig. 3 SIMS profiles of O^-.
After 1 h of immersion in de-aerated 0.02M NaCl solution.

1: unimplanted steel

2: molybdenum ion implanted steel (2.5×10^{16} ions.cm^{-2}).

phous layer. Conversely in sulphate medium, the decay of C_2 would be interpreted by an increase of the water content during a long term exposure to a non aggressive anion. Therefore it is suggested that C_2 is a faradaic pseudo-capacitance standing for the interactions between film and solution. Shift of oxidation state of the film cations associated to counter-ions exchange with the solution and/or intercalation mechanism can be responsible for this charge storage. This hypothesis was tentatively checked by performing measurements in a dehydrating solution (absolute ethanol + $Zn(NO_3)2$) on electrodes previously immersed for 1 or 100 h in an aqueous solution. Impedance spectra were recorded after one hour immersion; no significant evolution was observed after subsequent exposures. The Cole-Cole plot exhibits a single low frequency dispersion (Fig. 4). In the explored high frequency domain, no dielectric response was observed in this anhydrous medium. The parameters used for simulating the impedance data in both aqueous media and the ethanolic one are given in Table 1.

The C_2 values in the anhydrous medium are higher than when measured in the initial aqueous solutions. Again this can be attributed to the dehydration in alcoholic solution of the layer grown earlier in the presence of water. However ageing the film in the chloride medium already induced a dehydrated film: consequently subsequent exposure to alcoholic solution sparingly modifies the C_2 values (Table 1). Similarly, the large C_2 values found in alcoholic solution with electrodes aged in sulphate medium (Table 1) support the assumption of a progressive hydration in this later medium. Moreover the C_2 values obtained with Mo-implanted electrodes (Table 1) are lower in NaCl solution and higher in alcoholic medium than those found with non-implanted electrodes (Table 1). This is in agreement with the high hydration degree, mainly as hydroxylated Cr(II) species, of the outer part of the passive film formed on Mo implanted electrodes [11] preventing ionic exchange with chloride solution.

The whole body of experiments is fully consistent with the view of a passive layer which degree of hydration has a strong influence in its ability to exchange charges as indicated by the value of its low frequency capacitance value. Strongly hydroxylated, corrosion resistant layers exhibit a lower capacity, i.e. little exchange whilst dehydration by exposure to ethanol or chloride leads to higher capacitance. The change is even greater when dehydration takes place in initially more hydrated structures like in the presence of implanted Mo.

4. Conclusion

Impedance of 304 SS passivated in various solutions

Table 1 Fitting parameter variations with the electrolytic medium and exposure time. Unimplanted and Mo-implanted stainless steel samples (1 cm²)

Conditions	Aqueous medium					Alcoholic medium		
	E_{rest} (mV/ECS)	C_{01} (μF)	α_1	C_2 (μF)	α_2	C_{01} (μF)	C_2 (μF)	α_2
Unimplanted steel	NaCl (0.02 M)							
1 hour	- 33	11.8	0.30	30.0	0.60	7.0	56.0	0.72
100 hours	- 90	12.5	0.40	32.0	0.65	6.0	32.0	0.71
Unimplanted steel	Na_2SO_4 (0.007 M)							
1 hour	- 76	9.6	0.28	26.0	0.54	5.0	55.0	0.80
100 hours	- 6	10.0	0.30	20.0	0.74	4.5	80.0	0.70
Mo-implanted steel	NaCl (0.02 M)							
1 hour	+ 12	4.8	0.55	9.0	0.65	7.0	125.0	0.77

Table 2 Influence of composition and concentration of the electrolyte on fitting parameters. Stainless steel samples (1 cm²). 1 h exposure

Concentration	E_{rest} (mV/ECS)	C_{01} (μF)	α_1	C_2 (μF)	α_2
NaCl 0.02 M	- 33	11.8	0.30	30.0	0.60
NaCl 0.5 M	- 26	15.8	0.49	46.0	0.70
Na_2SO_4 0.007 M	- 76	9.6	0.28	26.0	0.54
Na_2SO_4 0.2 M	+ 23	14.1	0.55	22.6	0.40

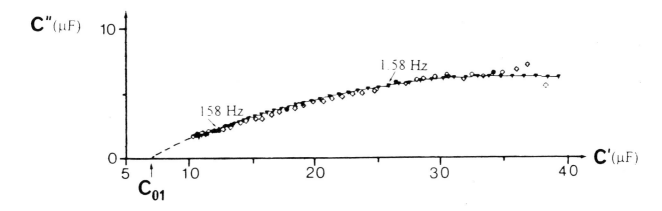

Fig. 4 Cole-Cole capacitance curves obtained in ethanolic medium +$Zn(NO_3)_2$ (80 g.L⁻¹). Stainless steel samples previously immersed in aqueous medium.
Experimental curve ◇ ◇ ◇ ◇
Fitting curve ▼—▼
$C_{01} = 7 \mu F$; $C_2 = 56 \mu F$; $R = 2.10^4 \Omega$; $a_2 = 0.72$.

when processed in the Cole-Cole representation indicates that the passive layer exhibits at high frequencies a dielectric behaviour related to the response of the whole film, most often described as a bi-layer structure. This suggests that relaxation times associated to the two components are not distinguished in our ex-

perimental conditions. At lower frequency a faradaic capacitance related to the ability of the film to exchange charges with the solution. Higher value of this capacitance are related to easier exchanges in dehydrated films. The likely relationship between this low frequency behaviour and the corrosion resistance deserves further detailed investigation. In particular the passive electrode impedance behaviour as a function of applied potential should provide useful information. Work is in progress in this area and will be reported in a forthcoming paper.

5. Notation

$$C = C_\infty + \frac{C_0 - C_\infty}{1 + (j\omega\tau)^{(1-\alpha)}} : \text{dielectric Cole-Cole Law}$$

C_0 : static capacitance ($\omega \longrightarrow 0$)
C_∞: high frequency limiting dielectric capacitance

For this work :

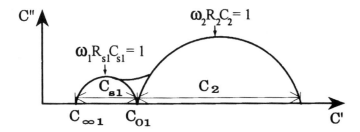

$C_{s1} = C_{01} - C_{\infty 1}$
C_2 : faradaic pseudo-capacitance

References

1. I. Olejford and B. Brox, Passivity of Metals and Semiconductors (Ed. by M. Froment), Elsevier Pub.Co., Amsterdam, p.561, 1983.
2. I. Olejford, B. Brox and J. Jevelstam, J. Electrochem. Soc., **132**, 2854, 1985.
3. A. Di Paola, Electrochim. Acta, **34**, 203, 1989.
4. A. Di Paola, D. Shukla and U. Stimming, Electrochim. Acta, **36**, 345, 1991.
5. P. C. Searson and R. M. Latanision, Electrochim. Acta, **35**, 445, 1990.
6. U.Stimming, Electrochim. Acta, **31**, 415, 1986.
7. R.Devaux, thesis, University of Lyon I, 1992.
8. K. S. Cole and R. H. Cole, J.Phys. Chem., **9**, 34, 1941.
9. R. W. Revie, B. G. Baker and J.O'M.Bockris, J. Electrochem. Soc., **122**, 1460, 1975.
10. L. D. Burke and M. E. G. Lyons, J. Electroanal. Chem. and Interfacial Electrochem., **198**, 347, 1986.
11. E. De Vito, thesis, University of Paris 6, 1992.

20

Passive Characteristics of Amorphous Ni–P Alloys

A. KROLIKOWSKI

Institute of Solid State Technology, Department of Chemistry, Warsaw University of Technology, ul. Noakowskiego 3, 00-664 Warsaw, Poland

Abstract

Anodic behaviour of electrodeposited amorphous Ni–P alloys (17–28 at.% P) was studied in neutral and acidic media using potentiodynamic and potentiostatic methods. These alloys exhibited a suppression of anodic dissolution suggesting passivation. This suppression was fairly insensitive to additions of chloride ions. The proposed model of the anodic process of amorphous Ni–P was based on different rates of active dissolution of P and Ni. Due to the inhibition of P oxidation the dissolution process is controlled by diffusion of Ni through the P rich surface layer. This model explains the observed insensitivity of the anodic behaviour of amorphous Ni–P to chloride ions.

1. Introduction

Even small additions of phosphorus destroy the passivity of nickel in acidic solutions and dramatically rise the anodic currents [1,2]. The critical content of P which prevents the passivation in 0.1N H_2SO_4 was approximated to be in the range 0.5–1.6 at.% [1]. On the other hand, Ni–P alloys with P content of *ca.* 20% exhibit a better corrosion resistance than pure Ni [3–5]. These P rich alloys show a suppression of dissolution in the potential range where Ni dissolves actively [6,7]. The interpretation of this behaviour is still a matter of discussion. The suppression of anodic dissolution was regarded as passivity involving the adsorption of hypophosphite ions [7] or the formation of a nickel phosphide film [6] on the alloy surface. Habazaki *et al.* suggested the accumulation of elemental phosphorus on the alloy surface which acts as a barrier against dissolution [8]. These interpretations do not take into account the structural aspect. Recent works revealed, however, that the above mentioned retardation of anodic dissolution is attributable to the amorphous structure of P rich Ni–P alloys [9–11].

This paper examines the anodic behaviour of amorphous Ni–P alloys in neutral chloride solution and gains a closer insight into the nature of the suppression of anodic dissolution of these materials.

2. Experimental

Ni–P alloys with P content ranging from 17 to 28 at.% were electrodeposited from sulphate–hypophosphite solutions [12] on copper foils. These films were 30–40 μm in thickness. The amorphous structure of the alloys was confirmed by X-ray diffraction.

Anodic behaviour of these samples was studied by potentiodynamic and potentiostatic polarisation methods. Sample preparation and design of the electrolytic cell were similar as those described elsewhere [13]. Measurements were performed mainly in deaerated 0.1N NaCl solution at $20 \pm 2°C$. The reference electrode was a saturated calomel electrode and all values of potential are given versus this electrode.

Anodic potentiodynamic curves were measured at a potential scan rate of 1 mV.s^{-1}. Before the measurements were started the samples were prepolarised cathodically (–0.7 V). Potentiostatic measurements were carried out by stepping the potential from the rest potential to selected anodic potentials, the change of current with time was monitored.

3. Results and Discussion

Anodic polarisation curves of amorphous Ni–P alloys (P $\geq$ 17%) in 0.1N NaCl are presented in Fig. 1. Curves of alloys with less P, which do not show amorphous structure [10], are inserted for comparison. Only the amorphous samples exhibit an evident current arrest between *ca.* –0.15 and 0.25 V. Increasing chloride concentration does not accelerate the dissolution in this potential region (Fig. 2), in agreement with [14]. No visible evidence of pitting was found; the surface remained intact (lustrous). Quite a similar suppression of anodic dissolution was observed for amorphous Ni–P in acidic media (Fig. 2.) These results suggest that the retardation of dissolution is attributable to P rich Ni–P alloys only and the nature of this phenomenon diverges from the classical concept of passivity.

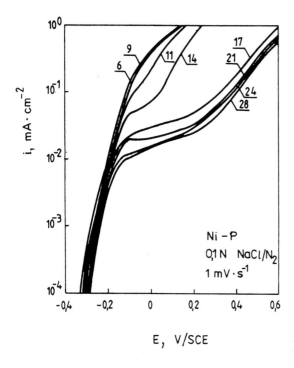

Fig. 1 *Anodic polarisation curves of Ni–P alloys in 0.1N NaCl solution. The numbers denote P content in the alloy in at.%. Alloys structure: crystalline (P < 10%), amorphous (P > 15%), transitional region (10–15% P) [10].*

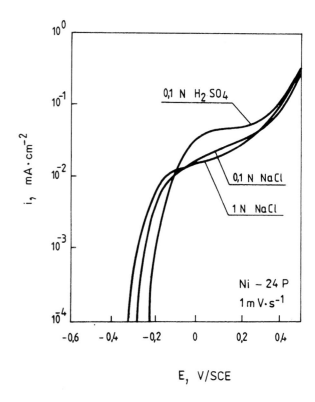

Fig. 2 *Comparison of anodic polarisation curves of Ni–24P in various solutions.*

Potentiostatic measurements, performed at potentials from the range of the current arrest, showed a continuous decay of dissolution current of amorphous Ni–P alloys (the results obtained for –0.15 V are shown in Fig. 3. Analysis of these decay curves indicates that the current is approximately in inverse proportion to the square root of polarisation time. This relation can be interpreted in terms of diffusion of a faster dissolving component of the alloy through the developing surface layer of a hardly dissolving component, according to [15]:

$$1/i = V(\Pi t)^{0.5}/nFaD^{0.5} \qquad (1)$$

where: V is the molar volume of the alloy; n, a and D - charge number of the electrode reaction, bulk mol fraction and diffusion coefficient of the dissolving component, respectively. Preferential dissolution of Ni from amorphous Ni-P was well documented [6, 7, 14]. Conceivably, eqn (1) may describe the dissolution current of Ni limited by diffusion through the P rich surface layer.

Before the potentiostatic test Ni–P samples were kept at the corrosion potential. Under this condition the surface enrichment in P may also occur (similar nature of dissolution of amorphous Ni–P was found at

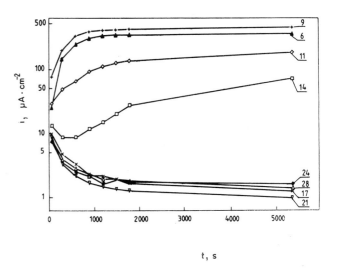

Fig. 3 *Dissolution current density vs time for Ni–P alloys in 0.1N NaCl at –0.15 V. The denotations as in Fig. 1.*

the corrosion potential and in the potential range of the current arrest [10]). Thus the potentiostatic results (Fig. 3) were fitted to the modified form of eqn (1) [8]:

$$1/I^2 = \Pi(t_o+t)V^2/(nFa)^2D \qquad (2)$$

where t_o is the time corresponding to the formation of the diffusion layer prior to the start of potentiostatic run. Figure 4 shows an example of such fitting. Estimated parameters of diffusion process are summarised in Table 1. The values for t are less than the time of exposure at the corrosion potential (90 min), conceivably, due to slower dissolution of Ni at E_{corr} compared with the anodic potential: –0.15 V. The estimates for the thickness of the diffusion layer:

$$d = (\Pi D(t_o+t))^{0.5} \qquad (3)$$

coincide well with the depth of the P rich layer (*ca.* 15 nm) determined by AES for Ni–20P in 0.2N HCl at 0 V [14]. Polarisation characteristics shown in Ref. [14] are very similar to those obtained in this study.

These results indicate that the retardation of anodic dissolution can be related to the slow oxidation of P from amorphous Ni–P. This effect may originate, at least in part, from structural and chemical homogeneity of this alloy. In our opinion, however, contribution of other features of amorphous Ni–P that differentiate it from crystalline Ni–P, namely: higher electrical resistivity, paramagnetism and chemical binding between the alloy components, should also be taken into consideration. Significant change in anodic oxidation of P on crystalline and amorphous Ni–P alloys was recently established [16].

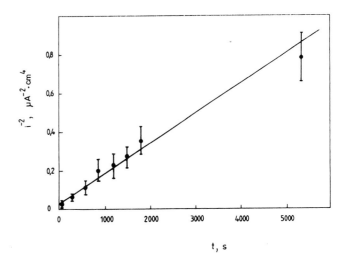

Fig. 4 Time dependence of reciprocal of the square of anodic current density for Ni–17P in 0.1 N NaCl at –0.15 V.

Anodic characteristics of amorphous Ni–P are scarcely affected by P content. This point confirms the primary importance of structural aspect.

4. Conclusion

This work shows clearly that the anodic behaviour of amorphous Ni–P diverges from the classical passivity concept. The model based on different rates of active dissolution of P and Ni is postulated (Fig. 5). Accordingly, due to the inhibition of P oxidation on amorphous Ni–P, a P rich zone is formed at the alloy surface and the dissolution process is controlled by diffusion of Ni through this P rich zone. This model is relevant

Table 1 Parameters of diffusion of Ni from amorphous Ni–P alloys polarised at –0.15 V in 0.1 N NaCl

	Diffusion coefficient cm^2/s^{-1}	Parameter t_o s	Diffusion layer thickness cm	Correlation coefficient (R)
Ni–17P	4×10^{-17}	5×10^{2}	8×10^{-7}	0.980
Ni–21P	3×10^{-17}	7×10^{2}	7×10^{-7}	0.987
Ni–24P	7×10^{-17}	1×10^{3}	1×10^{-6}	0.965
Ni–28P	7×10^{-17}	9×10^{2}	1×10^{-6}	0.968

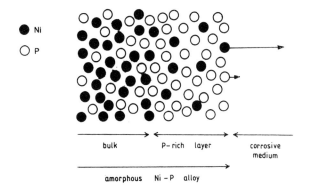

- ● Ni
- ○ P

bulk   P–rich   layer   corrosive medium

amorphous   Ni – P   alloy

Fig. 5 Schematic model of the anodic dissolution of amorphous Ni–P alloys.

to amorphous Ni–P in acidic and neutral media in the potential range from the corrosion potential up to *ca.* 0.25 V. The model explains the observed insensitivity of the anodic behaviour of amorphous Ni–P to chloride ions and the time dependence of the anodic current in potentiostatic tests.

5. Acknowledgement

The author is grateful to P. Butkiewicz for his experimental assistance in this work. Thanks are also due to the Batory Foundation for assistance in meeting the conference costs.

References

1. P. Marcus and O. Oda, Mem. Sci. Rev. Metall,. **76**, 715, 1979.
2. K. Masui, T. Yamada and Y. Hisamatsu, J. Met. Finish. Soc. Japan, **32**, 410, 1981.
3. A. Brenner, D. E. Couch and E. K. Williams, J. Res. Nalt. Bur. Stand., **44**, 109, 1950.
4. A. Kawashima, K. Asami and K. Hashimoto, J. Non-Cryst. Solids, **70**, 69, 1985.
5. A. Kawashima, Y. P. Lu, H. Habazaki, K. Asami and K. Hashimoto, Corr. Engng., **38**, 643, 1989.
6. G. Salvago and G. Fumagalli, Met. Finish., **85**, 31, 1987.
7. R. B. Diegle, N. R. Sorensen, C. R. Clayton, M. A. Helfand and Y. C. Yu, J. Electrochem. Soc., **135**, 1085, 1988.
8. H. Habazaki, S.-Q. Ding, A. Kawashima, K. Asami, K. Hashimoto, A. Inoue and T. Masumoto, Corros. Sci., **29**, 1319, 1989.
9. Z. Longfei, L. Shoufu and L. Pengxing, Surf. Coat. Technol., **36**, 455, 1988.
10. A. Krolikowski, M. Pelowska and P. Butkiewicz, Metall. Foundry Eng., **18**, 189, 1992.
11. A. Krolikowski and P. Butkiewicz, Electrochim. Acta, **38**, 1993, 1979.
12. R. S. Vakhidov, Elektrokhimiya, **8**, 70, 1972.
13. A. Krolikowski, Proc. 10th Int. Cong. Metallic Corrosion, Madras 1987, vol. II, 1169, 1987.
14. R. B. Diegle, N. R. Sorensen and G. C. Nelson, J. Electrochem. Soc., **133**, 1769, 1986.
15. E. Heusler and D. Huerta, Proc. Symp. Corrosion, Electrochemistry and Catalysis of Metallic Glasses. eds. R. B. Diegle and K. Hashimoto, Proc. Electrochem. Soc., 88–1, p.1, 1988.
16. U. Hoffmann and K. G. Weil, Corros. Sci., **34**, 423, 1993.

21

Passivity of Chromium in Concentrated and Anhydrous Solutions of Sulphuric Acid

B. STYPULA AND J. BANAS

Academy of Mining and Metallurgy, Institute of Foundry Engineering Kraków, Poland

Abstract

The electrochemical investigations of chromium were performed by means of potentiostatic and chronopotentostatic techniques in deaerated H_2O-H_2SO_4, CH_3OH-H_2SO_4, DMF-H_2SO_4 and formamide-H_2SO_4 solutions. X-ray photoelectron-spectroscopy (XPS) was used to study the structure of passive layers.

Chromium undergoes the oxide passivation in the whole range of acid concentration in investigated system. The mechanism of passivation is depended on electrolyte 'structure' especially on H^+/H_2O ratio. In the solutions with the ratio $H^+/H_2O < 1/4$ the passivation takes place with the participation of water molecules. The anodic films obtained in this solutions consist mainly of oxy-hydroxide chromium III, but it contain sulphur species on the lower valences S^0 and S. We suppose that the passivation process proceeds with the participation of undissociated acid molecules and/or HSO_4^- ions.

1. Introduction

The anodic passivation of chromium was the object of numerous publications[1–21] in connection with the importance of this metal in production of corrosion resistant alloys. Most investigations concerning the mechanism of passivation of chromium have been performed in diluted aqueous solutions [1–17]. In these solutions the passivation process proceeds with participation of water molecules. Water is a source of oxygen ions for oxide formation and stimulates the creation of polinuclear amorphous oxy-hydroxide Cr(III) films [12–14, 16–17, 20].

In environments with a small activity of water, i.e. in concentrated aqueous solutions and in anhydrous organic solutions of acids the mechanism of passivation is still not explained. This paper tries to explain the similarities and differences in the formation of passive films on chromium in aqueous and anhydrous organic solutions of sulphuric acid.

2. Experimental

The investigations have been carried out on pure polycrystalline chromium (99.9%). The solutions were prepared with anhydrous solvents—methanol, N, N–dimethylformamide, formamide containing non more then 0.02% water (the quantity of water was controlled by Karl-Fisher method)—and with anhydrous sulphuric acid.
The electrochemical experiments were performed with an interrupting potentiostat including automatic IR-compensation. an aqueous calomel electrode was used as reference electrode in aqueous solutions. In anhydrous organic environments the Ag/AgCl electrode in organic solvent was applied. The investigation were performed in the deaerated solutions by means of potentiostatic technique. X-ray photelectron spectroscopy (XPS) measurements of passive films on chromium were carried out using ESCA-3 spectroscope from Vacuum Generators with a $AlK_{\alpha 1,2}$ excitation (486.6 eV). The indepth analysis of anodic layer was performed by removing subsequent layers by sputtering with Ar^+ ions of 1.4 keV energy.

3. Results and Discussion

3.1 Aqueous solutions of sulphuric acid

Figure 1 demonstrates the influence of electrolyte structure of aqueous sulphuric acid solutions on anodic behaviour of chromium. Three ranges of acid concentrations can be distinguished in the diagram. The range 'a' corresponds to the diluted solutions, in which the molar ratio acid/water is lower than 1/4 and free water molecules (not bound in hydratation shell of hydrogen ions) are present in the medium. The passivation of chromium proceeds in this range with the participation of water molecules according to total reaction [22,23]:

$$Cr+ 2H_2O = CrOOH +3H^+ +3e \qquad (1)$$

and the Okuyama mechanism

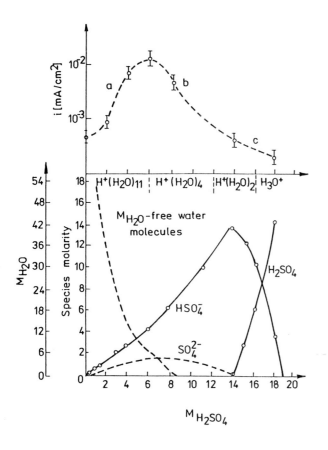

Fig. 1 *The influence of electrolyte structure of aqueous sulphuric acid solutions on anodic dissolution of chromium (E = 0.8 V, t = 1 h).*

$$Cr + H_2O = CrOH_{ad} + H^+ + e \qquad (1a)$$

$$CrOH_{ad} + H_2O = CrOOH + 2H^+ + 2e \qquad (1b)$$

The XPS measurements of surface layer on chromium in 4M H_2SO_4 (range a) presented in Fig. 2 demonstrate that the anodic passive film consists mainly of oxy-hydroxides of Cr(III) with a small hydration degree. The XPS spectra of the anodic layer formed in 18M H_2SO_4 show that in concentrated solutions (range C) the passive film is also composed of oxy-hydroxides of Cr(III) (Fig. 3). This film also contains a small amount of sulphur in a lower oxidation state (S^0, S^{-2}). The presence of these species is an evidence that acid molecules undergo reduction in the passivation process.

In concentrated sulphuric acid solutions (range b and c) 'free' water molecules are absent. Hence the passivation of chromium with participation of water molecules (eqn 1) is difficult. The HSO_4^- ions and H_2SO_4 molecules are most likely a source of oxygen ions in the passivation process. This possibility is supported by the linear dependence of the passivation potential on the logarithm of HSO_4^- concentration in range c (Figs 4 and 5). the above facts and composition of passive film allow us to propose the mechanism of passivation with the participation of hydrosulphate anions and undissociated acid molecules:

for range b,

$$Cr + HSO_4^- = CrHSO_{4ad} + e \qquad (2a)$$

$$CrHSO_{4ad} = CrOOH + SO_2 \qquad (2b)$$

for range c,

$$Cr + H_2SO_4 = CrHSO_{4ad} + H^+ + e \qquad (3a)$$

$$CrHSO_{4ad} = CrOOH + SO_2 \qquad (3b)$$

3.2 Organic solution of sulphuric acid

In completely anhydrous organic solutions the sulphuric acid shows oxidising behaviour and therefore

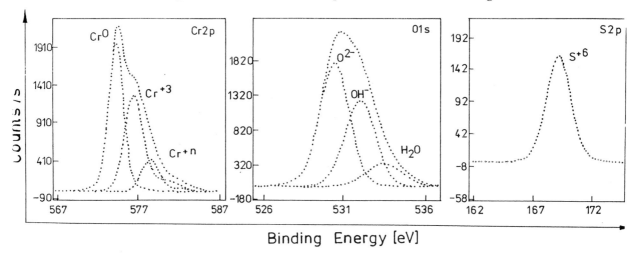

Fig. 2 *XPS spectra of passive film formed on chromium in 4M H_2SO_4 (E = 0.8 V, t = 1 h).*

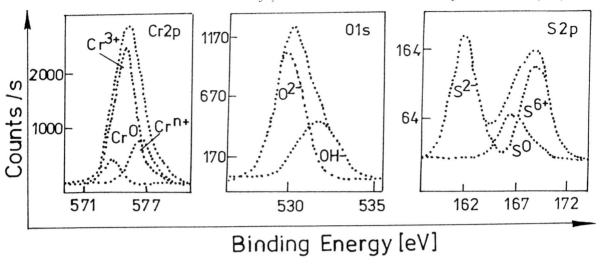

Fig. 3 XPS spectra of passive film formed on chromium in 18M H$_2$SO$_4$ (E = 0.8 V, t = 1 h).

the passivation action of chromium surface. Figures 6 and 7 present polarisation curves of chromium in H$_2$SO$_4$–CH$_3$OH and H$_2$SO$_4$-N,N-dimethylformamide. In these solvents, showing a small dielectric constant (ϵ_{CH_3OH} = 32.63, ϵ_{DMF} = 36.71) sulphuric acid is not completely dissociated. The dissociation constant K$_1$ of H$_2$SO$_4$ in N, N-dimethylformamide is equal to 9.5×10^{-4} (20°C).

The passivation ability of sulphuric acid molecules was observed also in N, N–dimethylformamide–HCl system. Figure 8(a) and 8(b) compare the influence of acid and water concentration on passivation of chromium in the above mentioned system. The passivation ability of sulphuric acid is analogous to the passivation ability of water.

Figure 9 shows the anodic polarisation of chromium in H$_2$SO$_4$–formamide solution in a solvent with a high dielectric constant (ϵ = 109.5) in which the acid is completely dissociated. The passivation process in this system can be connected with the oxidative ability of HSO$_4^-$ ions. The assumption is confirmed by the linear dependence of the passivation potential of chromium on the logarithm of HSO$_4^-$ concentration (Fig. 10). The XPS analysis of anodic passive layer, formed on chromium surface in HSO$_4^-$–formamide system, shows that passive film consist of Cr(III) oxy-hydroxide with a small amount of sulphide ions (Fig. 11). The presence of S^{-2} ions is an evidence that H$_2$SO$_4^-$ ions are reduced in the passivation process.

The similar anodic properties of chromium in aqueous concentrated and anhydrous organic solutions of sulphuric acid, as well as the similar composition of anodic passive film in both systems (the presence of sulphur compounds of lower valency) demonstrate that the mechanism of passivation in these systems is the same and proceeds with participation of HSO$_4^-$

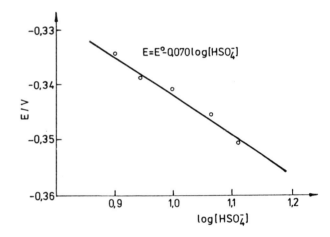

Fig. 4 Dependence of passivation potential of chromium on log [HSO$_4^-$] in H$_2$SO$_4$–H$_2$O solutions (range 'b').

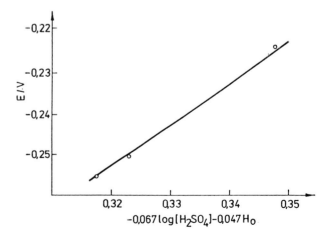

Fig. 5 Dependence of passivation potential of chromium on acidity H^0 and log [HSO$_4^-$] in H$_2$SO$_4$–H$_2$O solutions (rang 'c').

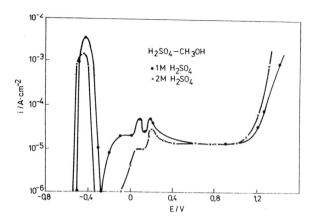

Fig. 6 *The anodic polarisation of chromium in CH$_3$OH–H$_2$SO$_4$ solutions.*

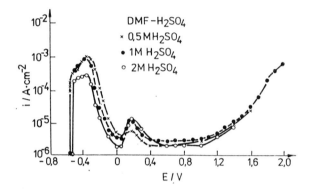

Fig. 7 *The anodic polarisation of chromium in N, N–dimethyloformamide–H$_2$SO$_4$ solutions.*

ions and H$_2$SO$_4$ molecules according to the reactions 2 and 3.

4. Conclusions

The XPS analysis of passive films on chromium demonstrates that anodic layer formed in anhydrous organic solutions and in aqueous concentrated solutions of sulphuric acid consists of Cr(III) oxy-hydroxide with a small amount of sulphur compounds of lower valency (S^0 and S^{-2}). In these solutions passivation of chromium proceeds via an anodic reaction between chromium surface atoms and HSO$_4^-$ or H$_2$SO$_4$ molecules. Hydrosulphate ions and sulphuric acid molecules show the oxidising ability and participate in the formation of oxide–hydroxide layer.

5. Acknowledgements

This work was supported by the Polish Committee for Scientific Research , Project No. 3 11989101.

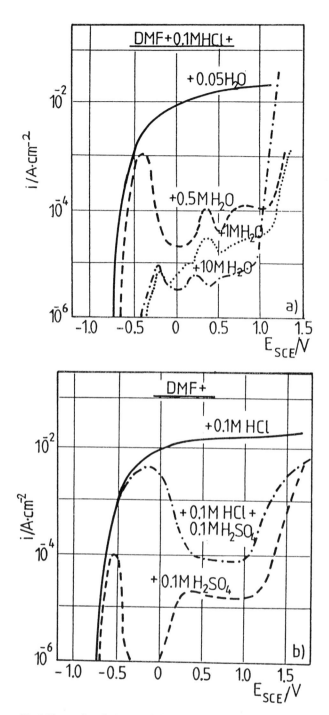

Fig. 8 *The anodic polarisation of chromium in N, N-dimethylformamide–0.1M HCl solutions: (a) the influence of water concentration; (b) the influence of H$_2$SO$_4$ concentration.*

References

1. R. D. Armstrong, M. Henderson and H. R. Thirsk, J. Electroanal. Chem., **35**, 119, 1972.
2. M. S. Basioury and S. Haruyama, Corros. Sci., **17**, 405, 1977.
3. M. Okuyama, M. Kawakami and K. Ito, Electrochim. Acta, **30**, 757, 1985.
4. H. Weidinger and E. Lange, Z. Electrochem., **64**, 468, 1960.

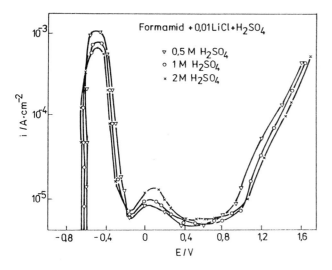

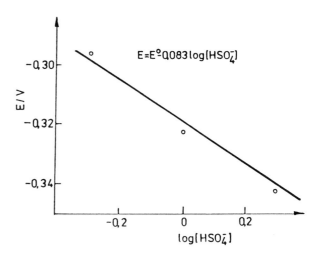

Fig. 9 The anodic polarisation of chromium in formamide–H₂SO₄ solutions.

Fig. 10 Dependence of passivation potential of chromium on log [HSO₄⁻] in formamide–H₂SO₄ solutions.

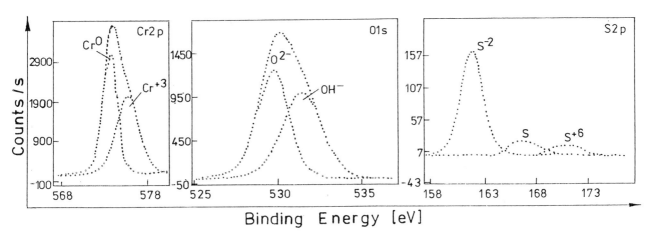

Fig. 11 XPS spectra of passive film formed on chromium in formamide–H₂SO₄ solution (E = 0.8 V, t = 1 h).

5. M. D. Armstrong and M. Henderson, J. Electrochem. Chem., **32**, 1, 1971.

6. Th. Heuman and H. S. Panesar, J. Electrochem Soc., **110**, 628, 1963.

7. K. E. Heusler, Proc. 4th Int. Symposium on Passivity, Princeton, NJ, 1977, (eds) R. F. Frankenthal and J. Kruger.

8. R. Beck, H. Schulz and B. Jansen, Electrochim. Acta, 31, 1131, 1986.

9. M. A. Genshaw and R. S. Sirobi, J. Electrochem. Soc., **118**, 1158, 1971.

10. M. Seo, R. Saito and N. Sato, J. Electrochem. Soc., **127**, 1909, 1980.

11. S. Haupt and H. N. Strehblow, J. Electrochem. Chem., **228**, 365, 1987.

12. J. H. Gerrestsen and J. H. W de Wit, Corros. Sci., **30**, 1057, 1990.

13. N. K. Chorievwa and A. M. Suchotin, Elektrochimia, **18**, 20, 1982.

14. A. M. Suchotin, M. N. Slepakow, P. Kosstikow and P. S. Stryknonow, Electrochimia, **18**, 285, 1982.

15. B. E. Wilde and F. G. Hodge, Electrochim. Acta, **14**, 619, 1969.

16. L. Bjornkvist and J. Olefjord, Corros. Sci., **32**, 231, 1991.

17. T. P. Moffat and R. M. Latanision, J. Electrochem. Soc., **139**, 1869, 1992.

18. G. T. Seaman, J. R. Myers and R. K. Saxer, Electrochim. Acta, **12**, 885, 1967.

19. N. D. Green, C. R. Bishop and M. Stern, J. Electrochem. Soc., **108**, 836, 1961.

20. G. Okamoto, Corros. Sci., **13**, 471, 1973.

21. B. Stypula, Metallury and Foundry Engineering, **16**, 119, 1990.

22. B. Stypula, Metallurgy and Foundry Engineering, **18**, 301, 1992.

23. B. Stypula, J. Banas, Electrochim. Acta, in press.

24. J. Banas, B. Mazurkiewixz and B. Stypula, Electrochim. Acta, **37**, 1069, 1992.

Influence of Electrochemical Formation Conditions on the Stability of Passive Layers Formed on Fe17%Cr Stainless Steel in Sulphuric Acid

F. LECRAS, G. BARRAL AND S. MAXIMOVITCH

Centre de Recherche en Electrochimie Minérale et Génie des Procédés - URA CNRS 1212, INPG-ENSEEG, 1130 rue de la Piscine, Domaine Universitaire, BP 75, 38402 Saint Martin d'Hères Cedex, France

Abstract

The stability of passive layers formed on stainless steel in sulphuric acid media, depends on the alloy composition and also on the layer forming conditions. Many results are available in literature but they often relate to different experimental conditions and are for this reason difficult to compare. Our aim was therefore to study on a given substrate and with the same test the influence of forming conditions on the stability of passive layers in H_2SO_4. Our results include an AES examination of the concentration profile of the elements.

1. Experimental

The substrate is an AISI 430 ferritic stainless steel, containing Ti additions (360ppm), drawn from an industrial cold-rolled sheet, whose chemical composition is given in Table 1. Measurements are performed on a rotating disk electrode (Ω = 2000 rpm) in a 0.1M H_2SO_4 aqueous solution (pH = 1.1) de-aerated by bubbling with Ar. The electrodes are 4mm dia. disks, cut in 1mm sheet, mounted with conductive resin on a conductive brass support and coated with a low-shrink resin.

1.1 Layer formation

Before the formation of layers, the electrodes are polished with abrasive paper 1200 and rinsed with distilled water. A, B, C layers are electrochemically formed in 0.1M H_2SO_4. The electrodes are immersed in the solution at the rest potential $E_{I=0}$. The different treatments are summarised in Fig. 1.

Layer A: oxidation to potential E_a ($E_a > E_{I=0}$) for time t_a with respect to $E_{I=0}$.
Layer B: initial polarisation at potential E_c for time t_c, then oxidation to potential E_a for time t_a.

Layer C: initial polarisation at potential E_c for time t_c, then oxidation by potential sweep of 1 mV/s from E_c to E_a.
Layer E: air formed layers. After polishing and rinsing, the layers are formed by exposure to air for time t_{air}.
Layer F: layers formed chemically in 65% HNO_3. After polishing and rinsing, the layers are chemically formed by immersion in 65% HNO_3 for time t_{im}.

1.2 Measurement of depassivation time

Figure 2(a) represents the potentiodynamic polarisation curve of Fe–17Cr in 0.1M H_2SO_4 at a sweep rate of 1 mV/s. The potentials are referred to the satured sulphate electrode (SSE = 665 mV/NHE). The active dissolution range between –950 and –800 mV shows a maximum I_p at –880 mV. The passive range extends from –800 to 500 mV. The transpassive region is ob-

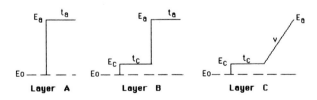

Fig. 1 Potential program for electrochemical formation of passive layers in 0.1M H_2SO_4.

Table 1 Composition (wt%)

Cr	Ni	Mn	Si	Ti	C	S	N
17.4	0.10	0.44	0.36	0.36	0.025	0.004	0.014

served beyond this potential. During the return decreasing sweep, the passive layer is not destroyed and we do not observe a reactivation peak for this sweep rate in the potential range where it was initially obtained.

The tests proposed in literature to measure passive layer stability [1] are based on the time needed for the destruction of the layer: measurement of the height of the reactivation peak during a slow decreasing sweep or measurement of the depassivation time under potentiostatic conditions in the vinicity of the current peak I_p. We used this second method which allows us to characterise the stability of layers formed under various conditions (chemical or in air) without introducing further modification arising from potential sweep during the test. The behaviour of a depassivating curve is plotted in Fig. 2(b): the initial dissolution current density is very low and in some cases, for short times, a negative current which may correspond to several reactions is observed (hydrogen evolution on

the passive layer, reduction of this layer or of other species. . .). The passive layer is gradually destroyed and the anodic current increases up to the steady state value corresponding to the active metal dissolution rate at this potential. The depassivation time τ is measured for the value $I_p/2$.

2. Electrochemical Result

We studied the influence of potential E_a and t_a for type A layers. For $t_a = 10\,min$, the depassivation time varies little with E_a (Fig. 3). Increasing time t_a at the potential $E_a = 400\,mV$, increases the stability of the layers but the depassivation time remains in the order of 10min even after 17h of polarisation (Fig. 4).

The influence of initial time t_c of polarisation at potential E_c treatment of the electrode is studied for type C layers (Fig. 5). The anodic polarisation E_a is fixed at 400 mV for all these experiments. When the duration of the initial polarisation at E_c increases, a

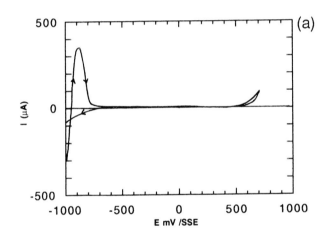

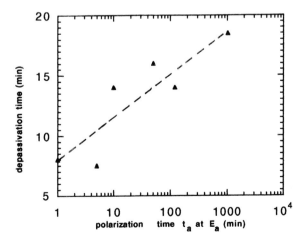

Fig. 3 *Depassivation time vs oxidation potential E_a for layers A.* $t_a = 10\,min.$

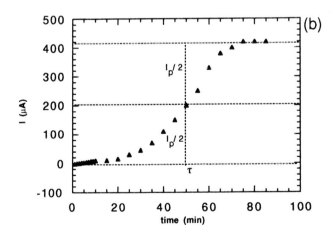

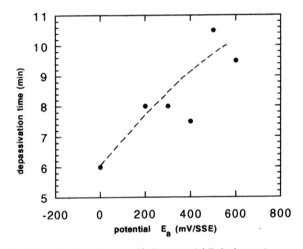

Fig. 2 *(a) Potentiodynamic polarisation curve for Fe17Cr in H_2SO_4 0.1M at $v = 1\,mV/s$; (b) Example of measurement of the depassivation time at $V = -880\,mV$.*

Fig. 4 *Depassivation time vs oxidation time t_a for layers A.* $E_a = 400\,mV.$

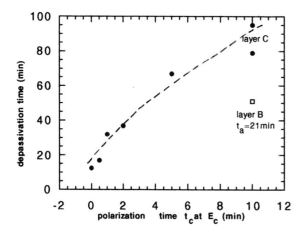

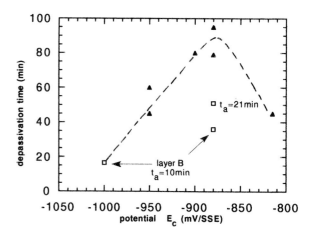

Fig. 5 *Depassivation time vs the time t_c of the initial polarisation at E_c for layers C and B.*
$E_c = -880$ mV, $E_a = 400$ mV; $t_a = 10$ min and 21min for layers B.

Fig. 6 *depassivation time vs the initial polarisation potential E_c for layer C and B.*
$tc = 10$ min; $E_a = 400$ mV; $t_a = 10$ min (layers B).

hard increase of τ with a factor *ca.* 10 is observed. It would seem that the surface of the metal would be modified during the active dissolution.

For the same initial activation treatment (–880 mV during 10 min), the oxidation mode is important too: comparing layers B and C for the same total oxidation time ($t_a = 21$ min corresponding to the time needed for the oxidation sweep), we observed that layers C are more stable than layers B.

We studied also the influence of the initial potential of polarisation E_c. Results are plotted in Fig. 6 for $t_c = 10$ min. The depassivation time has a maximum situated at –880 mV. On the other hand, a polarisation in the hydrogen evolution range (–1100 and –1000 mV) does not increase the stability of the layer.

In order to compare the influence of time for all the prepared layers, τ is plotted in Fig. 7 on a logarithmic scale (vs t_a for layers A, vs t_c for layers B and C, vs t_{air} for layers F and vs t_{im} for layers G).

Layers C obtained with the same oxidation time and different initial polarisation times t_c at –880 mV, have high τ values, which increase rapidly with time. For layers A the increase of τ with t_a is slower. The increase of τ with t_{im} is important for layers G. Layers F formed in air are less stable and their thickening or stabilisation is slow: after on month, their depassivation time is only 6 min. However, we observed a dispersion of results which seemed to depend on the quantity of water during the initial period of layer formation: after 17 h, layers formed in wet conditions are more stable than layers formed in dry conditions. This is in agreement with the stability of layers prepared in wet annealing atmosphere [1].

The results obtained for layers A, F and G, show that stability increases with the chemical oxidation

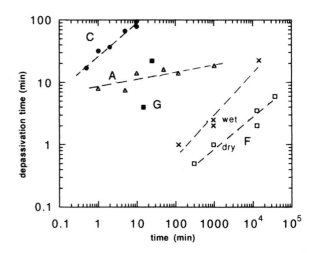

Fig. 7 *Depassivation time vs time for layers prepared in various conditions.*
Δ *layers A: t_a, $E_a = 400$ mV ;* ● *layers C: t_c, $E_c = -880$ mV, $E_a = 400$ mV;*
□ *layers F: t_{air} (dry);* ■ *layers G: t_{im};* × *layers F: t_{air} (wet).*

time (t_{air}, t_{im}) or electrochemical oxidation time (t_a). We can think that the stabilisation is related to 'ageing' phenomenoms like thickening, dehydration and segregation of some elements (Cr).

The slow variation of the depassivation time with t_a for layers A and also the difference of stability of layers B and C suggest that formation of the layer in severe electrochemical conditions, 'fix' the initial properties of the layer with few possibilities of stabilisation or rearrangement.

Concerning layers C, we first supposed that the layer reinforcement when polarising in the dissolution range was due to the destruction of the layer formed in the air after polishing. With this hypothesis, this layer should remain after direct oxidation and the electrochemical layer would be constructed on it, but would be rapidly destroyed when held at potential E_c. However, we observed that when E_c is situated in the hydrogen evolution range, the corrosion resistance of layers is not increased: the reinforcement seems more related to the presence of the dissolution current, since the phenomenon increases with the polarisation time and the optimum is observed at the current peak.

The effect of a polarisation at a potential where the alloy is dissolved, would be due to a variation of the superficial composition of the alloy [2] or the 'pre-layer' existing at the surface [3] by anodic segregation of chromium. However, these hypothesis seem contradictory with the steady state current of dissolution at E_c, except if the modification are a lowly variation of composition, or concern only a limited part of the area. In order to check this hypothesis, we measured the profiles of the different elements in the layer using Auger spectroscopy.

3. AES Analysis

The analysed samples were first subjected to the following treatments:

1. Type A layer with $E_a = 400$ mV and $t_a = 10$ min.

2. Type B layer with $E_c = -880$ mV $t_c = 10$ min and $E_a = 400$ mV $t_a = 10$ min.

3. Type C layer with $E_c = -880$ mV $t_c = 10$ min and $E_a = 400$ mV.

4. Type C layer with $E_c = -880$ mV $t_c = 0$ min and $E_a = 400$ mV.

The C, O, Cr and Fe profiles were measured during the etching of the surface with a rate of 5–10 Å/ min. The atomic percent of these elements are plotted in Fig. 8 for the four studied samples. Two types of profiles are observed owing they have been polarised 10 min at −880 mV (samples 2 and 3) or not (samples 1 and 4).

Samples 1 and 4 show the usual behaviour of passive layers [1]: the carbon contamination rapidly decreases. For samples 2 and 3, the high concentration of carbon on a large thickness seems to be related to the initial polarisation: all the layers are prepared in the same conditions with the same solution and the contamination should be the same. The high concentration of carbon on large thickness could be explained by the increase surface and rugosity after the initial treat-

ment. However, we did not observe an increase of dissolution current which was constant during the polarisation at Ec.

More probably, carbon seems to be fixed at the active metal surface during the prepolarisation and could have several origins: formation of reduction products from residual CO_2 or organic materials contained in the solution: CO_2 is highly soluble in acid medium and some small quantities could remain if bubbling of argon is not sufficient. Furthermore, CO_2 and many organic systems have thermodynamic potentials close to 0 (for example $Eo_{CO_2/CH_3OH} = 0.015$ V/NHE at pH = 1) and the reduction of CO_2 has been shown to occur on catalytic oxides [4]. In the potential range of active dissolution (−800 to −900 mV/ESS = −0.135 to −0.235 V/ENH), it is not excluded to reduce them on the active surface of metal or first layer of oxide.

The profile of iron shows a continual increase and the thickness of the layer is in agreement with that calculated from the oxygen profile. For chromium, the concentration increases rapidly up to a plateau: there is no maximum of concentration contrary to what is generally observed for air formed layers [1]. However, the Cr/Fe plot shows the relative superficial enrichment in chromium which seems higher and on a larger thickness for layers 2 and 3. This is true even if we calculate the thickness of the enriched layers from the oxygen peak. This important chromium concentration on a large thickness could explain the longer depassivation time measured for layers B and C.

4. Conclusions

The behaviour of passive layers, studied in a given medium, depends on the layer forming conditions. The first results on electrochemically formed layers A an C and air formed layers suggest that initial conditions are determinant for the further stability of the layers. A short prepolarisation in the active state modify the ratio Cr/Fe on a larger thickness and increases the stability of the layer. At this potential, it also seems possible to fix some species like carbon but additional studies will be necessary to determine its binding state by ESCA and its importance in the stability of the layer. The development of electrochemical treatments to modify passive layers seems then promising.

5. Acknowledgements

The authors gratefully acknowledge the contribution of C. Senillou for AES analyses and J. C. Joud and B. Baroux for helpful discussion.

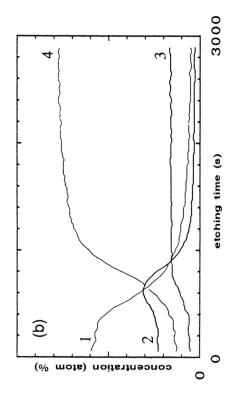

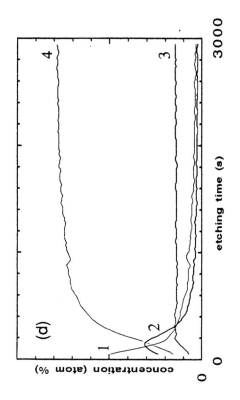

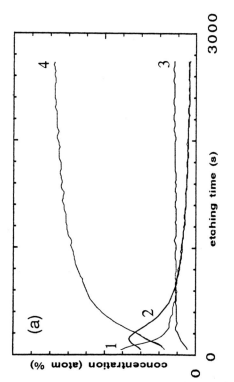

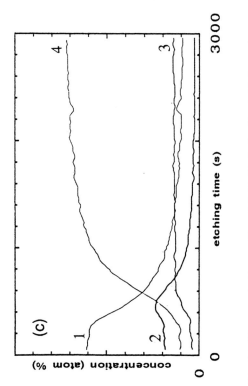

Fig. 8 AES concentration profiles of C, O, Cr, Fe for the studied samples
(a) Sample 1; (b) Sample 2; (c) Sample 3; (d) Sample 4.
1C-2 O-3Cr-4Fe.

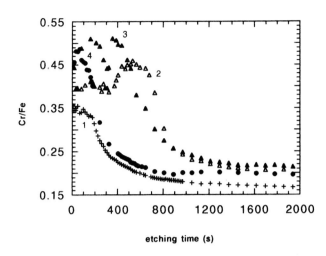

Fig. 9 Profile of Cr/Fe ratio for the studied samples.
1: sample 1; 2: sample 2; 3: sample 3; 4: sample 4.

References

1. D. Gorse, J. C. Joud and B. Baroux, Corros. Sci., **33**, 1455, 1992.
2. I. Olejord and C. R. Clayton, ISIJ International, **31**, 134, 1991.
3. D. D. MacDonald, J. Electrochem. Soc., **139**, 3434, 1992.
4. A. Bandi, J. Electrochem. Soc., **137**, 2157, 1990.

23

Identifiability and Distinguishability of Passivation Mechanisms by Electrochemical Impedance Spectroscopy and Electrogravimetric Transmittance

F. Berthier, J.-P. Diard* and B. Le Gorrec*

Laboratoire de Metallurgie Structurale, U.A. C.N.R.S. 1107, Universite de Paris XI, 91405 Orsay, France
*Ecole Nationale Supérieure d'Electrochimie et d'Electrométallurgie, Centre de Recherche en Electrochimie Minérale et en Génie des Procédés, U.A. C.N.R.S. 1212, Domaine Universitaire, B.P. 75, 38402 Saint Martin d'Hères, France

Abstract

A wide variety of mechanisms can be put forward to describe electrochemical dissolution and passivation mechanisms. The concepts of identifiability and distinguishability are used to compare dissolution and passivation mechanisms. The dissolution–passivation mechanisms of metal are seldom steady-state identifiable, however impedance spectroscopy provides the possibility of determining the values of parameters. The question arises, from a theoretical point of view, if ElectroGravimetry Transmittance (EGT), which should allow to determine atomic weight of adsorbate, leads by itself to a single set of parameter values for a given mechanism, or if it is necessary to couple up this technique with impedance spectroscopy. These concepts are applied to reaction mechanisms proposed for the interpretation of experimental data in the field of metal dissolution and passivation.

1. Introduction

A wide variety of reaction mechanisms, some of which are very complicated, have been proposed to describe metal passivation reactions. However when these mechanisms include a large number of steps and intermediate species, the significance of the measured values of the kinetic parameters, and thus the utility of the mechanism, may be questionable. In an article on the anodic dissolution of iron, Keddam et al. describe the procedure for modelling electrochemical reactions by impedance spectroscopy and conclude that: 'During the simulation calculation, it was found very difficult to establish other sets of kinetic parameters, and the rate constants change reasonably with the solution pH, but it is not possible to ascertain that the values given in the results were the only possible set' [1,2]. More recently, in a study on the dissolution and passivation of iron in neutral acetate buffer solutions, Takahashi et al. [3] encountered the same problem: 'Moreover, because of the large number of kinetic parameters, it is likely that an equally good fit to the data could have been obtained with a quite different set of values for the model parameters. Thus, no interpretation of the model parameters will be made.'

These questions partly explain the recent trend favouring the use of in situ methods for the physical of the species involved in metal passivation reactions rather than kinetic studies. However a priori methods for the investigation of models exist that can answer the questions posed in the literature [4, 5].

To establish a model, we first propose a mathematical structure containing certain parameters and then we determine their values by comparing theoretical expressions with experimental results. The values of the parameters cannot necessarily be determined. For noise-free ideal measurements and correct model, this depends uniquely on the mathematical structure of the model. An identifiability study tells us whether it is theoretically possible to determine the parameter values based on ideal synthetic data. If two different sets of parameters values can be determined for a given set of experimental results, these values are clearly meaningless. An identifiability study eliminates, for a given model, experimental methods that are not capable of measuring the parameter values. By studying the structural identifiability of reaction mechanisms before attempting to measure their kinetic parameters, it is possible to determine beforehand whether the measurements to be carried out will contain sufficient information to actually identify the mechanism. By assuming perfect experimental data, it is possible to determine whether a given mechanism can accurately reproduce experimental data for more than one parameter vector [6]. If only one solution exists, the model is said to be Structurally Globally Identifiable

(SGI); if a finite number of solutions can be determined, the model is said to be Structurally Locally Identifiable (SLI); if no finite number of solutions can be determined, the model is said to be Structurally Non Identifiable (SNI).

Different mechanisms can be compared objectively by studying their structural distinguishability. Two mechanisms are said to be Structurally Distinguishable (SD) if there are no parameters values that give the same output for a given input. Whatever the quality of measurements, two structurally non-distinguishable (ND) mechanisms can reproduce experimental results in a strictly identical manner.

These concepts are applied to reaction mechanisms proposed for the interpretation of experimental results obtained by steady-state investigations, Electrochemical Impedance Spectroscopy (EIS) and ElectroGravimetric Transmittance (EGT) [7–11] measurements in the field of metal dissolution and passivation.

2. Identifiability of Reaction Mechanism

Let us consider the mechanisms M1 proposed [12] to describe the dissolution and passivation of metals:

$$M, s \xrightarrow{\underline{K}_{o1}} M^{2+} + 2e^- + s \qquad \underline{K}_{o1} = k_{o1} \exp(2\alpha_{o1} fE); \ f = F/(RT)$$

$$M, s + A^{2-} \underset{K_{r2}}{\overset{K_{o2}}{\Leftrightarrow}} MA, s = 2e^- \qquad \begin{aligned}\underline{K}_{o2} &= k_{o2} A^{2-} \exp(2\alpha_{o2} fE) = k_{02} \exp(2\alpha_{o2} f\underline{E}) \\ \underline{K}_{r2} &= k_{r2} \exp(-2\alpha_{o2} f\underline{E});\end{aligned}$$

Assuming that the interfacial depletion in A^{2-} is negligible, the steady-state current density equation is given by:

$$i_f = 2F\Gamma \frac{\underline{K}_{o1}\underline{K}_{r2}}{\underline{K}_{o2} + \underline{K}_{r2}} = 2F\Gamma \frac{k_{o1} \exp(2\alpha_{o1} fE)}{1 + (k_{o2}/k_{r2})\exp(fE)}$$

Figure 1 shows a bell-shaped log i_f vs E curve, typical of metal dissolution and passivation phenomena, calculated for an arbitrary set of parameter values. The study of the steady-state current density equation shows that the only parameters which can be determined are α_{o1}, the product Γk_{o1} and the ratio of kinetic constants k_{o2}/k_{r2}, regardless of the experimental conditions such as the value of the pH or temperature. The mechanism is not steady state SGI.

Faradaic impedance and electrogravimetric transmittance equations are given by:

$$Z_f(s) = \frac{1}{2fi_f} \frac{\underline{K}_{o1} + \underline{K}_{o2} + s}{\alpha_{o1}\underline{K}_{r2} - \alpha_{r1}\underline{K}_{o2} + (\alpha_{o1} + \underline{K}_{o2}/\underline{K}_{o1})s}$$

$$\frac{\Delta m(s)}{\Delta E(s)} = \frac{fi_f}{F} \frac{A_e(\alpha_{r1}\underline{K}_{02} - \alpha_{o1}\underline{K}_{r2}) + (A_a\underline{K}_{o2}/\underline{K}_{o1} - A_e\alpha_{o1})s}{s(\underline{K}_{o2} + \underline{K}_{r2} + s)}$$

where $s = j\omega$

The set of parameters to be determined is defined according to $\sigma_{M1} = (\Gamma, k_{o1}, k_{o2}, k_{r2}, \alpha_{o1}, \alpha_{o2})$ when the species concentration A^{2-} and the temperature T are known.

The mechanism is said to be structurally globally identifiable by impedance spectroscopy if: $\forall E: Z_f(\sigma'_{M1}) = Z_f(\sigma_{M1})$ if and only if $\sigma'_{M1} = \sigma_{M1}$. Finding the set of parameters σ'_{M1} solving the equation $Z_f(\sigma'_{M1}) = Z_f(\sigma_{M1})$ for all potentials E is equivalent to solving the following system of equations:

$$i_f(\sigma'_{EIS}) = i_f(\sigma_{EIS})$$

$$\underline{K}'_{o2} + \underline{K}'_{r2} = \underline{K}_{o2} + \underline{K}_{r2}$$

$$\alpha'_{o1}\underline{K}'_{r2} - \alpha'_{r1}\underline{K}'_{o2} = \alpha_{o1}\underline{K}_{r2} - \alpha_{r1}\underline{K}_{o2}$$

$$\alpha'_{o1} + \underline{K}'_{o2}/\underline{K}'_{o1} = \alpha_{o1} + \underline{K}_{o2}/\underline{K}_{o1}; \forall \underline{E}$$

This system gives a unique solution such as (σ'_{EIS} = σ_{EIS} (Appendix 1). The mechanism M1 is SGI by EIS.

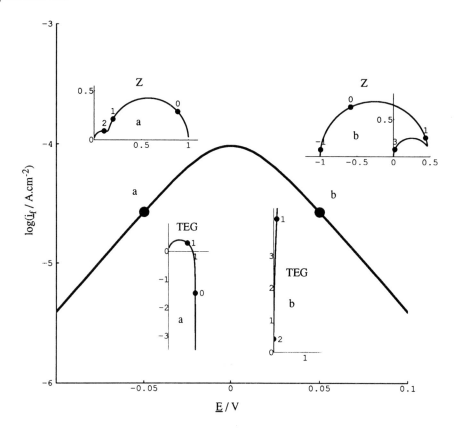

Fig. 1 *Steady-state log i_f vs potential curve, Nyquist plot for the impedance Z and the EGT.*
Z: - Im Z / / Rp/ = f(Re Z / / Rp/); EGT: – (1// Rm/) Im($\Delta m/\Delta E$) = f(1// Rm/) Re($\Delta m\Delta E$); k_{o1} = 1 s^{-1}; k_{o2} = 1 $mol^{-1}.cm^3. s^{-1}$; k_{r2} = 1 s^{-1}; r = 10^{-9} $mol.cm^{-2}$;
(α_{o1} = 0.5; α_{o2} = 0.3; A_e = 45 $g.mol^{-1}$; A_a = 100 $g.mol^{-1}$; C_{dc} = 10^{-5} $F.cm^{-2}$).

The mechanism is said to be SGI by EGT if and only if $\forall E$:$[\Delta m(s) / \Delta E(s)](\sigma'_{EIS}) = [\Delta m(s) / \Delta E(s)](\sigma_{EIS})$ if and only if $\sigma'_{M1} = \sigma_{M1}$ that is equivalent to solving the following system of equations:

$$i_f(\sigma'_{EGT}) = i_f(\sigma_{EGT})$$

$$A_e(-\alpha'_{o1}(\underline{K}'_{o2}+\underline{K}'_{r2})+\alpha'_{o2})=A_e(-\alpha_{o1}(\underline{K}_{o2}+\underline{K}_{r2})+\alpha_{02})$$

$$A'_a\underline{K}'_{o2}/\underline{K}'_{o1}-A_e\alpha'_{01}=A_a\underline{K}_{o2}/\underline{K}_{o1}-A_e\alpha_{o1}$$

$$\underline{K}'_{o2} + \underline{K}'_{r2} = \underline{K}_{o2} + \underline{K}_{r2}; \ \forall \underline{E}$$

This system gives a unique solution such as σ'_{EGT} = σ_{EGT}. The mechanism M1 is SGI by EIS. Figures 1 and 2 present the graphs of log i_f versus electrode potential for the following sets of parameters s = (Γ, k_{o1}, k_{o2}, k_{r2}, α_{o1}, α_{o2}, A_a) and σ' = (Γ, $\Gamma k_{o1}/k'_{o2}$, k_{r2}, k'_{o2}/ k_{o2}, α_{o1}, α_{o2}, A_a). The two curves are identical. The calculated EIS and EGT diagrams are plotted on these figures for two electrode potentials. The diagrams differ for the two sets of parameters, removing the ambiguity of steady-state kinetic parameter measurements.

3. Distinguishability of Reaction Mechanisms

Let us consider the mechanisms M2 and M3 proposed to describe the dissolution and passivation of metals [12]:

$$M, s \overset{K_{o1}}{\underset{K_{r1}}{\Leftrightarrow}} P, s + 2e \qquad \begin{aligned} \underline{K}_{r1} &= k_{r1} \exp(-2\alpha_{r1}f\underline{E}) \\ \underline{K}_{o1} &= k_{o1} \exp(2\alpha_{o1}f\underline{E}) \end{aligned}$$

$$M, s \overset{K_{o2}}{\rightarrow} Y, s + 2e \qquad \underline{K}_{o2} = k_{o2} \exp(2\alpha_{o2}f\underline{E})$$

$$Y, s \overset{K_{o3}}{\rightarrow} Y + 2e + s \qquad \underline{K}_{o3} = k_{o3} \exp(2\alpha_{o3}f\underline{E})$$

The mechanism M2 consists of a passivation step taking place in competition with a metal dissolution reaction comprising two first order steps, where Y designates the intermediate adsorbate of the metal dissolution reaction. The mechanism M3 which is written:

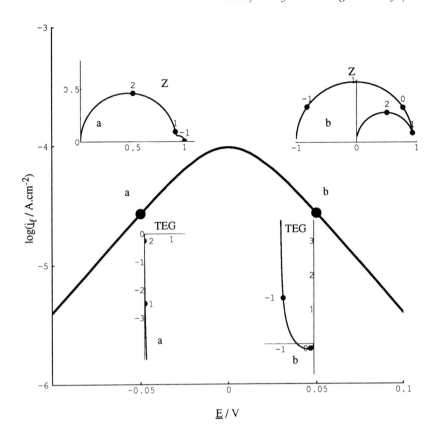

Fig. 2 Z: Steady-state log i_f vs potential curve, Nyquist plot for the impedance Z and the EGT.
Z:– ImZ// R_f = f(ReZ// Rp/); EGT:– (1// Rm/) Im(Δm/ΔE) = f(1// Rm/) Re(Δm/ΔE); k_{o1} = 10 s^{-1}; k_{o2} = 01 mol^{-1} $cm^3.s^{-1}$; k_{r2} = 0.1 s-1;
r = 10^{-10} $mol.cm^{-2}$; (α_{o1} = 0.5; α_{o2} = 0.3; A_e = 45 $g.mol^{-1}$; A_a = 100 $g.mol^{-1}$; C_{dc} = 10^{-5} $F.cm^{-2}$).

$$M, s \Leftrightarrow X, s + 2e \quad \underset{K_{r1}}{\overset{K_{o1}}{}}$$

$$\underline{K}_{o1} = k_{o1} \exp(2\alpha_{o1}f\underline{E})$$
$$\underline{K}_{r1} = k_{r1} \exp(-2\alpha_{r1}f\underline{E})$$

$$X, s \Leftrightarrow Q, s + 2e \quad \underset{K_{r2}}{\overset{K_{o2}}{}}$$

$$\underline{K}_{o2} = k_{o2} \exp(2\alpha_{o2}f\underline{E})$$
$$\underline{K}_{r2} = k_{r2} \exp(2\alpha_{r2}f\underline{E})$$

$$X, s + A \rightarrow X, s + B + 2e \quad \overset{K_{o3}}{} \quad \underline{K}_{o3} = k_{o3} \exp(2\alpha_{o3}f\underline{E})$$

is a mechanism involving an adsorbate X produced by the first step and which can be converted into a passivating agent Q in the second step or be used as a catalyst in the final step.

We have shown that these two mechanisms are steady-state SNI and SGI by IES [13,14]. Under steady-state conditions, the mechanism M3 is non-distinguishable from the mechanism M2, i.e. it is possible, given a set of parameter values for mechanism M2, to determine a set of parameter for mechanism M3 that produce the same steady-state curve. We have also shown that M2 and M3 mechanisms are distinguishable by EIS [13,14].

It can be shown that the mechanisms M2 and M3 are distinguishable by EGT and that they are Structurally Globally Identifiable since the EGT of mechanism M2 is expressed by:

$$\Delta m(s)/\Delta E(s) = K(a_0 + a_1 s + a_2 s^2) / [s (b_0 + b_1 s + s^2)]$$

where:
$a_o = A_e[\underline{K}_{o1} \underline{K}_{o3} (1 - \alpha_{o2}) - \underline{K}_{o3} \underline{K}_{r1} \alpha_{o2} - \underline{K}_{o2} \underline{K}_{r1} \alpha_{o3}]$
$a_1 = \underline{K}_{o1} A_p (1 + \alpha_{o3} + \underline{K}_{o3}/\underline{K}_{o2}) - A_e[(\underline{K}_{o1} + \underline{K}_{o2} + \underline{K}_{r1}) \alpha_{o3}$
$\quad + \underline{K}_{o3} \alpha_{o2}] + A_Y[K_{o1} (1 + \alpha_{o3} - \alpha_{o2}) + \underline{K}_{r1}(\alpha_{o2} - \alpha_{o3})]$
$a_2 = \underline{K}_{o1}/\underline{K}_{o2} A_p - A_e \alpha_{o3} + A_Y(\alpha_{o2} - \alpha_{o3})$
$b_o = \underline{K}_{o3} \underline{K}_{r1} + \underline{K}_{o2} \underline{K}_{r1} + \underline{K}_{o1} \underline{K}_{o3}$
$b_1 = \underline{K}_{o1} + \underline{K}_{o2} + \underline{K}_{o3} + \underline{K}_{r1}$

and that of mechanism M3 by:

$$\Delta m(s)/\Delta E(s) = K(n_0 + n_1 s) / (d_0 + d_1 s + s^2)$$

where:
$n_o = A_Q(\underline{K}_{o1} \underline{K}_{o2} + 2 \underline{K}_{o2} \underline{K}_{r1}) + A_X(- \underline{K}_{o1} \underline{K}_{o2} + \underline{K}_{r2} \underline{K}_{r1})$
$n_1 = \underline{K}_{o2} A_Q + A_X(- \underline{K}_{o2} + \underline{K}_{r1})$
$d_o = \underline{K}_{o1} \underline{K}_{r2} + \underline{K}_{r2} \underline{K}_{r1} + \underline{K}_{o1} \underline{K}_{o2}$
$d_1 = \underline{K}_{o1} + \underline{K}_{o2} + \underline{K}_{r1} + \underline{K}_{r2}$

Figures 1 and 2 provide graphs of log $\underline{i}_f$ vs electrode potential calculated using the theoretical expressions for each of the mechanisms. The two steady-state current density, potential curves are identical even though the EIS and EGT graphs are different.

4. Conclusion

The concepts of identifiability and distinguishability can be used to answer certain questions concerning the utility of kinetic models of electrode reactions. Dynamic methods (EIS and EGT) can render identifiable mechanisms that are not steady-state identifiable. Distinguishability is an objective criterion for comparing different mechanisms and their validation methods. It can be used to compare the potentiality of experimental methods used to test the validity of reaction mechanisms. Before proposing a reaction mechanism, whether based on electrical methods or any other experimental method, it is therefore essential to first study its structural properties.

5. Acknowledgements

Thanks are due to E. Walter (Laboratoire des Signaux et Systèmes, Ecole Supérieure d'Electricité) for fruitful discussion.

References

1. M. Keddam, O. R. Mattos and H. Takenouti, J. Electrochem. Soc., **128**, 257, 1981.
2. M. Keddam, O. R. Mattos and H. Takenouti, J. Electrochem. Soc., **128**, 266, 1981.
3. K. Takahashi, J. A. Bardwell, B. MacDougall and M. J. Graham, Electrochim. Acta, **37**, 477, 1992.
4. E. Walter, Identifiability of State Space Models, Springer-Verlag, Berlin, 1982.
5. E. Walter, Y. Decourtier, J. Happel and J-Y. Kao, AIChE J., **8**, 1360, 1986.
6. E. Walter, L. Pronzato, Identification de modèles paramétriques à partir de données expérimentales, Masson, Paris, 1993.
7. S. Bourkane, C. Gabrielli and M. Keddam, J. Electroanal. Chem., **26**, 471, 1988.
8. S. Bourkane, These, Paris 6, 1989.
9. S. Bourkane, C. Gabrielli and M. Keddam, Electrochim. Acta, **34**, 1081, 1989.
10. C. Gabrielli, M. Keddam and H. Takenouti, Proc. 4th E.I.S. Forum, C. Gabrielli ed., Montrouge, 1990, 31.
11. C. Gabrielli, M. Keddam and H. Takenouti, Corros. Sci., **31**, 129, 1990.
12. D. Schuhmann, J. Chim. Phys., **60**, 359, 1963.
13. F. Berthier, J.-P. Diard, P. Landaud and C. Montella, Proc. 5th E.I.S Forum, C. Gabrielli ed., Montrouge, 1991, 91.
14. F. Berthier, J.-P. Diard, P. Landaud and C. Montella, J. Electroanal. Chem., to be published.

Appendix 1

Analysis of the impedance spectroscopy identifiability of the mechanism M1

Look for the solutions of the following system of equations:

(a) $$(1+ \underline{K}'_{o2}/\underline{K}'_{r2})/[4Ff\Gamma'(\alpha'_{o1} \underline{K}'_{o1}+ \underline{K}_{o2})] = (1+ \underline{K}_{o2}/\underline{K}_{r2})/[4Ff\Gamma(\alpha_{o1} \underline{K}_{o1}+ \underline{K}_{o2})]$$

(b) $$1 + \underline{K}_{o2}'/\underline{K}_{o1}' + \underline{K}'_{r2}/\underline{K}'_{o1} = 1 + \underline{K}_{o2}/\underline{K}_{o1} + \underline{K}_{r2}/\underline{K}_{o1}$$

(c) $$\alpha_{o1} \underline{K}_{r2}'/\underline{K}_{o2}'-\alpha'_{rl} = \alpha_{o1} \underline{K}_{r2}/\underline{K}_{o2}-\alpha_{rl}$$

(d) $$\alpha_{o1}'/K_{o2}' + 1/\underline{K}'_{o1} = \alpha_{o1}/\underline{K}_{o2} + 1/\underline{K}_{o1} \qquad \forall \underline{E}$$

The first equation (a) is equivalent to two equations:

$$\forall \underline{E}$$

$$\underline{K}'_{o2}/\underline{K}'_{r2} = \underline{K}_{o2}/\underline{K}_{r2}$$

$$\Gamma'(\alpha'_{o1} \underline{K}'_{o1} + \underline{K}'_{o2}) = \Gamma(\alpha_{o1} \underline{K}_{o1} + \underline{K}_{o2})$$

It is thus possible to write:

$$\forall \underline{E}$$

$$\underline{K}'_{o2} = \underline{K}_{o2} \underline{K}'_{r2}/\underline{K}_{r2}$$

and (b):

$$K'_{o1} = \underline{K}_{o1} \underline{K}_{r2}/\underline{K}_{r2}$$

Equation (c) has a single solution:

$$\alpha'_{o1} = \alpha_{o1} \text{ and } \alpha'_{rl} = \alpha_{rl}$$

Equation (d) becomes:

$$\forall \underline{E}$$

$$(\alpha_{o1}/\underline{K}_{o2})\underline{K}_{r2}/\underline{K}'_{r2} + 1/\underline{K}_{o1} = \alpha_{o1}/\underline{K}_{o2} + 1/\underline{K}_{o1}$$

which is equivalent to:

$$\forall \underline{E}$$

$$\underline{K}'_{r2} = \underline{K}_{r2}$$

Thus $\sigma' = \sigma$. The mechanism is structurally globally identifiable.

24

Electrochemical Dissolution of $Fe_{60}Co_{20}Si_{10}B_{10}$ Amorphous Alloy and h.e.r Investigation on the Anodically Activated Alloy

M. Eyraud, F. Bellucci* and J. Crousier

Lab. de Physico-Chimie des Matériaux, Université de Provence, 13331 Marseille cedex 3, France
*Department of Materials and Production Engineering, University of Naples, Piazzale Tecchio 80125, Naples, Italy

Abstract

The electrochemical and electrocatalytic properties of $Fe_{60}Co_{20}Si_{10}B_{10}$ (Virtrovac 7600, G14) in deaerated 1M KOH solution at 25°C was investigated. Cyclic voltammograms for polycrystalline cobalt, iron and G14 alloy suggest that the first step in the anodic dissolution of the alloy is the dissolution of cobalt by a dissolution precipitation mechanism. Then the iron dissolution takes place. Poor electrocatalytic activity was exhibited by the amorphous alloy in the as-quenched state, while enhanced electrocatalytic activity was observed after anodic oxidation *in situ* at constant current density. The catalytic activity generally increased by increasing the oxidation current. This result has been attributed to an increase of the active surface of the catalyst after the anodic treatment rather than to an enhancement of the electronic properties of the surface.

1. Introduction

One of the most investigated electrochemical reactions is the hydrogen evolution reaction (h.e.r.) from both acid and alkaline solutions. Several methods have appeared in the literature dealing with suitable cathodes for this reaction as pointed out in a recent review by Trasatti [1] In spite of the effort put in this field, there is a need for new materials and/or for new physic and/or mechanical modifications of cathodes to enhance their catalytic activity. In the last decade amorphous alloys gained interest mainly for their corrosion resistance properties. In the past few years, however, attention has been devoted to these materials as cathodes for the h.e.r. [2–5]. In addition, it has been shown that these alloys need to be activated by using HF or via an electrochemical treatment (anodic oxidation) to exhibit good catalytic properties. After acid or electrochemical activation, however, an increase of the exchange current density, i_o, was observed, while the Tafel slope remains constant and equal to that of the untreated alloys. These observations led to the conclusion that the activation process affects the surface of the catalyst rather than the electronic state of the surface. Excellent cathodic properties were observed using the $Fe_{60}Co_{20}Si_{10}B_{10}$ amorphous alloy after an anodic treatment at 70°C in 30 wt % KOH [4]. Good catalytic properties have been claimed using this alloy in the as-quenched state [3, 6].

The aim of this paper is to further support the view that *in situ* anodic treatments enhance the catalytic activity of the G14 amorphous alloy. In addition, an attempt will be given to relate the catalytic activity of the anodically activated alloys to the electrochemical properties of the amorphous alloys in the same environment as determined by a cyclic voltammetric study of the polycrystalline iron, cobalt, and the G14 alloy.

2. Experimental

A classic three-electrode electrochemical cell was used. The counter electrode was a platinum sheet, placed in a separated compartment, and all the potentials were recorded with respect to a saturated sulphate electrode (SSE). The equipment consisted of a computer controlled (IBM-AT) PAR model 273 potentiostat, capable of both controlling the experiments, collecting and plotting the data. The amorphous alloy investigated in this paper, $Fe_{60}Co_{20}Si_{10}B_{10}$ (Vitrovac, G14) was kindly furnished by Vakuumschmelze GmbH, Hanau (D). The amorphous ribbons were used as-quenched, i.e. without any mechanical polishing. Before each experiment the electrodes were degreased with alcohol and rinsed several times with bi-distilled water, and finally cleaned in an ultrasonic bath. The bright face of the alloy was used as electrode, the dull face being masked by a high water resistant lacquer. The electrodes were immersed in the test solution without drying. The electrolyte solution was 1M KOH solution, deaerated by bubbling pure Argon gas for 2 h before the experiment. Cathodic polarisation curves on the as-quenched amorphous alloy were performed after the samples were polarised at –2000 mV for minutes to reduce any air-formed oxide. The same procedure was adopted for the alloy after anodic galvanostatic experiments under different anodic charging currents, i_{ac}, for two minutes and before cyclic polarisation.

3. Results and Discussion

3.1 Cyclic voltammetry

Several studies have been devoted to the anodic dissolution of iron in alkaline solutions [7–13]. Most of the studies used cyclic voltammetry and the voltammograms obtained in every papers were similar. After experiments by Raman laser is has been shown that at the end of the reduction sweep, Fe^{2+} species existed on the iron electrode as $Fe(OH)_2$ and therefore that the reduction did not reach the metallic iron state, the essential constituent of the passivating film being Fe_3O_4 [12]. Cobalt has been less extensively studied [14,15], due to its very complex anodic behaviour. Even if some voltammograms for FeCoBSi amorphous alloy in KOH solution are available [3, 4, 6l, it was necessary to investigate thoroughly its electrochemical behaviour by comparing with the voltammograms from polycrystalline iron and cobalt, obtained in the same experimental conditions, in the aim to determine a dissolution mechanism.

Figure 1 shows the stabilised voltammograms for both crystalline iron and cobalt. For cobalt, the charge under the cathodic peak is not in keeping with the charge under the anodic peak which indicates poor reduction of the species formed by anodic polarisation. The voltammogram for iron shows a main anodic peak, follows by a current plateau, and two well-formed cathodic peaks.

A series of triangular potential sweeps were performed on the G14 alloy. Figure 2 regroups the 1st, 5th and 10th cycles. The 1st cycle presents four anodic peaks referred to as A_1, A_2, A_3 and A_4, this last peak being just before the increase in current due to oxygen evolution. The cathodic scan presents a lot of small peaks among them only three are noted, C_1, C_2 and C_3 because they appear in the further cycles. For the 5th cycle, in the anodic scan, the peak A1 is replaced by a current plateau, the peak A_2 is higher and a new peak B_2 arises; the cathodic scan presents two well-formed peaks C2 an C_3. The 10th cycle shows almost the same trace, but the peak A_2 was smaller and the peak B_2 much higher. In the 25th cycle, the peak A_2 no longer exists. All the peaks, anodic or cathodic, increase by increasing the number of cycles as seen in Fig. 3 where

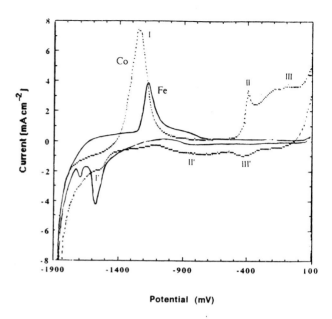

Fig. 1 Cyclic voltammograms for polycrystalline Fe and Co. Scan rate 20 mVs^{-1}.

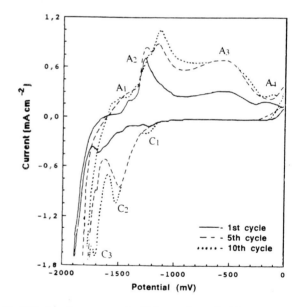

Fig. 2 Cyclic voltammogram for G14 amorphous alloy. Scan rate 20 mVs^{-1}.

10th and 25th cycles were presented. The amorphous sample changes colours during the sweeps. After the 5th cycle no change was observed. After the 10th cycle it became gilded, and after the 25th cycle it was dark brown. On further cyclic polarisation, the voltammograms are identical with the 25th cycle. To relate anodic and cathodic peaks, a series of voltammograms were performed by changing the anodic reversal potential. The voltammograms are

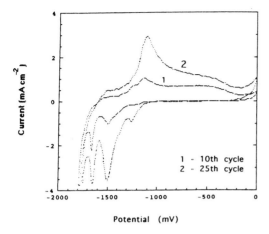

Fig. 3 Cyclic voltammogram for G14 amorphous alloy. Scan rate 20 mVs^{-1}.

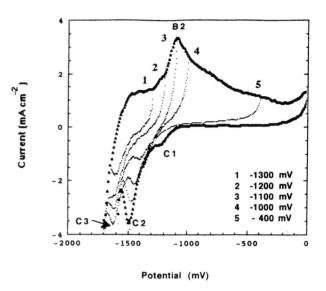

Fig. 4 Reversing potentials on stabilised voltammogram for G14.

presented in Fig. 4. The cathodic peak C$_1$ appears only when the anodic reversing potential, was higher than –400 mV. The peak C$_2$ is related to peak B$_2$ because the higher the anodic reversing potential the higher the peak C$_2$. The peak C$_3$ is the reduction peak of species formed during the plateau A$_1$. The most interesting feature is the excellent reproducibility of the going and returning part corresponding to hydrogen evolution, as soon as the voltammogram is stabilised, that means that the electrode surface is stabilised, and there is **no** influence of the anodic reversing potential on the hydrogen evolution overpotential.

By comparing the cyclic voltammograms for G14 with the ones for Fe and Co, some remarks can be presented:

- In the 1st cycle, peak A$_2$ appears as the main peak. The voltammogram is similar with the one obtained for Co, i.e. main anodic peak at the same potential and small cathodic peak.
- In the 10th cycle, peak A$_2$ is a shoulder and the main peak is peak B$_2$ and the voltammogram is very similar to the one obtained for Fe.

These two remarks make it possible to put forward a dissolution mechanism of the amorphous alloy: the first cycle is Co dissolution and the 5th cycle shows equal dissolution of Fe and Co. From the 10th cycle only Fe dissolution occurs.

3.2 Anodic oxidation
The anodic dissolution of the G14 alloy was carried out at different current densities for 2 min and are characterised by an E–t curve with different plateaus depending on the values of i$_{ac}$. Figure 5 shows some results for i$_{ac}$ values equal to 0.3, 2 and 3 mAcm^{-2}. Only a clear plateau at E = –2000 mV was observed for the i$_{ac}$

values investigated in this paper. For i$_{ac}$ equal to 3 mAcm^{-2} a sharp variation of the curvature in the E-t plot can be seen for E = –300 mV. These values of the potential (–1200 and –300 mV) are very close to the potentials at which passivity occurs as will be shown by the cyclic voltammetric study.

3.3 Cathodic polarisation
Cathodic polarisation curves were obtained under potentiostatic conditions at low scan rate on the as-quenched and anodically oxidised amorphous alloys. The curves follow a typical Tafel behaviour from which the Tafel slope, b, and the exchange current density, i$_o$,

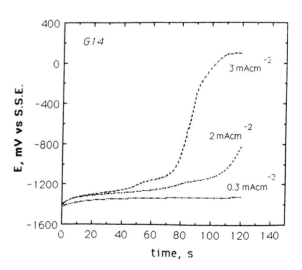

Fig. 5 Potential–time curve for G14 after different anodic treatments.

were evaluated. The latter values are reported in Fig. 6 as a function of i_{ac} together with the exchange current density of polycrystalline Pt, Fe, Ni, and Co in 1 M KOH at 25 °C as taken from the literature [3,5]. The values of i_o for the amorphous alloy in the as-quenched state is also reported. The values of i_o and of b (ranging between –110 and –135 mV) for the G14 alloy are in good agreement with those reported in the literature [3]. As can be seen from Fig. 6, the G14 alloy exhibits an increase of i_o by increasing the anodic oxidation current followed by a decrease of i_o for high values of i_{ac}. The best performance with this alloy was thus achieved after an anodic treatment with an oxidation current density of 0.3 and 2 mA cm^{-2}, respectively. A similar behaviour has been reported by Brossard *et al.* [4] for the G14 alloy in 30 % by weight of KOH at 70 °C. The maximum value of i_o was obtained after an anodic oxidation treatment 1 mA cm^{-2} [4].

Results obtained with the G14 alloy at high temperature[4], suggest that the high catalytic activity exhibited by this alloy after anodic oxidation, is due to the formation of crystalline Fe_3O_4 on the surface whose cathodic reduction leaves an active porous Fe layer on which the h.e.r. occurs. No change in the Tafel slope was reported suggesting that the enhanced catalytic activity exhibited by this alloy after anodic treatment, is due to a surface rather than to an electronic effect. A similar conclusion can be drawn also from the data of Juttner *et al.* [3], even if the low value the Tafel slope (95 mV) exhibited by this alloy in the as-quenched state at 25 °C (*ca.* –96 mV), would suggest a combined catalytic and surface effect.

Anodic treatment ($\mu A\ cm^{-2}$)	G14	
	$-\log i_o$	$-b$
as-quenched	5.86	97
10	5.70	133
50	5.80	118
100	5.32	112
300	5.20	120
500	4.55	135
700		
1000		
2000	4.56	137
3000	4.96	111
10^4		
10^5		
Polycrystalline materials	$-\log i_o$	$-b$
Platinum	4.62	120
Iron	5.00	135
Nickel	8.29	62
Cobalt	8.80	65

Fig. 6 Log of the exchange current densities i_o in $\mu A\ cm^{-2}$ and Tafel slope b in mV, for h.e.r G14, after anodic treatment for 2 min.

Results obtained in this paper (see Fig. 6) show that even at 25 °C the *in situ* anodic oxidation leads to an increase in the activity of the amorphous alloy as compared to that in the as-quenched state. The effect of the anodic charging current on the catalytic activity depends on the value of i_{ac}. For i_{ac} equal to 300 μA cm^{-2} the catalytic activity of the G14 alloy is comparable to that of crystalline Pt. This 'critical charging current density' is very close to the critical current for passivation of this alloy as can be seen from the cyclic voltammogram. It must be pointed out that changes of the amorphous sample colour during the anodic treatment were clearly observed only after the anodic treatment at the 'critical charging current density' at which the samples was covered by dark brown oxide. This oxide was not fully reduced in the subsequent cathodic pretreatment (–2000 mV for 10 min) before the cathodic polarisation curve was carried out. Catalytic activity of G14 as observed in this paper, can thus be attributed to the surface extent and to the electronic properties of the oxide-hydroxide layer rather than to the properties of an iron active surface. In fact the values of the Tafel slope are in the range 110–135 mV that is very close to the value of 120 mV observed for pure Fe and mild steel [1]. It can therefore be concluded that the enhanced catalytic activity exhibited by the G14 after anodic treatment, is due to a surface rather than an electronic effect. The decrease of the catalytic activity observed with the G14 after an anodic treatment with 3 mA cm^{-2}, is similar to that reported in the literature [4]. This result can be explained on the basis of following consideration. After an anodic treatment with 3 mA cm^{-2} final potential equal to + 98 mV was reached. As can be seen from the cyclic voltammogram (Fig. 2) at high values of the anodic potential a restructuration of the passive layer occurs leading to the formation of a compact rather than porous layer. This result leads, in turn, to a reduction of the catalytic surface, and thus of the value of i_o.

Characterisations of the alloy surface after electrochemical activation are in progress. The first results obtained by Auger analysis show the presence in the first layers of patches of iron or Si. Iron and Si were not detected together, cobalt was never detected. This analysis is in keeping with the proposed mechanism of dissolution.

4. Conclusion

Results obtained in this paper suggest that the first step in the anodic dissolution of the G14 amorphous alloy is the dissolution of cobalt, then the iron dissolution takes place. The catalytic activity versus the h.e.r in alkaline medium can be enhanced after an anodic treatment. The increased catalytic activity has been attributed to the formation of a porous layer on the amorphous surface, rather than to an enhancement of the electronic properties of the amorphous alloy.

References

1. S. Trasatti, in Advances in Electrochemical Science and Engineering, Vol. 2, Ed. by H. E Gerisher and C. W. Tobias, VCH Publishers New York, 1–86, 1992.
2. M. Enyo, T. Yamazaki, K. Kai and K. Suzuki, Electrochim. Acta, **28**, 1573, 1983.
3. Hailemichael Alemu and K. Jûttner, Electrochim. Acta, **33**, 1101, 1988.
4. J. Y. Huot, M. L. Trudeau, L. Brossard and R. Schulz, J. Electrochem. Soc., **138**, 1316, 1991.
5. K. Lian, D. W. Kirk and S. J. Thorpe, Electrochim. Acta, **36**, 537, 1991.
6. G. Kreysa and B. Hakansson, J. Electroanal. Chem., **201**, 61, 1986.
7. L. Ojefors, J. Electrochem. Soc., **123**, 1691, 1976.
8. R. S. S. Guzman, J. R. Vilche and A. J. Arvia, Electrochim. Acta, **24**, 395, 1979.
9. D. D. Macdonald and B. Roberts, Electrochim. Acta, **23**, 557, 1978.
10. D. D. Macdonald and B. Roberts, Electrochim. Acta, **23**, 781, 1978.
11. Z. Szklarska-Smialowska, T. Zakroczymski and C. J. Fan, J. Electrochem. Soc., **132**, 2543, 1985.
12. A. Hugot-Le Goff, N. Boucherit, S. Joiret and J. Wilinski, J. Electrochem. Soc., **137**, 2684, 1990.
13. D. Geana, A. A. El Miligy and W. J. Lorenz, J. Appl. Electrochem., **4**, 337, 1974.
14. H. Gomez Meier, J. R. Vilche and A. J. Arvia, J. Electroanal. Chem., **134**, 251, 1982.
15. R. D. Cowling and A. C. Riddiford, Electrochim. Acta, **14**, 981, 1969.

25

Growth/dissolution Processes of Anodic Films Formed on Titanium under α-irradiation, Studied by Coupling Electrochemistry and Inductively Coupled Plasma Atomic Emission Spectrometry

T. SAKOUT, J. C. ROUCHAUD, M. FEDOROFF AND D. GORSE

CECM-CNRS, 15 rue Georges Urbain, 94407, Vitry/Seine, France

Abstract

This work is concerned with the influence of α-radiation on the growth and dissolution processes of anodic oxide films formed on titanium in 1M H_2SO_4. The experimental procedure consists in bombarding a thin titanium foil with alpha particles emitted by an [241]Am source. The surface not in contact with the source is exposed to the corrosive solution: in these conditions, the anodic oxide film forms at the electrode surface under a constant flux of α-particles. Due to the energy loss in the foil, the energy of the more energetic α-particles attaining the anodic film varies between 1.3 and 2 MeV, depending on the foil thickness. By comparison with the film formation process on a 'bulk' titanium electrode without irradiation, either in the presence of the pre-existing native oxide film, or under conditions where it has been removed by polishing, the following results have been obtained by coupling electrochemistry (I(V) plots at 1 m Vs^{-1} scan rate) and Inductively Coupled Plasma/Atomic Emission Spectroscopy:

(i) For $E_\alpha \sim 1.6$ MeV, α-radiation favours both the electrochemical and chemical dissolution of the electrode during the anodic scan, the chemical dissolution being in some cases more than one order of magnitude larger than observed without irradiation on a surface free from its native oxide.

(ii) In contrast, with increasing the maximum incident energy at the metal/film interface up to $E_\alpha \sim 2$ MeV, the titanium electrode recovers progressively its behaviour without irradiation. The titanium concentration in solution approaches the level measured at a not irradiated titanium surface free from its native oxide.

1. Introduction

Not much is still known on what concerns the problem of radiation effects on (uniform or localised) corrosion from a basic point of view. The different factors (direct: nature and energy of the projectile, dose rate and dose..., or indirect: radiolysis products...) possibly influencing the aqueous corrosion still need to be identified properly. Moreover, the specific influence of radiation in this field has been a matter of debate for a long time. One explanation for this state of the art is that the phenomenon of irradiation-assisted corrosion may result from the interaction of cumulative and dynamic radiation effects on both the material substratum (metal or metallic alloy) and the corrosive aqueous solution in contact [1].

In order to go further in the understanding of radiation effects on aqueous corrosion, simplifications regarding the irradiation and electrochemical tests conditions and the corroding material are required. It is the purpose of the present work to determine the influence of irradiation by alpha-particles on the aqueous corrosion of a pure titanium electrode.

In a previous publication, α-radiation effects on passive films formed on a 304 austenitic stainless steel were studied [2]. Then titanium was taken in place of an industrial alloy, since the electrochemical growth and dissolution processes of the anodic oxide films on Ti, as well as the chemical and structural properties of the film have been widely studied in the past fifteen years [3].

As for the radiation source, the alpha-particles emitters, here [241]Am, are the only one easy to manipulate safely in the laboratory, allowing to investigate a wide energy spectrum, since the α-particles are emitted at 5.49 MeV, before being degraded by both the thin protective coating on the source and the titanium foil.

It was noted previously in a classical electrochemical study (using cyclic voltammetry), that irradiation by α-particles is able either to reactivate or to repassivate a titanium electrode in 1M H_2SO_4 at room temperature [4], depending on the energy of the a-particle reaching the titanium/film interface after travelling through Ti, in otherwise identical irradiation conditions (dose and dose rate), that is to say in an electrochemical environment for which such behaviour was never observed without irradiation.

On the other hand, reduction of the anodic film as a result of both chemical and electrochemical dissolution, at open circuit and under potential control, has been observed by some authors [5]. Likewise, the change in film growth and Ti dissolution efficiency with applied voltage and current density has been studied for titanium in 1M sulphuric acid, by coupling electrochemistry and atomic absorption spectrophotometry [6].

We pursue now our investigation of the phenomenon of radiation-enhanced corrosion on the same system as previously investigated (α-radiation, Ti foil, 1M H_2SO_4, room T) by coupling electrochemical I(V) tests with a quantitative determination of the titanium content in the solution obtained by Inductively coupled plasma atomic emission spectrometry (ICP/AES), in order to correlate the total dissolution rate to the electrochemical measurements. This method was chosen because it allows direct determination of titanium in the solution. The expected detection limit for this element could reach 1 $\mu g.L^{-1}$, while flame atomic absorption spectroscopy leads to a limit at least ten times higher [7].

2. Experimental

The experimental (electrochemical and irradiation) conditions have been fully described previously [1, 2]. The working electrode, schematised in Fig. 1, consists of a thin titanium foil of thickness $\leqslant 15\mu m$, contacting an Americium source (^{241}Am). The ensemble is sealed in a Pyrex tube, leaving a Ti surface of area S ~ 1.5 cm² (the exact S values are noted in Table 1) exposed to the 1M H_2SO_4 solution (V 200 cm³) at room temperature. In these conditions, the Ti electrode is bombarded by the back-face by the α particles and the source is isolated from the electrolyte. Due to energy loss in the metal roil, the energy of the α particles (E_α) reaching the metal/film interface (m/f) depends on the foil thickness. The energy of the α-particle, equal to 5.49 MeV when emitted in the ^{241}Am source, is degraded to values ranging from 1.3 to 2 MeV at the titanium/film interface, after passing through both the 2 μm thick protective coating (on the source) and titanium thicknesses going from 13.5 to 11 μm respectively. Only an estimated energy E_α of α-particles penetrating the solution can be calculated, since the energy loss in the native oxide is rigorously unknown (and is a part of the

problem under study). Under the present irradiation conditions, in all cases, the solution is irradiated through a few μm in depth, the maximum in electronic stopping power (dE/dx) is located 'far' from the f/s interface [8]. A platinum electrode and a saturated calomel electrode are used as counter and reference electrode respectively.

For the thin Ti foils, only gentle polishing with 1 μm alumina paste is made in order to avoid hole formation. A dummy source and a 15 μm Ti foil are used for testing the electrochemical behaviour of the system without irradiation. On the contrary, 'bulk' Ti electrodes are polished to mirror.

The experimental procedure is the following: the solution is degassed by N_2 bubbling for 2 h prior to introduction of the specimen in the main compartment of the electrolytic cell. Then the open circuit potential (OCP) is recorded before beginning the potential scan at a rate of 1 mV s^{-1}, from OCP up to 4 V/SCE. In this paper, we are concerned only by the anodic range below oxygen evolution, in order to avoid changes in the growth regime occurring as oxygen evolves.

Fractions of 10 mL of the solution were taken at several intervals during the potential scans for Ti determination. They were submitted to analysis by ICP/AES. We used an Atomscan 25 sequential spectrometer from Thermo-Jarrell-Ash, equipped with a 2 kW generator, working at 27.12 MHz, and a three quarter meter crossed Czerny-Turner monochromator. After optimisation in order to have the highest signal to background ratio, we used the following parameters: Power: 1150 W, photo-multiplier voltage: 1000 V, integration time: 10 s, observation height: 19 mm, slit height: 3 mm, nebulizer pressure: 40 psi, peristaltic pump rate: 1.5 mL.mn^{-1}, wavelengths: 323.452 and 336.121 nm. Solutions of known concentrations of titanium in 1M H_2SO_4, as also blanks with the same acid concentration, were used for standardisation.

The lowest detection limit reached in these experiments is 2 $\mu g.L^{-1}$. This value varies from one set of

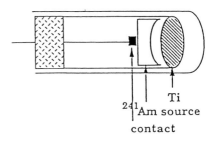

Fig. 1 Schematic drawing of the working electrode showing the ensemble Americium source (^{241}Am), titanium foil and the electrical contact.

Table 1 Titanium concentration in 1M H_2SO_4 electrolyte (el) in mg.L^{-1} measured by ICP/AES for different formation potentials during anodisation of titanium at 1mV.s^{-1} scan rate at room T: for unirradiated bulk Ti, a thin Ti foil, and different α-irradiated titanium thicknesses (the concentrations are corrected for the quantities removed by sampling)

potential V / SCE	Ti(el) (mgL^{-1}) Ti bulk without irradiation S =1.2cm^2 V =210cm^3	Ti(el) (mgL^{-1}) d_{Ti}=15μm unirradiated S =1.56cm^2 V =210cm^3	Ti(el) (mgL^{-1}) E_α = 2 MeV irradiated S =1.56cm^2 V =210cm^3	Ti(el) (mgL^{-1}) E_α=1.6 MeV irradiated S =1.95cm^2 V =170cm^3	Ti(el) (mgL^{-1}) E_α = 1.3 MeV irradiated S =1.43cm^2 V =170cm^3
-0.62	0.007 ± 0.002				
-0.55	0.0086 ± 0.0016				
-0.46	0.011 ± 0.002				
-0.2	0.013 ± 0.003				
0.0		0.0071 ± 0.0005	0.006 ± 0.002		
0.3		0.01 ± 0.0005	0.013 ± 0.002		
0.4	0.021 ± 0.001		0.015 ± 0.003		
0.46		0.0094 ± 0.0017			
0.6			0.0177 ± 0.0008		0.126 ± 0.002
0.8	0.020 ± 0.002	0.0132 ± 0.0005		0.175 ± 0.003	
0.9			0.016 ± 0.003		
1.0	0.027 ± 0.001	0.015 ± 0.003			0.064 ± 0.002
1.2			0.0187 ± 0.0006	0.121 ± 0.004	0.069 ± 0.002
1.6	0.031 ± 0.001			0.125 ± 0.004	
1.7		0.0215 ± 0.0008	0.034 ± 0.003		0.075 ± 0.002
1.8	0.037 ± 0.001			0.145 ± 0.004	
2.0	0.042 ± 0.003	0.023 ± 0.002	0.0333 ± 0.0015		0.081 ± 0.002
2.5				0.178±0.002	
3.0					0.105 ± 0.003

experiments to another, due to differences in the stability of the blank signal.

From the Ti measured concentrations, the real quantity of dissolved metal was calculated after correction for the quantity of element removed by each 10 ml sampling. The absence of a significant loss of titanium by sorption on the walls of the vessels was controlled.

3. Results and Discussion

The intensity-potential plots recorded under various irradiation conditions and without irradiation on a 'bulk' Ti electrode and on a thin 15μm thick Ti foil respectively (in the latter case, using the same experimental arrangement than under irradiation) are represented in Figs 2 and 3 (left scale: I in mA cm^{-2}). On the same figures (right scale: Q in mC cm^{-2}) are shown the variations of the anodic charge density with increasing formation potential (Q_{anodic}) and, for comparison, the measured concentrations of titanium converted into units of charge density ($Q_{ICP/AES}$) assuming that titanium dissolves as Ti^{4+}. However, in the presence of either the native oxide (or anodic oxide layers), the titanium dissolution and the electrochemical dissolu-

tion of the native oxide (possibly also as $Ti^{2+/3+}$, which is not accounted for in the estimated $Q_{ICP/AES}$) both contribute to $Q_{ICP/AES}$, while they participate with opposite signs to the charge transfer at the film/solution interface. The chemical dissolution of the native film also contributes to $Q_{ICP/AES}$. On the other hand, Q_{anodic} represents the sum of the film formation and electrochemical dissolution contributions, neglecting the other possible reactions [6]. We are thus aware of the fact that any comparison between these two quantities must be considered cautiously. The possible mechanisms of complexation and solvatation of the titanium cation in the α-irradiated solution are outside the scope of this paper.

The titanium concentrations detected in the electrolyte are reported in Table 1 as a function of the thickness of the titanium foil, for various formation potentials during the anodic scan. The corresponding E_α values (estimated on the basis of the electronic energy losses [8]) of the more energetic α-particles reaching m/f after passing through the titanium foil, for the three thicknesses tested are also indicated. Columns 1 and 2 refer to a bulk Ti electrode 'freshly' polished to mirror and to a gently polished thin Ti foil respectively, i.e. under conditions for which the native oxide is respectively removed (1) and present (2) at the electrode surface. In details, the results are the following:

3.1 Anodic film growth at a titanium electrode without irradiation

The I(V) plot of Fig. 2(a), recorded on a 'bulk' titanium electrode polished using successively finer grades down to 1 μm alumina paste, exhibits the typical features of a freshly polished electrode:

(i) the current density passes through zero at –0.62 V/SCE, the open circuit potential after ten minutes of immersion in the sulphuric solution reaching typically –0.69 V/SCE, a value approaching that for which the refractive indices of 'film free' titanium were found by ellipsometry [9],

(ii) an anodic peak of current density is visible, going through a maximum at *ca.* –0.5 V/SCE, consistent with an intense electrochemical titanium dissolution and responsible for the gradual increase of $Q_{ICP/AES}$ in this potential region. Then the current levels off rapidly to a 'plateau level' corresponding to the region of film thickening at (assumed) constant efficiency.

We note on Fig. 2(b) that a Ti electrode covered with its native oxide continues to dissolve slowly at a reduced rate in the region of film thickening up to 2 V/SCE, the so formed oxide layers being 'non barrier'.

3.2 Anodic film growth at a titanium electrode under α-radiation

It has been shown recently [4] that the efficiency for film growth is significantly modified under α-radiation, provided that the energy of the projectile reaching the metal/film interface be in a restricted range centred at $E_\alpha \sim 1.6$ MeV. This result was based on an electrochemical study performed on titanium targets of various thicknesses, which were chosen on both sides of the maximum of the curve representing the electronic stopping power of titanium versus range calculated for alpha-particles [1, 2].

In the present work, we concentrate on the role of the energy of the particle reaching the anodic oxide film, after having passed through the Ti foil (and the native oxide), which apparently affects considerably both the growth and dissolution processes, in a restricted energy range between roughly 1.3 and 2 MeV.

However, due to the broadening of the energy spectrum of the source, all particles with sufficient energy to reach the film/solution interface could also influence the corrosion mechanism.

We observe that α-radiation apparently reactivates the electrode surface, since an anodic peak attributed to titanium dissolution appears for $E_\alpha \sim 1.3$ MeV

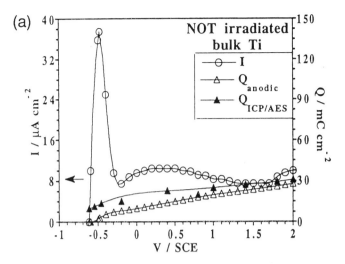

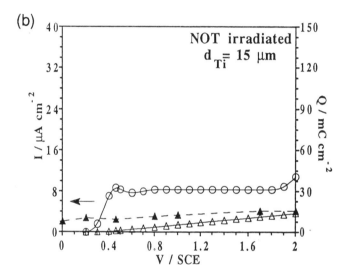

Fig. 2 Left scale: Intensity-potential plot recorded in the anodic region below oxygen evolution (open circle) and right scale: anodic charge passing through the interface versus potential (open triangle), charge deduced from ICP/AES measurements (black triangle), making the hypothesis that titanium dissolves as Ti^{4+} ($Q_{ICP/AES}$) for:

(a) bulk titanium polished to mirror, without irradiation,

(b) thin titanium foil (15μm) covered with its native oxide, without irradiation.

(Fig. 3(c)), and becomes clearly visible on the thinner irradiated 12.5 μm thick Ti foil for which $E_\alpha \sim 1.6$ MeV (Fig. 3(b)).

However, with increasing E_α up to 2 MeV, the electrochemical behaviour of the not irradiated electrode is apparently and progressively recovered (Fig. 3. (a)), that is to say under irradiation conditions for

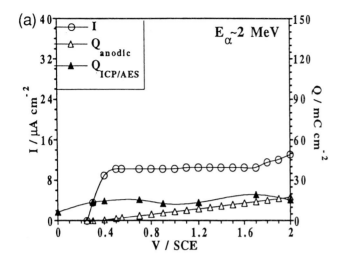

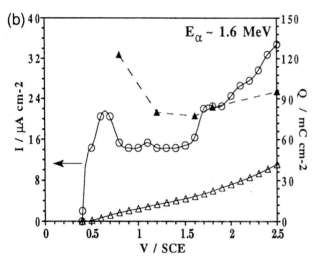

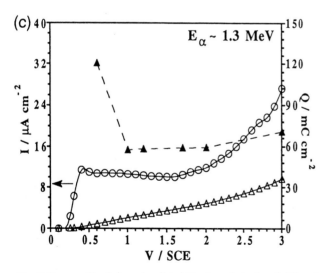

Fig. 3 Same as Fig. 2 for α-irradiated titanium electrodes of different thicknesses:

(a) $d_{Ti} = 11mm$ so that E_a 2 MeV at m/f.

(b) $d_{Ti} = 12.5mm$ so that E_a 1.6 MeV at m/f.

(c) $d_{Ti} = 13.5mm$ so that E_a 1.3 MeV at m/f.

which the alpha-particles penetrate more deeply the solution. However, it could be noted that in the latter case, the maximum in electronic stopping is the furthest away from the film/solution interface, by comparison with lower energies. At present, there is no available explanation for this behaviour.

In parallel with the electrochemical dissolution, and occurring in the same range of energies, an important chemical dissolution of the electrode is observed on Fig. 3(b) and 3(c) For E_α in the range of 1.6 MeV, this chemical dissolution is at most one order of magnitude larger than the electrochemical dissolution. This effect of irradiation is more marked in the potential region from V(I = 0) up to about 1V/SCE. A reprecipitation process at the electrode surface could possibly explain this result (see Fig. 3(b) and (c).

In contrast, as observed in Fig. 3(a) ($E_\alpha \sim 2$ MeV), for all potentials $\geqslant$ OV/SCE, the concentration of titanium in solution is found in between that detected without irradiation for an electrode polished to mirror or not (compare column 3 to column 1 and 2 on Table 1). Work is in progress in order to understand the apparently passivating influence of α-irradiation in this range of energies, by comparison with the lower energies considered above.

4. Conclusion

By coupling electrochemical measurements with an analysis of the composition of the electrolyte (titanium dissolved during anodization) by Inductively Coupled Plasma Atomic Emission spectrometry (ICP/AES), we have investigated the growth/dissolution processes of anodic oxide films which form on titanium under α-irradiation in 1M sulphuric acid at room temperature.

An important chemical dissolution of titanium, depending on the energy of the α-particles bombarding the metal/film interface, is evidenced in this work. This effect accompanies the electrochemical dissolution observed by using cyclic voltammetry in a previous study.

These chemical and electrochemical dissolution processes show a maximum efficiency for α-particles with energy $E_\alpha \sim 1.6$ MeV.

References

1. D. Gorse and T. Sakout, Materials under Irradiation, (ed) A. Dunlop, Trans. Tech. Publications, pp.451–466, 1993.
2. D. Gorse, B. Rondot and M. da Cunha Belo, Corros. Sci., **30**, 23, 1990.
3. Lj. Arsov, M. Froelicher, M. Froment and A. Hugot-Le Goff, J. Chim. Phys., **72**, 275, 1975; J. F. McAleer and L. M. Peter, J. Electrochem. Soc., **129**, 1252, 1982.

4. T. Sakout and D. Gorse, Proceedings of the 10th European Corrosion Congress, Barcelona, Spain, 5–8 July, 1993.

5. D. J. Blackwood, L. M. Peter and D. E. Williams, Electrochimica Acta, **33**, 1143, 1988.

6. J. L. Delplancke and R. Winand, Electrochimica Acta, **33**, 1551, 1988.

7. W. Slavin, Spectroscopy International, **4**, 22, 1992.

8. L. C. Northcliffe and R. F. Schilling, Range and Stopping Power Tables For Heavy Ions, Nuclear data Tables, Academic Press, A7, 233, 1970.

9. Lj. D. Arsov, Elecrtrochimica Acta, **30**, 1645, 1985.

26

Characterisation of Passive Layers of Bronze Patinas (Cu–Sn Alloys) in Relation with the Tin Content of the Alloy

L. ROBBIOLA*, C. FIAUD AND A. HARCH

Laboratoire d'Etude de la Corrosion, Ecole Nationale Supérieure de Chimie de Paris, 11 rue P. et M. Curie, F - 75005 Paris, France
*Laboratoire d'Electrochimie Analytique et Appliquée

Abstract

In order to develop our knowledge of the relations between the high degree of protectiveness of passive layers and their composition, corroded archaeological bronzes of Bronze Age (1500–950 BC) have been characterised by different methods of analysis. Results reveal that the passive layers have a bilayer structure and are due to the selective dissolution of copper from the copper solid solution of the alloy to the electrolyte. For all of the objects, the ratio $(Sn/Cu)_p/(Sn/Cu)_a$, with p for the outermost passive layer and a the alloy, has about the same value. Sn concentration within outermost passive layers is discussed in relation with the theoretical implications of the Kirchheim model.

1. Introduction

Corrosion studies on passive films are mostly concerned with processes occurring during short periods of time, i. e. often less than a hundred of hours. This paper is focused on the study of ancient metallurgical materials which have reached a stationary state after several hundreds of years.

Often, antique bronzes (Cu–Sn alloys) reveal 'uniform' surfaces which preserve the original surfaces, i.e. decor or polishing traces are still visible. This case of passive surface, known in the literature as 'noble patina' [1, 2], is a typical example of the protection of an alloy for more than several millennia.

Little is known about the nature of these passive surfaces and the processes of their formation. External aspect, composition and microstructure of passive layers in natural conditions will be discussed in this paper from results obtained on a corpus of Bronze Age objects (1500–950 BC). Then the results will be related to the Kirchheim model describing passive films formed on homogeneous alloys in the quasi stationary condition.

2. Experimental Method

2.1 Description of the samples

The samples are 13 archaeological artefacts—pins, bars, needles, rings—dated from about 1500 to 950 BC and discovered in an important Bronze Age metallurgical centre (Fort-Harrouard, France). They have been described in detail elsewhere [3]. The archaeological soil is representative of a moderately aggressive oxygenated soil [4]. All these objects were not restored. The objects are Cu–Sn alloys with a Sn content from 4 to 14% by weight. They contain minor elements such as Ni (from 0.1 to 1% by weight), As (0.1–0.35%), Fe and Ag and Sb (0.03-0.25%), Zn and Co(0.003–0.01). Three objects have a lead content of more than 1%. However, lead is not miscible in the alloy and does not modify the nature of the copper solid solution.

2.2 Characterisation of the surfaces

The passive surfaces were analysed by energy dispersive spectrometry (EDS) on areas of *ca.* 0.1 to 0.01 mm^2. Each artefact was introduced directly into the chamber of a scanning electron microscope (SEM) without any preparation to keep it intact. In another part of the work, corrosion products were scraped-off from the surface before being analysed by X-ray diffraction and by infrared spectrometry.

Three objects, considered as representative of the whole corpus, with different tin contents, were sectioned transversally and longitudinally for metallographic examinations and characterisation of the internal microstructure. They were homogeneous single phase alloys with an annealing structure and contained numerous copper sulphide inclusions due to the ores used [5].

3. Principal Component Analysis (PCA)

The PCA method was applied in order to interpret the EDS analyses of the patinas. The method allows [6], first, a description of the variations of the p analysed chemical elements (here Si, P, Cl, Fe, Ni, Cu and As) and, secondly, a representation of the whole analytical results in a different reference system with p-coordinates defined as the main axes. Each D_i axis (i = 1, .., p) is defined by an eigenvector U_i which corresponds to the eigenvalue λ_i of the diagonalised correlation matrix between the chemical elements. So the p eigenvectors (principal components) are expressed as a linear combination of the previous variables (Si, P, ...). The eigenvectors U_i are classified by decreasing order of the eigenvalues to obtain the best representation of the results for the first principal axes.

In our case, the analytical results were pretreated first by dividing each analytical point by its Sn value to have independent values, and then by scaling each variable by the standard deviation of the analysed points. This procedure allows to give an equal weighing for each value in the data matrix.

4. Results

The passive 'surface' of the bronzes appears as a shiny lustrous surface (Fig. 1) keeping the limit of the original surface. It is characterised by a two-layer structure (Fig. 2):

- an homogeneous outermost layer, around 5 to 40 μm deep, about the size of a metallic grain, and with different possible colours (green, dark grey, blue or bright grey). Copper sulphide inclusions from the alloy are still visible in the outermost layer. Outer crusts, sometimes observed on this layer, result from the precipitation of copper cations in the soil.

- an internal grey–brown layer, whose thickness may vary between 1 and 30μm. Intergranular attack is occasionally encountered at the boundary of the alloy.

From the microstructural examinations, it can be concluded that the passive structure has grown from the original surface to the alloy without apparent volume change.

A typical example of the distribution of the chemical elements is given in the Fig. 3.

The outermost layer is rich in tin, oxygen and soil elements (mainly Si, Al, P, Ca, Fe and Cl). The soil elements show significant variations in composition for the same object, and also from one object to another. Only the Sn-content in the outermost layer was found roughly constant in a given patina, yet it varied from one sample to another. Ratios (Sn/Cu) were then defined for the outermost layer $(Sn/Cu)_p$ and for the bulk alloy $(Sn/Cu)_a$. The ratio between the two quantities,

$$\beta = \frac{\left(\dfrac{Sn}{Cu}\right)_p}{\left(\dfrac{Sn}{Cu}\right)_a} \tag{1}$$

was found to be $\beta = 18 \pm 4$ for all of the samples. Similar values had been obtained in the literature [7–9]. So it appears that a general behaviour exists for objects buried in moderate aggressive soils (loam or dry or calcareous soil) for different metallurgical and historical periods (from 1000 BC [8, 9] to 1000 AD [7]).
The internal layer contains lower tin amounts than the outermost layer and no soil components. In this layer, the Sn/Cu varies strongly from one point to another.

Little information could be obtained from X-ray diffraction and infrared spectrometry results. The out-

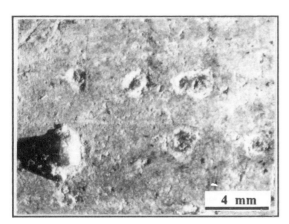

Fig. 1 Macroscopic view of a passive patina with some repassivated pits. Bronze Age bar (1250–1050 BC).

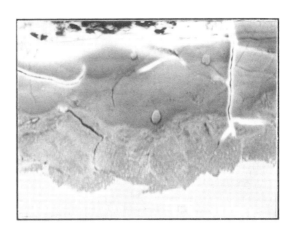

Fig. 2 Transversal microstructure of the passive layers. Inclusions within the passive layers and the alloy are copper sulphide inclusions (SEI).

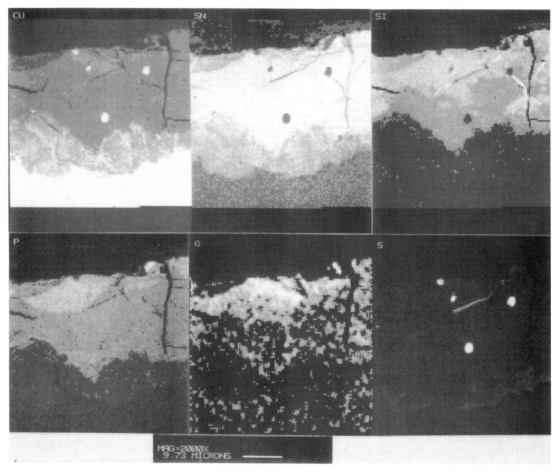

Fig. 3 X-ray maps of the right-centre of the Fig. 2 (left to right, top to bottom: Cu-Sn-Si-P-O-S).

ermost layer was amorphous and contained principally phosphate, silicate and hydroxyl groups under a non identified form. The products of the internal layer were difficult to isolate and no significant result could be interpreted. For both layers, it could only be assumed that the tin compounds are mainly amorphous hydrated tin oxides. It is known that tin (IV) oxide is thermodynamically stable over a large range of pH and potential values [10]. The copper compounds could not be identified as well-classified products, such as malachite, chrysocolle, libethenite..., as often encountered for corrosion copper products in bronzes [1,2,7,8].

The PCA on the outermost layers analysed by EDS is given in Fig. 4 for the two first principal components (72% of the PCA representation). PCA allows to attribute a relative weight to each participant (Si, P, ...) to the growth of the layer. If the chemical species are in the same region of the diagram, they can be supposed to have the same contribution to the growth of the passive layer. Figure 4 reveals different groups of chemical elements. The group of the incorporated soil elements (Si, P, Fe and Cl) are in the same region of the diagram and are strongly correlated, i.e. these ele-

ments have a similar contribution to the elaboration of the outermost layer. For the constituents of the alloy, copper is unrelated as well as As and Ni, showing that the basic element of the alloy does not have the same contribution as the minor or traces elements in the formation of the protective layer.

The presentation of the results in the Fig. 4 shows different classes of analytical points which correspond more or less to the amounts of soil elements (Si, P, Cl, Fe) in the patinas. We can remark that the classes are related to the different colours of the outermost layer. The largest amounts of soil elements are encountered in the blue patinas, whereas bright grey surfaces have a higher copper content than the others. The exact nature of the species in which the soil elements are incorporated has not been investigated and remains an open question.

Consequently, there exist very important variations in soil species from one kind of patina to another. The soil compounds maintain the electroneutrality but anions and minor contents or traces of metallic cations do not directly interfere in the relative enrichment of the tin cations in the passive layers. The reaction between the soil products and the metallic cations is

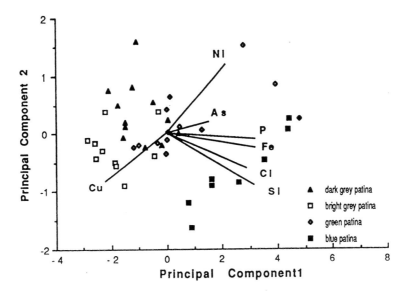

Fig. 4 Representation of the chemical variables and of the analysed points of the outermost passive layers. Principal components analysis.

not the rate determining step in the attainment of the passive state.

5. Discussion

The excellent corrosion resistance of tin bronzes in moderately aggressive natural conditions can be attributed to the formation of highly resistant tin compounds. Hydrated tin oxides ($SnO_2.xH_2O$) are known to form amorphous gel-like compounds. Furthermore, three important points appeared from the results:

(1) The pseudomorphic replacement of the surface alloy by the passive layers,

(2) The same β value (eqn 1) for all the patinas, and

(3) the large variation of soil components in the passive outermost layers.

Consequently, the high Sn amount in the passive layers can only be explained by Cu selective dissolution (decuprification). The soil components (mainly silicates and phosphates) do not seem to affect directly the process of the selective dissolution of the copper and the relative tin enrichment: their incorporation into the patina could be a second step in the formation of the passive surface. It can be assumed that the process leading to the oxidation of tin and copper is:

$$Sn + 4\,H_2O = Sn(OH)_4 + 4\,H^+ + 4e^- \qquad (2)$$

$$Sn(OH)_4 = SnO_2.xH_2O + (2{-}x)\,H_2O \qquad (2')$$

$$2\,Cu + H_2O = Cu_2O + 2\,H^+ + 2e^- \qquad (3)$$

$$Cu_2O + 2H_2O + 1/2\,O_2 = 2Cu(OH)_2 \qquad (3')$$

The process involved is in fact an internal oxidation of tin and copper: the relative Sn enrichment in the outermost passive layer could be due to both the higher mobility of copper cations and the greater stability of the tin oxide in the layer.

For homogeneous binary alloys in the stationary state, the approach of Kirchheim offers to explain the atomic fraction of the components in the passive layer and in the alloy [11], assuming that the composition of the subsequent layers depends on the diffusivities of the cations. In the passive state the average atomic fractions of the components can be estimated from the following formula for a binary alloy A_xB_{1-x} [12]:

$$\frac{X_p}{1-X_p} = \frac{X_a}{1-X_a}\frac{D_B}{D_A}\exp\left(\frac{\alpha(Z_B - Z_A).s.F.E'}{R.T.}\right) = \frac{X_a}{1-X_a}\beta \qquad (4)$$

where X_p is the atomic fraction of the component in the passive layer and X_a, in the alloy, D and Z are respectively the diffusivity and the valency of the cation components A or B. α is the transfer coefficient, s the jump distance of the cation and E' the electric field within the film.

In the case of bronzes, A = Sn and B = Cu. Figure 5 shows the results of measurements of the average composition of the copper atomic fraction in the passivating patina $X_{Sn,p}$ as a function of that in the alloy

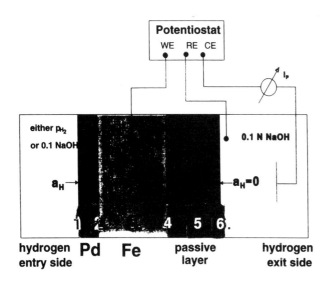

Fig. 1 Scheme of the processes in the electrochemical permeation experiment.

1. Adsorption of hydrogen atoms on the Pd surface; 2. Absorption of hydrogen atoms by the iron; 3. Diffusion of hydrogen atoms through the iron as screened protons; 4. Hydrogen transition through the interface Fe/ passive layer; 5. Hydrogen migration through the passive layer; 6. Hydrogen transition through the interface passive layer/electrolyte.

The electrochemical hydrogen permeation technique was used. The processes which happen at the sample are shown in Fig. 1.

The anodic permeation current is measured at the exit side (**6**). The dependence of permeation flux on temperature was measured with hydrogen activity $a_H = 1$ at the entry side of the permeation cell (**1, 2**), obtained by contacting this surface with flowing hydrogen (g) at $P_{H_2} = 1$ bar. The dependence on applied anodic potential, hydrogen activity, and sample thickness was measured with an electrochemical double cell introduced by Devanathan [5]. Here, the hydrogen activity is given by cathodic polarisation and hydrogen evolution controlled by the Volmer-Tafel-Heyrovsky mechanism (**1, 2**). Deaerated 0.1N NaOH solution was used as electrolyte.

From instationary measurements hydrogen diffusivities and concentrations in the passive layer were evaluated by means of the trapping theory [6].

3. Results

3.1 Hydrogen permeation coefficient

The hydrogen permeation coefficient Φ_H was measured for a hydrogen pressure of 1 bar ($a_H = 1$) at the entry side.

The following equations for the temperature dependence were obtained:

Pure iron:
$$\Phi_H^{Fe} = (3.6 \pm 1.4) \cdot 10^{-7} \exp\left(\frac{(35.5 \pm 0.9)\text{kJ} / \text{mol}}{RT}\right) \frac{\text{molH}}{\text{cm} \cdot \text{s}}$$

Pure iron with passive layer:
$$\Phi_H^{Fe/passive} = (18.9 \pm 28.1) \cdot 10^{-7} \exp\left(\frac{(42.0 \pm 3.0)\text{kJ} / \text{mol}}{RT}\right) \frac{\text{molH}}{\text{cm} \cdot \text{s}}$$

Figure 2 shows a plot of these equations and the measured data.

To calculate the hydrogen permeation coefficient for the passive layer separately, it was assumed that the reciprocal of the permeation coefficients related to layer thickness behave like permeation resistances. The following equation is obtained from Ohm's and Kirchhoff's law:

$$\frac{d_{Fe/passive}}{\Phi_H^{Fe/passive}} = \frac{d_{Fe}}{\Phi_H^{Fe}} + \frac{d_{passive}}{\Phi_H^{passive}}$$

The result was for the hydrogen permeation coefficient of the passive layer was:

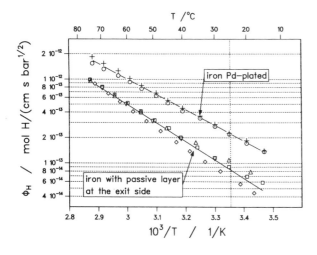

Fig. 2 Hydrogen permeation coefficient ΦH as a function of temperature; iron passivated or palladium coated.

$$\Phi_H^{passive} = (4.5 \pm 5.8) \cdot 10^{-11} \cdot \exp\left(\frac{(42.6 \pm 2.7)kJ / mol}{RT}\right)\frac{molH}{cm \cdot s}$$

The values at 25°C are listed below in Table 1.

Table 1 Hydrogen permeation coefficient at T = 25°C

	Fe	Fe + passive layer	passive layer
thickness	2 mm	2 mm	4 nm
$\Phi_H/\frac{molH}{cm \cdot s}$	$(3.6 \pm 0.1) \cdot 10^{-13}$	$(0.8 \pm 0.2) \cdot 10^{-13}$	$1.6 \cdot 10^{-18}$

3.2 Hydrogen charging by cathodic polarisation

First the permeation current was measured at constant hydrogen activities at the entry side varying the anodic potential at the exit side of the membrane, the potential has been varied in the range of –200 and + 400 mV$_{SHE}$. There was no significant dependence of the stationary hydrogen permeation current on the anodic potential. In case of a negligible influence of the passive layer on the permeation current density, its dependence on membrane thickness should be the same as for unpassivated iron, i.e. according to Fick's 1st law:

$$i_p \propto {}^1/_d$$

If, however, the passive layer controls the hydrogen permeation rate, no dependence of the permeation current density on the sample thickness (≡ iron thickness) is to be expected.

Figure 3 shows a plot of the permeation current density vs reciprocal sample thickness. It appears that there is no clear rate control of hydrogen permeation across the passive layer, but as well the thickness of the iron membrane as the passive layer are both rate determining.

The hydrogen permeation currents obtained from Pd-plated iron, passivated iron (by anodic polarisation), and chemically polished iron are shown in Fig. 4. It demonstrates the influence of the surface state on the stationary permeation current densities in dependence of the cathodic current density at the entry side. The passivated iron samples show a lower hydrogen permeation current; the linear dependence of the $\sqrt{|i_C| - i_P^s}$ with the Volmer-Tafel mechanism is found

only for Pd-coated iron. The passive layer formed by chemical polishing hinders hydrogen permeation much stronger than a passive layer formed by anodic polarisation in sodium hydroxide solution does.

The break-through time, t_b, differs for

- the uncharged sample (no hydrogen has been in the passive layer);

- the charged sample (the break-through time after a change in cathodic current, i.e. hydrogen activity at the exit side to a higher value.

In Fig. 5 the t_b-values obtained are plotted in dependence of the sample thickness. There appears to be a big trapping effect of the passive film binding some hydrogen almost irreversibly. Arriving hydrogen atoms will remain at these traps until total occupation. Therefor the break-through is retarded. After saturation of the passive film with hydrogen the permeation rate increases or decreases more quickly corresponding to a change in the hydrogen activity at the entry side. Applying the trapping theory [6,7], a rough estimation of the hydrogen diffusion coefficient

$$D_H = \frac{d^2}{20 \cdot t_b} \quad (d = \text{sample thickness})$$

is possible. Further, with respect to a balance of the molar numbers of hydrogen in the different layers,

$$n_H^{Fe/passive} = n_H^{Fe} + n_H^{Passive}$$

it is possible to evaluate the hydrogen concentration and diffusivity for the passive layer:

$$c_H^{passive} = \frac{V_{Fe}}{V_{passive}} \cdot \left(\frac{\Phi_H^{passive}}{d_{Fe}^2 / (20 \cdot t_b)} - c_H^{Fe}\right); \quad D_H^{passive} = \frac{\Phi_H^{passive}}{c_H^{passive}}$$

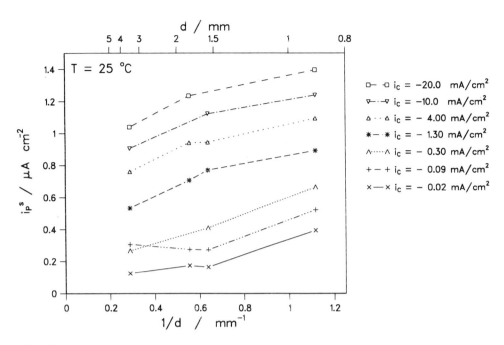

Fig. 3 *Steady-state hydrogen permeation current density as a function of iron membrane thickness and cathodic current density.*

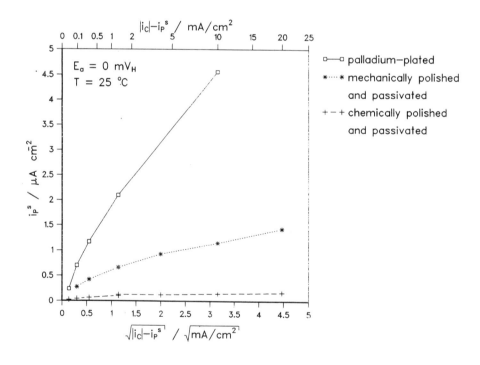

Fig. 4 *Influence of the surface conditions at the exit side on the hydrogen permeation current density as a function of the cathodic current density.*

Using the break-through times from Fig. 5, the diffusion coefficient results in $2 \cdot 10{-}16 \mathrm{cm^2/s}$ and the hydrogen concentration in *ca.* 10 mmolH/cm³. This value for the hydrogen concentration corresponds to 1/3 mole hydrogen per mole Fe_2O_3, if this oxide is assumed to be present.

3.3 Impedance measurements

The system used for the impedance spectroscopy measurements consisted of a frequency response analyser and a potentiostat, both controlled by a microcomputer. Impedance spectroscopy measurements were carried out over a frequency range of 0.01 to 9000

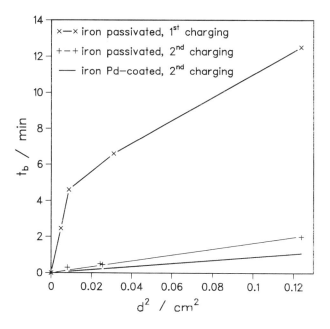

Fig. 5 Effects of iron membrane thickness and different surface films on the breakthrough time of hydrogen t_b.

Hz (stationary measurements at constant potential). Three sample states were compared:

- cathodically polarised iron (without passive layer and without hydrogen content);

- anodically polarised iron (with passive layer and without hydrogen content);

- anodically polarised iron saturated with hydrogen and with simultaneous hydrogen permeation.

A simple R-RC equivalent circuit was chosen to fit the data obtained. Values for capacities and resistances of the passive films in different sample states are listed in Table 2.

The capacity of the electrochemical double layer is much larger for unpassivated iron samples than for passivated ones. There seems to be no change in film capacity due to hydrogen.

Passivated iron samples show large polarisation resistances. The R values showed a big scatter, but there is an indication that they are smaller for samples saturated with hydrogen than for passivated samples without hydrogen. This points to an increased hydrogen content of the passive layer after hydrogen charging. An increasing hydrogen concentration of the passive layer should lead to an increasing concentration of Fe^{2+} ions which causes a lower resistance.

4. Discussion

The following hydrogen transport mechanism is assumed: In iron hydrogen atoms diffuse as screened protons [7]. Then the electron transfer reaction takes place at the interface iron/ passive layer, i.e. the electron flows to the potentiostat, while the proton moves through the passive film by means of activated jump diffusion [8] (see Fig. 6). Figure 7 (overleaf) shows a schematic plot of assumed volta potential (a), hydrogen activity (b) and concentration (c) across the composite membrane at steady-state hydrogen permeation. In iron there is only a low value of hydrogen concentration ($10–9$ mol.cm^{-3}) with small cH gradient. In the passive film, in contrast, high cH values (10^{-3} mol.cm^{-3}) and a high cH gradient are assumed to be expected.

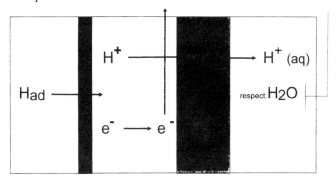

Fig. 6 Schematic representation of the hydrogen permeation mechanism.

Table 2 Resistance and capacity of the passive films in different states

surface state	capacity	resistance
unpassivated	1.2 mF	200 Ω
passivated	75 μF	100-500 kΩ
passivated and hydrogen saturated	71 μF	50-100 kΩ

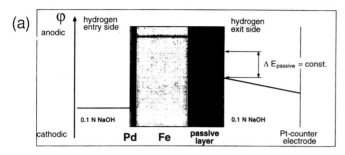

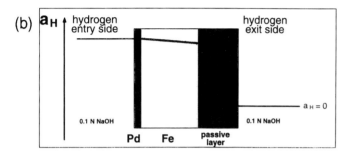

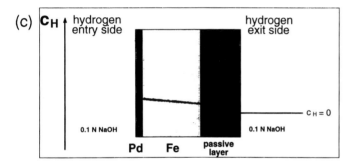

Fig. 7 Schematic representation of (a) the Volta potential φ, (b) the hydrogen activity a_H, and (c) the hydrogen concentration c_H across the iron sample with Pd layer and passive film.

References

1. J. Kruger, Corros. Sci., **29**, 149, 1989.
2. B. Cahen and C. T. Chen, J. Electrochem. Soc., **129**, 18, 474, 921, 1982.
3. Bloom and Glodenberg. In: H. H. Uhlig. In: R. W. Staehle and H. Okada, Passivity and Its Breakdown on Iron and Iron Base Alloys, 1976, Houston, Texas, Nat. Assoc. of Corr. Eng.
4. E. Riecke, Arch. Eisenhuttenwes., **49**, 509, 1978.
5. M. A. V. Devanathan and Z. Stachurski, J. Electrochem. Soc., **110**, 886, 1963.
6. A. McNabb and P. K. Foster, Trans. Met. Soc. AIME, **227**, 618, 1963.
7. E. Riecke and K. Bohnenkamp, Z. Metallkde, **75**, 76, 1984.
8. S.-I. Pyun and R. A. Oriani, Corros. Sci., **29**, 485, 1989.

28

Hydrogen and Iron Transport Through Passive Oxide Films on Electropolished Iron

*P. BRUZZONI AND E. RIECKE**

Comisión Nacional de Energía Atómica, Buenos Aires, Argentina
*Max-Planck-Institut für Eisenforschung GmbH, Düsseldorf, Germany

Abstract

The degassing rate of hydrogen from hydrogen-containing iron is strongly influenced by the nature of the oxide film which covers the iron surface, and usually much lower than oxide-free, Pd-plated iron.

This phenomenon has been investigated through hydrogen permeation experiments with gas-phase charging and electrochemical detection at 298 K T < 348 K. The exit surface of the permeation membrane was electropolished.

The hydrogen flow through the oxide film formed on this electropolished surface decreases as the time and temperature of the pre-polarisation treatment increases. This latter treatment is usually aimed at achieving a low background current. This decrease in the hydrogen permeation current is attributed to a continuous increase of the oxide film thickness L_{ox} in the range 2 nm < L_{ox} < 6 nm.

A correlation has been found between the so-called background current, which is mainly related to the oxide film growth rate or iron transport through the film, and the hydrogen permeation current. This is consistent with a mechanism of ion migration of both H and Fe under the influence of a strong electric field, in which the ion flow rate depends more or less strongly on the oxide film thickness, according to the ion charge Z_H and Z_{Fe}. The ratio Z_H/Z_{Fe} has been calculated assuming this model, yielding 0.27 to 0.3.

Space charge effects could explain some apparent deviations in the results, and provide a different picture of the role played by hydrogen in rendering passive films more protective.

1. Introduction

Hydrogen diffusion through transition metal oxides has been recently studied by means of the electrochemical hydrogen permeation technique at room temperature [1–4]. This investigations have the following aims:

1. To characterise materials which behave as barriers for hydrogen and can be easily deposited on metallic substrates;
2. To evaluate unwanted surface effects which appear while studying hydrogen permeation through metallic phases; and
3. To attempt to obtain information on the role played by hydrogen in the passivity of metals.

In the present work, hydrogen permeation through the oxide film grown on electropolished iron surfaces has been studied.

2. Permeation Cell and Membrane Preparation

The electrochemical hydrogen permeation cell with gas phase charging is shown in Fig. 1. The permeation membrane is located in a Teflon holder between charg- ing (left) and detection (right) cells. Pure H_2 (P = 1 bar) was used in the charging chamber. The electrolyte in the detection cell was 0.1 N NaOH, the anodic potential was set to +0.2 V (NHE), and the temperature ranged from 25 to 75°C.

Permeation membranes were cut from a cold rolled sheet of zone refined iron with the following contents of impurities in μg/g: C 60, Si 50, Mn 9. The membrane thickness 30, Al < 30, N < 10, O 40 and B 17. The membrane thickness was *ca.* 140 μm. The membranes were mechanically polished (SiC) up to #1000, heat treated (H_2, 1073 K, 6 hours) and degassed (vacuum, 573 K, 1 h).

The usual permeation membranes are covered on both sides by a thin film (*ca.* 60 nm) of electroplated Pd. The resistance of this Pd film to hydrogen permeation is believed to be negligible for membrane thickness larger than 100 μm. The Pd film on the entry side allows to reach local equilibrium between hydrogen in the gas phase and hydrogen in the metallic phase close to the entry surface. The Pd film on the exit side avoids the formation of iron oxide, a barrier for H permeation, on this surface. Hydrogen flow through such membranes is only limited by diffusion through the bulk metal.

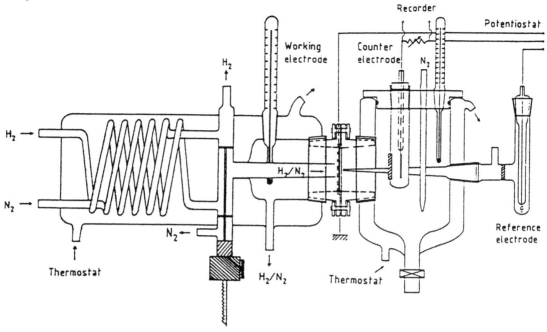

Fig. 1 Electrochemical hydrogen permeation cell with gas phase charging.

The Pd film was deposited according to the technique described by Pyun and Oriani [1]. Its quality was tested on Pd/Fe/Pd membranes through steady state permeation measurements with H_2 (g) charging at p = 1 bar. Permeation coefficients 0 to 15% lower than literature values [5,6] were obtained. Regarding unsteady state behaviour, the time to reach 90% of steady state flow was *ca.* 120 s, much larger than expected for pure, annealed iron. However, the response of the Pd film was reasonably fast to achieve the purpose of the present experimental technique, namely: to obtain hydrogen close to equilibrium with H_2 (P = 1 bar) at the metal/oxide interface in a time short enough to ensure that the passive film resistance to the hydrogen flow changes only slightly.

The permeation membranes used to study the hydrogen flow through the passive film had a Pd film on the entry side, but the exit side was not covered with Pd. Instead, this surface was electropolished according to the method of Sewell *et al.* [7]. After electropolishing, a *ca.* 1.5 nm thick oxide film develops in contact with air [8]. This film grows further as the exit surface is polarised in the detection cell.

3. Experimental Technique

In the usual hydrogen permeation technique with electrochemical detection, the exit side of the membrane is polarised during several hours at an anodic potential before introducing hydrogen. This procedure is in order to attain a very low background anodic current of some nA.cm^{-2}. This treatment is sometimes carried out at a temperature higher than the test temperature.

Curve 1 of Fig. 2 shows a typical log–log plot of the anodic current vs time in absence of hydrogen. Shortly after imposing the anodic potential, the current is high and decreases sharply. The current decreases at a slower rate as the polarisation time increases. The log-log plot shows that no constant value is attained. This has been observed even after 5 days of anodic polarisation.

This behaviour can be explained by a continuous growth of the oxide film with negligible dissolution rate and iron transport through the film by diffusion enhanced by a high electric field. This mechanism will be discussed below.

The fact that the oxide thickness continuously increases has led us to change the experimental technique: hydrogen has been introduced even in the presence of a high background current, as soon as 5 min after imposing the anodic potential. In this way, different oxide film thicknesses ranging from 2 to 6 nm could be tested. The thickness was estimated from the electric charge passed in absence of hydrogen, assuming the formation of either γ-Fe_2O_3 or Fe_3O_4 (0.18 equivalents/cm^3). The hydrogen current i_H was obtained by subtracting the background (iron) current i_{Fe} (curve 1) from the total current i_T, measured while hydrogen was present in the membrane (curve 2).

$$i_H = i_T - i_{Fe} \qquad (1)$$

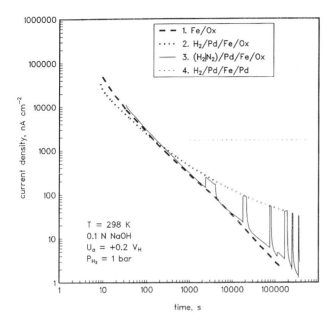

Fig. 2 Typical plot of the current evolution since the beginning of the anodic polarisation of electropolished iron at 25°C. (1). Iron free from hydrogen; (2). Iron in equilibrium with hydrogen (P_{H2}= 1 bar); (3). Iron, alternatively with and without hydrogen; (4). Hydrogen current density for a Pd-plated exit surface, calculated from literature data on hydrogen permeation coefficient in iron.

Curve 3 has been obtained by alternating N_2 and H_2 in the charging compartment. In this way, i_T and i_{Fe} are obtained from a single membrane. Coincidence with curves 1 and 2 is observed. Curve 4 is the expected i_H for a Pd/Fe/Pd membrane. It has been plotted to show in which extent hydrogen flow is reduced by the passive film on the exit side.

This modified technique has two advantages for studying the hydrogen permeation in anodic oxide films:

1. The role of polarisation time in alkaline solutions is recognised and can be investigated.

2. Films of lower thickness with high background currents can be studied.

4. Results and Discussion

The passive film is thought to be an spinel iron oxide of composition ranging from Magnetite (Fe_3O_4) to γ-Fe_2O_3 [9]. It has been suggested that hydrogen can occupy vacant cation positions in this spinel lattice, thus producing structures of the type Me_5HO_8 [10].

It is proposed that hydrogen moves in this film as an ion, by jumping to vacant cation sites. A high electric field is present in the film under anodic polarisation. The field will favour the flow of cations from the metal to the electrolyte. A similar mechanism is proposed for the iron movement.

As a first approach, it can be assumed that no space charge effects are present and that the field is high enough to make reverse flow negligible. Then, the following equations are obtained (j = H, Fe):

$$i_j = [(Z_j \, F \, D_j \, c_j)/a] \exp[Z_j \, F \, \Delta U \, a/(2 \, R \, T \, L_{ox})] \qquad (2)$$

where i = current density, Z = ion charge, F = Faraday's constant, D = diffusion coefficient in the oxide, c = ion concentration in the oxide close to the metal/oxide interface, a = jump distance, ΔU = voltage drop across the oxide film and L_{ox} = oxide film thickness. C_H is close to equilibrium with the charging environment (P = 1 bar), due to the low resistance of the metallic phase to the hydrogen flow in comparison with the oxide phase.

By combining these equations, a relationship is found between hydrogen current and iron current:

$$\left[\frac{i_H \, a}{Z_H \, F \, D_H \, C_H}\right] = \left[\frac{i_{Fe} \, a}{Z_{Fe} \, F \, D_{Fe} \, C_{Fe}}\right]^{(Z_H/Z_{Fe})} \qquad (3)$$

which predicts a slope (Z_H/Z_{Fe}) in a plot of log (i_H) vs log (i_{Fe}). This relationship fits in well with the experimental results (Fig. 3) except for very low film thickness (lowest time and highest background current). The slope (Z_H/Z_{Fe}) ranges from 0.27 to 0.3.

Space charge effects [11] can account for the deviations at low film thickness. Simple calculations using Poisson's equation show that the electric field is no longer homogeneous for ion concentrations above 10^{-6} mol cm^{-3}. A numerical calculation procedure has been developed to predict the ion flows taking space charge into account. It comprises the following steps:

1. Calculate the potential field assuming an initial concentration profile along n+1 potential wells (n= 5 to 20), to satisfy Poisson's equation;

2. Calculate the iron and hydrogen flow through the n potential barriers using a general transport equation involving both electric field and concentration gradient driving forces;

3. Recalculate the concentration profile after a small time interval;

4. Repeat this procedure until steady state has been reached.

Figure 4 shows a sample calculation, where the parameters have been chosen arbitrarily in order to fit the experimental data (Table 1). The observed bending of the plots at low thickness is predicted by this model.

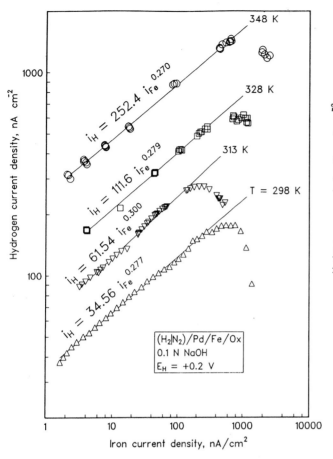

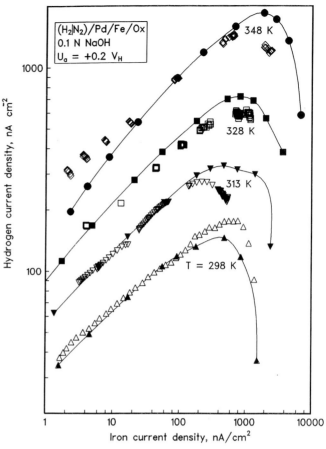

Fig. 3 Double logarithmic plot of i_H vs i_{Fe}, as defined in text. Best-fit equations of the type $i_H = A \cdot i_{Fe}^b$ were obtained excluding the highest iron current density points.

Fig. 4 Comparison between experimental (empty symbols) and calculated (filled symbols) values of i_H and i_{Fe}. The latter were obtained assuming a mechanism of ion diffusion in the presence of an electric field and with space charge effects.

Table 1 Parameters for the calculation of iron and hydrogen currents with space charge effects; $D = D_o \exp(-W/RT)$. The other parameters are defined in the text. Chosen CH values are proportional to H solubility in iron

PARAMETER	HYDROGEN	IRON
Z(ion charge)	1	3
D_o[cm^2/s]	4.50E−09	2.70E−04
W[kJ/mol]	45.00	92.50
conc.[mol/cm^3]	4.00E−05	1.00E−06 (348 K)
conc.[mol/cm^3]	2.40E−05	1.00E−06 (328 K)
conc.[mol/cm^3]	1.60E−05	1.00E−06 (313 K)
conc.[mol/cm^3]	1.00E−05	1.00E−06 (298 K)
Potential drop[V]	0.8	
jump dist.[cm]	3.00E−08	
diel.constant[As/(Vcm)]	8.85E−13	

The calculations show that hydrogen present in the film as H$^+$ causes a marked reduction of the iron flow, i.e. the corrosion rate, simply because of the repulsion forces between particles with charges of the same sign. This would mean that hydrogen present in the film improves passivity.

Other possible explanation for the deviation found at the initial stages of the film growth process is that the hydrogen concentration in the metal phase beneath the oxide film is lower than the equilibrium value with H2 (P = 1 bar), because of two possible reasons:

1. hydrogen flow is high enough to cause a considerable concentration drop in the metal phase;

2. the film thickness changes so fast in those initial stages that there is not enough time to reach the equilibrium hydrogen concentration at the metal/oxide interface due to the diffusion transients in the Fe membrane and in the Pd film.

5. Conclusions

1. The passive film on iron is not stable and grows continuously under anodic polarisation in alkaline solution. The hydrogen permeation rate through it changes accordingly.

2. This new hydrogen electrochemical permeation technique with high background current allows us to study the hydrogen transport through the passive film on iron for a wide oxide film thickness range.

3. The proposed mechanism for the hydrogen movement through the passive film on iron involves diffusion of hydrogen ions enhanced by a high electric field. A similar mechanism is proposed for the movement of the iron ions.

4. Space charge effects are likely to be present and can account for deviations from the high field transport model at low film thickness.

5. On the role played by hydrogen in passivity, the present results suggest that passivity could improved by hydrogen hindering iron ion migration through space charge effects.

References

1. Su-Il Pyun and R. A. Oriani, Corros. Sci., **29**, 485, 1989.
2. R.-H. Song, Su-Il Pyun and R. A. Oriani, J. Electrochem. Soc., **137**, 1703, 1990.
3. P. Bruzzoni and R. Garavaglia, Corros. Sci., **33**, 1797, 1992.
4. K. Schomberg. Diplomarbeit. Universität Dortmund, 1992.
5. E. Riecke, Werkst. Korros., **32**, 66, 1981.
6. J. Kumnick and H. H. Johnson, Acta Metall., **25**, 891, 1977.
7. P. B. Sewell, C. D. Stockbridge and M. Cohen, Can. J. Chem., **37**, 1813, 1959.
8. P. B. Sewell, C. D. Stockbridge and M. Cohen, J. Electrochem. Soc., **108**, 928, 1961.
9. M. J. Graham, J. A. Bardwell, R. Goetz, D. F. Mitchell and B. MacDougall, Corros. Sci., **31**, 139, 1990.
10. M. C. Bloom and L. Goldenberg, Corros. Sci., **5**, 623, 1965.
11. A. T. Fromhold, Jr. and J. Kruger, J. Electrochem. Soc., **120**, 723, 1973.

SURFACE TREATMENTS

AND

FILM MODIFICATIONS

29

Electrode Reaction Inhibition by Surface Modification and Localised Corrosion Resistance of Stainless Steels in Aqueous Solution

Y. C. Lu and M. B. Ives

Corrosion Laboratory, Institute for Materials Research, McMaster University, Hamilton, Ont., L8S 4M1, Canada

Abstract

The effects of alloying elements on the localised corrosion resistance of austenitic stainless steels are considered in terms of anodic and cathodic reaction inhibition. Surface analysis has confirmed that when local attack takes place nitrogen enhances the anodic segregation of beneficial alloying elements, such as chromium, and significantly prevents the transpassive dissolution of molybdenum. Alloyed chromium, molybdenum and nitrogen all improve the localised corrosion resistance of stainless steels by inhibiting the anodic processes. However, in order to further increase localised corrosion resistance, it is also necessary to constrain the cathodic reaction.

It is shown that cerium ion implantation of UNS S31603 stainless steel is very effective in inhibiting the cathodic electrode reaction involved in metallic corrosion. As a result of the inhibition of the kinetics of electrode processes, cerium treatment improves the localised corrosion resistance, and especially crevice corrosion resistance. This is supported by the results of electrochemical measurements in aerated 0.6M NaCl + 0.1M Na$_2$SO$_4$, and by accelerated corrosion tests.

1. Introduction

In 1967, Fontana and Greene [1] presented a model that describes many of the known aspects of that form of localised corrosion occurring in crevices. The model identifies the initiation and later development stages in the crevice corrosion of stainless steel in neutral chloride solution. In the initiation stage, due to restricted diffusion to the occluded site, oxygen becomes depleted inside the crevice, and the separation of anode and cathode develops, resulting in localised corrosion. Furthermore, during the development of crevice corrosion, the propagation of localised attack is stabilised by the large cathode-to-anode area ratio which exists. The large freely exposed surface acts as a cathode (with oxygen reduction supplying the current), while the crevice area acts as anode. Thus, extremely rapid localised dissolution of the metal can occur in the creviced area. Since corrosion depends on both anodic and cathodic reactions it can be reduced by inhibiting either.

There are different ways of inhibiting the electrode kinetics. In the current context consideration is confined to the effect of alloying elements on the localised corrosion resistance of austenitic stainless steels through surface modification.

2. Anodic Inhibition

Austenitic stainless steels have been used in environments where corrosion resistance is extremely critical. This resistance is attributed to the passive film formed on the surface protecting the thermodynamically active substrate from environmental attack. Chromium is the major film-forming element of stainless steels. However, overall localised corrosion resistance is enhanced by additional alloying elements of which molybdenum and nitrogen are particularly effective.

Molybdenum is found to be very effective in improving the pitting resistance of steels in chloride-containing media. Sugimoto and Sawada [2] have shown that the addition of molybdenum to stainless steels decreases the current density in the active region by two or three orders of magnitude in acidic solutions. Several studies have indicated that the addition of nitrogen can lead to improvements in passivation and pitting resistance [3–5]. The greatest effect of nitrogen has been observed when molybdenum is present, suggesting a possible synergism between molybdenum and nitrogen [6–8]. Sugimoto and Sawada [2] concluded that the passive film consists of a solid solution of Mo^{6+} in chromium oxyhydroxide. Due to the apparent thermodynamic stability of Mo(VI) oxide in acidic solutions, and the fact that Mo^{6+} may form a solid solution with CrOOH, they suggested that molybdenum is suppressed from dissolving transpassively and therefore contributes to the overall resistance to chloride attack. Hashimoto *et al.*[9] have suggested that the main beneficial effect of molybdenum is to decrease the rate of dissolution from active sites by the formation and retention of molybdenum oxyhydroxide or molybdate. Sakashita and Sato [10] have suggested that when molybdate ions adsorb onto a relatively thick membrane of an anion-selective hydrated iron oxide, they can alter ionic transport through the film by inducing ionic rectification. Lu and Clayton [11] used X-ray photoelectron spectroscopy (XPS) to study passive films formed on molybdenum bearing stainless steels, and further supported this theory.

Surface analysis has provided evidence of enhanced surface segregation, resulting in surface films enriched in beneficial elements [12,13]. The results from the Auger (AES) analysis, shown in Fig. 1, indicate that passive films are rich in chromium compared to the alloy substrate. However, the nitrogen-bearing alloy UNS N08367(AL6XN) retains more molybdenum than the low nitrogen version, UNS N08366(AL6X), at high applied potential (250mV vs SCE) in 1M HCl, and enhances the anodic segregation of beneficial alloying elements, such as chromium, when local attack takes place. These beneficial effects may be attributed to the formation of fine surface nitrides [14]. As indicated in Figs 2–4, chromium and molybdenum nitrides are more stable than pure metal in acidic solution. Surface nitriding may constrain the anodic dissolution of iron by more than two orders of magnitude in a neutral saline solution. This effect may hinder the auto-catalytic acidification process of pit formation in seawater. In acidic solution, iron nitride is less stable than chromium and molybdenum nitrides. This may enhance the enrichment of chromium and molybdenum during repassivation at a local breakdown site. It has been also shown [15] that nitrogen significantly prevents the transpassive dissolution of molybdenum. This effectively retains molybdenum in the passive surface, and thereby improves the localised corrosion resistance of stainless steels. In summary, chromium, molybdenum and nitrogen all improve the localised corrosion resistance by inhibiting the anodic processes, and probably in a synergistic manner.

3. Cathodic Inhibition

To date, the commercial alloy development approach has been to increase the percentage of chromium, molybdenum and nitrogen, in order to increase localised corrosion resistance. But crevice corrosion remains a problem in these alloys. Crevice corrosion and pitting corrosion have quite different initiation stages. For crevice corrosion, an oxygen concentration cell is established due to the depletion of oxygen or other oxidant in the crevice site. The enhanced cathodic reduction of oxidant adjacent to a localised attack site intensifies the localised corrosion. In order to prevent crevice corrosion, it is necessary to constrain the cathodic reactions always associated with metal dissolution.

Cerium was first reported to increase the localised corrosion resistance of aluminium alloys [16–18], possibly by inhibiting the cathodic reduction of oxygen. However, little detailed research has been done on the inhibition of the cathodic electrode reactions and its role in improving the localised corrosion resistance of stainless steels.

The improvement in localised corrosion resistance of stainless steel by cerium ion implantation was evaluated by electrochemical measurements and a standard crevice corrosion spool test [19]. Rotating disc electrodes were employed to study the cathodic electrode process and its inhibition by cerium implantation on UNS S31603 stainless steel in a solution of 0.6M NaCl + 0.1M Na_2SO_4. This was compared with an earlier study made on gold discs [20]. Figure 5 shows the disc current vs. potential for implanted and unimplanted stainless steel samples. It is clear that on the unimplanted disc the cathodic current is rotation speed dependant. However the cathodic reactions on the cerium-implanted electrodes are greatly restrained and seem not to depend on rotation speed. As discussed previously [20], in this test solution the cathodic reaction at potentials > –1200 mV on gold are only related to the reduction of oxygen. However, on a stainless steel disc, it was found [19] that there is also current contributed from the reduction of surface metal oxide and/or hydroxide. The magnitude of this current is independent of rotating speed. The current contributed from the reduction of oxygen on UNS S31603 is found to be a linear function of the square root of the rotational angular velocity of the disc and follows the Levich relationship [21], implying that the reduction of oxygen on the unimplanted steel is controlled by mass transport processes. The cathodic reduction of oxygen is greatly inhibited by cerium ion implantation and appears to be limited by charge transfer processes.

In acidic solutions, one of the major cathodic reactions involved in metallic corrosion is the reduction of protons. The kinetics of this reduction at a range of pH values were reported previously [20]. The cathodic reduction of protons on UNS S31603, both as-received and cerium-implanted, was measured in a deaerated 0.1M Na_2SO_4 + 0.6M NaCl solution at pH 3.50. In this

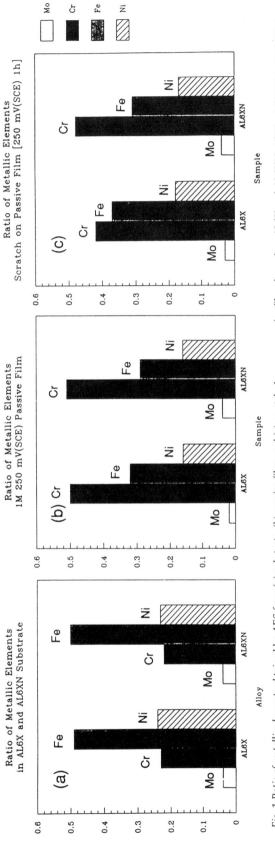

Fig. 1 Ratio of metallic elements obtained by AES from (a) substrate, (b) passive films and (c) scratched areas on passive films formed on AL6X and AL6XN at 250 mV(SCE) in 1M HCl for 1 h followed by scratching.

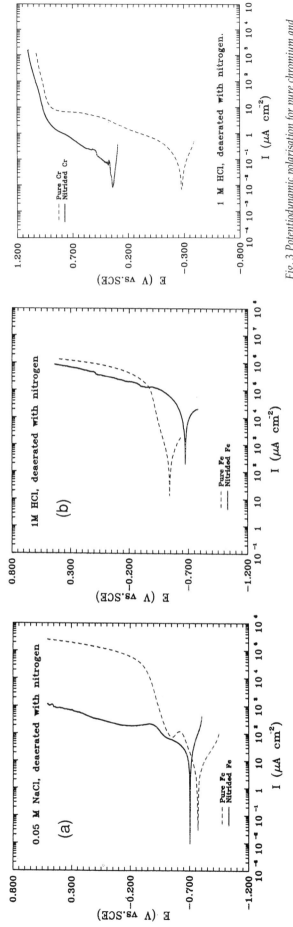

Fig. 2 Potentiodynamic polarisation for pure iron and nitrided iron in deaerated (a) 0.05M NaCl (pH 7.5) and (c) 1M HCl solution.

Fig. 3 Potentiodynamic polarisation for pure chromium and nitrided chromium in deaerated 1M HCl solution.

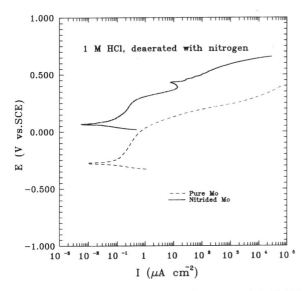

Fig. 4 Potentiodynamic polarisation curves of pure Mo and nitrided Mo in deaerated 1M HCl solution.

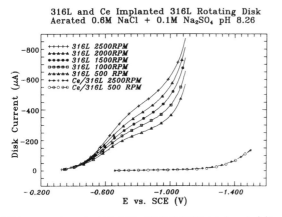

Fig. 5 Disc current vs potential plot for UNS S31603 stainless steel discs, unimplanted and cerium ion implanted, cathodically polarised in an aerated 0.1M Na₂SO₄ + 0.6M NaCl solution (pH = 8.26) at different disc rotating speeds.

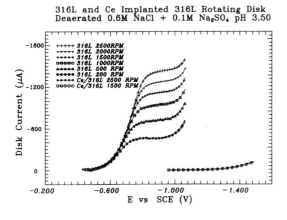

Fig. 6 Disc current vs potential plot for the rotating discs in a deaerated 0.1M Na₂SO₄ + 0.6M NaCl solution (pH = 3.50) at different speeds.

solution in the absence of surface oxides/hydroxides, the cathodic current at potentials more positive than – 1200 mV is entirely due to hydrogen reduction. Figure 6 shows the current vs potential for UNS S31603 discs cathodically polarised at a range of rotating speeds. It is clear that the cathodic current contributed from the reduction of hydrogen ions for the stainless steel disc is rotation-speed dependant. The cathodic current on cerium-implanted discs is restrained greatly and appears not to depend significantly on rotating speed. whereas on the unimplanted surfaces, the reduction reactions are controlled by mass transport. However, cerium ion implantation greatly inhibits this reduction, as it does the reduction of oxygen.

Cerium has a much higher affinity for oxygen than have the transition metal alloying elements in stainless steels. The highly stable cerium oxide may therefore form a barrier on the stainless steel surface. However, with a surface concentration of only 1.24 at.%, one cannot expect the cerium to form a continuous barrier covering the whole surface. It is possible that the cerium oxide may 'poison' only the active sites at which the cathodic reaction is normally catalysed. These active sites could be the relatively sparse kink and step sites on the crystalline surface [22], which have been shown to be readily inhibited, for example, during the dissolution of alkali halides [23]. This makes the reduction of the oxygen take place under higher overpotential and with more than two orders of magnitude smaller current density than that on an unimplanted steel electrode.

The stable cerium oxide may also be expected to block anodic dissolution. This is demonstrated in Fig. 7, where the anodic polarisation of unimplanted and implanted steels are compared. In addition to reducing the passive current by at least one order of magnitude, the implantation has increased the breakdown potential by at least 250 mV.

A parallel ASTM G48 B crevice corrosion test also demonstrates that the cerium-implanted sample had no crevice corrosion after 24 h of testing at room temperature, while an untreated sample was readily attacked by the 10% ferric chloride solution. A field test is also in progress in Florida [24]. Preliminary observations from that test confirm the beneficial effect of cerium in improving the crevice corrosion resistance of stainless steels in tropical sea water.

4. Conclusions

Nitrogen enhances the anodic segregation of beneficial alloying elements, such as chromium, when local attack takes place and significantly prevents the transpassive dissolution of molybdenum. The effects of alloying elements, chromium, molybdenum and

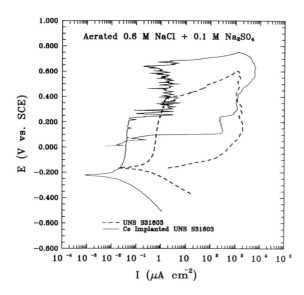

Fig. 7 Potentiodynamic polarisation curves of unimplanted and cerium-implanted UNS S31603 steels in aerated 0.1M Na$_2$SO$_4$ + 0.6M NaCl solution.

nitrogen improve the localised corrosion resistance of stainless steels by inhibiting the anodic processes. However, in order to further increase localised corrosion resistance, it is also necessary to constrain the cathodic reactions.

Cerium ion implantation on UNSS31603 stainless steel is very effective in inhibiting both the cathodic and anodic electrode reaction involved in metallic corrosion. The action of the cerium is to form the thermodynamically-stable cerium oxide, which, even in very small quantity, reduces the cathodic and anodic reactivity, presumably by blocking the reactive surface sites, such as kinks in atomic ledges. As a result of the inhibition of the kinetics of electrode processes, cerium treatment improves the localised corrosion resistance, and especially crevice corrosion resistance, of stainless steels. This is supported by results obtained by both electrochemical measurements and by accelerated corrosion tests.

5. Acknowledgement

This work was supported by grants from Chemetics International Company Ltd., The University Research Incentive Fund (Province of Ontario), and the Natural Sciences and Engineering Research Council of Canada.

References

1. M. G. Fontana and N. D. Greene, Corrosion Engineering, McGraw-Hill, New York, 41, 1967.
2. K. Sugimoto and Y. Sawada, Corros. Sci., **17**, 425(1979).
3. J. Eckenrod and C. W. Kovack, ASTM STP 679, PA: American Society for Testing and Materials [ASTM], Philadelphia, 1977, p. 17.
4. K. Osozawa and N. Okato, in Passivity and its Breakdown on Iron and Iron Based Alloys, p. 135. NACE, Houston, TX (1976).
5. J. E. Truman, M. J. Coteman and K. R. Pirt, Brit. Corros. J., **12**, 236, 1977.
6. A. J. Sedriks, Intl. Metal. Reviews, **28**, 306, 1983.
7. R. Bandy and D. Van Rooyen, Corrosion, **39**, 227, 1983.
8. O. I. Lukin *et al.*, Zashet. Met., **115**, 545, 1979.
9. K. Hashimoto, K. Asami and K. Teramoto, Corros. Sci., **19**, 3, 1979.
10. M. Sakashita and N. Sato, in 'Passivity of Metals' R. P. Frankenthal and J. Kruger, Editors, p. 479, The electrochemical Society, Corrosion Monograph Series, Princeton, N.J., 1978.
11. Y. C. Lu, and C. R. Clayton, J. Electrochem. Soc., **132**, 2517, 1985.
12. Y. C. Lu, R. Bandy, C. R. Clayton and R. C. Newman, J. Electrochem. Soc., **130**, 1774, 1983.
13. C. R. Clayton, L. Rosenzweig, M. Oversluizen and Y. C. Lu, in 'Surfaces, Inhibition and Passivation', E. McCafferty and R. J. Brodd, eds, Electrochem. Soc., Pennington, NJ, 1986, p. 323.
14. Y. C. Lu, M. B. Ives and C. R. Clayton, 'Synergism of Alloying Elements and Pitting Corrosion Resistance of Stainless Steels', Corros. Sci., in press.
15. Y. C. Lu and M. B. Ives, Corros . Sci., **33**, 2, 317, 1989.
16. D. R. Arnott, B. R. W. Hinton and N. E. Ryan, Corrosion, **45**, 12, 1989.
17. F. Mansfeld, S. Lin, S. Kim and H. Shih, J. Electrochem. Soc., **137**, 78, 1990.
18. H. Shih, Y. Wang and F. Mansfeld, Corrosion '91 paper No. 136, NACE, Houston, 1991.
19. Y. C. Lu and M. B. Ives, Corros. Sci., **34**, 1173, 1993.
20. M. B. Ives, Y. C. Lu and J. L. Luo, Corros. Sci., **32**, 91, 1991.
21. V. G. Levich, Physicochemical Hydrodynamics, Prentice Hall, Inc. Englewood Cliffs, N. J., 1962.
22. W. F. Burton, N. Cabrera and F. C. Frank, Proc. Roy. Soc., **243A**, 299, 1950.
23. M. B. Ives, Ind. Eng. Chem., **S7**, 34, 1965.
24. M. B. Ives, in 'Applications of Stainless Steels '92, Kristiantads Boktryckeri AB, Stockholm, Sweden, **1**, 436, 1992.

30

Modification of Passive Zinc by Alloying Elements

*W. Paatsch and W. J. Lorenz**

Bundesanstalt für Materialforschung und -prüfung (BAM), D-12 200 Berlin, Germany
*Universität Karlsruhe, Germany

Abstract

The semiconductor properties of passive films formed in alkaline aqueous solutions on Zn, Zn–Co, and Zn–Ni and Zn–Fe substrates were studied by electrochemical and photoelectrochemical methods. Substrate doping with Co, Ni and Fe leads to a homogeneous distribution of these elements in the passive oxide layer. The density and energy distribution of shallow donor states were determined from deviations from the ideal Mott-Schottky behaviour observed by electrochemical impedance measurements and from optical quantum efficiency experiments. The doping elements Co, Ni and Fe seem to generate deep donor levels, as indicated by impedance measurements.

1. Experimental

Zn and low-alloyed Zn films of 30 μm thickness were electrodeposited on a steel substrate at T = 298 K, using a boric acid weak electrolyte of pH = 5.5, containing $CoCl_2$ or $NiCl_2$. Zn–Fe was deposited using alkaline solutions. According to microprobe analysis the alloying metals are homogeneously distributed in the Zn-matrix. All potentials are referred to the saturated calomel electrode (SCE) as the reference electrode.

The passive layers on Zn and the alloyed substrates were performed at 0.6 V < E_{SCE} < 1.0 V for 2 h. The depth distribution of Co, Ni and Fe within the doped passive oxide films was investigated by X-ray photoelectron spectroscopy (XPS). Details of the experimental setup and the electronic equipment used for the electrochemical dc and ac measurements (cyclic voltammetry, galvanostatic transients, and the electrochemical impedance spectroscopy (EIS)) have been described elsewhere [1]. Photoelectrochemical experiments were carried out using chopped light techniques [2].

2. Results and Discussion

2.1 Electrochemical measurements

The electrochemical behaviour of the different substrates in alkaline solution is characterised by cyclic voltammograms in Fig. 1. The only significant difference between Zn and the Zn alloys can be found in the transition from the passive to the transpassive state characterised by an additional anodic peak being similar to transpassive behaviour of pure Co and Ni [3,4].

Electrochemical impedance measurements were carried out in the potential range –0.5 V < E_{SCE} < 0.6 V. The upper limit was chosen to avoid interferences of the faradaic process occurring at higher anodic potentials (Fig. 1). Thus all samples studied were

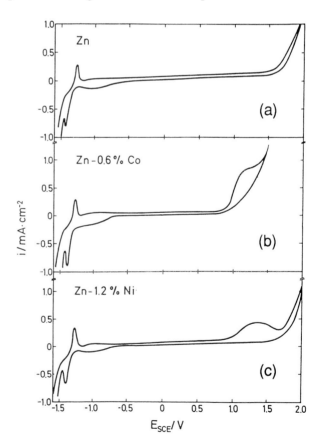

Fig. 1 Cyclic voltammograms of (a) Zn, (b) Zn–0.6% Co and (c) Zn–1.2% Ni in aerated 0.1M NaOH at T = 298 K. $dE_{SCE}/dt = 1.0$ mV s^{-1}.

prepolarised at $E_{SCE} = 0.6$ V for 2 h in order to establish a quasi-steady state corresponding to a nearly constant thickness and composition of the passive layer.

The capacitances of the passive films at different potentials were determined from the imaginary part of the impedance data at $f = 4 \times 10^3 s^{-1}$. The impedance data are nearly frequency-independent in the range $10^{-3} s^{-1} \leqslant f \leqslant 10^4 s^{-1}$. The values of C obtained from the experiments were found to be about one order of magnitude lower than typical C_H values. Therefore, C_{sc} values were calculated from the experimental C data corrected by a constant and typical value of $C_H = 20 \times 10^{-6} F\, cm^{-2}$.

Figure 2 shows the well known Mott-Schottky plot for passive films on Zn and Zn–Co substrates. Similar results are received for Zn–Ni and Zn–Fe samples. It can be seen that the C_{sc} values follow the Mott-Schottky relation in the investigated potential range for the passive films on pure Zn and Zn–Co alloys below 0.6% Co. At higher cobalt contents, however, the data deviate from the linear Mott-Schottky relation at $E_{SCE} > + 0.4$ V which is more or less the same for passive Zn–Ni

and Zn–Fe and which is not due to anodic faraday reactions. The mean flatband potential of E_{FB} -0.75 V agrees with literature data [5]. The apparent donor concentrations of the order of magnitude of $10^{20} cm^{-3}$ are relatively high but typical for many semiconducting passive films. The electronic properties of the n-type semiconducting passive films on Zn–Co, Zn–Ni and Zn–Fe alloys are obviously not ideally compatible with the simple Mott-Schottky relation. The observed deviations can be explained by assuming deep electronic states that become ionised only at higher potentials [6,7].

2.2 Photoeletrochemical measurements

Photocurrent–potential curves at constant photon energy of $h\nu = 3.5$ eV of anodised Zn, Zn–Co, Zn–Ni and Zn–Fe substrates showed an onset of the photocurrents at potentials for E_{SCE} of -0.7 V, in agreement with the EFB values obtained by extrapolation of the linear portion of the Mott-Schottky plots. Furthermore, photovaltage measurements of anodised Zn and Zn-alloy substrates carried out galvanostatically at a constant photon energy of $h\nu = 3.5$ eV showed an onset of the photovoltage for E_{FB} of -0.7 V. The negative sign of the observed photovoltage, E_{ph}, indicated n-type semiconducting properties of the different passive layers.

If the quantum efficiency Q is plotted as $(Qh\nu)^2$ vs $h\nu$ according to [8] and [9] the band gap energy E_g is given by the plot with the intercept of the photon energy axis. For the single crystal ZnO, the band gap energy $E_g = 3.2$ eV is independent of the potential (Fig. 3). The results for passive films formed on unalloyed and alloyed Zn substrates can also be represented by a $(Qh\nu)^2$ vs $h\nu$ plot, as shown in Fig. 3. Straight lines are obtained exhibiting intercepts, $h\nu_o = E_g$, which, however, vary with the electrode potential. In contrast to crystalline ZnO the anodically formed passive layers

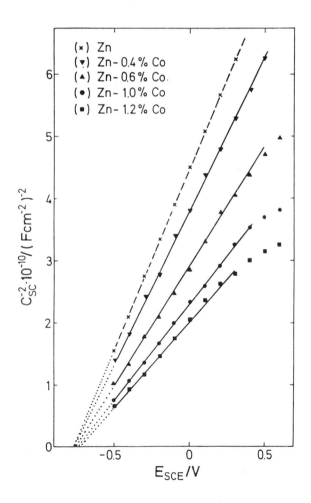

Fig. 2 Mott-Schottky plots of passive layers on Zn and Zn–Co substrates.

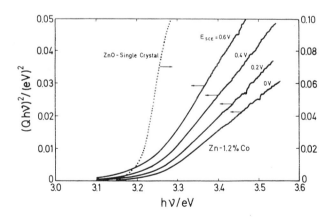

Fig. 3 Determination of the band gap energy of ZnO single crystal (dotted curve) and the passive layers on Zn–1.2% Co substrate (full curves) depending on the electrode potential ($0\, V < E_{SCE} < 0.6\, V$).

on Zn, Zn–Co, Zn–Ni and Zn–Fe substrates show higher E_G values which decrease with increasing potential approaching the value $E_g = 3.2$ eV at about E_{SCE} + 1.0 V. The higher band gap energy of the anodically formed passive layers on unalloyed and alloyed Zn substrates may be explained by structural changes compared to the ZnO single crystal structure. Due to coulometric as well as XPS depth profile measurements the thickness of the anodic oxide layers is *ca.* 100 Å and supposed to be quasi amorphous. The decrease of the band gap energy of all anodised samples with increasing band-bending can be understood in terms of the energy distribution schematically represented in Fig. 4. A decrease of the Fermi energy produces a depletion of electronic states below E_F^o.

Accordingly, the energy of the optical transitions, $h\nu_1$ from the valence band to the first unoccupied levels is decreased when the Fermi energy E_F is shifted down into the band gap with increasing $\Delta\Phi sc$. Deep energy levels within the band gap located more than 0.7 eV below flatband energy may be generated by the doping elements Co and Ni as calculated after [8, 9]. However, they could not be observed by optical means. The transition from the valence band to these deep states, which become partly ionised due to higher band-bending have either a too-small optical absorption coefficient or a two-small quantum yield to be observed with the experimental setup used in this work.

The corrosion inhibition function of the alloy d-metal components can be qualitatively explained by a dopant–valency–interaction model[10]. The corrosion current is reduced when the charge carriers are eliminated. These are mobile, negatively charged Zn cation vacancies. The segregated alloy elements, according to XPS measurements, can combine with such vacancies and scavenge further vacancies by forming immobile ion pairs.

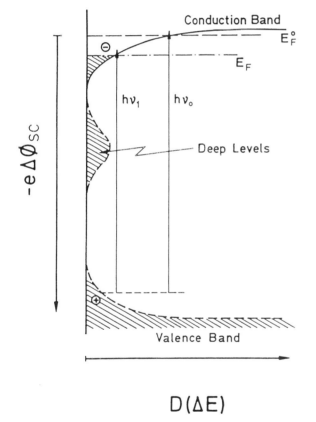

$$D(\Delta E)$$

Fig. 4 Schematic state distribution; $\Delta\Phi_{sc} = E_{SCE}$ *(electrode potential vs SCE) -* E_{FB} *(flatband potential),* $D(\Delta E)$ *= density of states,* E_F *= Fermi energy,* E_F^o *= Fermi energy under flatband condition.*

References

1. J. Hitzig, K. Jüttner, W.J. Lorenz and W. Paatsch, J. Electrochem. Soc., **133**, 888, 1986.
2. W. Paatsch, J. Phys., Paris, **38**, C 5-151, 1977.
3. H. Gomez Meier, J. R. Vilche and A. J. Arvia, J. Electroanal. Chem., **138**, 367, 1982.
4. R. S. Schrebler Guzman, J. R. Vilche and A. J. Arvia, J. Electrochem. Soc., **125**, 1578, 1978.
5. B. Pettinger, H. R. Schoppel, T. Yokoyama and H. Gerischer, Ber. Bunsenges. Phys., Chem., **78**, 1024, 1974.
6. M. H. Dean and U. Stimming, Electroanal. Chem., **228**, 135, 1987.
7. M. H. Dean and U. Stimming, Corros. Sci., **29**, 199, 1989.
8. W. W. Gartner, Phys. Rev., **116**, 84, 1959.
9. W. Kautek, H. Gerischer and H. Tributsch, J. Electrochem. Soc., **127**, 2471, 1980.
10. W. Kantek, M. Sahre and W. Paatsch, Electrochim. Acta, in press.

31

Modification of Passive Films on 304L by Cyclic Polarisation Influence on Film Stability in Acidic Environments

C. Duret–Thual and F. Barrau

IRSID Unieux, Pont du Sauze, Z. I. du Parc 42490 Fraisses, France

Abstract

Polarisation cycles were applied to 304L electrodes in different solutions Na_2SO_4 and $(NH_4)_2SO_4$ at two pH(3 and 7). The modified passive layers resulting from the different oxidation-reduction cycles were studied, in particular their corrosion behaviour in acidic environments. The electrochemical response of the surface during the polarisation cycles was significantly different as a function of the cation present in the electrolyte. Impedance measurements were conducted in the treatment solution after different potentiostatic exposures (from 2 to 20 and 70 h). An influence of the initial electrochemical treatment was still detected after 20 h of exposure. The stability of the modified passive layer was studied in deaerated 2M H_2SO_4 at room temperature using a reactivation test (cathodic scan, 3 V/h from –200 mV/SSE). All the surfaces electrochemically treated exhibited an improved behaviour as compared to the as-polished condition. The cyclic treatments performed at pH 3 in Na_2SO_4 appeared as the more beneficial with respect to corrosion resistance.

1. Introduction

Oxide layers with electrocatalytic properties may be developed on metallic electrodes by simple cyclic polarisation [1, 2]. In this study, polarisation cycles were applied to 304L electrodes in different electrolytes, with the aim of examining:

- the evolution of the passive layer characteristics resulting from these electrochemical treatments,
- the corrosion resistance of these layers with respect to their stability in acidic environments.

2. Experimental

2.1 Material and surface preparation

All the experiments were performed using the same heat of cold rolled 304L the composition of which is given in Table 1. The specimens were square sheets of $20 \times 20 \times 1$ mm dimensions. Surface preparation involved grinding with SiC abrasive paper (grade 1200) then polishing with diamond paste to 1 μm finish. All the electrochemical experiments were carried out after a minimum air exposure of 16 h after specimen preparation.

2.2 Electrochemical treatments

Two solutions were used for the electrochemical treatments, 0.1M ammonium sulphate and 0. 1M sodium sulphate at different pH. They were prepared with deionized water and RP NORMAPUR reagents ($\geq$ 99.5%). The pH was adjusted by addition of concentrated H_2SO_4, NH_4OH or NaOH.

Electrochemical treatments mainly consisted of polarisation cycles or potentiostatic exposures (–200 mV against the saturated sulphate electrode SSE) in the abovementioned solutions. The polarisation cycles were applied at room temperature from –2000 to +300 mV/SSE with two sweep rates 20 and 100 mV s^{-1}, –2000 mV being the end potential value. After twenty cycles, the specimens were removed from the electrochemical cell, rinsed and sonicated in deionised

Table 1 Composition (wt%)

C	Mn	Si	S	P	Ni	Cr	Mo	Cu	Al	Ti
0.018	1.549	0.385	0.005	0.018	9.26	18.38	0.058	<0.025	<0.003	<0.010

water then dried using acetone and a flash of warm air (T ≤ 35∞C) to prevent water condensation. EIS experiments were conducted in the same solution as the polarisation cycles, so in these cases the specimens were kept in the same electrochemical cell. Electrode surfaces were always in the vertical position in the cell.

Prior to the electrochemical treatments (potentiostatic or polarisation cycling), the specimens were held at –2000 mV/SSE for 5 min in order to get an initial standard surface. Previous descriptions of this kind of cathodic treatment showed that not all oxides present were removed (especially Cr oxides) [3].

2.3 Evaluation of modified passive films

To evaluate the modifications of the passive layer induced by the electrochemical treatments, different techniques involving both *ex situ* and *in situ* experiments were used. The *in situ* studies were performed by Electrochemical Impedance Spectrometry (EIS). After the polarisation cycles, EIS spectra were recorded at –200 mV/SSE after a stabilisation of 2 h at this potential. The effect of a prolonged exposure at –200 mV/SSE was also investigated. To examine the change of behaviour induced by the polarisation cycles, we compared the EIS spectra obtained after similar time exposure at –200 mV with or without initial electrochemical treatment. Experiments were performed using an EG&G 273A potentiostat and a Schlumberger 1250 frequency response analyser. EIS was conducted between 65 kHz and 2 mHz at an amplitude of 10 mV rms.

Chemical *ex situ* analyses were carried out on a few samples by GDOS (Glow Discharge Optical Spectrometry) and XPS (X-Ray Photoelectron Spectroscopy). Qualitative data on both film chemical composition and thickness were obtained in this study. The samples investigated were subjected to the electrochemical treatment and then exposed to air before analysis. They were compared to the as polished specimen exposed to air.

To study the behaviour of electrochemically modified electrodes in acidic environment, a reactivation test was conducted in deaerated sulphuric acid. After deaeration of the cell and of the electrolyte in a separate vessel, the solution was introduced, the specimen being polarised at the starting potential (–200 mV/SSE). Before beginning the potential sweep (3 V. h⁻¹) in the cathodic direction, the specimen was polarised 2 mins at –200 mV/SSE. The acid concentration (2M) was adjusted to obtain for the reference sample (as polished specimen) a significant reactivation peak. All the tests were performed at room temperature.

3. Results and Discussion

3.1 Polarisation cycles

Cyclic polarisation was applied to 304L specimens in the four solutions (ammonium and sodium sulphate pH 3 and 7) with two sweep rates 100 and 20 mV s⁻¹. The three parameters investigated affected the voltammograms obtained to various degrees.

The cation and sweep rate effects are displayed in Figs 1 and 2, showing the sample behaviour during the last cycle (20th). The cation influence is quite important since it induces changes in the anodic peaks and in the current densities involved. Characteristic values (potential, current) from the voltammograms are gathered in Table 2.

In the presence of Na⁺, two anodic peaks are generally seen around –1000 and –500 mV/SSE whereas in the presence of NH4⁺, only one broad anodic peak is present in the same potential range. In one condition (pH 3, 20mV.s⁻¹), the voltammograms exhibited one broad anodic peak in the two media (Fig. 1(b)). A general feature of the curves obtained in the sodium sulphate solutions is also that the hydrogen evolution reaction takes place at lower potentials.

Depending on pH and sweep rate, different evolutions of current densities were observed. At pH 3, higher anodic current densities (anodic potential

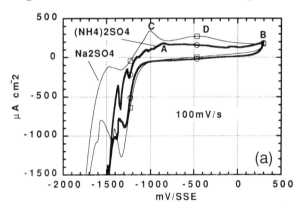

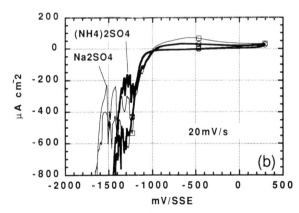

Fig. 1 Cyclic voltammertry in pH 3 solutions (20th cycle).

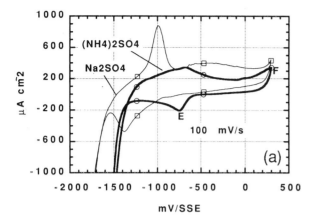

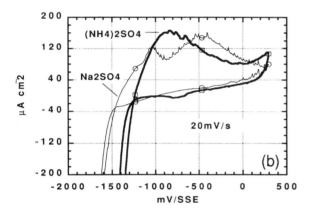

Fig. 2 Cyclic voltammetry in pH 7 solutions (20th cycle).

scan) are associated with the presence of Na+. This is particularly the case at 20 mV.s^{-1} where the maximum anodic current density increases with the number of cycles whereas it decreases in the presence of NH4+. A very disturbed zone is observed in the cathodic potential range (1700 to –1300 mV/SSE), showing current oscillations in the two scanning directions. These current variations are more pronounced in the sodium sulphate solution and at the lower sweep rate. The reason for these variations has not been yet elucidated. However H$_2$ evolution and pH variations at the electrode surface could be partly responsible for these phenomena. At the higher pH, the general shape of the voltammograms in the anodic potential scan is the same but the involved current densities are higher (see Table 2) and increase with the number of cycles contrarily to the results obtained at pH 3. In the reverse scan, reduction peaks (around –750 mV/SSE) are present at the 20th cycle especially in the presence of NH4^{+}.

The occurrence of these different peaks and their identification will be discussed taking into account experiments conducted in the same solutions, at different scanning rates and in an extended potential range (–2000, +1000 mV/SSE). The behaviour of pure

iron and chromium will be also considered.

At first it should be mentioned that after the surface cathodic 'cleaning' (–2000 mV/SSE), the surface is not free from chromium oxide [3]. Moreover during the successive passivation–reduction cycles it is likely that the Fe-rich mixed oxide–hydroxide layer is not completely reduced especially in the conditions of pH 7 and high scanning rate, as shown by [4] at pH 8.5 and [5,6] at a higher pH in the case of pure iron. A thickening of the passive layer then results from such polarisation cycles.

Some useful information was deduced from the experiments conducted at high potentials with respect to the contribution of the Cr (III)/Cr (VI) redox behaviour. In the different solutions used in this study, the oxidation Cr (III)/Cr (VI) takes place in the potential range + 650 mV/SSE and a conjugated reduction peak is observed around 0 mV/SSE. In agreement with [7], we found that the reduction peak was associated with a change of Cr (VI) to Cr (III) species occurring in the film. The fact that the addition of Cr (VI) to the solution did not produce any modification of the voltammogram supported this idea. With respect to the oxidation of Cr to Cr (III), several papers [8,9] describe the mechanism of pure chromium oxidation in acidic solutions. In our experimental conditions, an active/passive transition cannot be observed even at pH 3 as mentioned by [8]. As nickel selectively dissolves and does not contribute to a significant extent to the constitution of the passive film [10], the transition Ni/Ni (II) may participate to the anodic current in the potential range –700, –800 mV/SSE as deduced from the voltammogram obtained for pure Ni (Table 2). The anodic peaks present in the voltammograms (–2000, +300) are then to be associated mainly with iron redox transitions.

A rather complete description of the cyclic voltammetry diagrams for pure iron in various neutral or alkaline media has been proposed recently by several authors [4,5]. The comparison of these results with those obtained by [11] on Fe–Cr and on the alloy 316L by [7] supports the following preliminary conclusions:

- the broad anodic peak A (Fig. 1(a)) observed mainly in ammonium sulphate corresponds to different oxidation steps leading to the formation of Fe (II) and Fe (III) oxides and hydroxides with probably the presence of FeOOH in an external layer; the increase in current B (Fig. 1a) observed before the Cr (III) to Cr (VI) transition comes from subsequent oxidation of Fe(II) species present in the film in Fe (III) as mentioned by [7]. The two well differentiated oxidation peaks C and D (Fig. 1a) detected in the sodium sulphate solution probably indicate the formation of Fe (II) soluble species for the more cathodic one, the second and more broad peak D originating from the other oxidation steps as in the previous solution;

Table 2 Characteristics of the polarisation cycles

	mV s⁻¹	Anodic scan				Cathodic scan	
	mV s^{-1}	mV/SSE	µA cm^{-2}	mV/SSE	µA cm^{-2}	mV/SSE	mV/SSE
NH$_4^+$ 304L pH3	100			-850	180	-1300	
	20			-650	32		
Na$^+$ 304L pH3	100	-1000	340	-500	270	-1325	
	20			-550	70		
NH$_4^+$ 304L pH7	100			-680	345		-750
	20			-840	170		-800
NH$_4^+$ Fe pH7	20	-1120		-720		-1320	-920
NH$_4^+$ Ni pH7	20			-770			-900
Na$^+$ 304L pH7	100	-990	880	-450	400	-1380	
	20	-1050	128	-500	160	-1370	

- the peak E at –750 mV/SSE (Fig. 2(a)) in the cathodic scan is conjugated with the current increase F at +200 mV (Fig. 2(a)) and thus is related to Fe (III) to Fe (II) reduction, possibly in the film. The last more cathodic peak then indicates subsequent reduction of species probably coming from the solution.

3.2 Electrochemical impedance spectroscopy

Our experimental procedure involved a potentiostatic hold of the specimens at –200 mV/SSE before recording the first EIS spectra. Some information on film evolution could then be derived from the current density change during this time period, depending on the prior electrochemical treatment. The observed general trend is that higher quantities of charge are involved in the passive layer evolution for specimens previously subjected to cyclic polarisations. It is noticed that more than 2 h of potentiostatic hold are required to obtain a 'stable' passive film. This was confirmed by EIS spectra recorded after different time periods.

Impedance data were obtained systematically after 2 and 20 h of ageing at –200 mV/SSE. Nyquist plots vs the investigated parameters are represented in Fig. 3. They are characteristic of the behaviour of passive electrodes and they can be described as portions of more or less depressed semi-circles. In a few cases a deviation from this behaviour was observed at low frequencies after 2 h ageing and attributed to an evolution of the passive interface during the impedance

measurement. The diagrams exhibited significant differences as a function of pH, cations present in the solution and exposure at –200 mV/SSE. Using the EQUIVCRT program [13], low frequency resistance's were calculated from the experimental data and their variations with the experimental conditions were the following:

- at pH 3, in the NH4+ containing solution, similar trends were noted between 2 and 20 h exposure, the Rlf (low frequency resistance) values being in the order 'cyclic 100 mV.s⁻¹ > 20 mV.s⁻¹ > –200 mV potentiostatic'. In the presence of Na+ no significant differences are detected after 2 h for the different treatments investigated.

- at pH 7 the same Rlf ranking is obtained in all conditions ('cyclic 100 mV.s⁻¹ > –200 mV potentiostatic > 20 mV.s⁻¹'). Rlf corresponding to the NH4+ containing solution are higher than those obtained in the presence of Na$^+$. After 20 h of potentiostatic exposure, the influence of the prior electrochemical treatment was still observed. Experiments carried out after 70 h however confirm that this effect no longer exists.

3.3 Reactivation tests in deaerated 2M sulphuric acid

These tests were carried out after air exposure of the specimens. In the 'as-polished' conditions, the reactivation curves (2M H$_2$SO$_4$, cathodic scan from –200 mV/SSE, 3V.h⁻¹) presented two peaks at –770 and –850 mV/SSE with maximum current densities in the order

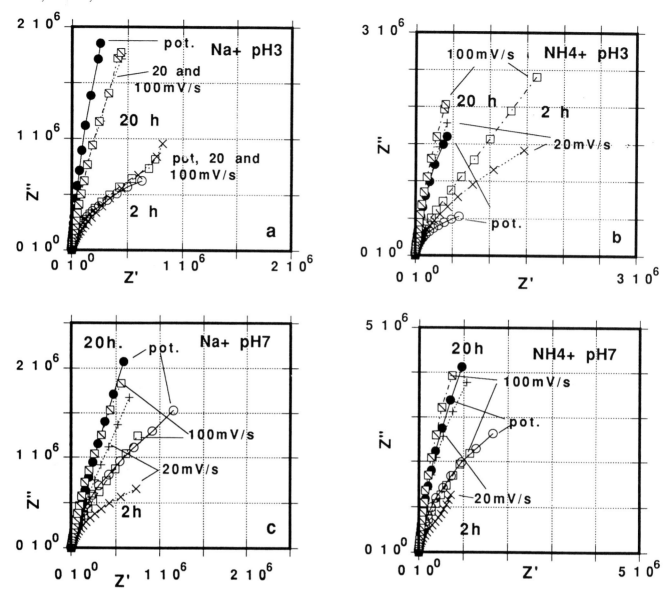

Fig. 3 Nyquist plots obtained in sodium and ammonium sulphate solution after 2 and 20 h at –200 mV/SSE with or without prior electrochemical treatment (pot. = –200 mV/SSE not previously cycled) Z' = real part (V) , Z"= imaginary part (V) (in all diagrams first point 65 kHz, last point 2.5 nHz, 10 points per decade).

of respectively 80 to 160 $\mu A.cm^{-2}$. The specimen surface was tarnished in these conditions. All the electrochemical treatments led to an improvement of film stability as shown in Table 3; most often no surface degradation was apparent after the test. The different specimens were compared on the basis of the peak area referring to a base line defined for each reactivation curve. The calculated quantities of charge are therefore minimised with respect to the real value. The highest resistance to the reactivation test was obtained by cyclic polarisation at pH 3. It is noteworthy that all specimens electrochemically treated exhibited a major reactivation peak located around –850 mV/SSE, the second one (–770 mV) being most often insignificant.

The occurrence of the two peaks are related to the structure and composition of the passive layer. In more diluted sulphuric acid (0.5 and 1M) at the same scanning rate, only the first peak is detected at –770 mV/SSE.

3.4 Surface analyses

GDOS analyses were essentially performed on specimens treated by cyclic polarisation at pH 7 in the ammonium sulphate solution. The thickness of the layers developed after treatment did not differ significantly from those obtained in the as-polished conditions. No nitrogen enrichment was observed in the superficial layer by this method.

Table 3 Results from the reactivation test in 2M deaerated sulphuric acid

pH/ion/treatment	E$_1$ peak (mV/SSE)	i$_1$ (μA cm^{-2})	E$_2$ peak (mV/SSE)	i$_2$ (μA cm^{-2})	Q total mC cm^{-2}
As polished	-760 to -770	44 to 78	-845 to -850	86 to 174	13.1 to 18.2
7/Na$^+$/-200	-780	5	-876	116	5.2
7/NH$_4^+$/-200	-773	6	-869	62	2.4
3/Na$^+$/20					0
3/Na$^+$/100			-863	1 to 4	0.02
3/NH$_4^+$/20					0
3/NH$_4^+$/100	-733	6	-867	15	1.1
7/Na$^+$/20	-783	2	-824	2.8	0.13
7/Na$^+$/100	-788	3	-839	5 to 13	0.4 to 0.8
7/NH$_4^+$/20	-760	5			0.4
7/NH$_4^+$/100	-762	3	-850	13	0.8

XPS qualitative analyses were conducted on samples treated at pH 3. Higher thicknesses were found for passive layers grown by cyclic polarisations compared to the as-polished condition. Differences in the chemical composition of the films were also detected and should be specified in the future.

4. Conclusion

Cyclic polarisations induced passive film modification and improved the film behaviour in deaerated sulphuric acid. An important influence of the cation was found since it affects the electrochemical behaviour of the surface as shown in the voltammograms. Such effects were ascribed to adsorption processes [14] on pure iron. In this case, similar internal oxide layers were obtained for both Na+ and NH$_4^+$ cations but the outer oxyhydroxyde layers differed in thickness [14]. In the present study it appeared from preliminary results of surface analyses that passive layers grown by successive reduction cycles are thicker than those obtained in as-polished conditions. No obvious correlation between impedance data and the results from the reactivation test was observed, especially when the behaviour of 'potentiostatic' specimens was compared to that of the 'cycled' state. The layer thickness is probably an important parameter to consider in the reactivation test. The presence of nitrogen in the passive film has to be carefully studied, especially in the case of electrochemical treatments conducted in NH^{4+} containing solutions, since this element could play a role in modifying the hydrogen evolution reaction kinetics as mentioned by [15] for Fe–N alloys. Nitrogen effects on the film stability in acidic conditions should also be taken into consideration. More investigations are required particularly chemical analyses, to give a more accurate interpretation of the data generated in this study. Work is in progress in this area.

Electrochemical treatments such as cyclic polarisation cycles thus provide a meaningful way to modify passive films and possibly increase their resistance to specific environments. Such studies also contribute to a better understanding of passivation processes.

5. Acknowledgements

The authors gratefully acknowledge the contribution of D. Costa, G. Boutin for XPS and GDOS analyses and N. Dowling for helpful discussion.

References

1. M. R. Gennero de Chialvo and A. Chialvo, Electrochimica Acta, **36**, 13, 1963–1969, 1991.
2. F. Rouina, Thèse Institut National Polytechnique de Grenoble, February 1990.
3. P. Marcus and I. Olejord, Corros. Sci., **28**, 6, 589–602, 1988.
4. C. A. Melendres, M. Pankuch, Y. S. Li and R.L. Knight, Electrochimica Acta, **37**, 15, 2747-2754, 1992.
5. Z. Szklarska-Smialowska, T. Zakroczymski and C. J. Fan, J. Electrochem. Soc., **132**, 11, 2543–2548, 1985.
6. A. Hugot, Le Goff, J. Flis, N. Boucherit, S. Joiret and J. Wilinski, J. Electrochem. Soc., **137**, 9, 2684–2690, 1990.

7. N. Ramasubramanian, N. Preocanin and R. D. Davidson, J. Electrochem. Soc., **132**, 4, 793–798, 1985.

8. L. Björnkvist and I. Olejord, Corros. Sci., **32**, 2, 231–242, 1991.

9. T. P. Moffat and R. M. Latanision, J. Electrochem. Soc., **139**, 7, 1869–1879, 1992.

10. E. de Vito, Thèse Universite Paris VI, November 1992.

11. D. Thierry, D. Persson, C. Leygraf, D. Delichere, S. Joiret, C. Pallotta and A. Hugot-Le Goff, J. Electrochem. Soc., **135**, 2, 305–310, 1988.

12. J. H. Gerretsen and J. H. W. De Wit, Corros. Sci., **31**, 545-550, 1990.

13. B. A. Boukamp, Equivalent Circuit (EQUIVCRT.PAS) Users manual, 2nd Edition, 1989.

14. S. Juanto, R. S. Schrebler, J. O. Zerbino, J. R. Vilche and A. J. Arvia, Electrochimica Acta, **36**, 7, 1143–1150, 1991.

15. V. Brusic, G. S. Frankel, B. M. Rush, A. G. Schrott, C. Jahnes, M. A. Russak and T. Petersen, J. Electrochem. Soc., **139**, 6, 1530–1535, 1992.

32

Surface Modification of Stainless Steel by an Alternating Voltage Process

F. Mansfeld, S. H. Lin and L. Kwiatkowski

Corrosion and Environmental Effects Laboratory (CEEL), Department of Materials Science and Engineering, University of Southern California, Los Angeles, CA 90089-0241, USA

Abstract

In the Alternating Voltage Passivation Process (AVPP) a potential square wave is applied for a certain length of time at a constant DC bias potential E_o in the passive region. The effects of the process parameters E_o, pulse length P, amplitude A and ratio R_o of the duration of the anodic to the cathodic portion of the pulse in 0.05M H_2SO_4 on the passive properties of type 304 stainless steel have been evaluated using factorial design experiments. The passive properties have been determined by recording the time to reactivation t_p during open-circuit potential decay and the critical current density i_{crit} and the passive current density i_p recorded in 0.1M H_2SO_4 after open-circuit potential decay. For the optimum process parameters a significant increase of t_p and large decreases of i_{crit} and i_p have been determined. Surface analysis by Auger electron spectroscopy suggests that AV passivation increases the thickness of the passive film and leads to chromium enrichment.

1. Introduction

It has been shown recently that surface modification by an electrochemical process—Alternating Voltage Passivation Process (AVPP)—applied to stainless steels (SS) produces surfaces with improved corrosion resistance [1, 2]. In the AVPP, a potential square wave is applied for a certain length of time at a constant bias potential E_o. The main process parameters are pulse amplitude A, pulse width P, E_o and ratio R_o of the duration of the anodic to the cathodic portion of the pulse (Fig. 1). The objective of the research presented here is an evaluation of the effects of AVPP process parameters on the protective properties of the surface layers of stainless steel. The results of this evaluation will be applied in further research to improve the passive properties of low-Cr stainless steels. The present investigations have been limited to studies in acidic media. Further research is planned to evaluate the resistance of modified surface layers on stainless steels to localised corrosion in chloride-containing media.

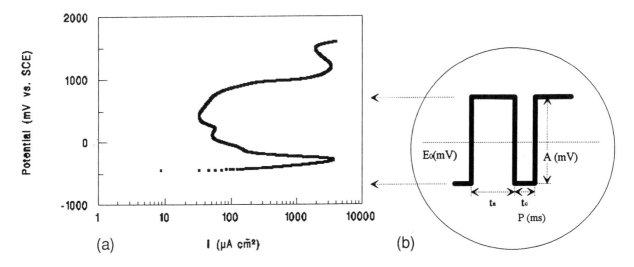

Fig. 1 Schematic illustration of the measurement principle: (a) anodic polarisation curve for 304 SS in 0.05M H_2SO_4; (b) location of the square wave in comparison with the polarisation curve, where A, P, $R_o = t_a : t_c$, and E_o denote pulse amplitude, pulse width, anodic to cathodic polarisation time ratio and DC bias potential, respectively.

2. Experimental Approach

Flat disks of 304 SS were abraded, rinsed in distilled water in an ultrasonic bath and cathodically polarised in 0.05M H_2SO_4 at –750 mV vs SCE for 5 min. After this cleaning step and rinsing in distilled water, the electrode was immediately immersed in 0.05M H_2SO_4 and polarised at E_o with a superimposed potential square wave. All solutions were open to air. A factorial design experiment was applied to determine the optimum values of A, P, E_o and R_o (see Fig. 1). All AV passivation experiments were performed with a PAR potentiostat model 173 combined with a PAR Universal Programmer model 175. Electrochemical methods were chosen to characterise the corrosion resistance of the modified passive layers. The decay of the open-circuit potential after AV passivation was recorded in 0.1M H_2SO_4 in order to determine the layer stability. An anodic polarisation curve was then recorded ($10 \, mV.s^{-1}$) starting at the corrosion potential E_{corr} reached in the previous open-circuit decay experiment. The values of the time to activation t_p, the critical current density for passivation i_{crit} and the passive current density i_p were the criteria of the stability of the modified passive film. These experiments were carried out using a PAR model 173 potentiostat with a 276 GPIB interface card which was controlled by an IBM XT computer.

In order to compare the electrochemical behaviour of the modified surfaces in the presence of a redox reaction, cathodic polarisation curves as well as cyclic polarisation studies were performed in a borate–boric acid buffer at pH = 8.4. In the latter case a potential sweep rate of $100 \, mV.s^{-1}$ was applied.

Auger electron spectroscopy (AES) was used to determine concentration profiles of Fe, Cr, O and S in the modified surface layers.

3. Results and Discussion

3.1 The effects of process parameters

According to the procedure described elsewhere [2], the direction of changes of i_{crit}, i_p and t_p as a function of A and P were determined in a 3^2 factorial design experiment using combinations of {A} = 450, 900, 1350 mV and {P} = 50, 100, 150 ms. The results obtained by the method of least squares fitting are presented in Fig. 2. Optimum values of E_o and R_o have been established earlier from the analysis of 2^2 factorial design experiments, where it was found that E_o and R_o should be fixed at E_o = 300 mV and R_o = 4:1. For a fixed pulse width P an increase in amplitude A decreases i_{crit} and i_p (Fig. 2(a) and (b)). This effect is especially significant for A changing from 450 mV to approx. 950 mV. In the range of A changing from 950 to 1350 mV i_{crit} is low and not very sensitive to the changes of A and P, while i_p

increases. The time to reactivation t_p increases with increasing A up to 950 mV and decreases in the range 950–1350 mV (Fig. 2(c)). For a fixed value of A an increase of P affects all three parameters, but its influence seems to be not as significant as the effect of A. From these results one may conclude that the amplitude A of the alternating voltage (AV) exerts a larger effect on the protective properties of the surface layers than the pulse width P. However, as it has been discussed elsewhere [2], there is no simple relationship between improvement of the passive layer stability and AV amplitude alone due to significant interaction effects between A and P.

The results in Fig. 2 provide the A and P values necessary for achieving values t_p, i_{crit} and i_p which describe the most stable passive layers achieved by AVPP. Considering average effects, {A,P} = {1050 mV, 90 ms} were chosen as optimum parameters which were applied in the experiments described below.

3.2 Protective Properties of Passive Layers Produced by AVPP

The open-circuit potential decay curves obtained for an untreated sample, after DC passivation and after optimised AVPP (Fig. 3 (a)) as well as the anodic polarisation curves (Fig. 3(b)) show significant improvements of the oxide film stability after AVPP. The efficiency of AVPP can be judged by the efficiency ratios $R_{tp} = t_p/t^o_p$, $R_{ic} = i^o_{crit}/i_{crit}$ and $R_{ip} = i^o_p/i_p$ where t^o_p = 33 s, i^o_{crit} = 3700 $\mu A.cm^{-2}$ and i^o_p = 40 $\mu A.cm^{-2}$ are the results recorded for the untreated 304 SS. For the results shown in Fig. 3 R_{tp} = 100, R_{ic} = 80 and R_{ip} = 4. It is important to note that the values of i_{crit} and i_p were determined after open-circuit potential decay. The large decrease of i_{crit} seems to suggest that the passive film formed by AVPP dissolved only a few weak spots during open-circuit potential decay and therefore less anodic charge was needed for repassivation.

In preliminary investigations it has been observed that surface modification by AVPP affects the hydrogen evolution reaction [1]. Another important reaction especially in neutral or alkaline environments which exerts an influence on the corrosion behaviour of the metal is cathodic reduction of oxygen. Cathodic polarisation curves in aerated borate buffer for the untreated sample as well as after the passsivation treatment show clearly that the rate of oxygen reduction is considerably faster on the untreated SS 304 than after DC or AV passivation (Fig. 4). In the latter case the lowest cathodic currents in the entire potential range were observed.

One reason for the beneficial effect of AVPP is the greater thickness of the film in comparison with DC or untreated sample [3]. An increase of the thickness of the passive film on modified surfaces can be deduced

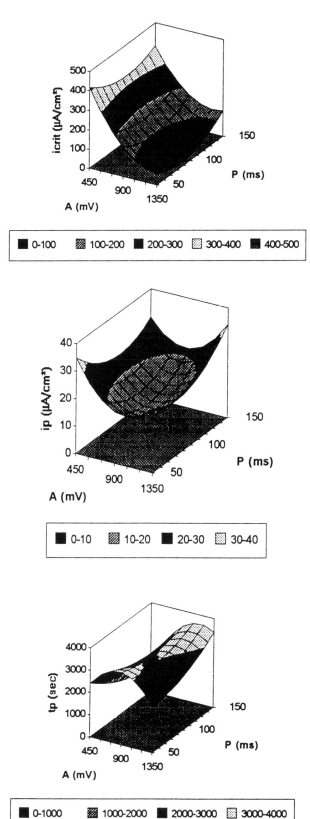

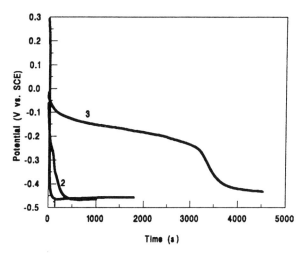

Fig. 3(a) Open-circuit potential decay for 304 SS exposed to 0.1M H₂SO₄; curve 1: as-received; curve 2: after DC passivation at E = 500 mV for 20 min; curve 3: after AVPP for 20 min with {A,P} = {1050 mV, 90 ms}, E₀ = 300 mV, R₀ = 4: 1.

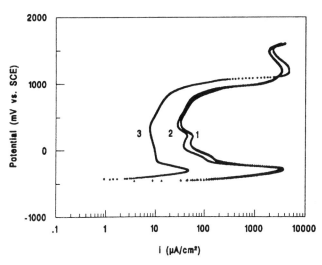

Fig. 3(b) Anodic polarisation curves in 0.1M H₂SO₄ after potential decay (Fig. 3(a)).

Fig. 2 Results of 3² factorial design in terms of functional dependence of i_crit, i_p and t_p on A and P at E₀ = 300 mV and R₀ = 4: 1. Time of treatment 20 min.

from Fe and O elemental profiles determined with AES (Fig. 5). An increase of the passive film thickness due to AVPP was found independently from cathodic reduction curves in deaerated borate of pH = 8.4, in which the largest cathodic charge was consumed for the reduction of the passive film formed on 304 SS by AVPP [4]. Another important result shown in Fig. 5 is the smaller amount of S in the passive film after AVPP as compared to dc passivation. AES measurements indicate Cr enrichment in passive films after AVPP (Fig. 6). Qualitative conclusions about the changes in the chromium content or stoichiometry after AVPP may also be drawn from cyclic polarisation curves carried out in deaerated borate buffer (pH = 8.4) (Fig.

185

33

Modification of Passive Films Composition due to Water Ageing

B. Legrand, C. Goux, C. Senillou and J. C. Joud

Laboratoire de Thermodynamique et Physico-Chimie Métallurgiques, URA CNRS 29, ENSEEG, BP 75, 38402 Saint Martin d'Hères Cedex, France

Abstract

The effect of water ageing on the composition and thickness of passive films formed on Fe–Cr–Si alloys by heat treatment in water containing hydrogen atmospheres has been investigated using Auger Electron Spectroscopy.

Depth intensity profiles were obtained using ion sputtering and were converted into concentration profiles using either the relative sensitivity method or the 'Sequential Layer Sputtering' model.

The influence of the annealing atmosphere and of the silicon concentration on the behaviour of the passive film during water ageing has been studied.

In particular it is shown that a large Si content (around 1at.%) can delay the chromium segregation in the passive film which is normally obtained.

1. Introduction

The main object of the present study is to investigate the in depth composition of passive films according to several parameters such as composition, ageing time, heat treatment in order to obtain the maximum chromium segregation. A special attention is paid on the precise location of the chromium segregation.

Three different materials are studied: a $FeCr_{17}$ alloy, a $FeCr_{17}Si_1$ alloy and an industrial ferritic stainless steel containing 17%Cr as major element and several minor elements among which 0.3%Si. To obtain the passive films, cold worked specimens are annealed during 15 mn at 820°C in two different H_2O/H_2 containing atmospheres, following the conditions described in [1]. Atmosphere dewpoints are either $\sigma = -67$°C ($P_{H_2O} = 4.10^{-6}$ atm, 'Dry annealing') or $\sigma = -58$° C ($P_{H_2O} = 1.3.10^{-5}$ atm. 'Wet annealing'). After annealing, specimens are aged in pure distilled water for several days to several months.

2. Surface Analysis

Auger electron spectroscopy is performed using a Riber Spectrometer fitted out with a C.M.A. and a Channeltron electron multiplier. The primary energy of the electron beam is 2 keV, the sample current is *ca.* 1 μA and E dN(E)/dE spectrums are obtained with a modulation tension of 4 V.

We investigate the evolution of the in depth composition using Xe-ion sputtering with a differential pumping gun. The acceleration voltage is 1 kV and the Xenon pressure is 10^{-5} mmHg. The defocalized ion beam area is *ca.* 4 mm × 4 mm. We evaluate the sputtering rate to approx. 5–10 Å.mn^{-1} (calibration on metallic oxides).

Depth intensity profiles are converted into concentration profiles using either the relative sensitivity method or the S. L. S. model.

2.1 Quantification using the Relative Sensitivity Method

This method supposes that the peak-to-peak amplitude for the Auger signal corresponding to an element is proportional to the amount of this element [2].

In this framework, the atomic concentration of the element is:

$$X_i = \frac{h_i / S_i}{\sum\limits_j h_j / S_j} \qquad (1)$$

where:
• The index j represents one characteristic peak for each element in the sample.
•h_i is the peak-to-peak amplitude for the derivated Auger characteristic peak corresponding to the ith element. We use for Fe the LVV peak at 703 eV, for Cr the LVV peak at 529 eV, for Si the LVV peak at 92 eV, and for SiO_2 the LVV peak at 76 eV.
•S_i is the sensitivity factor corresponding to this signal. As it depends on the experimental analysis conditions, S_i has been determined in a previous work [3]. In

our study, the composition of the passive film (Figs 3, 5, 6) is represented by an atomic percentage of metallic element according to eqn (2):

$$\%Fe = \frac{X_{Fe}}{X_{Fe} + X_{Cr} + X_{Si}} \qquad (2)$$

The relative sensitivity method is very easy to apply and gives good qualitative results, but it supposes that the composition is the same in every layer studied by Auger analysis and does not take into account the sputtering crater geometry.

2.2 Quantification using the Sequential Layer Sputtering Model

The sequential layer sputtering model [4, 5] is based on the statistical nature of the sputtering process and supposes a layered structure of the sample.

It describes the contribution of the layer n to the surface after a sputtering time t by a Poisson distribution:

$$\theta_n(t) = \frac{(t/t_o)^{n-1}}{(n-1)!} \cdot \exp(-t/t_o) \qquad (3)$$

where t_o is the time necessary to remove the equivalent of one monolayer.

According to this model, the intensity of the derivated Auger peak for an element is:

$$I(t) = I^o \cdot \sum_{n=1}^{N} \sum_{m=0}^{M} \theta_n(t) \cdot X_{n+m} \cdot \exp(-m/\lambda) \qquad (4)$$

where:
• λ is the mean free path of the Auger electrons in monolayer.
• X_{n+m} is the molar concentration in the n+m layer
• I^o is a normalisation factor relative to this element and to the experimental conditions. I^o is determined by a calibration in the matrix where the concentration of each element is well known.

The summation over m takes into account the contribution of all the layers below the nth one (see Fig. 1) and the summation over n takes into account all the layers which contribute to the surface after t mn of sputtering.

Practically, we evaluate a first value of $I^o.X(n)$ for each layer n using the procedure described by Hoffmann [5] according to the relation (5).

$$I^o.X(t) = I(t) - \lambda \cdot \frac{dI(t)}{dt} \qquad (5)$$

Then, we calculate I(t) with these values. We compare it with the experimental result and we adjust $I^o.X(n)$ until the calculated intensity profile seems as

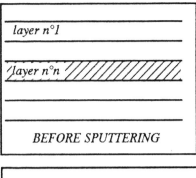

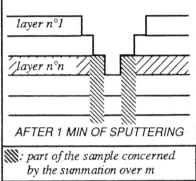

Fig. 1 Decomposition of the passive film according to the Sequential Layer Sputtering model.

similar as possible to the experimental one (see Fig. 2, p.190). Then, in order to compare the S L S model to the relative sensitivity method, we calculate the composition according to eqn (2).

Actually, this work has only been done for FeCr17 alloys because the evaluation of I^o, easy for Fe and Cr is more difficult for Si which is at a very small concentration in the matrix, and for SiO_2 which is not in the matrix.

3. Experimental Results

As indicated in the previous paragraph, the S L S model has only been applied in the limited case of pure Fe–Cr alloys. So, we first present the whole experimental results obtained with the relative sensitivity method.

3.1 Effect of ageing and of the annealing atmosphere

A first series of measures was performed on FeCr17 alloys in order to study the effect of water ageing. Figure 3 shows that, for films formed in 'dry' hydrogen atmosphere, neither composition nor thickness are changing during the first month of ageing but that a long term ageing (8 months) induces an important chromium segregation in the outermost part of the film and an increase of its thickness. The uncertainty about the experimental results is roughly 5 at.%.

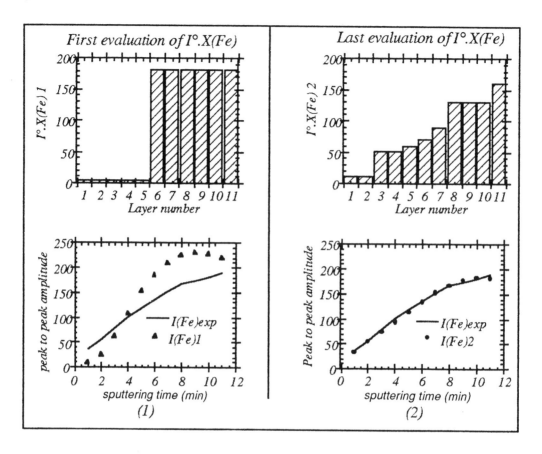

Fig. 2 *Example of deconvolution of an intensity profile using the Sequential Layer Sputtering model.*
(1): first evaluation of I°.X(Fe); (2): last evaluation of I°..X(Fe).

For films formed in 'wet' hydrogen atmosphere, the same evolution is observed, but after only one month of ageing. Moreover, after that rapid evolution, composition and thickness seem to be constant during a longer ageing and they reach limits which are similar to those obtained in the case of 'dry' annealing.

The annealing atmosphere has however no significant effect on the metallic composition of the film observed before ageing. Moreover, the initial oxygen content (see Fig. 4, p.192) is not drastically changed between the two treatments. The residual level, observed in the bulk part of the sample, is due to a redeposition effect.

So, we can presumably assume that it is a modification of the film structure (oxide, hydroxide) which allows the acceleration of the chromium segregation induced by ageing.

3.2 Effect of silicon
The effect of silicon on the film composition is studied analysing $FeCr_{17}Si_1$ alloys. In this case, the Si content is the sum of a metal and oxide contribution. Figure 5 (p.192) shows that, at this concentration, silicon segregates and that the chromium segregation is decreased and delayed relatively to $FeCr_{17}$ alloys.

A previous study performed on industrial stainless steel (see Fig. 6, p.192) containing 17%Cr and 0.33%Si did not exhibit these effects [6].

So, we can suppose that there exists a 'critical' concentration (somewhere between 0.33% and 1%) above which the effect of silicon on chromium segregation is noticeable.

3.3 Comparison between the S L S model and the Relative Sensitivity Method
In Fig. 7 (p.193), we compare the chromium concentration profiles obtained with the relative sensitivity method an with the S L S model for an $FeCr_{17}$ alloy dry or wet annealed and aged during respectively 27 or 26 days.

In comparison with the relative sensitivity method the chromium segregation is not located in the outermost part of the film but slightly deeper and that there is a chromium depletion near the passive film-matrix interface.

These observations have been checked on all the samples of $FeCr_{17}$ alloys (wet or dry annealed) corresponding to the various times of ageing.

Furthermore we can tentatively apply the S L S model for oxygen, in particular to see the oxygen

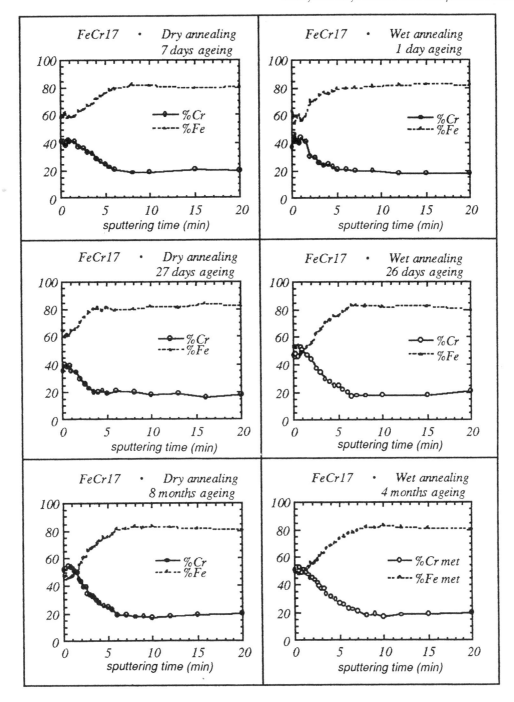

Fig. 3 Evolution of the passive film composition during water ageing for FeCr$_{17}$ alloys: Role of the annealing atmosphere composition.

profile (for which I° is difficult to determine) with the assumption that the bulk oxygen content is always the same. For the same sample indicated in Fig. 7, Fig. 8 (p.194) shows the evolution with the layer number of the atomic fraction of Fe and Cr completed with the ratio X(O)/X(O)bulk. The evolution of the atomic fraction of Cr clearly indicates a segregation of this element at roughly 4 layers in depth with a larger tail in the case of wet annealed samples. The maxima of the curves representative of the ratio X(O)/X(O)$_{bulk}$ are

slightly shifted to the outer part of the passive film relatively to the chromium curves. This fact can suggest a duplex nature of the passive layer in good agreement with the results of Marcus [7].

Moreover, we observe a constant and very low level of iron at the outer part of the film suggesting a possible dissolution of this element in water during ageing following the same process as that described by Kirchheim for samples under polarisation [8].

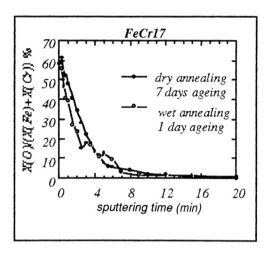

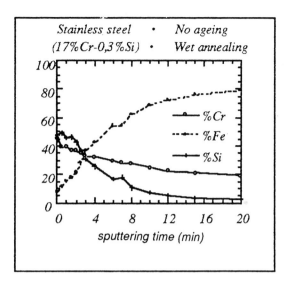

Fig. 4 *Influence of the annealing atmosphere on the in depth oxygen contents in the passive film before ageing.*

Fig. 6 *Passive film composition before water ageing for an industrial stainless steel containing 17%Cr and 1%Si.*

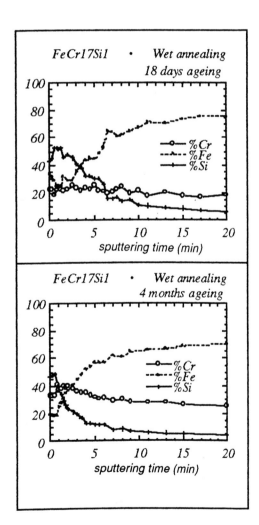

Fig. 5 *Evolution of the passive film composition during water ageing for FeCr₁₇Si₁ alloys: Effect of silicon.*

4. Conclusion

The main object of the present study was to investigate the in depth composition of passive films formed on $FeCr_{17}$ and $FeCr_{17}Si_1$ alloys after dry or wet annealing at 820°C.

The composition of the annealing atmosphere does not change the metallic segregation before water ageing. Conversely, we observe an acceleration of the chromium segregation induced by water ageing in the case of wet annealed specimens. So, we can conclude to an important effect of the annealing conditions on the structure of the passive film.

The influence of silicon has also been investigated. The main conclusion is the existence of an important initial silicon segregation which delay the chromium one for samples which silicon content is above a critical concentration (0.33% < C < 1%).

The application of the S L S model to the exploitation of the Auger intensities for $FeCr_{17}$ alloys allows to localise more precisely the enrichments of chromium and oxygen and the depletion of iron.

We observe shifted distributions of chromium and oxygen element, which is consistent with a duplex structure of the passive film.

The low and constant iron level at the outer part of the film suggest that the passive film evolution during water ageing is not only controlled by chromium oxidation at the film surface but also by iron dissolution in water.

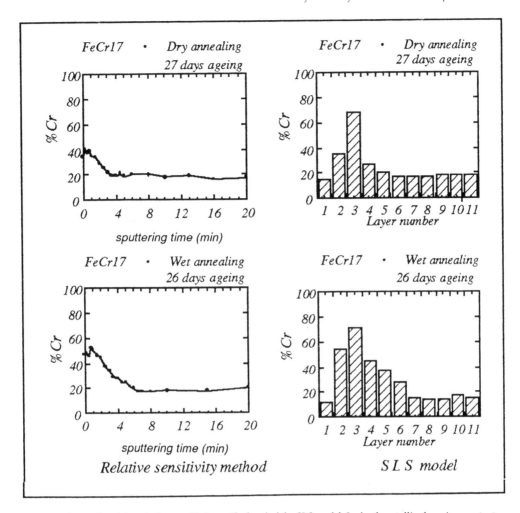

Fig. 7 Comparison between the results of the relative sensitivity method and of the SLS model: In depth metallic chromium contents.

References

1. D. Gorse, J. C. Joud and B. Baroux, Corros. Sci., **33**,1455, 1992.
2. L. E. Davis, N. C. MacDonald, P. W. Palmberg, G. E. Riach and R. E. Weber, Handbook of Auger Electron Spectroscopy, 2nd edition, 1976.
3. J. Wasselin, thesis, INP Grenoble, 1984.
4. J. M. Sanz and S. Hofmann, Surface and Interface Analysis, **8**, 147–157, 1986.
5. S. Hofmann, Progress in Surface Science, **36**, 1991.
6. C. Goux, D. Liu, D. Gorse, J. C. Joud and B. Baroux, Material Science Forum, 111–112, 1991.
7. P.Marcus, this symposium, 1993.
8. R. Kirchheim, B. Heine, H. Fischmeister, S. Hofmann, H. Knobe and U. Stolz, Corros. Sci., **29**, 899–917, 1989.

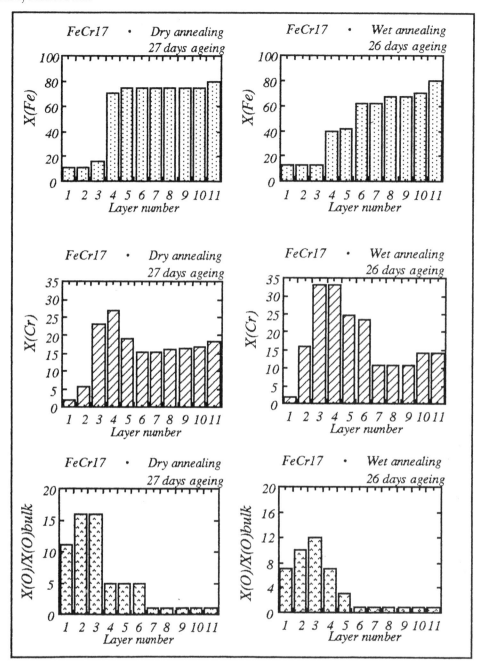

Fig. 8 Atomic concentrations of Fe, Cr O in the passive films determined using the SLS model.

34

Chromate-free Surface Treatment: MolyPhos —A New Surface Conversion Coating for Zinc. Optimising the Treatment by Corrosion Testing

G. Bech-Nielsen, I. Chorkendorff* and P. T. Tang†

Chem. Dept. A, The Technical University of Denmark, DK-2800 Lyngby, Denmark
*Physics Department
†Centre for Advanced Electroplating

Abstract

As an alternative to chromating of electroplated zinc a new treatment has been developed. The composition of the treatment solution (molybdate/phosphate ratio), the temperature and duration of the treatment were optimised using the Taguchi statistical planning and evaluating the corrosion resistance obtained in the various conditions by simultaneous CMT (Corrosion Measurements by Titration) and EC (Electrochemical Measurements).

Protective layers formed on zinc and also on cathodically polarised electroless nickel were analysed by Auger analysis. The Mo/P ratio is higher in the protective layer than in the treatment solution, and molybdenum is apparently present in a low oxidation state.

1. Introduction

Electroplated zinc is nearly always given a corrosion protecting chromate treatment. However, growing concern about use of chromate solutions, considering occupational health and environmental effects, makes it desirable to find alternative methods for corrosion protection, involving much less toxic compounds. In several studies seeking alternatives to chromate the chemically related molybdates and tungstates and also hetero-polyions of these have been tested, however, so far with limited success [1–3].

2. Experiments

We have made examinations of the corrosion protection afforded by a great number of treatments, where molybdates, tungstates and several combinations of these with other elements were used. The most promising results were seen with molybdate combined with phosphate, a combination involving much less toxic components than chromate. The most common method for industrial evaluation of corrosion protection is the Neutral Salt Spray Test, but the more recently introduced Prohesion Test [4] is claimed to give results with a more reliable acceleration factor. However, for faster and more quantitative evaluation of the protective quality of various treatments we used continuous CMT measurements (Corrosion Measurements by Titration) and electrochemical measurements (EC) at intervals [5, 6]. These measurements were carried out in 3% NaCl solution, saturated with CO_2-free air at pH = 5.000. In order to identify optimum treatment parameters (time, temperature, concentrations of molybdate and of phosphate) Taguchi statistical planning of experiments [7] was employed. Over the range of experimental conditions two compositions of solutions were identified, giving low corrosion rates similar to those for chromated samples. In one the molar ratio of Mo/P was 0.33, in the other it was 0.66, and the two solutions have been called 'MolyPhos 33' and 'MolyPhos 66', respectively. All corrosion measurements were followed for at least one hour, after which measurements by the two methods indicated a constant, common rate of corrosion. An example is shown in Fig. 1. Samples treated according to the optimum procedures were also tested by Salt Spray Tests and Prohesion Tests, with chromated samples as reference material. In salt spray the chromated samples were clearly superior to the MolyPhos treated ones, but in the Prohesion Test the two treatments were equally good. A sample of zinc plated from an acidic bath was a little more resistant, when treated with MolyPhos than when chromated.

3. Further Investigations

In chromating it is obvious that chromate acts as an oxidant, and some zinc is oxidised. A similar role for

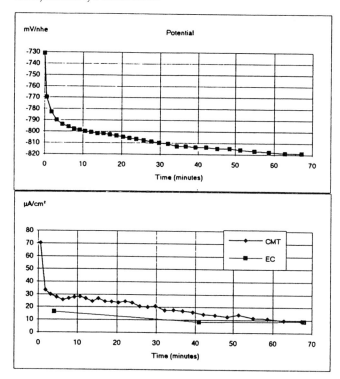

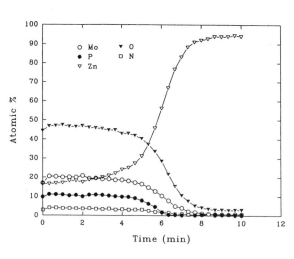

MolyPhos 66 on Zn

1 min ≈ 180 atomic layers

Fig. 3 Auger depth profile of the layer formed on zinc by treatment with a solution containing molybdate and phosphoric acid. Molar ratio Mo/P = 0.66. A very low carbon signal has been omitted.

Fig. 1 Corrosion testing of zinc treated with a solution of molybdate and phosphoric acid. Molar ratio Mo/P = 0.66. The test was made using a rotating disc electrode in a 3% NaCl solution, saturated with CO_2 free air, pH = 5.000. The upper part of the diagram shows the corrosion potential. In the lower part both continuous recording of CMT measurements and individual EC measurements are shown.

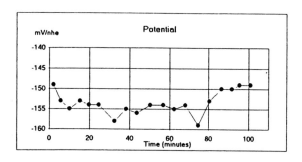

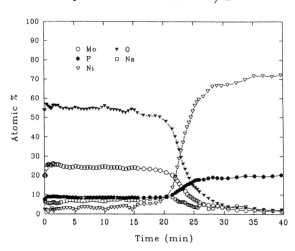

MolyPhos 132 on Ni/P

1 min ≈ 47 atomic layers

Fig. 4 Auger depth profile of the layer formed on electroless Ni (Ni/P) by cathodic treatment with a solution containing molybdate and phosphoric acid. Molar ratio Mo/P = 1.32.
A very low carbon signal has been omitted.

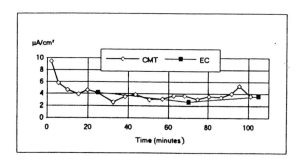

Fig. 2 Corrosion testing of electroless Ni treated with a titanate-molybdate solution. Corrosion measurements of the same sample prior to treatment: CMT: 45 μAcm^{-2}, EC: 11 μAcm^{-2}.

molybdate might be expected, but not in the way known for phosphomolybdate being mildly reduced to produce a blue colour. No hydrogen evolution is seen during treatment, though the potential of zinc is from *ca.* –800 (at the start) to –750 mV/nhe (after some 2 min, the preferred treatment time at 60°C). In the MolyPhos 33 solution (pH = 2.1) a faint blue colour is seen after treatment of many samples in succession, but not in the MolyPhos 66 solution (pH = 4.6).

In order to obtain more information about the properties of the coating formed by the MolyPhos treatment some modifications were made. A sample of zinc was treated at room temperature in a MolyPhos solution, and no film was formed. Then the sample was polarised cathodically, and a coating was formed, providing equally good corrosion protection as those produced in the normal way. Similarly a sample of electroless nickel was polarised cathodically in a MolyPhos solution. A decorative, pink layer was formed, and the sample showed highly improved resistance to corrosion in the moist SO_2 chamber [8] as well as according to CMT and EC measurements. A treatment of electroless nickel with a solution containing a combination of titanate and molybdate (which was not successful for treatment of zinc) gave the most efficient protection of the metal noted so far, see Fig. 2. Fundamental problems with corrosion of pure nickel and also of electroless nickel in aerated solutions have been noted [5], so the agreement between CMT and EC measurements was of special interest.

Apparently, it did not matter, whether a layer was formed due to reduction of molybdate by an amount of zinc being oxidised, or whether molybdate was reduced in a cathode reaction; the essential properties of the layer were the same.

Then various layers were examined by Auger analysis. Two depth profiles are shown in Figs 3 and 4. Figure 3 shows the Auger profile on zinc, plated from a cyanide bath and treated with MolyPhos 66. A nitrogen signal is assumed to arise from additives in the plating bath. In view of a very small carbon signal (omitted in the diagram) the nitrogen is hardly likely to have arisen from an organic compound. However, an estimate of the stoichiometry of the layer cannot be made with any confidence. The diagram in Fig. 4 is more suited for the purpose. The part of the layer examined after some 5 to 10 min of sputtering has a composition, which may represent most of the layer. The nickel signal is very low, and phosphorus coming from the metal phase may be disregarded in view of being the minor component in the metal. So by attributing an oxidation state of +5 to phosphorus at these depths and the normal oxidation states of +1 and –2 to sodium and oxygen one arrives at an oxidation state of + 2.3 to +2.5 for molybdenum. This is certainly no

accurate figure; small corrections, which may have to be applied, will cause a noticeable change in the result. However, it can be taken as evidence of a substantial reduction of molybdate. At the same time it is seen that the layer formed by reaction of zinc with a solution with a Mo/P ratio of 0.66 results in a layer with a Mo/P ratio of 2.0, Fig. 3. In Fig. 4 the layer was formed by reaction with a solution having a Mo/P ratio of 1.32, and in the layer the ratio is 2.7. A stable phosphate complex with molybdenum in an oxidation state of +3 is known, having 2 alkali ions, 2 Mo atoms and 4 phosphate groups [9]. The Na/P ratio of 0.8 seen in Fig. 4 is of the corresponding order of magnitude. Then, however, a major part of the molybdenum should appear in different compounds, probably oxides.

XPS examination of the molybdenum signal was attempted, but spectra conforming with data for low valency species (including zero) were not found. The effects of sputtering may be responsible for this. In the outmost layers molybdate was identified prior to sputtering.

Additional examinations will be needed, both to clarify more safely the conditions for obtaining good protection with MolyPhos treatment and to establish the characteristics and the compositions of the layers formed by the treatment.

References

1. G. D. Wilcox and D. R. Gabe, Br. Corros. J., **22**, 254, 1987.
2. U. Buttner, J. L. Jostan and R. Ostwald, Galvanotechnik, **80**, 1589, 1989.
3. UK Patent Application, GB 2070073 A (Kobe Steel Ltd., Japan).
4. F. D. Timmins, J. Oil Col. Chem. Assoc., **62**, 131, 1979.
5. G. Bech-Nielsen in 'Electrochemical Methods in Corrosion Research IV', O. Forsen ed., Materials Science Forum, Vols **111–112**, 525, Trans Tech Publications, Switzerland, 1992.
6. G. Bech-Nielsen and T. Dorge, SurFin '91 Technical Conference, Toronto, Canada, Proceedings p. 955, 1991.
7. G. Taguchi and S. Konishi, 'Orthogonal Arrays and Linear Graphs', American Suppliers Institute, Dearborn, Michigan, USA, 1987.
8. ASTM Standard G 87–84.
9. A. Bino and F. A. Cotton, Angew. Chem., **91**, 496, 1979.

35

Passive Film Evolution on Stainless Steels with the Medium Composition: Electrochemical Studies in the Presence of Chloride and Molybdate Ions

L. Farah, C. Lemaitre and G. Beranger

Université de Technologie de Compiègne, LG2mS, URA 1505 du CNRS, B.P. 649, 60206 Compiègne Cedex, France

Abstract

In order to obtain information concerning the mechanisms of corrosion in presence of the passive film of stainless steels, and in the changes of the properties of this film due to medium modifications, we have performed a.c. impedance measurements in several media containing sodium chloride and sodium molybdate, to separate the conditions where adsorption, transport, or diffusion can be obtained. The static potential values were chosen in three domains: close to the rest potential, at a potential where the passivity is stable or in transpassivity conditions. The results obtained have shown that the limiting step of the corrosion process was not the same at the three potentials: close to the rest potential, the corrosion is limited by a diffusion process; when the passive film is stable, the limiting step is due to the existence of a capacity, and the passivity breakdown shows short circuits in the film. The influence of the added molybdate ions is not identical in the pitting domain of concentrations or in the passivating one: the changes in the passive film modify the thickness and the properties of the passive film.

1. Introduction

By using stochastic measurements of the pitting potentials of stainless steels, previous works in our laboratory [1] have shown that the value of this parameter depends on the medium composition. It appears that, in neutral chloride containing solutions, the presence of inhibitive species, such as chromate or molybdate ions, modifies the evolution of the pitting potential when a critical ratio of inhibitive and aggressive species is obtained [2, 3]: a change in the passive film behaviour is observed, with a 'pitting domain' of concentrations, and a 'passivating' one . That can be explained by a change in the limiting step of the corrosion process, e.g. adsorption, electrostriction, diffusion, transport [4–6]. In this study, our aim was to correlate these previous results with information obtained from a.c. impedance measurements about the above mentioned processes, in different neutral chloride solutions containing several amounts of sodium molybdate, with concentrations chosen to obtain different mechanisms.

2. Experiment

The alloy is an austenitic stainless steel, AISI 316 L, all the samples are cut from industrial cold rolled sheets in the solution treated state. The exposed area S of test disk specimens is equal to $0.785\,cm^2$ (10 mm dia.). Their surface is wet mechanically polished using SiC paper until the grade 1200 and with a diamond past until the 3 μm grade, then air aged in the ambient air for 24 h before the test.

The media were prepared with distilled water, and contained $0{,}2\,mol.L^{-1}$ sodium chloride with different amounts of sodium molybdate: 0–0.01–0.1–0.2 $mol.L^{-1}$. These concentrations were chosen because a previous work had established that there is a critical amount of sodium molybdate $0{,}1\,mol.L^{-1}$, when [NaCl] = 0.2 $mol.L^{-1}$ in deaerated conditions, for which a change is observed between pitting and inhibition of pitting [2, 3]. Thus, these media were deaerated by a nitrogen flux in the solution. For our cells, containing approx. 0.33L, the duration of that deaeration was 2 h, and the nitrogen flux was maintained on the solution during the measurements.

The polarisations curves were obtained from potentiokinetic tests, at a potential sweep rate of 120 mV min^{-1}. That rate was used to avoid a strengthening of the passive film during the measurement in passive region of potentials. The reference electrode was a saturated calomel electrode, and the counter electrode was in platinum, with an area of 1 cm^2.

For a.c. Impedance measurements, we have performed tests between 10000 and 0.01 Hz with 4 frequencies by decade. We have fixed several potentials corresponding respectively to the passive or the transpassive region of the polarisation curves, or to the rest potential. The alternative amplitude of potentials was equal to ± 25 mV.

3. Results

3.1 Polarisation curves

The four polarisation curves corresponding to each medium are reported in the Fig. 1. One can see that the rest potentials are obtained between –400 and –175 mV/SCE, and the pitting potentials can be determined between +136 and +1000 mV/SCE, with a great ennoblement due to molybdate ions presence in the media. The current densities in the passive region are decreasing with the increase of the molybdate ions concentration, but the obtained values are similar, between 2.10–6 an 10–5 A.cm^{-2}.

3.2 Impedance measurements

Results obtained without molybdate addition in the chloride solution and with 0.01 mol.L^{-1} of sodium molybdate addition are very similar, thus we have reported only the impedance diagrams without molybdate ions in the medium, in form of BODE diagrams (Fig. 2). On the other hand, the observations made with 0.1 and 0.2 mol.L^{-1} are the same, thus only results obtained with a 0.2 mol.L^{-1} addition are here reported (Fig. 3). It appears in Fig. 2 that the impedance modulus in the sodium chloride vs the logarithm of the frequency is not a line, in contrast with the cases corresponding to the inhibitive role of the molybdate ions in solution (Fig. 3). In this later one, the potential values of 400, 600 and 900 mV/SCE correspond to the passive region, but 1200 mV/SCE is a potential in the transpassive region: the corresponding impedance diagram is very different.

4. Discussion

The impedance of passive film is due to the excess of electric charges near the solid/liquid interface [7, 8], and the diagrams obtained can be considered as capacitive [9]. With the results obtained in our work, it is easy to determine the corresponding capacitance, for each frequency. If Z' and Z" are respectively the real and the imaginary part of the impedance Z, one has Z – Z' + j Z" with:

$$1/Z = (1/R) + j\omega C \tag{1}$$

thus:

$$(Z")^{-1} = \omega C \tag{2}$$

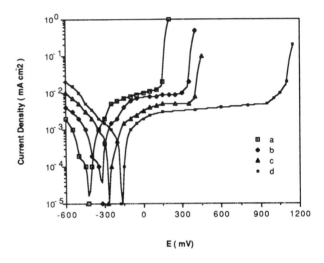

Fig. 1 Effect of the sodium molybdate concentration on polarisation curves of a AISI 304L stainless steel in 0.2 mol.L^{-1} NaCl media:
(a) without molydate; (b) sodium molybdate addition = 0.01 mol.L^{-1};
(c) sodium molydate addition = 0.1 mol.L^{-1}; (d) sodium molydate addition = 0.2 mol.L^{-1}.

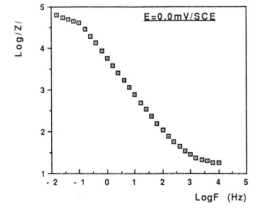

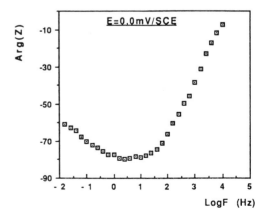

Fig. 2 BODE diagrams obtained with the AISI 304L stainless steel in 0.2 mol.L^{-1} NaCl.

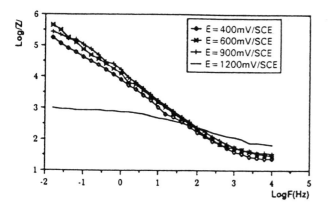 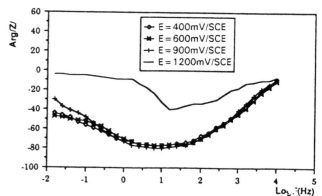

Fig. 3 BODE diagram of the stainless steel in NaCl solution + 0.2 mol.L⁻¹ Na₂MoO₄.

We have calculated the values of these capacities for each potential in each solution, and the results obtained with 0.1 and 0.2 mol.L⁻¹ of added sodium molybdate are reported in Figs 4 and 5 respectively. It appears in these figures that three potential domains have shown: they correspond, on the polarisation curves, to the cathodic, the passive, and the transpassive regions. It is possible, by using this method, to reproduce curves similar to polarisation curves, but concerning the electric capacity of the interface: when the passive film is stable, the capacity is minimum [7].

If one consider a surface area S of the passive film, the corresponding capacity is given by:

$$C = \varepsilon_o \varepsilon_r S / d \qquad (3)$$

where d is the thickness of the film, ε_o and ε_r are the dielectric constants. Thus, it is possible to obtain information on the thickness evolution of the passive film. The value of the capacity takes into account, with the oxide/hydroxide film capacity, the double layer, the charge into the passive film, and the charge of surface, but there is a decrease of this value if the concentration of molybdate ions increases: that is in accordance with an increase of the passive film thickness due to the molybdate presence in the medium.

Figures 4 and 5 show that the minimum value of the capacity is observable between 0 and +200 mV/SCE for an addition of 0.1 mol.L⁻¹ of molybdate ions, and between 0 and +1000 mV/SCE for an addition of 0.2 mol./L⁻¹: the thickness of the passive film depends on the concentration of the sodium molybdate in the solution, and the potential domain of passivity is increased by the addition of this species, as previously indicated [3].

The results obtained without molybdate (Fig. 2) or with little amounts of this species show that the BODE

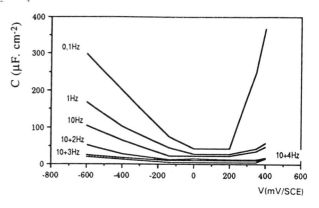

Fig. 4 Effect of frequency on the capacitance in 0.2M NaCl + 0.1M Na₂MoO₄.

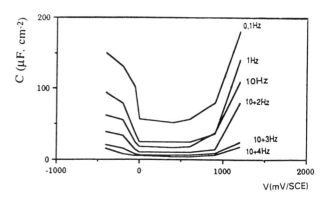

Fig. 5 Effect of frequency on the capacitance in 0.2M NaCl + 0.2M Na₂MoO₄.

diagram concerning the impedance modulus is not a line: that corresponds to a value of R different to zero in the eqn (1). Then R = Rt and a charge transfer exists: the limiting step of the corrosion process is charge transfer through the passive film [7]. When molybdate ions are present in sufficient quantities (Fig. 3), the BODE diagram is a line with a slope -1, and the impedance can be written with the eqn (2). In this case the process occurs without transfer, and is limited by diffusion [5]. In the transpassive region of potentials, pits are initiated, and short circuits of diffusion can play a role [10].

5. Conclusion

By using impedance diagrams, we have determined the evolution of the capacity of the interface solid/liquid vs the frequency, the potential and the molybdate amount in solution for stainless steels in neutral chloride media. The results obtained have permitted to correlate these results with observations made by using polarisation curves, and concerning the different domains of potential, and particularly the passivity one. It appears that the capacitance values are modified by a change in the amount of molybdate in the medium, and there is also a change in the limiting step of the corrosion process when the molybdate amount is sufficient *vis-a-vis* the chloride content to be, in the inhibition diagram previously established [3], in conditions of pitting inhibition. If the charge transfer is

shown in the pitting domain, the diffusion process is also occurring in the inhibition one. In order to confirm these observations, other tests are actually performed with rotating electrodes to separate clearly the diffusion conditions.

References

1. C. Lemaitre, B. Baroux and G. Beranger, C. R. Acad. Sci. Paris, t 308, série II, 171, 1989.
2. C. Lemaitre and G. Beranger, Werkstoffe und Korrosion, **40**, 229, 1989.
3. C. Lemaitre, B. Baroux, A. Abdel Moneim and G. Beranger, Corros. Sci., **31**, 585, 1990.
4. H. H. Strehblow and B. Titze, Corros. Sci., **17**, 461, 1979.
5. M. N. Mc Donald and D. D. Mc Donald, J. Electrochem. Soc. **134**, 1, 41, 1987.
6. N. Sato, Electrochimica Acta, **16**, 1683, 1961.
7. M. G. S. Ferreira and J. L. Dawson, J. Electrochem Soc., **132**, 760, 1985.
8. D. D. Mc Donald, J. Electrochem. Soc., **129**, 9, 1974.
9. I. Epelboin, Electrochemica Acta, **11**, 221, 1966.
10. K. Juttner, F. Mansfield and W. J. Lorentz, Eurocorr '87, Karlsruhe, Germany, Proc. DECHEMA Ed., 1987, p.543.

36

Passive films on Glassy FeCrPC Alloy—The Role of Phosphorus in Stability

B. Elsener and A. Rossi*

Institute of Materials Chemistry and Corrosion, Swiss Federal Institute of Technology, ETH Hönggerberg, CH-8093 Zurich, Switzerland

*Dipartimento di Chimica e Tecnologie Inorganiche e Metallorganiche, Universita di Cagliari, Via Ospedale 72, 1-09124 Cagliari, Italy

Abstract

The passivation of the amorphous alloy $FeCr_{10}P_{13}C_7$ has been studied by XPS surface analysis and electrochemistry. Passive films formed in 1N HCl and 1N H_2SO_4 at different potentials on mechanically polished samples of the alloy were studied. The thickness of the films increases linearly with the applied potential in both electrolytes. The passive films are enriched in chromium oxi-hydroxide, phosphates are incorporated in notable amounts. The composition of the substrate underneath the film is independent of the potential, the main feature is the marked enrichment in phosphorus (129.4 eV) compared to the bulk alloy. An intermediate species of phosphorus (131.7 eV) is found at the interface passive film/alloy, it shows a chemical state similar to elemental phosphorus. The outstanding corrosion resistance of this alloy may thus be attributed to this enrichment of phosphorous underneath the film, reducing to a great extent the ionic conductivity of the interface, and to the incorporation of phosphates in the passive film.

1. Introduction

Amorphous alloys containing phosphorus as the major metalloid have been found by different research groups to be highly corrosion resistant [1–3]. As summarised in our earlier work [4], XPS surface analytical studies on FeCr- and FeCrNi-based metallic glasses indicate a substantial enrichment of oxidised phosphorus P(+V) at 133 eV in the passive film. AES analysis of FeNiBP amorphous alloys reported a maximum of P underneath the film [5]. There is no general consensus about the oxidation state or the nature of the lower oxidation states of phosphorus observed in passive films on chromium- and nickel-based metallic glasses [2, 4–6]. Regarding the beneficial aspect of phosphorus, a specific synergistic interaction with chromium has been suggested [6], but evidence from work on FeNi based amorphous alloys shows that chromium does not necessarily have to be present to achieve a high corrosion resistance [7].

This combined XPS surface analysis and electrochemical study reports results on the passivity of the alloy $FeCr_{10}P_{13}C_7$ in 1N H_2SO_4 and 1N HCl. The discussion focuses on the role of phosphorus in enhancing the stability and the resistance against localised corrosion of phosphorus containing amorphous alloys.

2. Experimental

The amorphous alloys were produced by rapid quenching (RQ) carried out under controlled helium atmosphere at 0.8 bar. The resulting ribbons were typically 1.5 cm wide and ca. 30–40 μm thick. The wheel side was dull, the non-contact side bright. X-ray microprobe analyses, performed on both sides of the ribbons, confirmed that the ribbons have uniform composition.

'As received', mechanically polished (to 1 μm diamond paste) and mechanically polished and polarised samples were analysed. Anodic polarisation was started after 5 min exposure to the electrolyte at the open circuit potential. The samples were removed from the cell under applied potential, then rinsed with bidistilled water, dried in a nitrogen stream and transferred under nitrogen into the fast entry air lock of the spectrometer. The untreated part of the sample was masked by a gold ring.

XPS analyses were performed with an ESCALAB MkII spectrometer (Vacuum Generator Ltd., UK). The X-ray source was Al Kα (1486.6 eV) and run at 20 mA and 15 kV. The pass energy (fixed analyser transmission mode) was 20 eV. The instrument was calibrated according to [8], the binding energies are referred to the C1s signal at 285.0 eV. Details of background subtraction and curve fitting are given in [4]. The quantitative analysis was performed according to the three layer model (contamination, passive film, substrate), the values of the mean free pass and the cross sections used are the same as reported in [4].

3. Results

Immersion of 'as received' samples in deaerated hydrochloric or sulphuric acid results in OCP in the passive range (–50 to +170 mV SCE), the anodic polarisation curve shows a wide passive range. The mechanically polished samples instead show OCP in the active range (–360 mV SCE) and a distinct active/passive transition in a very narrow potential range is observed [4]. No pitting or crevice corrosion occurred in 1N HCl. The current/time curves for the potentiostatic passivation shows a linear relationship log i vs log t for 'as received' and mechanically polished samples at all passivation potentials, the current densities becoming higher at more positive potentials.

The high resolution X-ray photoelectron spectra for the O1s, Fe2p$_{3/2}$, CR2p$_{3/2}$ and P2p signals measured on a mechanically polished sample after passivation for 60 min in 1N HCl at +0.5 V vs SCE are shown in Fig. 1 with the background subtraction ac-

cording to [9] and curve fitting to determine the integrated peak intensities for further quantitative evaluation. The binding energies of Fe(met), Cr(met) and P(met) remain constant at 707.0±0.1, 574.1±0.1 and 129.4+0.1 eV respectively. The peak binding energies of oxidised chromium, iron and phosphate in the passive films increase with the applied potential [4]. The P2p spectra after anodic polarisation always show an additional intermediate state of phosphorus at 131.7 eV regardless of potential. This signal is not detected in 'as received' or air formed oxide films [10] but it has been identified in mechanically polished samples and analytical application of X-ray excited Auger electrons suggests that it may be elemental phosphorus [11]. The major component of the O1s spectra after passivation is the M-OH or O-P component at 531.4 eV.

The thickness t of the passive film increases linearly with the applied potential in both electrolytes (Fig. 2). The composition of the passive films as a function of the passivation potential is shown in Fig. 3. Oxidised chromium is considerably enriched in the passive film and continues to increases steadily up to 0.3 V SCE where 80% of the cations in the film are Cr

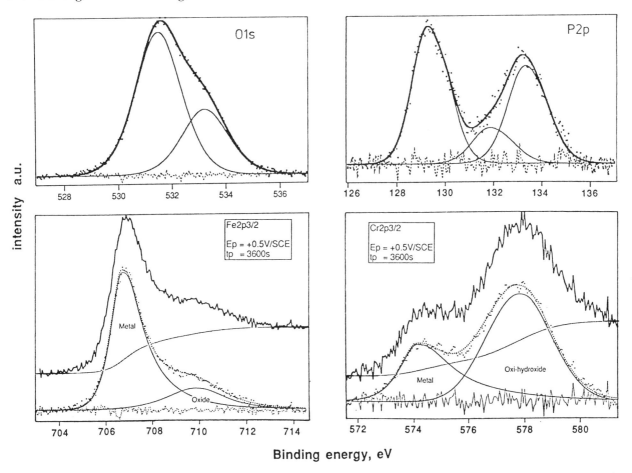

Fig. 1 *High resolution XPS spectra of O1s, P2p, Fe2p$_{3/2}$ and Cr2p$_{3/2}$ signals with background subtraction and curve fitting to determine the integrated peak intensities. Amorphous alloy Fe$_{70}$Cr$_{10}$P$_{13}$C$_7$ polarised for 1 h in 1 N H$_2$SO$_4$ at +0.5 V SCE.*

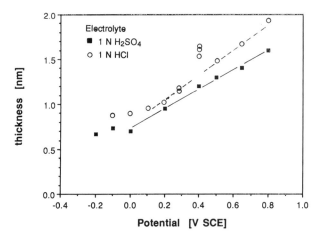

Fig. 2 *Thickness of the passive film formed in deaerated 1N HCl or 1N H₂SO₄ on mechanically polished samples of the amorphous alloy* Fe₇₀Cr₁₀P₁₃C₇ *(1 h of polarisation).*

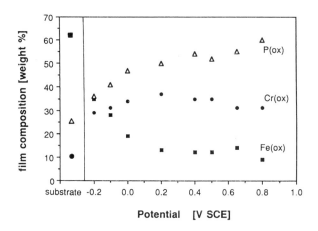

Fig. 3 *Composition of the passive film formed in deaerated 1N H₂SO₄ on mechanically polished and passivated samples of the amorphous alloy* Fe₇₀Cr₁₀P₁₃C₇ *(weight percent determined from the integrated XPS intensities according to the three layer model [4, 10]).*

(III), whereas the iron oxide content decreases. Considerable amounts of phosphates are incorporated into the passive film.

The composition of the substrate underneath the film is independent of the potential (Fe 62%, Cr 11%, P 27%). The main feature is the marked enrichment of phosphorus (at 129.4 eV) compared to the bulk alloy, which is more pronounced at higher potentials (higher film thickness) and/or at a take off angle of 60° [10].

4. Discussion

The excellent resistance against pitting corrosion of this and other amorphous alloys with phosphorus in aggressive media (up to 12 M HCl) has been attributed in literature to the enrichment of chromium oxy-hy-

droxide in the passive film [1, 12], to the formation of a bipolar passive film by incorporation of phosphates into the film [3] or to a synergistic interaction of chromium and phosphorus [6]. The strong chromium (III) oxy-hydroxide enrichment found after passivation of 'as received' samples [12] and is due to the selective dissolution of iron from the thick oxide film present after rapid solidification [13]. Instead the model of a bipolar passive film may apply to samples passivated after mechanical polishing that contain large amounts of phosphates in the passive film.

The common features of all the passivated samples are:

(i) a marked enrichment of phosphorus (at BE = 129.4 eV) in the substrate underneath the film observed for this alloy [4, 10, 13] and other phosphorus containing metallic glasses with a high corrosion resistance [5, 7],

(ii) the presence of an intermediate oxidation state of phosphorus [4, 6, 10, 13] located at the interface film substrate [4]. The chemical state of phosphorus with BE ≈ 129.4 eV corresponds to a partially covalent bound P in the alloy (similar to Cr₃P). The intermediate oxidation state has been identified as elemental phosphorus [11].

This double layered structure with phosphates (+V) in the passive film and elemental phosphorus at the interface film/substrate might be responsible for the high corrosion resistance of phosphorus containing amorphous alloys, reducing the ionic conductivity of the interface and thus improving its stability. Further evidence for this hypothesis give the following electrochemical results:

- At the same level of Cr(III) enrichment in the passive film the alloy with 13% phosphorus is much more resistant against localised corrosion than that with 13% boron [3, 10].
- the mechanism of passivation of amorphous iron–chromium alloys without phosphorus is the same as for Fe–Cr alloys or stainless steels whereas alloys with P do not show the expected dependence of i_{crit} on pH [14].
- the passive current density for this amorphous alloy and FeNiBP is practically independent of the solution pH in contrast to passive films on iron–chromium alloys or stainless steels [4, 10]
- the current transients measured during metastable pitting of this alloy show a much shorter lifetime than those on stainless steels and on the alloy with boron [15], indicating a very high repassivation rate of the P bearing alloy.

These facts may be explained by the presence of the P enriched layer at the interface metal/passive film and—in case of scratching or pitting—by the rapid direct oxidation of P to P(+V) and its incorporation as phosphate into the passive film. In addition, the strong covalent bonds metal-metalloid stabilise the alloy and inhibit the anodic dissolution. Amorphous alloys without P, i.e. $Fe_{70}Cr_{10}B_{13}C_7$, lack these positive effects and despite a similar enrichment of chromium oxi-hydroxide in the passive film are thus much less corrosion resistant [10].

5. Conclusions

The passive film on the amorphous alloy $Fe_{70}Cr_{10}P_{13}C_7$ formed on mechanically polished samples is rich in iron oxide at low potentials. At higher passivation potentials a chromium oxi-hydroxide enrichment (up to 80% of the cations) is found. The passivation occurs by active alloy dissolution and active/passive transition results in phosphates being incorporated into the passive film.

Phosphorus (129.4 eV) is enriched underneath the passive film. An intermediate species of P (131.7 eV) is present after passivation at the inner part of the interface film/substrate, chemically it is similar to elemental phosphorous.

The outstanding corrosion resistance of the alloy $Fe_{70}Cr_{10}P_{13}C_7$ and its high repassivation rate may be attributed to this enrichment of phosphorus underneath the film, reducing to a great extent the ionic conductivity of the interface, and to the incorporation of phosphates in the passive film.

6. Acknowledgements

The Authors wish to thank the Institute of Physics at the University of Basel for the production of the amorphous alloys. The work was financially supported by the Italian National Research Council (Comitato Tecnologico and Progetto di Chimica Fine e Secondaria).

References

1. K. Hashimoto, in Passivity of Metals and Semiconductors, ed. M. Froment, Elsevier, Amsterdam, p. 235
2. C. R. Clayton, M. A. Helfond, R. B. Diegle and N. R. Sorensen, Proc. of Symp. on Corrosion, Electrochemistry and Catalysis of metallic glasses, R. B. Diegle and K. Hashimoto eds, El. Chem. Soc. Vol. **88–1** p.134, 1988.
3. S. Virtanen, B. Elsener and H. Boehni, J. of Less-common Metals, **145**, 581, 1988.
4. A. Rossi and B. Elsener, Surface Interface Analysis, **18**, 1992, 499–504.
5. G. T. Burnstein, Corrosion NACE, **37**, 1981, 549.
6. T. P. Moffat, R. M. Lantanision and R. R. Ruf, J. Electrochem. Soc., **139**, 1992, 1013.
7. M. Janik-Czachor, ISIJ International, **31**, 1991, 149.
8. M. P. Seah, Surface Interface Analysis, **14**, 1989, 488.
9. P. M. A. Sherwood, 'Practical Surface Analysis' ed. Briggs and M. P. Seah, Appendix 3, p.445, J. Wiley, 1983.
10. B. Elsener and A. Rossi, Electrochimica Acta, **37**, 2269–2276, 1992.
11. A. Rossi and B. Elsener, 'Characterisation of surface films on FeCrPC alloys by XPS and X-ray excited Auger peaks', this volume, pp.12–16.
12. K. Asami, K. Hashimoto, T. Masumoto and S. Shimodeira, Corros. Sci., **16**, 909, 1976.
13. B. Elsener and A. Rossi, Metallurgy and Foundry Engineering, **18**, 175–187, 1992.
14. B. Elsener, S. Virtanen and H. Boehni, Electrochim. Acta, **32**, 927, 1987.
15. S. Virtanen, B. Elsener and H. Bohni, Proc. Symp. Transient Techniques, The Electrochem. Soc. Proc., Vol. **89–1**, 1989.

37

Modification of Ta/Ta₂O₅ Films by Platinum Deposits

O. KERREC, D. DEVILLIERS*, C. HINNEN†, P. MARCUS†, H. CACHET** AND H. FROMENT**

E.D.F. Département Etude des Matériaux, Centre de Recherche des Renardières, route de Sens, Ecuelles BP 1, 77250 Moret sur Loing, France
*Université P&M Curie, Laboratoire d'Electrochimie, 4 Place Jussieu, 75252 Paris Cedex 05, URA CNRS 430
†Ecole Nationale Supérieure de Chimie de Paris, Laboratoire de Physicochimie des Surfaces, 11 Rue Pierre et Marie Curie, 75231 Paris Cedex 05, URA CNRS 425
**Université Pierre et Marie Curie, Laboratoire de Physique des Liquides et Electrochimie, 4 Place Jussieu, 75252 Paris Cedex 05, UPR CNRS 15

Abstract

It is well known that the oxidation of tantalum substrates leads to the formation of a Ta_2O_5 coating resistant to corrosion. In this paper the so-called M. O. structure (Metal/Oxide) is prepared by the anodic oxidation of tantalum electrodes in sulphuric acid media. The M. O. E. system Ta/Ta_2O_5/Electrolyte presents a diode effect: only cathodic current is observed, corresponding to protons reduction.

The conduction properties are considerably enhanced if a M.O.M. structure (Metal/Oxide/Metal) is prepared in incorporating a noble metal for example. Two techniques were used to deposit platinum on Ta/Ta_2O_5 electrodes either from the electroreduction or from photodeposition in acidic solution containing platinum salts (Pt(II) or Pt(IV) ions). In both cases, surface analyses were performed by X-ray Photoelectron Spectroscopy and Scanning Electron Microscopy, in order to characterise the structure and the chemical states of the deposits. XPS analyses of the Pt4f core-level indicate that metallic platinum is present in both kinds of film. This is corroborated by SEM images showing the presence of metallic crystallites. I–E curves show that the diode effect is completely cancelled: the M. O. M. structure behaves like a platinum electrode. However, impedance spectra are different from those obtained with platinum electrodes: the oxide layer gives rise to a supplementary loop in the Nyquist diagram, attributed to tunneling conduction through the oxide.

In conclusion, M.O.M. structures may constitute a new kind of corrosion-resistant anodes; the deposit of an electronic conductor on a passivated layer generates a new type of structure in which the conduction is enhanced. This should be taken into account when studying the mechanism of corrosion process.

1. Introduction

The oxidation of 'valve' metals like tantalum leads to the formation of a compact passivating oxide film which protects the metallic substrate from corrosion. The so-called M. O. (Metal/Oxide) structures like Ta/Ta_2O_5 used as electrodes present a diode effect since cathodic reactions can be observed at high overvoltage although anodic reactions are totally blocked [1,2].

This diode effect is due to the lack of localised states at the oxide surface. Under anodic polarisation, electrons cannot be exchanged between the donor species present in solution and the Solid/Liquid interface because acceptor intermediate states do not exist at the oxide surface. On the contrary, under cathodic polarisation, electrons are provided by the metallic substrate and they can be transferred to acceptor oxidised species of the redox couple in solution.

In the case of a metallic electrode, the rate of electron transfer between the redox couple and the metal depends on the nature of the material and the redox couple. In the case of an electrode covered by an oxide layer, the rate depends also on the intrinsic properties of the film (thickness, impurity inclusions, charge carriers concentration, surface states density). One of the modification techniques consists in the creation of intermediate states at the oxide surface by deposit of a noble metal like platinum; for example, $Ti/TiO_2/Au$ structures have already been studied by Schultze and Elfenthal [3].

In this paper, the $Ta/Ta_2O_5/Pt$ structure preparation is described, as well as the factors involved in the remarkable enhancement of the electron transfer observed with these structures.

2. Experimental

2.1 Ta/Ta$_2$O$_5$/Pt structure preparation

Platinum metal was chosen because it has a very good electronic conductivity, excellent catalytic properties and high corrosion resistance. Several preparation methods were tested: electrochemical, photoinduction, sputtering and electroless but only the two first techniques gave mechanically stable deposits adherent to tantalum oxide substrates.

2.1.1 Electrochemical deposit

A cathodic galvanostatic pulse (i = – 3.7 mA cm^{-2} at 40°C) was applied to a Ta/Ta$_2$O$_5$ structure dipped in a solution containing a platinum salt. Two solutions were tested: Ammonium Hexachloroplatinum (IV) (NH$_4$)$_2$PtCl$_6$ 0.5 10^{-3} M in HClO$_4$ 1M (solution A) and Hydrogen Dinitrosulphatoplatinum (II) H$_2$Pt(NO$_2$)$_2$SO$_4$ 1.3 10^{-3} M in H$_2$SO$_4$ 10^{-2} M (solution B). In fact, only the second solution gave rise to an adherent and mechanically stable deposit. A rotating disk electrode was used and surfactants were added to enhance bubble evacuation but these two techniques led to non adherent platinum deposits.

2.1.2 Photoinduced deposit

For many years the photon-assisted technique has been studied for metal deposition [4–7]. It presents many advantages for oxide substrates because no electrical contacts are required and deposit is limited to the illuminated area. Tantalum disks covered by a Ta$_2$O$_5$ layer (band gap of about 4.5 eV) were dipped into a quartz cell containing a platinum salt in acidic solution. Both solutions A and B were used successively for each sample. Samples were irradiated by a Cermax Xenon illuminator (LX150UV, ILC Technology) providing *ca*. 1 W in the wavelength range 250–325 nm, creating electron-hole pairs. Most of the visible and infrared radiations of the lamp spectrum were eliminated by use of a cold mirror (ORIEL-375FV86-50). The spot diameter was 5 mm. Irradiation times were 6 minutes for each solution.

2.2 Deposit characterisation

2.2.1 Surface analysis

The platinum deposit morphology was determined by SEM and its valency states by XPS. XPS measurements were performed with a VG Scientific ESCALAB MARK II. An Al Kα X-Ray source (1486.6 eV) was used with a power of 600 W. The analyser was operating at 20 eV pass energy for high resolution spectra of Pt4f core levels and 50 eV for survey spectra. The analysed area was 10 mm^2. Analysis of pure platinum was carried out in the spectrometer to give the reference fitting parameters of the Pt4f metallic states. Reference data

for oxidised forms of platinum were taken in the literature [8].

2.2.2 Electrochemical analysis

The electrochemical properties of Ta/Ta$_2$O$_5$/Pt structures have been determined by I–E curves, voltammograms and impedance spectra and compared with those of a pure platinum electrode and of a Ta/Ta$_2$O$_5$ structure. Experiments were performed in sulphuric acid 0.5M containing FeSO$_4$/Fe$_2$(SO$_4$)$_3$ 10^{-2} M. A Z Computer Tacussel was used for impedance measurements in the 10^5–10^{-2} Hz frequency range. The amplitude of the sinusoidal signal was 5 mV.

3. Results and Discussion

3.1 Deposit formation

Several factors influence the deposition kinetics: firstly, the difference of the platinum oxidation state and the nature of the complexing anion; secondly, the pH of the solution. Indeed, two reactions are in competition: hydrogen evolution and platinum ions reduction. More the hydrogen evolution overvoltage is important (–60 mV/pH [1]), more the platinum nucleation is enhanced. The higher pH of solution B can explain its better behaviour. As soon as platinum crystallites are formed, electrodeposition is in competition with proton reduction. The faradaic efficiency of the first reaction decreases sharply when the platinum surface increases. We think that hydrogen bubbles mask the metallic active surface on which they are formed. So, proton reduction is blocked and platinum nucleation can continue on other sites. These different phases can be pointed out on Fig. 1; in the case of a Ta/Ta$_2$O$_5$ electrode, the important cathodic overvoltage is re-

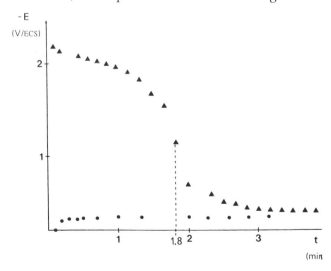

Fig. 1 Potential of the working electrode during cathodic deposition from a H$_2$Pt(NO$_2$)$_2$SO$_4$ solution. Galvanostatic technique: i = –3.7 mA cm^{-2}; t = 40°C. Working electrode: Ta/TaOx structure (▲); platinum (●).

lated to the first nucleation step. The overvoltage decreases with time and a point of inflexion is observed for t = 1.8 min with the applied experimental conditions, corresponding to the complete oxide surface covering. Then, the curves show a flat shape similar to that obtained with a pure platinum electrode on which hydrogen evolves easily.

3.2 Surface analysis

SEM micrographs reveal the morphology difference between the two kinds of deposit. Figure 2(a) shows that the surface is entirely covered with platinum in the case of an electrochemical deposit method. This massive deposit is the result of crystallites coalescence; their diameter is *ca.* 0.5μm. On the contrary, the quantity of platinum crystallites is smaller in the case of a photoinduced deposit. The nucleation seems only car-

(c) Photodeposition of Pt on Ta/Ta₂O₅(39 nm): long irradiation period.

Fig. 2 SEM micrographs of two kinds of platinum deposits.

(a) Electrochemical deposition of Pt on Ta/Ta₂O₅(39 nm).

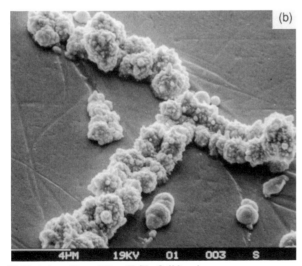

(b) Photodeposition of Pt on Ta/Ta₂O₅(39 nm).

ried out on surface defects, provided by mechanical polishing (Fig. 2(b)). In Fig. 2(c), the micrograph shows that a longer irradiation period does not give rise to new nucleations, but leads to an increase of the size of the existing crystallites.

The experimental XPS spectrum of the Pt4f region of an electrochemical deposit is presented on Fig. 3(a). It was decomposed into two elementary peaks, corresponding respectively to $Pt4f_{7/2}$ and $Pt4f_{5/2}$. An asymmetry of these peaks at low kinetic energies, due to the shake-off process, is observed. Fitted and experimental spectra are comparable with that of the pure platinum reference sample: $E_b(Pt^0_{7/2}) = 71 \pm 0.1$ eV and $E_b(Pt^0_{5/2}) = 74.1 \pm 0.1$ eV. Usually, the thickness of the deposit can be estimated from the Pt4f signal intensity, using the equation relative to an homogeneous layer on a substrate:

$$I_{Pt} = I_{Pt}^{\infty}\left[1 - \exp(-\frac{d_{Pt}}{\lambda_{Pt}^{Pt}\sin\theta})\right]$$

where I_{Pt} and I^{∞}_{Pt} are respectively the intensities emitted by the deposit and by a bulk Pt substrate at the corresponding take-off angle θ, d_{Pt}, the film thickness and λ_{Pt}^{Pt}, the attenuation length of the $Pt4f_{7/2}$ signal. By using, λ_{Pt}^{Pt}, = 17.3 Å as calculated from the Seah and Dench correlation [9] the equivalent deposited thickness should reach *ca.* 30 Å. However, the film thickness is very much underestimated because of the important porosity of the films.

Two successive photoinduced platinum deposits were realised on tantalum oxide, as described above.

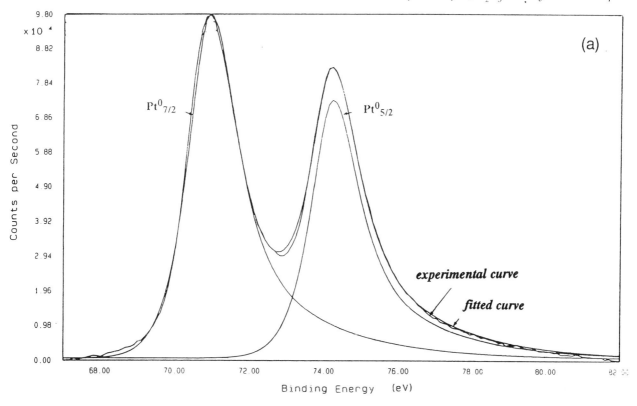

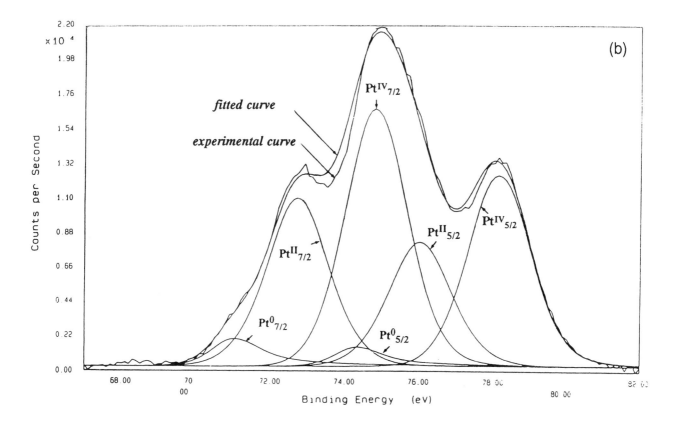

Fig. 3 X-Ray Photoelectron Spectroscopy: Pt4f region of the spectrum for a Ta/Ta$_2$O$_5$/Pt structure.
(a) electrochemical deposition; (b) photodeposition.

The experimental spectrum of the Pt4f region is presented on Fig. 3(b). It is very different from the spec-supplementary doublets are observed at higher binding energies, ($E_b = 72.7$ eV and $E_b = 74.8$ eV for the $Pt4f_{7/2}$ peaks), corresponding respectively to $Pt^{II}_{7/2}$ and $Pt^{IV}_{7/2}$ [8]. The metallic doublet remains visible at the binding energy of the metallic substrate showing (i) that the deposited film is very thin and (ii) that the charging effects are not involved. Pt peak intensities are smaller than those corresponding to the electrochemical deposit as can be seen in comparing Fig. 3(a) and 3(b), but not small enough to be consistent with the presence of small platinum crystallites. The assumption of a full coverage of the surface although not detected by SEM is reinforced by the fact that a colour change is visible to the naked eye: the purple coloured oxide became bright grey after the photoinduced deposit.

In addition, on the survey spectrum, a chlorine peak at $E_b = 199$ eV [8] was detected. This proves that Pt^{IV} can be attributed to non entirely reduced platinum salt deposit from solution A. In the same way, Pt^{II} would arises from a non complete reduction of the platinum salt of solution B.

The photodeposition method gives rise to an important nucleation on the tantalum oxide surface; however, the deposit growth is rapidly stopped and the amount of deposited platinum is lower than in the case of the electrochemical method.

3.3 Electrochemical analysis

The equilibrium potential E_{eq} of the $Ta/Ta_2O_5/$Electrolyte system is not defined. On the contrary, in the case of a $Ta/Ta_2O_5/Pt$ structure (prepared by photodeposition or electrochemical techniques) dipped into a solution containing a redox couple, E_{eq} is perfectly defined and is equal to the potential of a pure platinum electrode in the same solution. For a M.O.M'. structure, E_{eq} must not depend on the nature of the deposited noble metal M'.

3.3.1 I–E curves

The curves corresponding to an electrochemically formed $Ta/Ta_2O_5/Pt$ structure (geometric area = 0.16 cm^2) and to a pure platinum electrode (area: 0.0314 cm^2) both present the characteristic Z-shaped curves of reversible systems. Near E_{eq} the slope of the linear part of the I–E curve corresponding to the M. O. M'. structure is lower than in the case of platinum. This is due to the existence of a supplementary energy barrier with regard to the case of a Metal/Electrolyte interface. Contrary to a Ta/Ta_2O_5 structure, electrons can be transferred in both cathodic and anodic directions.

3.3.2 Cyclic voltammetry

Voltammograms of electrochemical and photoinduced deposits are presented on Fig. 4. In the case of electrochemical $Ta/Ta_2O_5/Pt$ and Pt electrodes, sev-

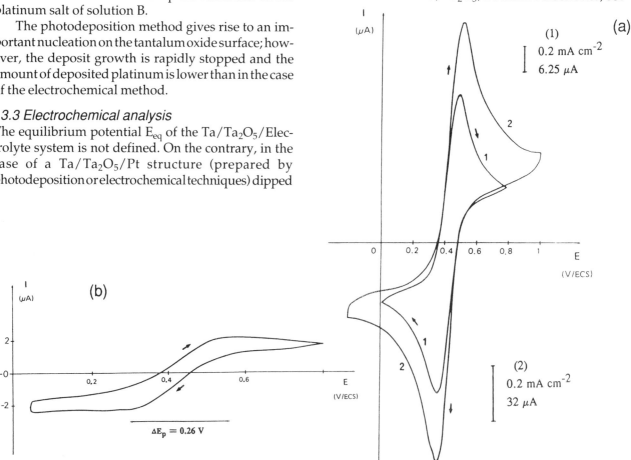

Fig. 4 Voltammograms ($n = 10mVs^{-1}$; $t=25°C$). Electrolyte: H_2SO_4 0.5 M + $FeSO_4/Fe_2(SO_4)_3$ 10^{-2} M.
(a) Comparison of a platinum working electrode [A = 0.0314 cm^2; curve (1)] and a Ta/Ta_2O_5(39 nm)/Electrodeposited Pt structure [A = 0.16 cm^2; curve (2)]. (b) Ta/Ta_2O_5(39 nm)/Photodeposited Pt structure.

eral cycles in the range –0.2 V to +0.8 V *vs* SCE were performed to activate the surface. Indeed, the potential difference ΔE_p between anodic and cathodic peaks decreases when the cycle number increases, and reaches a constant value. It is well known [10,11] that very thin oxide layers are electrochemically formed on the metallic platinum electrodes in acidic electrolytes. The same compounds may be formed in the H_2SO_4 solution containing the Fe^{3+}/Fe^{2+} redox couple. This oxide is easily reducible in acidic medium and this explains the gradual surface activation as and when reduction cycles are performed. Moreover, activation of the Ta/Ta$_2$O$_5$/Pt structure induced by platinum ions migration in the oxide surface layer cannot be involved because the decrease of ΔE_p is also observed for a pure platinum electrode. However, the stationary value of ΔE_p is higher for Ta/Ta$_2$O$_5$/Pt (140 mV) than for Pt (120 mV); on the contrary, the peak intensity (taking into account the geometric area) is lower: $i_{pPt} \approx 1.2$ mA cm^{-2} and $i_{pTa/Ta2O5/Pt} < 0.85$ mA cm^{-2}. A correlation can be made with the different values of the slope observed on I–E curves. In the case of the photoinduced deposit, $\Delta E_p \approx 260$ mV and the current intensity was much lower than that obtained with previous electrodes. This behaviour can be explained by the small quantity of metallic platinum and by the partially oxidised form of the platinum deposit revealed by XPS.

3.3.3 Impedance spectroscopy

These studies were performed on electrochemically prepared Ta/Ta$_2$O$_5$/Pt structures and Pt electrodes at the equilibrium potential of the Fe^{3+}/Fe^{2+} redox couple in H_2SO_4 0.5 M (0.68 V/SHE). As seen above, the catalytic properties of platinum arise with the activation time. Nyquist diagrams were plotted for several values of ΔE_p. Diagrams for M.O.M'. structures present two or three distinct regions (see Figs 5(a) and 5(b)). Each diagram exhibits a capacitive semi-circle in the high frequency region; the C_{ox} capacitance associated to the loop (0.55 μF cm^{-2}) is equal to that found for a Ta/Ta$_2$O$_5$ structure of the same thickness when no faradaic current flows [12]. This proves that the oxide layer is not perforated and that the model taking into account short circuits of 'Pt nails' in the oxide [13] cannot be considered. Indeed, in the latter case, Nyquist diagrams would be similar to that obtained for the Pt/Electrolyte system. Moreover, platinum insertion into the bulk oxide cannot be involved because in that case, the decrease of C_{ox} due to a diminution of the dielectric constant would be observed.

In the medium frequency range, another semi-circle is observed, which diameter R_{ct} increases with ΔE_p. It corresponds to the charge transfer step at the Electrode/Electrolyte interface. The associated

Helmholtz capacity value C_H is constant. R_{ct} and C_H values for the Ta/Ta$_2$O$_5$/Pt structure and the platinum electrode are given in Table 1. i_0 values are lower than for pure platinum electrodes.

In the low frequency range, if the charge transfer is not the limiting kinetic process (i.e. for $\Delta E_p = 200$ mV and $\Delta E_p = 260$ mV), a 45° straight line corresponding to the diffusion of solvated Fe^{3+} and Fe^{2+} electroactive species in solution is observed, as on a metallic electrode. Thus, the Ta/Ta$_2$O$_5$/Pt/Electrolyte system can be modelled by the equivalent circuit represented on Fig. 5(c). The supplementary loop in comparison with the usual Randles circuit modelling the Pt/Electrolyte system is attributed to the oxide layer capacitance C_{ox} in parallel with the oxide resistance noted R_{ox}. The capacity associated to the potential drop in the space charge layer present at the Oxide/Pt interface is not pointed out on the Nyquist diagrams. This is probably due to the fact that the space charge capacity C_{sc} is in series with C_{ox}; C_{sc} value is very high because of the very small thickness of the space charge layer probably induced by doping effect of platinum insertion in the outer oxide layer. Z_W is the Warburg impedance.

Several processes are involved: diffusion of the solvated electroactive species, electron transfer through the outer Helmholtz plane between these species and the energy levels at Electrode/Electrolyte interface and electron transfer through the solid phase. The remarkable modification of the behaviour induced by a platinum deposit is principally due to two factors:

– Firstly, in the case of the Ta/Ta$_2$O$_5$/Electrolyte system, there are no energy states at the oxide surface. Indeed, occupied states of the valence band and unoccupied states of the conduction band are separated by a wide gap devoid of surface states. As the Fermi level of the Fe^{3+}/Fe^{2+} redox couple is located in this forbidden region, electrons cannot be transferred according to the Gurney-Gerischer theory [14]. A platinum deposit creates an important quantity of surface states in the gap (see Fig. 6) and induces a very important R_{ct} diminution, in comparison with a Ta/Ta$_2$O$_5$ structure; R_{ct} values are similar to those obtained for platinum (see Table 1).

– Secondly, Ta/Ta$_2$O$_5$ and Ta/Ta$_2$O$_5$/Pt structures differ by the presence on the latter of an energy barrier associated to the charge carriers depleted region at the Oxide/Platinum interface, characterised by its thin thickness d_t and its height $\Delta E_t = 1.65$ eV. Electron transfer through this barrier is interpreted in terms of direct tunneling. R_{ox} corresponds to the 'tunneling resistance' ($R_{ox} \approx 80$ Ω for $d_{ox} = 39$ nm) and d_t to the tunneling distance. This barrier explains why smaller values of i_0 are obtained for Ta/Ta$_2$O$_5$/Pt structures in

Table 1 Results of electrochemical analysis for Pt or Ta/Ta$_2$O$_5$/Pt electrodes

electrode material	ΔE_p (mV)	R_{ct} (Ω cm^2)	C_H (μF cm^{-2})	i_0 (mA cm^{-2})
Pt	120	6.3	44	4.0
	210	24.2	47	1.4
Ta/Ta$_2$O$_5$/Pt	200	25.0	40	1.0
(d_{ox} = 39 nm)	260	38.4	41	0.7
(electrogenerated Pt)	320	384.0	33	0.06

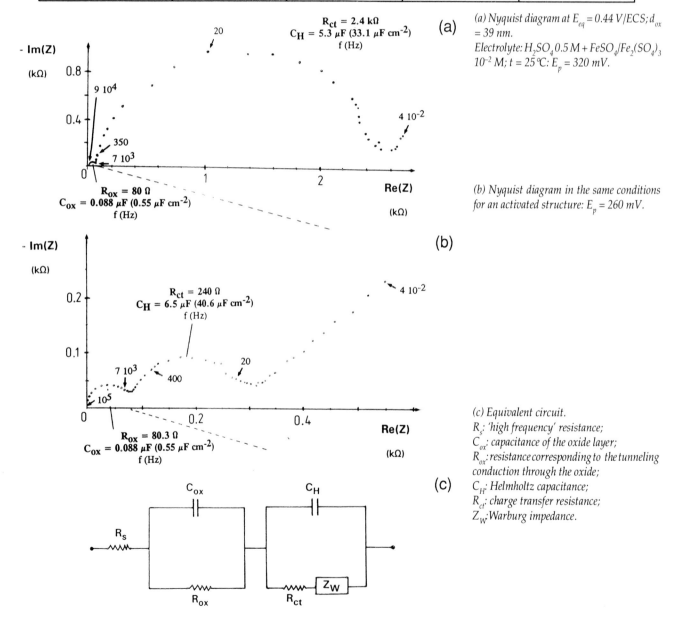

(a) Nyquist diagram at E_{eq} = 0.44 V/ECS; d_{ox} = 39 nm.
Electrolyte: H_2SO_4 0.5 M + FeSO$_4$/Fe$_2$(SO$_4$)$_3$ 10^{-2} M; t = 25 °C: E_p = 320 mV.

(b) Nyquist diagram in the same conditions for an activated structure: E_p = 260 mV.

(c) Equivalent circuit.
R_s: 'high frequency' resistance;
C_{ox}: capacitance of the oxide layer;
R_{ox}: resistance corresponding to the tunneling conduction through the oxide;
C_H: Helmholtz capacitance;
R_{ct}: charge transfer resistance;
Z_W: Warburg impedance.

Fig. 5 Electrochemical analysis: impedance spectroscopy for the Ta/TaO$_X$/Electrodeposited Pt/Electrolyte system.

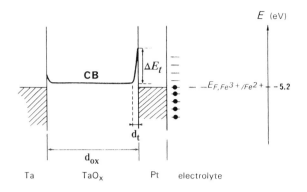

Fig. 6 Band diagram of the Ta/TaO_x/Pt/Electrolyte system.

CB: conduction band;
$E_{F\ Fe3+/Fe2+}$: Fermi level of the redox couple;
d_t: tunnel distance;
ΔE_t: height of the tunnel barrier.

comparison with those obtained with platinum. When a Ta/Ta₂O₅ structure is polarised, for which charge density at the oxide surface is negligible, the potential drop is localised within the oxide: band edges are fixed. On the contrary, for Ta/Ta₂O₅/Pt, the Oxide/Pt interface is completely degenerated because of the high surface state density. Electrons can be easily transferred and the electrode is reversible. The exchange current density depends on the energy difference between the electron affinity in the oxide and the deposited metal work function.

4. Conclusion

XPS and SEM techniques have revealed that photoinduced platinum deposits formed at the surface of thick tantalum oxides are very thin and partially oxidised. Those obtained by electrochemical method result from the coalescence of metallic crystallites.

The electrochemical behaviour of Ta/Ta₂O₅/Pt structures is similar to that of a metallic platinum electrode. The enhancement of electronic conductivity is interpreted in terms of an important diminution of the charge transfer resistance at the Electrode/Electrolyte interface.

The deposit of an electronic conductor on a passivated layer generates a new type of corrosion

resistant anodes. New catalytic Ta/Ta₂O₅/IrO₂ DSA's for oxygen evolution have been proposed [15,16]. Another study concerns the perfecting of Ti/TiO_x/PbO₂ electrodes used for example for the ozone preparation [17].

Moreover, the catalytic effect induced by electronic conductor deposition should be taken into account when studying the mechanism of the corrosion process of materials in which releasing products or impurities can be deposited.

References

1. V. A. Macagno and J. W. Schultze, J. Electroanal. Chem., **180**, 157, 1984.
2. J. W. Schultze and V. A. Macagno, Electrochim. Acta, **31**, 355, 1986.
3. J. W. Schultze and L. Elfenthal, J. Electroanal. Chem., **204**, 153, 1986.
4. F. Mollers, H. J. Tolle and R. Memming, J. Electrochem. Soc., **121**, 1160, 1974.
5. J. J. Kelly and J. K. Vondeling, J. Electrochem. Soc., **122**, 1103, 1975.
6. H. Cachet, M. Froment and A. Messad, J. Electroanal. Chem., **284**, 263, 1990.
7. J. Bruneaux, H. Cachet, M. Froment and A. Messad, Electrochim. Acta, **36**, 1787, 1991.
8. C. D. Wagner, W. M. Riggs, L. E. Davis, J. F. Moulder and G. E. Mullenberg 'Handbook of X-Ray Photoelectron Spectroscopy', G. E. Mullenberg (Ed.), Perkin-Elmer Corporation, Physical Electronics Division, Eden Prairie, Minnesota (1979).
9. M. P. Seah and W. A. Dench, Surf. Interface Anal., **1**, 2, 1979.
10. J. J. Van Benschoten, J. Y. Lewis, W. R. Heineman, D. A. Roston and P. T. Kissinger, J. Chem. Educ., **60**, 1983, 772.
11. N. R. de Tacconi, J. O. Zerbino, M. E. Folquer and A. J. Arvia, J. Electroanal. Chem., **85**, 213, 1977.
12. O. Kerrec, Thesis, Université Pierre et Marie Curie, Paris, 1992.
13. J. W. Schultze, Materials Chem. and Phys., **22**, 417, 1989.
14. R. Memming, 'Charge Transfer Processes at Semiconductor Electrodes', J. Electroanal. Chem., **11**, 1, 1979.
15. C. Comninellis and G. P. Vercesi, J. Applied Electrochem., **21**, 136, 1991.
16. C. Comninellis and G. P. Vercesi, J. Applied Electrochem., **21**, 335, 1991.
17. J. E. Graves, D. Pletcher, R. L. Clarke and F. C. Walsh, J. Applied Electrochem., **22**, 200, 1992.

38

The Effect of Pretreatments on the Corrosion Behaviour of Aluminium

H. J. W. Lenderink and J. H. W. de Wit

Delft University of Technology, Laboratory of Materials Science, Division of Corrosion Technology and Electrochemistry, Rotterdamseweg 137, 2628 AL Delft, The Netherlands

Abstract

Electrochemical techniques (impedance measurements and recording of polarisation curves) were used to study the corrosion behaviour of aluminium and aluminium substrates covered with a chromate conversion layer, in neutral solutions. In addition the conversion layers after exposure at different potentials have been investigated by Scanning Electron Microscopy (SEM). SEM shows that the porous chromate conversion layer has been converted to a thinner and more compact oxide layer. This is also confirmed by impedance measurements. In particular the high frequency time constant could be related to the presence of an oxide film on blare aluminium or to the chromate conversion coating. Beside the change of the morphology of the chromate conversion coating, its composition also modifies. Especially at high anodic potentials, chromate ions migrate from the film which results in a less protective layer.

1. Introduction

Filiform corrosion is a special type of localised corrosion. It occurs on certain metals, for example aluminium, underneath protective organic coatings. It is characterised by a thread-like track of corrosion products which may have considerable length but little width or depth [1]. The filiform consists of a slowly moving active head, in which the corrosion reactions occur, and an inactive tail of corrosion products. Filiform corrosion usually starts at a defect in the paint film. Examples of such defects are defects due to mechanical damage, an edge which is insufficiently covered by the paint or defects which are the result of long time of exposure to an aggressive environment (blistering of the paint).

Before aluminium substrates are painted, they are generally pretreated, for example by anodising or chromatising. The resulting oxide layers usually diminish the sensitivity for filiform corrosion. In addition they considerably decrease the growth rate in case filiform corrosion damage occurs.

Certain mechanisms have been proposed for filiform corrosion [1–3]. Hoch [1] proposed a mechanism in which he considered the head of the filiform as anodic and the area under the coating in front of the head as cathodic. The substrate in the head is forced to dissolve and therefore the adhesion of the paint with the substrate fails. Another mechanism has been proposed by van der Berg *et al.* [2]. They considered the main part of the filiform head to be anodic. However a small area of the head located in the tip of the paint–substrate is considered to be cathodic. At that spot oxygen is reduced and hydroxide ions are produced (pH increases locally considerably) and therefore the paint–substrate adhesion fails. In both mechanisms there is separation of anodic and cathodic areas.

Despite much research the actual mechanism is still unclear. In literature one prefers usually the first mentioned mechanism for painted aluminium substrates without giving experimental or theoretical support. Nevertheless the question remains how conversion layers on aluminium behave in neutral, acid and alkaline environments and how this behaviour is related to the mechanism of filiform corrosion.

This article deals with the corrosion behaviour of the aluminium and aluminium substrates covered by a chromate conversion layer. In literature these conversion layers have been described, particularly during their formation on substrate materials such as iron [4,5], zinc [6], titanium [7] and aluminium alloys [8,9]. Surface analysing techniques such as XPS and SIMS [8,9] were used to determine the composition of the final formed layers. Although their morphology and chemical composition depend on the application techniques, these layers are supposed to consist of aluminium hydroxide-oxide containing Cr^{3+} and Cr^{6+} species. The Cr species are largely located in the outer layers of the film.

2. Experimental

Pure aluminium cylinders (99.995%, Johnson Matthey Ltd) with a diameter of 0.6 cm were first covered with a two component resin (Metacoat F495, Sikkens NL) and then embedded in another epoxy resin (Epikote 828 Shell and Loromin C260 Hardener, BASF). Only one side of the embedded aluminium cylinder was exposed to the electrolyte. The electrodes were abraded (grit 600), polished (1 μm), ultrasonically cleaned and finally rinsed with ethanol. The electrodes were mounted in a specially designed rotating disk electrode cell [10]. The electrodes were not anodised but only polarised to several anodic potentials during the measurements. This results in an oxide film of several nm thickness.

In addition a commercially chromated Al5050 alloy supplied by industry was used. These specimens were rinsed only with ethanol. A PMMA tube was carefully glued on the surface area with inert epoxy resin to provide a measuring cell.

The electrolyte solution used in the experiments contained 0.1M Na$_2$SO$_4$. The electrolyte solutions were prepared from analytical grade (Merck) salts and de-ionised water.

Before measuring the polarisation curves or performing impedance measurements, the electrode was left at open circuit to obtain a stable corrosion potential and then polarised to a chosen anodic potential. The impedance measurements were recorded when the current fluctuations were less than 2.5 % of the total current in 1 min. The amplitude of the perturbation was 10 mV and the frequency was varied from 99 kHz to 1 mHz. All measurements were performed in a Faradaic cage.

A Schlumberger 1286 Electrochemical Interface and a Schlumberger 1255 Frequency Response Analyser were used for the impedance measurements. The polarisation curve was recorded using a EG&G PAR 173 potentiostat. The electrode potential was scanned from –1500 mV to 500 mV with a rate of 1 mV/s. The polarisation curves were recorded at least twice. All potentials were measured against a saturated Hg/HgSO$_4$ (SSE) electrode (+660 mV vs NHE, T = 293K), and the potentials in this article are given with respect to this electrode.

The SEM used for investigating the surface morphology, was a JEOL 840A scanning microscope. A gold layer has been deposited onto the samples. The voltage was 10 kV. Some of the samples were bent to look at the oxide film in more detail.

3. Results and Discussion

As described above the conversion layer can be con-sidered as a layer of aluminium hydroxide/oxide, contaminated with Cr^{3+} and Cr^{6+} species. Figure 1 shows the conversion layer. It has a very porous char-acter. The figure is representative of both the exposed (at the corrosion potential for *ca.* 16 h) and not-exposed layer since the morphology did not change during exposure at the corrosion potential. The porous film is converted into a more compact layer if the sample is held at –250 mV for *ca.* 16 h (Fig. 2). Figure 3 shows in more detail that the film contains large pores and some remnants of the porous layer left on the compact film. This is clearly shown in Fig. 4 for the bent specimen. As a result of the bending process the compact layer has been cracked.

Impedance measurements also confirm the modi-fication of the conversion layer. Figure 5 shows the Nyquist plot for samples polarised at –250 mV after 400 and 1000 min exposure. The shape of the Nyquist plot is quite similar to that of a blank aluminium specimen (without conversion layer). The time con-stant (semi-circle) at high frequencies is attributed to the oxide film [11,12] or to the anodic aluminium dissolution reaction at the metal–oxide interface [13]. Our results show that this time constant is due to the oxide film as is also suggested by Frers [11] and Bessone [12]. When the oxide film is considered as a dielectric material, a linear relation between the potential and the inverse of its capacitance is expected and was also obtained (Fig.6).

From Fig. 6 the thickness d of the (anodised) oxide or conversion layer could be calculated as a function of the potential using equation:

$$d = \frac{\varepsilon_o \varepsilon_r A}{C_{ox}} \qquad (1)$$

in which ε_o is the dielectric constant of vacuum; ε_r is the relative dielectric constant of medium, A is the area and C_{ox} is the capacitance of the oxide or conversion layer.

A thickness of the (anodised) oxide film on the blank aluminium (without a chromate conversion coat-ing) of about 3nm has been calculated ($\varepsilon_{r,Al2O3}$ = 9.8, Fig. 6). This agrees with the thickness (10–50 Å/V), found in literature [14]. The calculated thickness of the chromate conversion layer is about 1 μm (ε_r = 8) and agrees with a rough estimation from the SEM pictures. This implies that the corresponding time constant in the impedance data can actually be attributed to the oxide film and not to the anodic dissolution reaction of aluminium at the metal–oxide interface as suggested by Brett [13].

During exposure to the electrolyte solutions at a electrodepotential of –250 mV a slow change of the morphology of the conversion layer with time is ob-

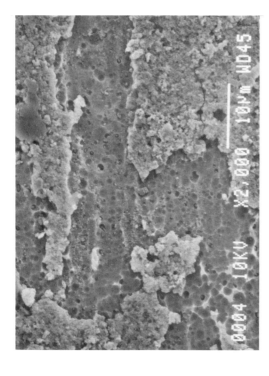

Fig. 1 SEM photo of the chromate conversion layer. The morphology is representative of the not exposed area as well as for the exposed surface (16 h at −1250 mV).

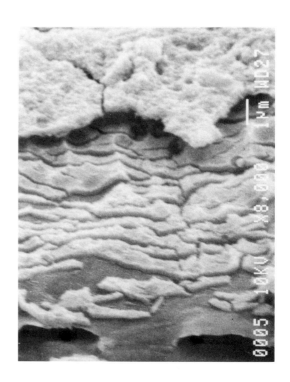

Fig. 2 SEM photo of the chromate conversion layer after exposure (−250 mV for 16 h).

Fig. 3 SEM photo of the conversion layer. (more detailed) (−250 mV for 16 h).

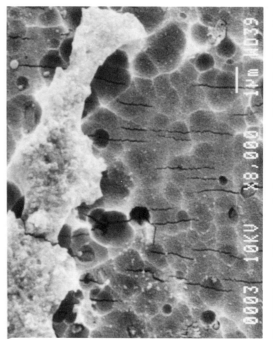

Fig. 4 SEM photo of the conversion layer. The specimen was bent 90° (−250 mV for 16 h).

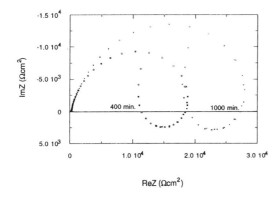

Fig. 5 Nyquist plot of chromate conversion coating on Al5050 alloy. Specimen is held for 400 and 1000 min at –250 mV.

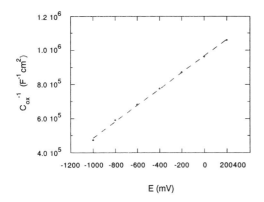

Fig. 6 The inverse of the capacitance of the oxide layer on bland aluminium(without conversion layer).

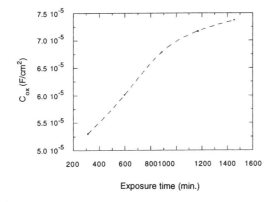

Fig. 7 Capacitance of the conversion layer as a function of exposure time.

served. The diameter of the semi-circle at high frequencies in Fig. 5 (i.e. the resistance of the conversion layer) increases with time. This implies that the layer is improving. In addition its capacitance increases as Fig. 7 shows. The value of the capacitance may increase due to the higher density of the oxide (larger ε_r) and / or due to the dissolution of the film which results in a

thinner film and therefore in a larger capacitance (SEM, Fig. 4).

The polarisation curves of the pure aluminium specimen and the chromated Al5050 alloy are shown in Fig. 8. Figure 8 shows that the passive current density of the chromated Al5050 alloy is initially about a factor 100 smaller than that of the pure aluminium specimen. This is due to the presence of the conversion layer on the aluminium alloy which impedes the anodic dissolution reaction of aluminium. The porous film may directly start to modify to a more compact layer as soon as the potential is scanned anodically. This prohibits the current to increase.

Beside the change in the morphology, the composition of the conversion layer is modified. Above anodic potentials of 0 mV the passive current density increases enormously due to the oxidation of the Cr^{3+} species present in the chromate conversion layer ($2Cr^{3+} + 7H_2O \Leftrightarrow Cr_2O_7 + 14H^+ + 6e$). The film only partially dissolves because the current density remains smaller than that of the bare aluminium alloy without conversion layer (10^{-6} A cm^{-2}). The protective property of the film becomes diminished. This can also be concluded from the current density (10^{-7} A cm^{-2}) of the reversed scan in the cyclic polarisation curves in Fig. 9 at potentials below 0 mV, where the influence of the conversion layer is still visible. In the subsequent scan (Fig. 10), the current behaviour is similar to that of the first reversed scan and thus confirms again the influence of the remaining conversion layer. When the vertex potential of a cyclic polarisation curve is kept below 0 mV, the low current density of *ca.* 10^{-8} A cm^{-2} could be maintained even in the subsequent scans. So, If the potential exceeds 0 V, the protective character of the film is reduced as a result of chromate dissolution.

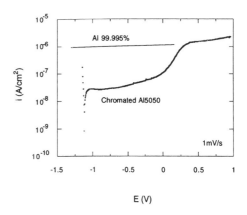

Fig. 8 Polarisation curves of pure aluminium and Al5050 alloy with conversion coating in a 0.1M Na_2SO_4 solution.

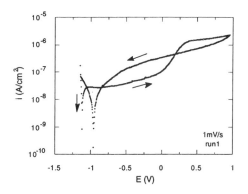

Fig. 9 Cyclic polarisation curves of Al5050 alloy with conversion coating; run 1.

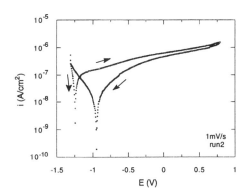

Fig. 10 Cyclic polarisation curves; run 2.

The impedance diagram of the chromate conversion layer shows a very complex shape (Fig. 11), particularly in the frequency range from 0.1 to 100 Hz. An analysis of the impedance diagram by the Boukamp 'Equivcrt' program of [15] gives several time constants. It is not easy to relate these time constants to the occurring reactions. Some general remarks however can be made. As mentioned earlier, the time constant indicated as (2) in Fig. 11 (left) can be ascribed to the conversion layer itself. The time constant (1) at low frequencies may be attributed to the dissolution process which becomes more pronounced at more anodic potentials. As can be observed in the impedance diagram (Fig. 11 (b)), the impedance doesn't show clearly the time constants at medium frequencies.

The polarisation curves of the chromate conversion layers show features which are relevant to the filiform corrosion process in the (anodic) head of the filiform, although we do not know which potential on the curve corresponds to that of the filiform head. The behaviour of the chromate conversion layers at low anodic potentials indicates that such layers may slow down the growth of the filiform. The layer slowly dissolves if the anodic potential in the filiform head is kept low.

4. Conclusion

From the SEM photos and the cyclic polarisation curves it can be concluded that the porous chromate conversion layer has been modified to a more compact layer by polarising the samples anodically up to 1000 mV (SSE). Beside the modification of the conversion layer, it's composition is also changed, particularly at potentials more positive than 0 mV. Cr^{3+} species from the film are oxidised and dissolved as chromate. Impedance measurements provide an useful additional information on the processes in the chromate conversion layer as a function of potential. However a detailed analysis is not yet possible due to the porous structure. Experiments will be done in the future to explain more details.

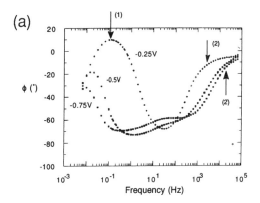

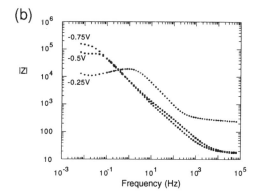

Fig. 11 Bode plot of Impedance measurement of Al5050 alloy in 0.1M Na_2SO_4 solution at –0.75, –0.5 and –0.25V, (a) phase angle ϕ and (b) Impedance |Z|. Two time constants are marked as (1) and (2).

5. Acknowledgement

This research is supported by the Netherlands Foundation for Chemical Research (SON) with financial aid from the Netherlands Technology Foundation.

References

1. G. M. Hoch and R. F. Tobias, Nace Corrosion '71, paper 19.
2. M. van Loo, D. D. Laiderman and R. R. Bruhn, Corrosion, **9**, 277–283, 1953.
3. W. van der Berg, J. A. W. van Laar and J. Suurmond, Adv. Organic Coat. Sci. Techn., 188–207, 1979.
4. Z. Szaklarska and R. W. Steahle, J. Electrochem. Soc., **121**, 9, 1146–1152, 1974.
5. R. A. Powers and N. Hackerman, J. Electrochem Soc., **100**, 7, 314–19, 1953.
6. A. Pirnat, L. Meszaros and B. Lengyel, Electrochim. Acta, **35**, 2, 515–522, 1990.
7. H. H. Uhlig and A. Geary, J. Electrochem. Soc., **101**, 5, 215–224, 1954.
8. M. F. A. Rabbo, J. A. Richardson and G. C. Wood, Corros. Sci., **16**, 689–702, 1976.
9. M. Koudelkova and J. Augustynski, J. Electrochem. Soc., **124**, 8, 1165–1168, 1977.
10. J. A. L. Dobbelaar, The use of impedance measurements in corrosion research. Thesis Delft University of Technology, 1990.
11. S. E. Frers, M. M. Stefenel, *et al.*, J. Appl. Electrochem., **20**, 996–999, 1990.
12. J. Bessone, C. Mayer, K. Juttner and W. J. Lorenz, Electrochim. Acta, **28** (2), 171–175, 1983.
13. C. M. A. Brett, Corros. Sci., **33** (2), 203–210, 1992.
14. H. J. de Wit, C. Wijenberg and C. Crevecoeur, J. Electrochem. Soc., **126** (5), 779–785, 1979.
15. B. A. Boukamp, User manual 'Equivcrt' Sofware .

39

The Effect of Hydrostatic Pressure on the Modification of Passive Films Formed on 316L Stainless Steel in Sea Water

A. M. Beccaria, G. Poggi, E. Tartacca and P. Castello

Istituto per la Corrosione Marina dei Metalli, Area di Ricerca del CNR, Genova, Italy

Abstract

The effect of the increase of the hydrostatic pressure on the corrosion behaviour of 316L stainless steel in sea water was studied in the pressure range from 1 to 300 atm. Electrochemical tests and XPS technique were used. It was observed that the resistance to localised corrosion decreases by increasing the hydrostatic pressure. The great pitting susceptibility at high pressure was linked to the change of the composition of the passive film being a Mo rich mixed (Cr, Fe) hydroxide the film formed at atmospheric pressure and a 'dry', depleted Mo, mixed (Cr, Fe) oxide the film formed at high pressure.

1. Introduction

Field tests carried out in natural sea water at different depths (1, 1000, 1500, 2500 m) [1, 2] show that the pitting susceptibility of 316 stainless steel increases by increasing depth. This increase was only ascribed to the change of the dissolved oxygen concentration (D.O.) in surface and in deep sea water since D.O. value is 7 ppm about in the surface and 2 ppm or less in the depth; the effect of the hydrostatic pressure was ruled out. Conversely, laboratory experiments carried out by changing the hydrostatic pressure and leaving unchanged all other parameters (D.O., T, etc.) show the effect of the hydrostatic pressure on the corrosion behaviour of many metals and alloys [3–7]. This influence has been attributed to the different nature of the corrosion layers formed at different pressures, owing to the change of some chemico–physical parameters, like ion activity, ion hydration degree, etc. to which free energies of the reaction of different corrosion compounds are linked. In fact, owing to the enhanced activity [8] of Cl^- ions and their penetrability into the passive layer, several metal oxides [4, 7] can be converted into soluble oxychlorides, thus forming pit initiation sites and owing to the decreased ion hydration degree the oxide/hydroxide ratio decreases thus modifying the protective power of the corrosion layer [3, 9]. Therefore the greater pitting susceptibility of 316 L stainless steel at high pressure would be also linked to the change of the composition of its corrosion layer, due to the effect of the hydrostatic pressure. As it is known [10–13] the passivating power of corrosion film is due to the formation of an oxyhydroxichloride which

contains Fe^{3+}, Ni^{2+}, Mo^{6+}, Cr^{3+} as cations with relatively high Cr content. The presence of OH groups on the steel surface is a necessary precursor for passive film formation [14]. At high pressure the concentration of OH^- in the passive film might decrease by depleting its corrosion resistance to localised attack.

In order to verify whether the hydrostatic pressure may modify the composition of corrosion film formed on 316L stainless steel and affects its resistance to localised corrosion attack, investigations were carried out at different hydrostatic pressures (1, 10, 100, 300 atm.) at 10°C in sea water at pH 8.2, with a D.O. content of 6.5 ppm.

2. Experimental Method

The tests specimens (cylinders of 0.5 cm dia. and 2.5 cm height for electrochemical experiments and sheets of 1.5×4.5×0.5 cm for free corrosion experiments) were obtained from 316L stainless steel plates having the following nominal percent composition: Cr 18.5; Ni 10.0; Mo 2.5; C 0.03; remainder Fe; austenitic structure. The specimens were polished up to 600 grade emery paper, then degreased with petroleum ether and with methanol before being immersed in sea water at pH 8.2 and at 10°C. The pH value remained almost unchanged. The experiments were carried out in pressure vessels pressurised with an hydropneumatic pump by using the corrosive solution as working fluid so that the dissolved oxygen concentration remained effectively unchanged during the pressure increase, ranging from 7.00 ppm at the beginning of the experiment to 6.0 ppm at the end, after 72h exposure time. The electrodes (arranged in the same manner as in the ASTM G5/72 cell) were: a working electrode made of 316L, two counter electrodes made of platinum, and an Ag/AgCl reference electrode. The tests pressure values were: 1, 10, 100, 300 atm. Electrochemical tests were carried out on specimens previously kept for 2h in sea water at various pressures. Corrosion current density and polarisation resistance were measured with polarisation potentiodynamic curves obtained with a scanning speed of 250 mVh^{-1}. The impedance was measured by perturbing the equilibrium potential of the specimen with 10 mV a.c. with a frequency decreasing from 10KHz to 10 mHz. The impedance values were plotted in Nyquist diagrams (ReZ vs −ImZ). Both passivation film breakdown potential (Ep) and pitting protection potential (Epp) were also measured. Measurements of passivation film breakdown and protection to pitting potential values were made with semi-stationary potentiostatic polarisation. By beginning from the equilibrium potential, the working electrode potential was increased in steps of 1 mV every 10 min up to Ep value, which was indicated by the sharp increase of the anodic current. When Ep was established, the working electrode potential was lowered in the same manner (1 mV/10 min) until the backward curve crossed the forward curve. Free corrosion tests were carried out on specimens previously immersed 72h at 1, 100, 300 atm.

The corrosion layer was characterised with XPS spectroscopy. X-Ray photoelectron spectra were obtained by using a vacuum generator L.t.d. Escalab 210 spectrometer equipped with three chambers (analyser, preparation chamber and fast energy lock). All X-Ray photoelectron spectra were recorded in fixed analyser transmission (FAT) mode 20 e.V. pass energy. Mg K$_{\alpha 1,2}$ radiation (1253.6 e.V.) was used at residual 10^{-9} torr pressure.

3. Experimental Results and Discussion

3.1 Electrochemical tests

Polarisation potentiodynamic curves (Fig. 1) show that the average corrosion rate of 316L stainless steel in sea water slightly increase by increasing pressure, being corrosion current density (i_{corr}) and polarisation resistance (Rp) values respectively equal to 0.195 µA cm^{-2} and 153 KΩ cm^2 at 1 atm and equal to 0.258 µA cm^{-2} and 124 KΩ cm^2 at 300 atm. Any modification of the electrochemical mechanism can be avoided, because the Tafel slopes of the polarisation curves remain rather unchanged when the hydrostatic pressure increases. The increase of the corrosion current density is probably due to the change of the composition of the passive layer leading to the smaller passivation domain, as confirmed by the feature of the

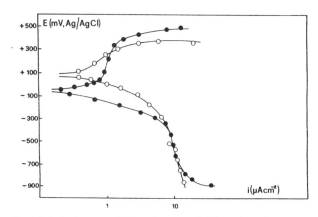

Fig. 1 Polarisation potentiodynamic curves of 316L specimens in sea water at −•− and at −o− 300 atm. 2h pre-exposure time.

anodic curve. Moreover the positive shift of the corrosion (Table 1) and rest (Fig. 2) potentials suggests that at the metal–solution interface the pH mean value would decrease while the mean Mn$^+$ activity would increase, thus confirming the possibility of localised corrosion phenomena shown from the pitting susceptibility measurements (Fig. 2).

In fact Fig. 2 shows that by increasing the hydrostatic pressure, the following parameters decrease:

- the breakdown film potential,
- the difference between Eeq and Epp potentials,
- the perfect passivity domain,
- the repassivation domain, decrease.

The impedance measurements confirm the decrease of the mean corrosion resistance of 316 L stainless steel shown by potentiodynamic polarisation although the Rω$_{\rightarrow 0}$ values (Table 2) at atmospheric pressure are similar to the R$_p$ values (Table 1) but smaller

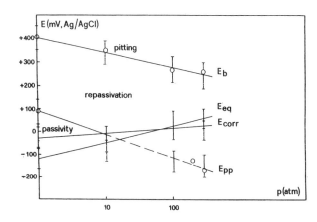

Fig. 2 Breakdown film potential (Ep), corrosion potential (Ecorr), equilibrium potential (Eeq), repairing film potential (Epp), vs hydrostatic pressure.

Table 1 Electrochemical parameters obtained from polarisation curves of 316L specimens in sea water after 2h of pre-exposure time

p atm	i_{corr} mA cm^2	E_{corr} mv, Ag/AgCl	R_p KΩ	B_a mV/dec	B_c mV/ dec
1	0.195	− 59.30	153	140	− 135
10	0.220	+ 0.09	140	150	− 135
100	0.240	+ 30.50	133	170	− 130
300	0.258	+ 51.90	124	175	− 127

Table 2 Rv$_{\to}$0 and electrode capacitance values obtained from impedance measurements of 316L in sea water after 2h of pre-exposure time

p atm	R_{Cu+O} KΩcm^2	C mFcm^{-2}
1	161.90	83.16
10	105.66	49.76
100	75.17	26.9
300	84.86	12.5

than those of R_p at high pressure. The feature of the impedance curves (Fig. 3) in the whole pressure range suggests a multiple adsorption/desorption complex process, with more than one time constant [15], being all semicircles below the real axis. The electrode capacitance values, ranging from 83 to 20 μF cm^{-2} (Table 3) suggest that the oxide are the main components of corrosion layers in the whole pressure range. Therefore the highest value of the capacitance at atmospheric pressure suggests the presence of compounds with a ionic bond, such as M–OH hydroxides. The straight lines of ReZ vs –JmZ in the lowest frequency range suggest a mass transport through the passive film easily due to the presence of several pores formed by localised corrosion attack. The slope of these straight lines increases by increasing the hydrostatic pressure, by confirming the greatest extent of localised corrosion at high pressure.

4. Free Corrosion Test—ESCA Analysis

XPS analysis (Table 3) shows that the passive film of specimens exposed in sea water at atmospheric pressure is mainly formed by a mixed, Mo rich, (Fe^{2+}, Fe^{3+}, Cr^{3+}, Ni^{2+}, Mo^{6+}) oxyhydroxide (Fig. 4). These results

are in good agreement with the literature data which ascribe to the presence of a Cr^{3+} hydroxide–Mo^{6+} hydroxide a good resistance of stainless steels against to localised corrosion [16, 18]. Cr^{3+} hydroxide forms a very compact films which are stable to the attack of Cl^- ions and Mo^{6+} decreases the activity of surface active sites through the formation on them of Mo-oxyhydroxide or Fe–Cr molybdate, thus forming an uniform passive films. The presence of bound water may increase the corrosion resistance of 316L stainless steel because, according to the Okamodo theory [19] the bound water may react to give various species or be deplaced by the other ions coming from surroundings, resulting in degradation of passive film. On the other hand, the bound water acts as the beneficial species to capture the dissolving metal ions with consequent formation of the new film resisting against further attack by surroundings. The composition of the passive film changes by increasing pressure, as XPS analysis (Table 3, Fig. 4) shows. In fact the bound water and Me–OH bonds amounts decrease, so that the passive film is mainly formed by mixed 'dry' oxides with a $Cr3^+$ content higher than that of the bare alloy and a Mo total content ($Mo^{4+} + Mo^{6+}$) lower than that of the bare alloy (Fig. 4), thus suggesting the preferential Mo dissolution through the formation of soluble Mo_xCl^{y-} complex. Table 3 moreover shows the presence of

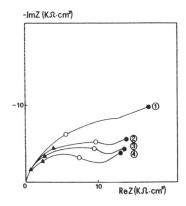

Fig. 3 Impedance curves of 316L in sea water after 2h of pre-exposure time at: 1 atm, curve (1) 10 atm, curve (2); 100 atm, curve (3); 300 atm, curve (4) in the frequency range from 10 mHz to 10 kHz. ● 10 mHz; ○ 100mHz; ▲ 1Hz.

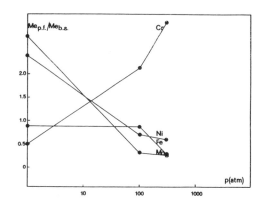

Fig. 4 Metal concentration in the passive film/metal concentration in the bar alloy vs hydrostatic pressure.

Table 3 Binding energies (e.V.), in round brackets, and percentages of Fe, FCr, Ni, Mo, O compounds in the corrosion layers formed on 316L stainless steel in sea water 72 h exposure time

Elem	P/atm	709.5	710.5	711.5	712.3						852.3	853.9	855.6
	1	15	58	—	27				1		40	—	60
Fe_{2p}	100	5	49	44	—			Ni_{2p}	100		50	40	10
	300	—	27	67	—				300		27	60	—
		$Fe(OH)_2$	αFe_2O_3	$\gamma FeO \cdot OH$							Ni_{met}	NiO	$Ni(OH)_2$
		575.6	576.5	577.0	578.0					529.6	530.5	532.0	533.2
	1	—	—	100	—				1	15	—	45	38
Cr_{2p}	100	29	53	—	17			O_{1s}	100	23	7	37	17
	300	15	55	—	29				300	29	9	35	13
		CrO_2	CrO_2O_3	$CrOH_3$	CrO_3					oxide	spinel	OH^-	H_2O
		227.6	230.7	231.6	232.2	232.7	235.1						
	1	—	—	—	—	48	35						
Mo_{3d}	100	32	—	29	—	—	—						
	300	44	29	—	18	—	—						
		Mo_{met}	$MoCl_4$	$FeMoO_3$	MoO_3	MoO_3	Mo-oxhyd						

some amounts of Mo-chlorides (probably $MoCl_4$) in the corrosion film. The presence of great amount of FeOOH non protective product [13] confirms the lower protection power of the corrosion film formed at high pressure. The lower concentration of OH^- groups and of bound water, explains the decrease of the electrode capacitance values by increasing pressure (Table 2). Moreover the (OH, H_2O) concentration can be correlated with the electrode capacitance values or the polarisation resistance (Fig. 5). As this figure shows, greatest the capacitance, greatest the polarisation resistance.

5. Conclusions

The increase of the hydrostatic pressure modifies the composition of the passive films, since:

- at atmospheric pressure the corrosion layer formed by a mixed oxy-hydroxide (Cr^{3+}, Fe^{2+}, Fe^{3+}, Ni^{2+}, Mo^{6+}) Mo rich, presenting good resistance to localised corrosion and showing a perfect passivity range of 200 mV at least.

- at high pressure the corrosion layer shows a decrease of OH groups and bound water concentration. The corrosion layer is mainly formed by Cr^{3+} rich, Mo^{6+} 'dry', depleted Mo oxides. $MoCl_4$ and FeOOH are also present. The resistance to localised corrosion decreases since the passivity domain disappears and the repassivation domain is lower than at atmospheric pressure.

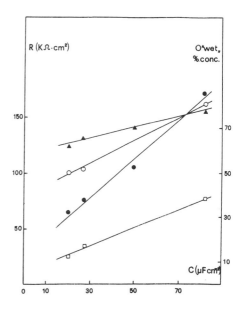

Fig. 5 Polarisation resistance Rp –▲–, $R_{\omega \to 0}$ Resistance –●–, total 0% 'wet' –○–, and partial 0% bond water ·–□–□– vs capacitance values.

6. Acknowledgement

This work was partially supported by the Cee-Mast 004 Project.

References

1. F. M. Reinhart, Corrosion of Materials in Hydrospace Technical Report 504. 11, 12, 107 US Naval Civil Engineering Laboratory. Cal 1966 Alexandria, Virginia.

2. F. M. Reinhart, Rep 834-US Naval Engineering Laboratory. Port Hueneme Corrosion of Metals and alloys in the Deep Ocean 1976. Alexandria Virginia.

3. A. M. Beccaria, P. Fiordiponti and G. attogno, Corrosion Sci., **29**,403, 1989.

4. A. M. Beccaria and G. Poggi, Corrosion, **42**, 470, 1986.

5. A. M. Beccaria and G. Poggi, Corros. Prev. and Control, **34**, 51, 1987.

6. A. M. Beccaria and G. Poggi, Brit. Corros. J., **20**, 188, 1985.

7. D. Festy and A. M. Beccaria, Corrosion '89, New Orleans April 17–21, Paper 292, Publ by NACE 1989.

8. H. Horne, Marine Chemistry, Wiley Interscience, New York, 73, 1969.

9. A. M. Beccaria and G. Poggi, Corrosion Sci., in press.

10. K. Asami, K. Hashimoto and S. Shimodaira, Corrosion Sci., **18**, 151, 1978.

11. J. E. Castle and C. R. Clayton, Corrosion Sci., **17**, 7, 1977.

12. K. Asami and K. Hashimoto, Corrosion Sci., **17**, 559, 1977.

13. I. Olefjord and H. Fischmeister, Corrosion Sci., **15**, 697, 1975.

14. T. P. Moffat and R. M. Latanision, J. Electrochem. Soc., **139**,1869,1992.

15. H. P. Hack and H. W. Pickering, J. Electrochem. Soc., **138**, 690, 1991.

16. M. A. Streicher, Corrosion, **30**, 77, 1974.

17. K. Sugimoto and Y. Sawada, Corrosion Sci., **17**, 425, 1977.

18. K. Hashimoto, K. Asami and K. Teramoto, Corrosion Sci., **19**, 3, 1978.

19. G. Okamoto, Corrosion Sci., **13**, 471, 1973.

40

Effect of Laser Surface Remelting on Passivity and its Breakdown on Al–Si Alloys

S. VIRTANEN AND H. BÖHNI

Institute of Materials Chemistry and Corrosion, Swiss Federal Institute of Technology, ETH-Hönggerberg, CH-8093 Zürich, Switzerland

Abstract

In order to study the role of composition and structure on the passivity and its breakdown, studies were carried out on laser surface modified Al–Si alloys. Surface remelting of Al–Si cast alloys results in a large structural refinement.

Studies in neutral solutions showed that the dissolution rate is strongly decreased by laser surface remelting. The anodic reaction in the passive range is significantly retarded, as can be seen from polarisation measurements. Thus a structural refinement due to the laser-treatment leads to a formation of a more protective passive film. The passive films were *in situ* characterised by photoelectrochemical methods. The results clearly show that the laser surface remelting modifies the composition and the structure of the passive film.

In Cl⁻-containing solutions the laser-remelted alloys show a remarkably higher resistance against pitting corrosion than the untreated alloys. The pitting potentials of the laser-remelted alloys exhibit values up to 400 mV more positive than in the case of the untreated alloys. The pits that are formed on the surface of the laser-treated surface are much smaller than on the substrate, indicating that initiation and growth of pits are strongly affected by the microstructure. If the Si content becomes too high (17%), the laser-treated surfaces exhibit more heterogeneities, which leads to formation of a less resistant passive film—especially in the interface regions of the neighbouring laser tracks. Even in this case the pit growth is retarded in the laser-treated surface probably due to a smaller size of the inhomogeneities compared to the substrate.

1. Introduction

Aluminium and its alloys are materials with a highly protective passive film but unfortunately they are susceptible to localised corrosion in presence of chlorides in the solution. The pitting corrosion behaviour of aluminium alloys is strongly dependent on the microstructure. Many alloying elements have a limited solid solubility in aluminium and the formed intermetallic precipitations can lead to a diminished localised corrosion resistance.

Studies on metastable aluminium alloys prepared by ion-implantation [1] or by rapidly quenching [2] have shown positive effects of many alloying elements on the localised corrosion resistance of aluminium, when present homogeneously in solid solution. Generally, a structural refinement due to rapid solidification has been found to increase the localised corrosion resistance not only of aluminium alloys but of stainless steels, as well [3–6].

During this work the effect of laser surface remelting, which induces a high structural refinement, on the passivity and its breakdown of a series of Al–Si alloys was studied. Surface modifications due to a laser treatment can strongly affect the corrosion behaviour. Laser surface melting of an AISI 304 stainless

steel has been found to improve the pitting resistance [7,8]. This effect was explained to be due to a more homogeneous distribution of the sulphides and carbides which act as pit nucleation sites. The aim of this work was to study the effect of both chemical composition and structure on passivity and its breakdown of Al–Si alloys.

2. Experimental

Cast substrates (Al–Si) were surface-modified by surface-remelting by a 1.5 kW CO_2 laser. Experimental details of the laser surface modifying are given elsewhere [9].

For the electrochemical measurements the samples were mechanically polished with abrasive paper, rinsed with ethyl alcohol and distilled water and dried. All measurements were carried out in deaerated solutions (N_2) at room temperature.

The polarisation curves and the impedance spectra were measured with a conventional electrochemical set-up. The photocurrents were generated by focusing chopped light of a 150 W Xe lamp onto the electrode which was held under potentiostatic control. For the photocurrent spectra a monochromator was placed in the light beam and the wavelength was scanned in the range of 800–250 nm. In all experiments the lock-in technique was used (chopper frequency 30 Hz). For the local photoelectrochemistry the Xe lamp was replaced by a Laser (HeCd) and the cell was mounted on a xyz-stage. A more detailed description of the experimental set-up for photoelectrochemical measurements is given elsewhere [10,11].

3. Results

Laser surface remelting of Al–Si cast alloys results in significant structural refinement [9,12]. The silicon particles situated at the interfaces between adjacent laser tracks are coarser than in the track centres. Alloys with silicon contents of 7 and 12% solidify with dendritic microstructures. Al–17Si alloy exhibits mixed microstructures consisting of eutectic regions and regions with primary silicon surrounded by eutectic. For pure aluminium no structural changes could be obtained by laser-remelting.

The effect of laser surface remelting on the dissolution rate of the alloys at the corrosion potential in a deaerated 0.1M NaCl solution is given by Fig. 1 which shows R_p-values determined from the impedance spectra. Since no pitting corrosion takes place under these circumstances the R_p-values characterise the protectiveness of the passive films against general dissolution. For all these alloys with silicon ranging from 7–17 % the dissolution rates of the laser remelted surfaces are significantly decreased. The dissolution rate of pure Al is only slightly reduced by the laser treatment.

The effect of the laser remelting on the polarisation curves in 0.1M NaCl is shown in Figs 2–4. The cathodic and the anodic current densities are in all cases diminished by the laser treatment. In the case of the Al–7Si and Al–12Si samples the laser treatment leads to re-

markable ennoblement of the pitting potential, whereas in the case of the Al–17Si sample the pitting resistance measured as the pitting potential is not affected by the laser treatment.

The sample surfaces show different type of attack for the substrate and for the laser treated surfaces. The pits formed on the surface of the substrate are much larger than those on the laser remelted surfaces (Fig. 5(a,b)). Typical for the laser-treated surfaces is furtheron that pits are preferably initiated at the interface zones of the laser traces (Fig. 5(b)). Potentiostatic measurements at a potential below the pitting potential on the alloy Al–17Si show that current peaks—corresponding to film breakdown events—become much smaller after the laser-treatment.

The passive films formed on the different samples were further characterised by photoelectrochemical methods. The band-gap energies were determined from the measured photocurrent spectra by replotting the spectra according to an indirect transition [13]. The band-gap energies are shown in Fig. 6 for the untreated and the treated samples after passivating them in 0.1M Na_2SO_4/N_2 at E = –400 mV SCE for 1 h. The optical band-gap energies of the untreated samples are in all cases very near the optical band-gap energy of the passive film on pure aluminium. The band-gap energy of the passive film on pure aluminium and on Al–7Si shows no changes after the laser-treatment, whereas the laser treatment shifts the band-gap energy of the alloys Al-12Si and Al–17Si to smaller values.

Figure 7 (p.228) shows the distribution of local photocurrents measured with the laser-spot-scanning apparatus of the laser remelted surface of Al–7Si and Al–17Si after passivating the samples in 0.1M Na_2SO_4 at E = –400 mV SCE. These figures represent typical laser-spot maps of 8 measurements at various surface

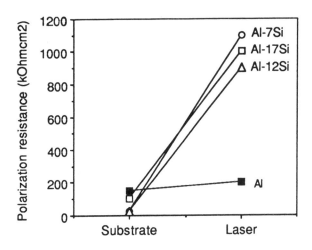

Fig. 1 R_p-values at the corrosion potential in 0.1M NaCl/N_2.

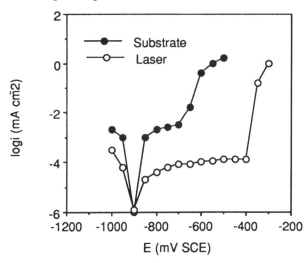

Fig. 2 Potentiodynamic polarisation curves of Al-7Si in 0.1M NaCl/N_2 (sweep rate 0.2 mV/s).

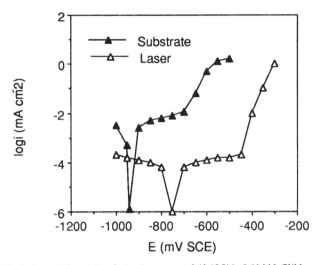

Fig. 3 Potentiodynamic polarisation curves of Al-12Si in 0.1M NaCl/N₂ (sweep rate 0.2 mV/s).

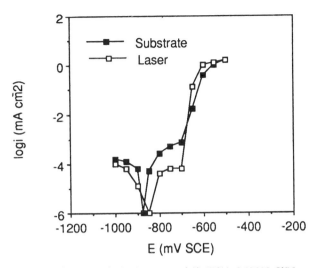

Fig. 4 Potentiodynamic polarisation curves of Al- 17Si in 0.1M NaCl/N₂ (sweep rate 0.2 mV/s).

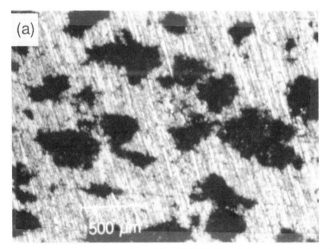

Fig. 5 Photograph of the surface of the Al-17Si after pitting corrosion in 0.1M NaCl/N₂ (a) without laser-treatment, (b) with laser surface remelting.

locations. The laser-treated Al–12Si alloy showed a very similar behaviour as the Al–7Si alloy. In all cases the laser tracks and the interface regions show differences in the photoresponse. In the case of the Al–17Si alloy these differences are more significant than in the case of the Al–7Si and Al–12Si alloys. The surface roughness of the laser-remelted surfaces was determined by atomic force microscopy at different surface regions. The scans very clearly show that the medium roughness of the different laser-remelted surfaces lies in the same range and hence the different photocurrent distributions are not merely due to differences in the surface roughness.

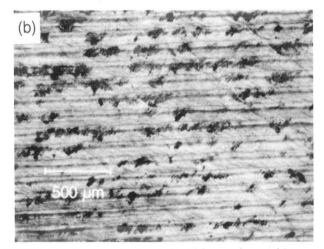

Fig. 6 Optical band-gap energies for passive films in 0.1M Na₂SO₄/N₂ at E = –400 mV SCE/1h.

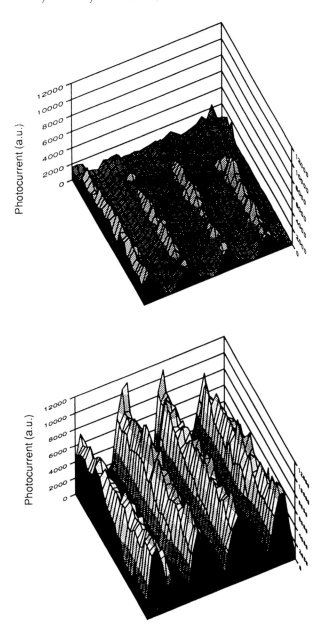

Fig. 7 Typical laser-spot-scanning images in 0.1M Na$_2$SO$_4$/N$_2$, E=–400 mV SCE (a) Al–7Si laser-treated, (b) Al–17Si laser-treated (Side length of the square = 450 μm, spot size = 8 μm, step size = 15 μm).

4. Discussion

The electrochemical measurements show that laser surface remelting, which leads to structural changes in the Al–Si alloys, can strongly modify their corrosion behaviour. In all cases shown here the homogenisation of the structure leads to a formation of a more protective passive film (decrease of the R$_p$-values). In case of the Al–7Si and Al–12Si alloy the pitting resistance is strongly increased by the laser remelting of the surface. Even in the case of the alloy Al–17Si, where the

pitting potential is not increased by the laser treatment, the pit growth is retarded by the laser surface remelting.

Changes in the band-gap energy values of the passive films are signs that the composition or/and the structure of the films has changed. Generally, an introduction of states into the band-gap due to a more defective passive film can lower the optical band-gap energy (E$_g$). The E$_g$-values for the substrates are practically independent of the Si-content—even though the heterogeneity of the alloy increases with increasing Si-content. This suggests that the passive film of these alloys is principally an Al-oxide film and its structure (defectiveness) either is not changed by increasing structural heterogeneity of the underlying material or that such changes in the structure are not reflected in the measured values of the band-gap energy. The passive film on Al–7Si shows the same photocurrent behaviour without and with the laser remelting. This suggests that the composition of the passive film is not significantly changed by the laser-treatment. The higher pitting resistance of the laser-treated alloy is therefore mainly due to a higher structural and chemical homogeneity of the laser remelted surface. Since the pitting potential of the laser-treated Al–7Si alloy is remarkably more noble than that of pure aluminium, it can be further concluded that Si, when homogeneously distributed, increases the pitting resistance of aluminium.

The decrease in the E$_g$-value of the passive film of the laser-remelted Al–12Si and Al–17Si alloy suggests that the composition of the passive film has changed—probably due to an increase of the Si-content. This can result from the fact that the laser-remelting increases the solid solubility of silicon in aluminium and that this silicon in solution can probably take part in the film formation. The finding that the band-gap energy of the Al–12Si alloy after the laser-treatment is decreased (i.e. Si-content of the films has increased), but the pitting potential of this alloy is about the same as that of the alloy Al–7Si indicates that an increase of the Si-content of the passive film does not lead to such an increase of the pitting resistance which would be reflected in the pitting potential.

Further, the poor pitting corrosion resistance of the laser-treated alloy Al–17Si suggests that in this case laser remelting is not able to produce a surface alloy with a high structural homogeneity. This is confirmed by the local photocurrent measurements. A comparison of Fig. 7(a) with 7(b) clearly shows that in the case of the Al–17Si alloy the heterogeneity of the surface is much larger than in the case of the Al–7Si alloy. Since this cannot be explained by varying surface roughness, these differences have to be due to variations in the composition or in the structure of the passive film formed on the laser zones and in the interface regions. Even though it is not possible without further experiments to determine the reason for these differences, the local variations in the photocurrent are a sign for an inhomogeneous passive film. Clearly the passive film formed on the interface regions is less resistant to pitting.

5. Conclusions

The results of the electrochemical and photoelectrochemical characterisation of the samples clearly show that laser surface remelting of Al–Si alloys without modification of the alloy composition can strongly improve the quality of the passive film. In all cases the dissolution rate is largely decreased by the laser-treatment. Alloys up to 12% Si exhibit a significantly higher pitting potential after the laser-treatment. This effect is mainly due to the structural refinement achieved by the laser treatment. If the Si content becomes too high (17%), the laser-treated surfaces exhibit more heterogeneities, which leads to formation of a less resistant passive film—especially in the interface regions of the neighbouring laser tracks. Even in this case the pit growth is retarded in the laser-treated surface probably due to a smaller size of the inhomogeneities compared to the substrate.

References

1. E. McCafferty, P. P. Trzaskoma and P. M. Natishan, in Advance of Localized Corrosion, Eds. H. Isaacs, U. Bertocci, J. Kruger and S. Smialowska, NACE, Houston, Texas, 1990, p. 181.

2. H. Yoshioka, S. Yoshida, A. Kawashima, K. Asami and K. Hashimoto, Corros. Sci. **26**, 795, 1986.

3. A. Saito and R. M. Latanision, in Proc. Int. Congr. on Met. Corr., Vol. **3**, p. 122, Toronto, 1984.

4. T. Tsuru and R. M. Latanision, J. Elecrochem. Soc., **129**, 1402, 1982.

5. P. C. Searson and R. M. Latanision, Corrosion, **42**, 1986, 161.

6. A. Kawashima and K. Hashimoto, Corros. Sci. **26**, 467, 1986.

7. J. B. Lumsden, D. S. Gnanamuthu and R. J. Moores, in Proc. Int. Symp. on Fundamental Aspects of Corrosion Protection by Surface Modification, Eds. E. McCafferty, C. R. Clayton, Proc.-Vol. **84–3**, p. 122, The Electrochemical Society Inc., Pennington, NJ, 1984.

8. E. McCafferty and P. G. Moore, in Proc. Int. Symp. on Fundamental Aspects of Corrosion Protection by Surface Modification, Eds. E. McCafferty, C. R. Clayton, Proc.-Vol. **84–3**, p. 112, The Electrochemical Society Inc.,Pennington, NJ, 1984.

9. E. Blank and M. Pierantoni, Technische Rundschau, **15**, 86, 1991.

10. P. Schmuki and H. Böhni, J. Electrochem. Soc., **139**, 1908, 1992.

11. P. Schmuki and H. Böhni, in Oxide Films on Metals and Alloys, Eds. B. R. MacDougall, R. S. Alwitt, T. A. Ramanarayanan, Proc.-Vol. **92–22**, p. 326, The Electrochem. Society, Pennington NJ, 1992.

12. S. Virtanen, H. Böhni, R. Busin, T. Marchione, M. Pierantoni and E. Blank, to be published in Corros. Sci.

13. E. J. Johnson, in Semiconductors and Semimetals, R. K. Willardson and A. C. Beer, Eds., Vol. 3, Chapter 6, Academic Press, New York, 1967.

41

Laser Pulsed Irradiation of Passive Films

R. Oltra, G. M. Indrianjafy and I. O. Efimov*

URA CRNS 23 Reactivite des Solides, Universite de Bourgogne, BP 138 21004 Dijon, France
*On leave from the Institute of Chemical Physics, 142432 Chernogolovka, Russia

Abstract

Channel flow double electrode (CFDE) and electrochemical quartz microbalance (EQCM) techniques combined with the pulsed laser depassivation were applied to investigation of passive films on Fe and Ni electrodes respectively. With CFDE it was shown that anodic dissolution contributes partially 15% to the total charge during the Fe electrode repassivation. With EQCM it was found that the thickness of the passive film on a Ni electrode reaches its maximum (~8 Å) after 2 h of passivation.

1. Introduction

The passivity of metallic materials in aqueous electrolytes can be disturbed by ionic breakdown or imposed mechanical breakdown of passive film. This breakdown followed by bare metal dissolution and passive layer rebuilt, gives rise to localised corrosion if the first process dominates or to rapid recovering of passive film in the opposite case. So information about metal dissolution and oxidation rates on bare metal surface are of great importance.

Two main problems arise here. The first one is that the duration of depassivation and the depassivated area must be known exactly. Intensive irradiation of passive surfaces (between 10 to 50 MW.cm^{-2}) with short-pulsed Laser (6 ns) was proved to be a useful tool for this purpose [1].

Induced by such radiation, deformations in subsurface layer of metal give rise to passive film breakdown followed by a current transient [2]. The second problem is that both metal dissolution and repassivation processes take place through with a charge transfer. So some additional assumptions should be made to separate their respective contribution to the total repassivation current.

Another approach is to combine current measurements on depassivated electrode with some other techniques. The aim of this work was to investigate the abilities of two of them—channel flow double electrode (CFDE) and electrochemical quartz crystal microbalance (EQCM)—when they are combined with the pulsed laser depassivation technique. These methods (CFDE and EQCM) are well-known in electrochemistry [3,4], but were not largely applied yet to depassivation–repassivation phenomena [5].

1.1 CFDE technique

CFDE is a hydrodynamic method to study electrodes kinetics. Two electrodes are placed in the electrolyte stream close to each other with an insulating gap (0.4 mm) between them (Fig. 1). A special bi-potentiostat designed by Keddam *et al.* [6] has been used to avoid ohmic coupling between them. It has been checked that this device operates also up to 500 Hz. After the laser depassivation of the upstream iron electrode metal, a part of the Fe^{2+} ions emitted into solution is oxidised at the downstream glassy carbon electrode. The flow rate has been fixed to 10^{-2} m.s^{-1}, so that the time delay between the current transient on the depassivated iron electrode and the transient response on the collecting electrode was around 40 ms.

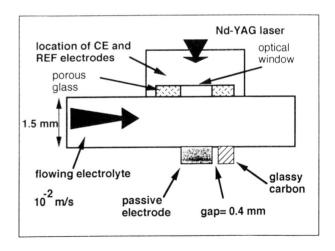

Fig. 1 Schematic setup of the CFDE technique combined with the pulsed laser depassivation technique.

CFDE technique is characterised by the collection efficiency (N), which in steady-state conditions is equal to the ratio between the currents on collecting and emitting electrodes. For the transient measurements, the same coefficient can be introduced as the ratio between the integrated currents, i.e. the amount of material, as shown for ring-disc electrodes by Albery [7]. The CFDE cell was calibrated by applying potential square pulses to the emitting electrode for a redox system (Fe^{2+}/Fe^{3+}) [8]. In the frequency range characteristic of the dissolution current transients due to the laser activation (up to 500 Hz), the collection efficiency was found to be constant and equal to 0.3. Being independent from frequency this value stands also for the total charge collection efficiency. We must note that (N) can be assumed to describe only faradaic process because the frequency range of double layer recharge is much higher.

The CFDE technique was applied to the depassivation analysis of iron passivated in a $HClO_4$ pH = 1 solution at +0.3 V/SSE. The potential of the collecting electrode was fixed at +0.8 V/SSE in order to collect Fe^{2+} species. Typical current transients have been defined as a function of laser energy (Fig. 2). For high laser energy a transient response was detected on the collecting electrode at t = 40 ms after the current pulse on the depassivated electrode, demonstrating that Fe^{2+} were released into solution. The faradaic balance between the currents related to dissolution and repassivation of freshly exposed metal has been estimated by comparing the total charge measured on the depassivated electrode (Q_{tot} = 370 μC) with the charge involved in the transient response at the collecting electrode (Q_{coll} = 8.5 μC) taking into account the collection efficiency N = 0.3. The contribution of anodic dissolution (Q_{diss}) to Q_{tot} can be expressed as function of Q_{coll}

$$Q_{diss} = (n_e/n_c N) \bullet Q_{coll} \tag{1}$$

where n_e = 2 corresponds to the charge transferred in Fe^{2+} generation and n_c = 1 corresponds to the oxidation of Fe^{2+} to Fe^{3+} on the collecting electrode.

Calculated from eqn (1), Q_{diss} = 57μC, it means that *ca.* 15% of the total charge in the depassivation–repassivation sequence is supplied by metal dissolution. This value can be underevaluated because the experimental collection effeciency must be lower than the theoretical value.

1.2 EQCM technique

Quartz resonator operating with one face in contact with solution acting simultaneously as working electrode is a sensitive tool for electrochemical processes investigation. When combining with laser

depassivation the following properties of EQCM should be taken into account:

- quartz resonant frequency is sensitive not only to the electrode mass changes, but also to the stresses in the deposited film and to the motion of the solution. The last two factors are important when electrode surface is irradiated by laser.
- laser depassivation is a local process, so differential quartz sensitivity should be used in frequency shift interpretation.

Local calibration of 6 MHz AT cut quartz crystal was performed by local deposition of Cu from a sulphuric solution ($CuSO_4$ 200 g.L^{-1}, H_2SO_4 52 g.L^{-1}) using a mask with a hole (0.05 cm dia.).

Differential sensitivity is 60 kHz.cm^2/mg at the centre of the crystal and decreases following the exp.($-1.7r^2 /R^2$) at the distance r from the centre (R = 0.65 cm is the gold electrode radius) in agreement with [9].

Commercial frequency counters are not suitable for laser depassivation kinetic measurements due to their slow time resolution (60 ms) and due to the noise generated by the laser power source.

So we used the direct measurements of small frequency differences, recording the quartz oscillations by the means of a digital oscilloscope (see appendix 1). A change of mass of *ca.* 100 ng gives rise to a frequency difference of *ca.* 6 Hz.

Laser depassivation experiments were performed on a Ni film (3.5μm) deposited on a Cu (3.5 μm) substrate covering the gold plate of EQCM. The laser energy was transferred through an optical fibre of 1.5 mm in diameter, which was immersed directly in the solution leading to a laser spot of the same diameter.

In the kinetic measurements, the Ni electrode was irradiated by a 6 ns laser pulse (0.9 J/cm^2) in water and sulphuric solutions (1M and 3M H_2SO_4) after passivation during 1 h at E=-0.1 V/SSE.

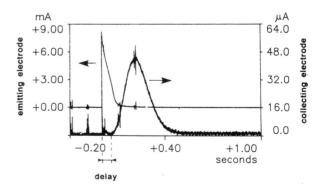

Fig. 2 CDFE current transients for a depassivation with laser intensity > 30 MW.cm^{-2} on:

(a) emitting and (b) collecting electrodes. The depassivated area was 5 mm^2.

During the first 0.5 ms, the change of frequency caused by stresses in metal was negative in all three media (Fig.3). Then the sign of Δf changed to the positive one due to the motion in solution induced by the mechanical action of the laser. The maximum value of about 100 Hz observed at 10 ms and the characteristic times of the frequency decrease (~ 0.2 ms) were the same in these media indicating the equal damping in liquids. The asymptotic behaviour of Δf was different: in water it diminished to zero, in corrosive media it tended to a nonzero value (5 Hz in 1M and 50 Hz in 3M H_2SO_4) caused by the mass-losses reacted to faradaic processes.

In faradaic balance measurements the time of polarisation in 1M H_2SO_4 at E = –0.1 V/SSE before laser depassivation was varied from 3 to 200 min. After depassivation the total mass losses calculated from the final frequency shift correspond to the sum of the initial passive film mass and the mass of metal dissolved during the current transient. The last one is much shorter than on Fe electrode(~ 10 ms) and can be attributed totally to anodic dissolution, because passive film breakdown proceeds without charge transfer (the passive film contains Ni in oxidised form) and the current due to the slow passive film rebuilt is small. Thus the mass-losses caused by anodic dissolution can be calculated from the integral of the transient. Subtracting this mass from the total one, we obtain the mass of the passive film as a function of passivation time, i.e. the kinetics of oxide film growth (Fig. 4).

From these data, assuming an average film density

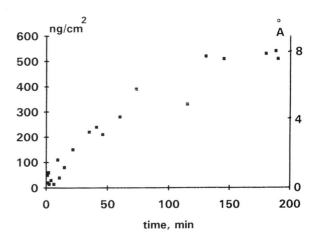

Fig. 4 Change of the thickness and the mass (per cm^2) of the passive film on a Ni electrode as a function of the passivation time in 1M H_2SO_4 at E = –0.1 V/SSE. The irradiated area was 14×10^{-2} cm^2.

as for NiO (6.7 g.cm^{-3}) one can conclude that the film thickness reaches its maximum (~ 8.2 Å) after 2 h of polarisation, in agreement with XPS data [10].

2. Conclusion

It was shown that mass and charge balance in repassivation process as well as the kinetics of passive film growth can be investigated using CFDE and EQCM techniques combined with the pulsed laser depassivation.

3. Acknowledgements

The authors thank M. Keddam (UPR15 CNRS Paris VI) for the CFDE equipment. I. O. Efimov thanks the 'Conseil Régional de Bourgogne' for his postdoctoral grant.

References

1. R. Oltra, G. M. Indrianjafy and J. P. Boquillon, Journal de Physique IV, **1**, 769, 1991.
2. V. A. Benderskii, I. O. Efimov and A. G. Krivenko, J. Electroanal.Chem., **315**, 29, 1991.
3. T. Tsuru, Materials Science and Engineering, **A 146**, 1, 1991.
4. D. A. Buttry and M. D. Ward, Chem. Rev., **92**, 1355, 1992.
5. M. Itagaki and T. Tsuru, Proc. 2nd EIS meeting, Santa Barbara USA, July 1992.
6. N. Benzekri, R. Carranza, M. Keddam and H. Takenouti, Electrochim. Acta, **34**, 1159, 1989.
7. W. J. Albery and M. L. Hitchman, 'Ring-Disc Electrodes', Oxford Univ.Press, 1971, p. 151.
8. R. Oltra, G. M. Indrianjafy, M. Keddam and H. Takenouti, to be published in Corros. Sci.
9. C. Gabrielli, M. Keddam and R. Torresi, J. Electrochem. Soc., **138**, 2657, 1991.
10. P. Marcus and J. Oudar, J. Microsc. Spectrosc. Electron., **4**, 63, 1979.

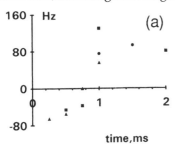

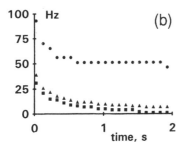

Fig. 3 Laser induced change of frequency of EQCM covered by a Ni film in water (■), 1M (▲) and 3M (●) H_2SO_4 at short (a) and long (b) time scales.

Appendix 1

High sensitivity measurement of the frequency shift of the EQCM

At t = 0, two oscillation curves were recorded on the digital oscilloscope with a time delay T between them. This time was chosen exactly to put these oscillations in phase. It implies that the number of oscillations in this time interval is integer.

After the disturbance of frequency due to the change of mass of the quartz, at the time t, oscillation curves were recorded again with the same time delay T between them. The number of oscillations with the new frequency was not integer in this interval, so a time shift τ arose between the equal phase points. The equation relating f_o, t, T and τ with the frequency shift f(t) (which is supposed to be a function of time) reads:

$$\tau = \int_t^{t+T} \Delta f(\xi) f_o^{-1} d\xi \qquad (3)$$

When t < T, $\Delta f(t)$ can be restituted from (3) by differentiating τ(T) dependence. For t >> T eqn (3) is reduced to: f(t) T

$$\tau = \frac{\Delta f(t)T}{f_o} \qquad (4)$$

A change of mass of *ca.* 100 ng gives rise to the frequency difference *ca.* 6 Hz. The required time for this shift measurements T equals 1ms if the oscilloscope time resolution T equals 1ns.

42

Irradiation Effects in Anodic Oxide Films of Titanium

*D. Gorse, T. Sakout and Y. Serruys**

CECM-CNRS, 15 rue Georges Urbain, 94407, Vitry/Seine, France
*SRMP-CEREM, CEA-Saclay, 91191, Gif/Yvette, France

Abstract

Ex situ studies of irradiation effects in anodic oxide films of titanium, performed by using implantation with rare gas or metallic ions in the 10–100 keV range (typically below 1 keV/amu) show that the electronic, optical and electrochemical properties of the film can be modified by this treatment, depending on the irradiation parameters. Particularly, ion implantation changes the stability of the anodic film with respect to the aqueous corrosion. On the other hand, it is shown in this paper that α-radiation provides the most suitable means for studying *in-situ* the influence of irradiation on growth and stability of anodic films which are formed at a metallic electrode. The principle of the mounting consists in irradiating with α-particles through the metal electrode: it allows to vary the energy of the α-particles bombarding the metal/film interface by changing the thickness of the target electrode. Particularly, in this study of the anodisation of titanium under α-radiation, two energy domains are evidenced:

(i) one favouring *both* the chemical and electrochemical dissolution of the Ti electrode with increasing energy E_α from 400keV up to 1.6MeV,
(ii) followed at higher energy (E_α up to 3 MeV in the present study) by a regime where the passivation of the Ti electrode is favoured, the ionic conductivity of the film formed under these irradiation conditions being apparently lower than without irradiation. The respective role of the electronic and nuclear stopping power in Ti and in solution for explaining such opposite electrochemical behaviours is discussed.

1. Introduction

For over thirty years, ion-implantation has been used for altering the properties of a wide range of materials. As for semiconductors and insulating or semiconducting oxides, ion implantation in anodic oxide films produces radiation damage and doping effects, responsible for changes in their electronic and optical properties [1].

Generally speaking, the slowing down of incident particles in the target electrode occurs by electronic excitations, ionisations and nuclear collisions, i.e. by electronic and nuclear stopping respectively. However, when a material is implanted in the near surface region, with heavy (rare gas or metallic) ions of low energy in the 10–100 keV range (typically below ~1 keV/amu), nuclear stopping is predominant. The damage and implantation profiles cannot be separated. Sputtering effects may occur (except with He+, the maximum sputtering yield being around 1 keV in TiO₂). However, an advantage of ion-implantation is that it allows to vary the defect production rate and distribution over the film thickness.

For example, anodic films on titanium at various stages of the anodisation process (and various thicknesses [2]) have been irradiated with He+ ions of energy ranging from 7 to 17 keV, increasing with advancing the stage of growth, in order to keep the damage profile in the inner part of the film. Under these conditions, it was shown by photoelectrochemistry that irradiation causes a broadening of the optical absorption edge and the appearance of sub-gap peaks of photoresponse, revealing the presence of radiation-induced defect states in the gap of the Ti anodic oxide, located at approximately the same energies below the conduction band edge than already observed with Ar-ion bombardment of various TiO_2 surfaces [3]. In addition, it was remarked that for the same fluences ($3 \cdot 10^{16}$–$1.2 \cdot 10^{17} He^+ cm^{-2}$) the radiation effects are more marked in the early stage of growth, on non-crystallised thin films than on thicker crystalline ones.

Ion implantation modifies the electrochemical behaviour of anodic films of titanium, as it is the case with other metals like Zr, Hf, Ta... [1, 4], affecting both the ionic and electronic conductivity of the film. Measurements of the repassivation charge after implantation reveal an increase in ionic conductivity due to irradiation. Moreover, the presence of radiation-induced defect states in the band-gap of the anodic oxide (which cannot be identified exactly until now, no calculation of the band structure being available) causes noticeable changes in the rate of electron transfer reactions (and gas evolution) occurring with redox species in solution, over a wide energy range penetrating deeply inside the bandgap, well beyond the band edges. However, these *ex situ* studies suppose that repassivation of the anodised electrode, occurring at immersion in the corrosive solution, is taken properly into account after irradiation. The consequences on the stability of these materials regarding the uniform or localised (aqueous) corrosion have been considered in the literature.

At the present time, there is a need for *in situ* electrochemical studies of the anodic oxidation process *under flux*. As a result of the above mentioned studies, changes in efficiency of film growth are expected under irradiation, and should depend on the particular conditions chosen, with respect to the ratio of electronic to nuclear stopping. Under conditions of predominant electronic stopping, using α-radiation, it was reported recently that the anodisation of Ti can be modified [1]. On the other hand, the problem of anodisation under visible or u.v. light irradiation is widely treated in the literature, where changes in the ionic transport properties have been related to the photogeneration of minority carriers, subsequent development of a space charge in the outer portion of the growing film and variation of the electric field strength. Moreover, one can find examples where some additional chemical changes in the film have been conjectured, since the observed irradiation effects were found permanent during further growth without u.v. light [5].

This paper is concerned with the effect of α-radiation on the electrochemical behaviour of a titanium electrode, with the purpose of determining the respective influence of the electronic but also of the nuclear stopping on the anodisation of titanium in the early stage of growth. We concentrate on both the film growth and dissolution processes, as a function of the irradiation parameters. It is shown that changes in the

electronic stopping are not sufficient to explain the observed results, and that the existence of a nuclear stopping power must be taken into account, except if some interaction between the radiolysis and corrosion products might be involved.

3. Irradiation Conditions

In an *in situ* study of anodic film growth under irradiation in corrosive environment (1M H_2SO_4), the choice of the radiation source is made critical by the presence of the aqueous phase, but somewhat imposed on purpose of limiting the irradiation of the solution [1]. Indeed, the chemistry of radiolysis products could be complicated by possible interactions with the corrosion products, close to the film/solution interface, by comparison with 'bulk' solutions or pure water. Until now, irradiation with heavy ions was not used in electrochemistry, since it requires the facilities of high energy machines, due to the short penetration depth of heavy ions through matter (for example, a typical range* of 0.5 μm is found for 1MeV argon ions in a copper target [6]). β or υ-radiations are more adapted for modifying the redox behaviour of solutions, their influence on the anodisation of the electrode being only deduced indirectly in that case.

α-radiation provides the most suitable radiation source for the problem under study: sealed sources, for example Americium (^{241}Am) emitting 5.49 MeV α-particles, are available, easy to manipulate in the laboratory by comparison with neutron or β sources due to the short range (~ a few μm) of the α-particles in matter, providing an intense flux of α-particles in a wide energy range (activity = 0.35 mCi). The energy spectrum of the source has its maximum slightly degraded (~5.4 MeV), and exhibits a broad tail towards low energies due to a 2μm protective metallic coating. At these energies, the range in water and titanium are *ca.* 34 μm and 17μm respectively. Thus, whatever the mounting, it will be uneasy to separate radiation effects in the substratum during anodisation and in solution. We choose the arrangement schematised in Fig. 1 for the working electrode: the ^{241}Am source is pushed into contact with a Ti foil of ~1.5 cm² area obturating a Pyrex tube (fitted to the source dimen-

---

* We recall some definitions. The particle-matter interaction, or the energy loss process for a projectile going through a target, is ruled by statistical laws. The stopping power is the average energy loss per unit path length in a material: $S(E) = \dfrac{dE}{dx}$, given in MeV cm⁻¹ in this paper.

The range of a particle is the distance it penetrates into a material before coming to rest. In this work, penetration depth or range are used alike. If the lateral deflections and the statistical fluctuations are disregarded, the range is approximated by,

$$R = \int_0^E \left(-\frac{dE}{dx}\right)^{-1} dE$$

sions). The electric contact is made on the back face of the radiation source.

In Fig. 2 is represented the electronic stopping power (accounting for the energy spectrum of the source) for α-particles as a function of the range in titanium and in water [7], for two different thicknesses of the titanium foil, 9 and 12.5 μm respectively, which yield two opposite electrochemical behaviours, as will be shown in the following. The energy deposited into the solution is only divided by a factor of *ca.* 2, going from $d_{Ti} = 9–12.5 \mu m$.

Comparing the range of the α particles to the maximum film thickness attainable with Ti, and taking into account the energy dispersion in the titanium foil, it is obviously impossible to stop α-particles of energy in the range of a few MeV over the film thickness: a flux penetrates deeply into the solution, making uneasy the discrimination between direct radiation effects and indirect ones due to the solution radiolysis. It is an

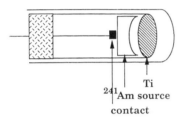

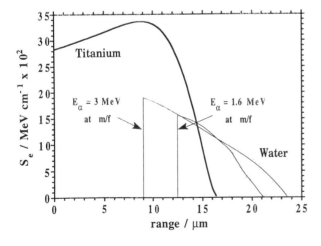

^{241}Am source
contact

Ti

Fig. 1 Schematic drawing of the working electrode showing the ensemble constituted by the Americium source (^{241}Am), the titanium foil and the electrical contact.

important problem to be solved, since until now mainly the role of the radiolysis products in modifying the corrosion properties of irradiated materials was studied in the literature. However, this arrangement makes a convenient compromise, since the energy deposited in solution can be varied easily by changing the Ti thickness, and reduced to a minimum as already emphasised [8].

Even if it the standard deviation on both thicknesses of the metallic coating and Ti target could be precisely estimated, the fact that the spectrum of the source is not known with sufficient accuracy (especially its low energy tail), does not allow a study of the $E_\alpha = 10–100$ keV range with the present mounting: it is a drawback since radiation damages in Ti anodic oxides are produced in this low energy range, as noted above. The *in situ* irradiation effects in this range should be interesting to compare with *ex situ* photoelectrochemical studies showing the influence of radiation damage on the electronic structure at various stages of film growth [1].

In the present case, the slowing down of the α projectile is mainly due to electronic excitations and ionisation occurring in both the growing film and in solution: in this aspect, the problem resembles that of anodic growth under illumination [5]. However, we shall see in the following that electronic stopping is not sufficient to explain the observed results.

4. Electrochemical Conditions

These are described elsewhere [8]. Only the working electrode deserves special attention. The potentials are given with reference to the saturated calomel electrode. Mechanical polishing with 1 μm alumina paste is performed on all specimens. The Open Circuit Potential (V_{OCP}) is measured after its stabilisation in solution. It attains –0.685 V/SCE ± 0.025 V on the bulk titanium electrode polished to mirror used as a standard, so approaching the value of the film free surface according to ellipsometry measurements [9]. Only gentle polishing in order to avoid hole formation is made on the thin Ti foils before the electrochemical tests under irradiation. Consequently, the average V_{OCP} reaches values between 0.125V and 0.4V/SCE, depending on the thickness of the Ti foil bombarded. This point will be discussed in a short coming paper.

5. Comparison of the Intensity–Potential Plots Obtained at a Ti Electrode Without or Under α-radiation

We compare in Figs 3 and 4 intensity–potential plots obtained at 1 mVs^{-1} sweep rate on a Ti electrode in 1M H$_2$SO$_4$ at room temperature, respectively without and

Fig. 2 Electronic stopping power for α-particles vs range in titanium for two different foil thicknesses 9 and 12.5μm, and the corresponding ones calculated in water taking into account the energy spectrum of the source after passage through titanium.

under α-irradiation. The cathodic limit is chosen above the potential of hydride formation. Only the early stages of growth are studied, the anodic limit being 4 V/SCE. Two cycles are shown, with exception of Fig. 4(a).

The role of the native oxide in modifying the electrochemical reactions at a Ti electrode is evidenced in Fig. 3: on the freshly polished electrode, an intense anodic peak is visible, attributed to the electrochemical dissolution of titanium. It does not appear on a thin foil which was polished as for the tests under irradiation, that is to say still covered with its native oxide.

Likewise, the influence of the energy of the α projectiles bombarding the metal/film interface on the electrochemical behaviour of the Ti target is visible in Fig. 4. Indeed, using the present mounting, mainly the maximum energy of the α-particles reaching the metal/film interface (m/f), E_α, and consequently the range and energy deposition in solution, vary when the Ti foil thickness is changed. However, it is remarkable that the energy loss in the liquid phase adjacent to the film/solution interface (integrated over a thin liquid layer, a few nm for example) does not vary significantly, within the precision of the nuclear data tables

[7]. However, significant differences in the I(V) plots are visible when varying Eα from 1.6 to 3 MeV. Under irradiation, in all cases, the electrode surface is initially covered with its native oxide.

Bombardment with E_α <1.6 MeV α-particles produces the reactivation of the surface: two anodic peaks are visible in the first cycle located at $V_1 = 0.8$ V/SCE and $V_2 = 1.2$ V/SCE, persisting on the 2nd and 3rd cycles (with a −0.3 V shift for V_1 on the 2nd cycle and following) and a third one appears on the 2nd cycle at $V_0 = -0.15$ V/SCE. But, on the average, the anodic charge passed during the potential sweep below oxygen evolution (~2 V/SCE) is conserved. This result is particularly visible on Fig. 5 showing the anodic charge (Q_{an}, in mC.cm^{-2}) calculated below 2 V/SCE as a function of the Ti target thickness for the 1st and 2nd cycles. It is also remarkable that under these particular irradiation conditions, Q_{an} attains the value obtained on bulk Ti polished to mirror ($Q_{an} = 26.4 \pm 3$ mC.cm^{-2}). We are aware of the fact that the reactivation processes are different in both cases (by comparison of Figs 3(a) and 4(a)).

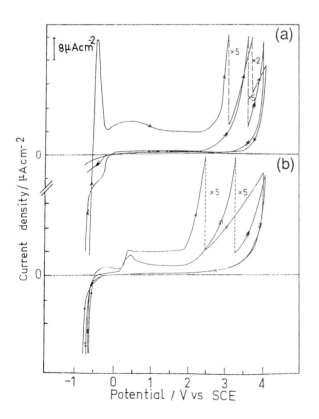

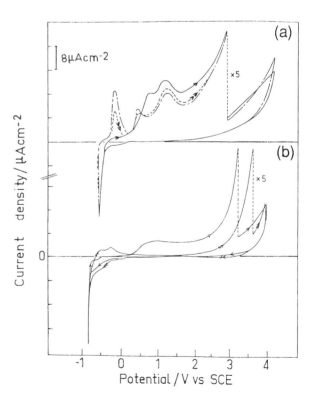

Fig. 3 *Intensity–potential plots recorded at 1 mVs^{-1} scan rate for titanium electrodes in 1M H$_2$SO$_4$ at room T without irradiation: (a) bulk freshly mechanically polished Ti electrode; (b) gently mechanically polished titanium foil, 15 μm thick, covered with its native oxide.*

Fig. 4 *Same as Fig. 3 for gently mechanically polished Ti electrodes bombarded with α-particles by the back face, showing the influence of the energy of the projectile: (a) d$_{Ti}$ = 12.5 mm so that E$_\alpha$ < 1.6 MeV at m/f, and (b) d$_{Ti}$ = 9 mm so that E$_\alpha$ < 3 MeV at m/f.*

43

Modification of Oxide Films by Ion-Implantation and its Influence on the Corrosion Behaviour*

J. W. Schultze, A. Michaelis, O. Karstens and Ch. Buchal†

Institut für Physikalische Chemie und Elektrochemie, Heinrich-Heine Universität Düsseldorf, 4000 Düsseldorf, Germany
†Institut für Schicht- und Ionentechnik, Forschungszentrum Jülich GmbH, 7500 Jülich, Germany

Abstract

The effect of radiation damage on the corrosion behaviour of Ti passive layers in a radioactive environment is simulated by 200 keV Ti^+ implantation into 40 nm TiO_2 layers.

Measurements were carried out on semi-implanted specimens to allow a direct comparison of implanted and not implanted regions under exactly the same experimental conditions.

The implantation generates a large amount of Frenkel-defects and a strong amorphization of the target, which yields a high conductivity of the implanted oxides enhancing e.g. electron transfer reactions (ETR). RBS spectra show that the stoichiometry is shifted towards titanium suboxides. The structure change also occurs in ellipsometric measurements by a strong change of the optical constants of the implanted layers. In contrast to non implanted oxides the implanted ones show a significant absorption. Photoelectrochemical measurements confirm this ellipsometrical result by the fact that the absolute photocurrent values are higher on the impl. area than on the non impl. one. Evaluation of photocurrent-potential curves also gives evidence for the high amorphicity of the implanted layers. UV-laser irradiation of the oxide layers with high power density yields no significant increase of photocorrosion on the implanted area.

1. Introduction

The modification of passive layers is of great interest for: understanding and prevention of localised corrosion, electrocatalysis, photoelectrochemistry, electronics and sensor development. Ion implantation is now a conventional technique for modifying the near surface properties of metals [1] as well as of oxides [2] or oxide films [3]. As an example Ti^+ ion implantation into TiO_2 passive layers on Ti is chosen to discuss the induced radiation damage effect on the corrosion behaviour. Ti^+ ions were used for implantation to rule out doping effects. Moreover, implantation energies of 200 keV were chosen. In this case the implanted ions stop only at the oxide/metal interface and the underlying metal, whereas the radiation damage is also spread over the oxide (see Fig. 1). Complementary to previous investigations on this system at our laboratory [4–6] now semi-implanted samples were used. This preparation allows the direct comparison of implanted and non-implanted sample areas under exactly the same experimental conditions. Thus a better quantification of the implantation effects is possible. Beneath electrochemical standard methods several optical *in situ* methods (laser scanning photocurrent method and micro-ellipsometry) are used to get complementary information and a high local resolution of electrode reactions.

The presented modification experiments were carried out to simulate the radiation damage in a radioactive environment. That is mainly caused by massive particles like hot atoms or alpha-radiation, but the hot electrons generated by beta- and gamma-radiation can also cause a small amount of Frenkel-defects. Therefore the presented results may also be relevant for the corrosion behaviour of container material for the disposal of HAW (Highly Active Waste). For this purpose Ti is a very promising material.

2. Experimental

2.1 Electrodes and surface processing (implantation)

The titanium (99.6%) plates (10 mm dia.) were first mechanically polished with emery paper 200, 600, 1200, and 2400, and finally electropolished applying a procedure given in [7] to get a smooth surface with open grain boundaries. All presented experiments were carried out in 0.5M H_2SO_4 (pH 0.3 at 25°C). The oxide films were formed by anodic potentiodynamic polarisation up to 20 V yielding an oxide thickness of

---

*Part of a Ph.D thesis of A. Michaelis

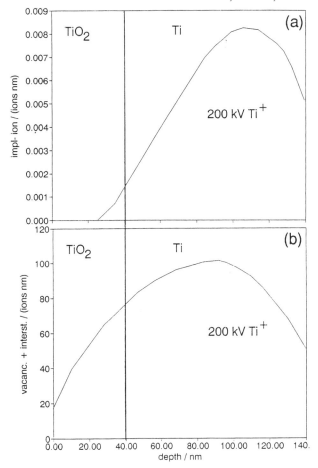

Fig. 1 Calculated distributions of the implanted ions (a) and of vacancies and interstitials (b) after 200 keV Ti+ implantation. All values are given per implanted ion and per nm. Calculations are performed by Trim-C code Ver. 3.3 [23].

ca. 40 nm. Electrochemical potentials are given with respect to the standard hydrogen electrode (SHE) throughout this paper.

For ion implantation, the samples were clamped against a copper heat sink in the target chamber of an Eaton NV 3204 ion implanter. One half of the sample surface was covered by an aluminium foil to get semi-implanted electrodes with a sharp implantation boundary (see inset of Fig. 6). Oil free vacuum is maintained at 10^{-5} Pa. The implanted dose was 10 at.% (*ca.* 10^{16} ions.cm^{-2}).

2.2 Optical methods and RBS
For the measurement of photocurrents at high local resolution an electrochemical *in situ* laser scanning apparatus is used. For this the beam of a frequency doubled Ar+-laser (Spectra Physics 2020) with a wavelength of 257 nm (4.8 eV) is focused onto the sample surface by use of a metallographic microscope (Leitz Metallux 3) with an UV-objective (OFR). This affords the addition of a 45°UV mirror to the microscope-tube. A minimal spot-size radius of *ca.* 1 μm can be obtained.

The construction allows a simultaneous illumination and observation (via a CCD-camera) of the surface. For simultaneous electrochemical measurements a three electrode configuration cell is used which is positioned under the objective on a x,y-stage (1μm step-size). The laser-photons generate electron/hole pairs in the n-semiconducting Ti-oxides yielding a photocurrent i_{ph} which information about the local electronic properties can be deduced from. The method developed by Butler *et al.* [8, 9] is described in detail in [10, 11].

Further the light of a Xe-high pressure lamp can be fed through the microscope. This allows the detection of photocurrent spectra with a local resolution of *ca.* 100 μm. For this a monochromator is added to the optical path between lamp and microscope.

For micro-ellipsometry a standard multiple angle of incidence setup with a rotating analyser is used [12].

RBS spectra were obtained by impinging a He+ beam of 1.5 mm dia. and 1.3 MeV energy onto the sample under 60° tilt angle. RBS data were evaluated for concentration profiles and oxide thickness measurements but not for crystallographic information.

3. Result and Discussion

3.1 RBS and Micro-Ellipsometry
Figure 2 shows the titanium step of RBS spectra of a 40 nm TiO$_2$ layer before and after 200 keV Ti+ implantation. Three effects can be observed:

(i) The slightly increased step width indicates a small broadening of the layer.
(ii) The increased step height means an increase of the Ti concentration, i.e. the formation of a suboxide TiO$_{2-x}$. This result agrees with cyclovoltammetry. The current-potential curves show the typical behaviour of Ti-suboxides (increased oxygen evolution at *ca.* 2 V).
(iii) The slope of the line at the interface metal/oxide shows a small decrease indicating a mixing in this region.

All these effects suggest a changed concentration profile of oxygen ions which are mixed and smeared out by recoil formation and preferential sputtering. These results are in qualitative agreement with multiple angle of incidence ellipsometric measurements (Fig. 3) on both sides (impl. and not impl.). Each point in Fig. 3 refers to one ellipsometric measurement at one angle of incidence in the range from 50 to 84°. By fitting of these points the optical constants (refractive index n and absorption-coefficient k) of the sample as well as the layer thickness d can be determined. The values of these quantities are given in Fig. 3, index 2 labels the constants of the Ti-substratum and index 1 those of the

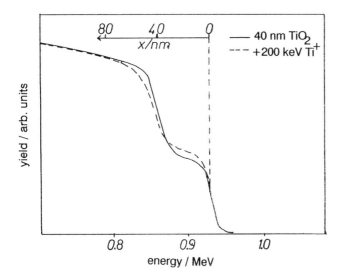

Fig. 2 RBS spectra of not implanted and 200 keV Ti^+ implanted TiO_2 passive layers. The Ti region is shown.

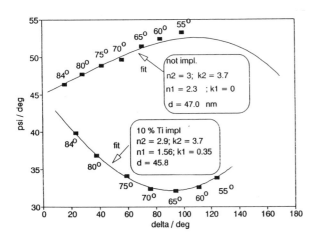

Fig. 3 Multiple angle of incidence ellipsometry on the implanted and not implanted sample area. Each delta-psi point refers to one ellipsometric measurement at one special angle of incidence (angles are indicated). The lines give the best calculated fit with the indicated parameters. Index 2 labels the constants of the Ti-substratum and index 1 of the oxide layer.

layer. A very strong change in the optical constants of the layers before and after implantation occurs, indicating a structure or stoichiometry change agreeing with the RBS-result. After implantation a significant absorption $k_1 = 0.32$ arises which is a hint for a more metal-like behaviour of the implanted oxide (higher conductivity). The higher conductivity, higher absorption resp. is the consequence of an enhanced amorphicity and a large amount of formed localised defect states in the mobility gap of the implanted region (hopping mechanism).

The measured small oxide thickness decrease of ca. 2 nm is not significant and can be due to the inhomogeneity of the modified layer causing an uncertainty in the thickness evaluation and is therefore not in contradiction to (i) REM micrographs and measurements with a laser-profilometer showed that the roughness of the surface is not increased after implantation. Therefore, it can be ruled out that the effect (i) is due to a roughness enhancement.

3.2 Electron Transfer Reactions

Electron transport from the electrolyte to the metal and vice versa through the oxide film depends on electron conductivity. In passive films on metals this can be investigated by electron transfer reactions (ETR). For this the outer sphere ETR Fe^{2+}/Fe^{3+} is used (0.05M in 0.5M H_2SO_4). TiO_2 is an n-type semiconductor due to oxygen vacancies which cause donor levels near the conduction band. On pure, undisturbed layers cathodic ETR take place only at high overvoltage. The anodic reaction is blocked because of the increased band bending. Figure 4 shows the ETR on the impl. and not impl. area. Implantation enhances the cathodic current by half an order of magnitude and enables anodic current showing the enhanced conductivity in agreement with the ellipsometrical result. This is an important fact for the corrosion behaviour of Ti in a radioactive environment, since radiation will increase the conductivity and decrease the overpotential of electrode reactions. Due to superimposed ionic processes the equilibrium potential U_0 is not correctly indicated.

The behaviour of the modified oxide is analogous to that of titanium suboxides generated for the 'LiGA-Process' [13, 14] in an etching solution of 0.5M H_2O_2

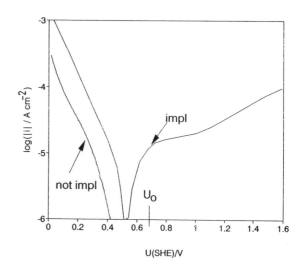

Fig. 4 Tafel plots of the Fe^{2+}/Fe^{3+} redox reaction (0.05M) in 0.5M H_2SO_4 on unimplanted and implanted TiO_2 layers of the semi-implanted samples. Currents below 1 μAcm^{-2} are not necessarily related to electron transfer reactions.

and 0.5M NaOH. These oxides, too, feature a high amorphicity and conductivity [15].

3.3 Photoelectrochemistry

In Fig. 5 the photocurrent spectra at different electrode polarisations on the impl. and not impl. area are shown. Evaluation of these curves (Gärtner evaluation [17]) yields a small decrease of the indirect bandgap from 3.4 eV (not impl.) to 3.2 (impl.). This decrease can be explained by a spread of energy levels into the region of the energy gap (Urbach tails), i.e. after implantation the lattice is disordered (no long range interaction) which means that a strong amorphication of the target has taken place. The large amount of defect states within the mobility gap of the implanted oxide can change the absolute photocurrent values as well. On one hand these states can lead to a decreasing photocurrent because they act as recombination centres, on the other hand an increasing photocurrent can occur due to an enhanced absorption [16]. From Fig. 5 it is evident that the latter case is dominant here. This is in good agreement with the ellipsometric result described before (high absorption coefficient kl on the impl. area). It should be emphasised that the evaluation of the absolute photocurrent values as done here, is possible only, if semi-implanted specimens are applied. This is one of the great advantages of the chosen sample preparation, i.e. for the first time quantification of absolute photocurrent values is possible.

Figure 6 shows photocurrent laser scans across the implantation boundary in dependence on the applied electrode potential. At 1V the photocurrent on the impl. region is smaller than on the not impl. one. Surprisingly, this relation inverts at higher potentials. To confirm this result photocurrent-potential curves of locodynamic experiments shown in Fig. 7 were

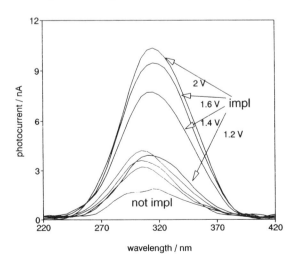

Fig. 5 Photocurrent spectra on the implanted and not implanted sample area for different indicated anodic polarisations (lock-in technique).

detected. For this the electrode was scanned in μm steps normal to the incident laser beam while the potential increases simultaneously like in a common potentiodynamic sweep. The advantage of such a locodynamic experiment is that the laser always hits a nearly fresh surface area, therefore effects of previously induced oxide growth can be ruled out. Agreeing with Fig. 6 a strong increase of the photocurrent for the impl. region occurs at *ca.* 1 V. The shape of these curves can in principle be explained by an anodic shift of the flat-band potential U_{fb} of the implanted layer. Nevertheless former investigations at our laboratory showed that U_{fb} is almost not affected by implantation [4, 18, 19]. Therefore, this effect is probably mainly caused by a spontaneous formation of laser induced oxide under the applied high laser power. Moreover the shape of the photocurrent-potential curves is affected by the crystal-structure of the layers. According to the Gärtner model [17] the photocurrents should show a square root dependence on the potential in the case of crystalline semiconductors. Only the curve of the non-implanted area looks like this. From literature it is known that those TiO_2 layers have a polycrystalline structure [20, 21]. The implanted area shows a more exponential increase which indicates a large contribution of the Poole-Frenkel effect [22]. This is again evidence for the strong amorphicity of the implanted layers.

3.4 Photocorrosion

To get direct information about the corrosion behaviour of the impl. and not impl. area, a photocorrosion test was carried out. For this the surface was illuminated by a focused laser beam (257 nm) with a power density of some kW/cm^2 at an applied anodic potential of 3 V. The large amount of generated electron/hole pairs under these conditions leads to a strong formation of laser induced oxide due to the changed potential distribution caused by the holes which accumulate at the oxide/electrolyte interface.

The amount of formed laser oxide is a measure for the susceptibility to photocorrosion. In a succeeding laser scan with low power density a large amount of formed laser oxide yields a small photocurrent at this location and vice versa. This is due to the E-field influence on the photocurrent, i.e. a thick oxide layer causes a lower potential gradient under potentiostatic conditions, which decreases the electron/hole pair separation in this case.

In Fig. 8 such a laser scan across the implantation boundary and the previously laser induced oxide spots is shown at an electrode potential of 0.8 V. The absolute photocurrent values at the marked laser induced spots are a little smaller on the impl. area than on the not impl. one, but that effect is not significant. So the

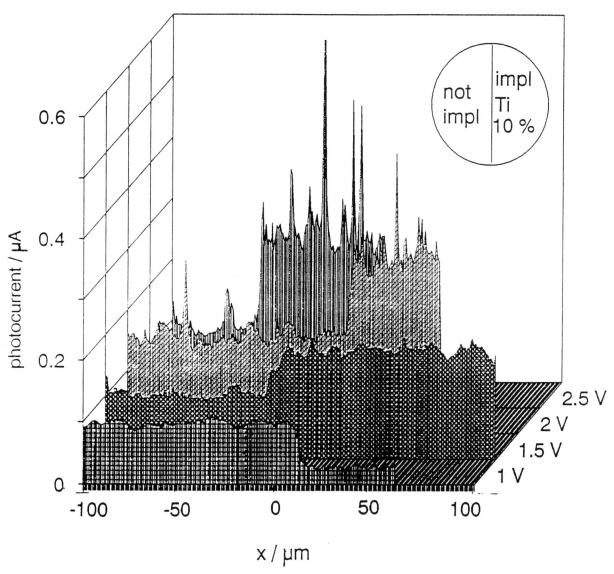

Fig. 6 *Photocurrent laser scans across the implantation boundaries in dependence on the applied anodic potential. Laser power ca. 20 μW, spot dia. ca. 3 μm. The inset shows the geometry of the specimen.*

conclusion can be drawn that the photocorrosion is not strongly enhanced by the implantation.

4. Conclusions

The radiation damage caused by 200 keV Ti$^+$ implantation of 40 nm TiO$_2$ layers is studied. The damage yields a drastic change of the structure and the conductivity of the passive films due to Frenkel pair generation and amorphization of the target. RBS spectra show that the stoichiometry is shifted towards titanium suboxides. The structure change can also be measured by ellipsometry, where a strong change of the optical layer constants occurs. The implanted oxides show a significant absorption k_1. Photoelectrochemical measurements confirm this ellipsometrical result by the fact that the absolute photocurrent values are higher on the impl. area than on the not impl. one. Evaluation of the photocurrent–potential curves also gives evidence for the high amorphicity of the implanted film.

Laser irradiation of the oxide films with high power density yielded no significant increase of photocorrosion on the implanted area.

Moreover, radiation induced amorphication and increasing conductivity will cause a faster adjustment of the equilibrium potential of the electrode. If one regards the susceptibility of Ti to hydrogen embrittlement at cathodic potentials, this fact can be advantageous for the corrosion behaviour because the free corrosion potential will shift towards the anodic region were Ti is protected by passive films.

5. Acknowledgement

The financial support of this work by the Bundesministerium für Forschung und Technologie is gratefully acknowledged.

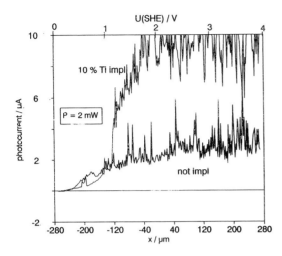

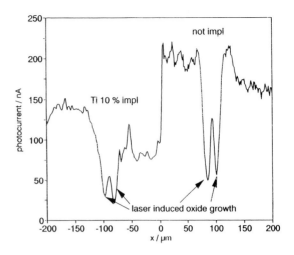

Fig. 7 Locodynamic measurement of photocurrents in dependence on the potential under laser illumination on the implanted and not implanted area (laser power = 2 mW, spot dia. 4 μm, wavelength = 257 nm, dU/dt = 4 mV/s, stepsize = 1 μm).

Fig. 8 Photocurrent laser-scanning scan across the implantation boundaries after laser induced oxide growth at the indicated location (laser power ca. 20 μW, spot dia. ca. 3 μm).

References

1. E. de Vito, M. Keddam and P. Marcus, this issue.
2. R. Kelly, J. Electrochem. Soc., **134**, 1667, 1987.
3. J. W. Schultze, B. Danzfuss, O. Meyer and U. Stimming, Mat. Sci. Eng., **69**, 273, 1985.
4. J. W. Schultze, L. Elfenthal, K. Leitner and O. Meyer, Electrochim. Acta, **33**, 911, 1988.
5. L. Elfenthal and J. W. Schultze, in K. Ebert and R.v. Ammon, Safety of the Nuclear Fuel Cycle, VCH Verlag, 1989.
6. M. Wolff, J. W. Schultze, J. L. Delplancke, Ch. Buchal and L. Elfenthal, J. Electroanal. Chem., **300**, 283, 1991.
7. Lj. D. Arsov, Electrochim. Acta, **30**, 1645, 1985.
8. A. M. Butler, J. Electrochem. Soc., **130**, 2358, 1983.
9. A. M. Butler, J. Electrochem. Soc., **131**, 2185, 1984.
10. J. W. Schultze, K. Bade and A. Michaelis, Ber. Bunsenges. Phys. Chem., **95**, 1349, 1991.
11. K. Bade, O. Karstens, A. Michaelis and J. W. Schultze, Faraday Discussion **94**, 1992, in press.
12. A. Michaelis and J. W. Schultze, J. Thin Solid Films, Proc. ICSE, 1993, in press.

13. E. W. Becker, W. Ehrfeld, P. Hagmann, A. Maner and D. Munchmeyer, Microelectronic Engineering, **4**, 35, 1986.
14. A. Maner, W. Ehrfeld and R. Schwarz, Galvanotechnik 79, 4, 1988.
15. A. Michaelis, A. Thies and J. W. Schultze, Dechema Monographien Vol. 125, VCH-Verlag, 459, 1992.
16. K. Leitner, Ph. D thesis, University of Düsseldorf, 1987.
17. W. W. Gartner, Phys. Rev., **116**, 84, 1959.
18. L. Elfenthal, J. W. Schultze and O. Meyer, Corros. Sci., **29**, 343, 1989.
19. J. W. Schultze, L. Elfenthal, K. Leitner and O. Meyer, Mat. Sci. Engng, **90**, 253, 1987.
20. J. L. Delplancke and R. Winand, Electrochim. Acta, **33**, 1539, 1988.
21. Ch. Buchal, J. L. Delplancke, M. Wolff, L. Elfenthal and J. W. Schultze, Radiation Effects and Defects in Solids, **114**, 225, 1990.
22. A. R. Newmark and U. Stimming, J. electroanal. Chem. **164**, 89, 1984.
23. J. F. Ziegler, J. P. Biersack and U. Littmark, The Stopping and Range of Ions in Matter—Vol. 1, Pergamon Press, Oxford, 1985.

44

Simulation of γ-Ray Effects on Passive Films by Laser Irradiation

A. Michaelis and J. W. Schultze

Institut für Physikalische Chemie und Elektrochemie, Heinrich-Heine-Universität Düsseldorf, 4000 Düsseldorf, Germany

Abstract

The γ-photoeffect on TiO_2 passive layers is simulated by *in situ* UV-laser irradiation with high power density at high local resolution. Micro-ellipsometric, micro-reflection spectroscopic and laser scanning examinations of the anodic passive film formation on single Ti-grains yield a strong dependence of the layer thickness on the underlying grain-structure. On grains with thicker initial oxide layers a smaller radiation damage (lower amount of laser induced oxide) occurs than on thin layered grains, indicating a poorer resistance of those grains to photocorrosion. Anisotropy–ellipsometry shows that the oxide induced by the radiation grows epitaxially but with a different crystal orientation than the underlying initial TiO_2. By multiple angle of incidence ellipsometry and micro-reflection-spectroscopy a significant absorption of the laser induced oxide can be detected due to enhanced electronic conductivity caused by a large amount of defect states within the bandgap of the formed oxide. Electrochemical investigations show that the resistance of this oxide to corrosion is very low. Nevertheless, the radiation caused no direct dissolution of the passive layer, e.g. no laser induced pitting occurred.

1. Introduction

Various possible mechanisms of corrosion induced by γ-irradiation of nuclear environments have been discussed in the literature [1, 2], but few experiments exist with special systems only [3, 4]. Homogeneous dissolution of passive films could be caused by products of radiolysis, whereas microscopic substratum effects (grains, grain boundaries etc.) could cause localised corrosion or oxide growth. For the investigation of possible localised corrosion mechanisms, two aspects are emphasised in this paper with titanium as an example:

(i) The role of microscopic properties of oxide films.
(ii) Reactions caused by electron/hole-pair formation which are induced by localised *in situ* UV-laser irradiation for simulation of the γ-photoeffect.
Passive Ti is used as an example, since it is discussed as an outer barrier material for the final disposal of highly active waste (HAW) [5] in salt mines such as those at Gorleben.

1.1 Interaction of γ-radiation with passive films and its simulation

The principle interaction mechanisms of γ-quanta with matter are summarised in Fig. 1. The Compton- and Photo-effect are most important because by these ef-

fects most of the energy is transferred. Primary, ionisation of matter and generation of hot electrons with kinetic energies in the MeV range takes place. The corrosion damage effect E_{damage} is a function of four sub-effects,

$$E_{damage} = f(E_{rad}, E_T, E_{destruction}, E_{ph})$$

These are radiolysis of the surrounding electrolyte E_{rad}, which produces short living radicals and oxidising agents like H_2O_2, temperature increase E_T (up to

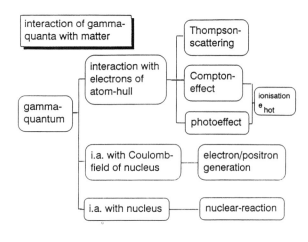

Fig. 1 Principle interaction mechanism of γ-quanta with matter (the abbreviation i.a. means interaction).

Part of a Ph.D thesis of A. Michaelis

250°C for HAW), structure change $E_{destruction}$ by generation of a small amount of Frenkel-defects and γ-photoeffect E_{ph}, which produces electron/hole (e/h) pairs in the n-type semiconducting TiO_2 layers. Only the first effect E_{rad} is caused primarily by the γ-quanta via direct ionisation. The other effects are caused secondarily by the relaxation of the hot-electrons. Although E_{rad} and E_T are very important for the corrosion behaviour, here especially E_{ph} is regarded.

A straightforward estimation shows that this effect can be simulated by UV laser-irradiation with high efficiency. If HAW with a dose rate of 10^3 gray/h is regarded, in a Ti passive layer an absorption of 10^8 $MeV.cm^{-2}h^{-1}$ takes place if a layer thickness of 100 nm with a density of 4 $g.cm^{-3}$ is assumed. That corresponds to a generation of *ca.* 10^{18} e/h pairs/s assuming a typical energy inversion rate of 10%. This would yield photocurrents in the nA range.

On the other hand with focused UV-laser irradiation (hv > E_{gap}), power densities in the $kWcm^{-2}$ range can be obtained yielding photocurrents higher than µA on much smaller areas. Therefore long periods of time of γ-irradiation can be simulated by UV-laser irradiation in laboratory time scales and at high local resolution if only E_{ph} is regarded.

For this the beam of a frequency doubled Ar^+-laser (257 nm) is focused onto the sample surface by use of a microscope with an UV-objective. A minimal spot radius of *ca.* 1 µm can be obtained yielding power densities up to some kW cm^{-2}. For simultaneous electrochemical investigations a three electrode configuration is used (see Fig 2(a)).

Observation of the sample surface via a CCD-camera shows that a grain depending oxide growth as well as oxygen evolution takes place. In combination with this laser scanning apparatus, micro-ellipsometry and reflection spectroscopy are applied to characterise layers on single grains. Especially micro-ellipsometry will be emphasised here. By reason of the anisotropy of Ti, information not only about layer thickness and optical properties can be deduced by ellipsometry, but also about epitaxy and crystal orientation. For this the method of anisotropy ellipsometry is presented. Herewith, important information about the local corrosion behaviour of Ti can be deduced.

2. Experimental

2.1 Optical methods

The experimental setup of the UV-laser scanning setup is described detailed in [6, 7]. For micro reflection spectroscopy an optical fibre can be added to the microscope of the laser raster device. Light reflected from the sample is fed to the spectrometer along the fibre where it is split into its spectral components. As light source the microscope lamp is used. By using an objective with a magnification of twenty a lateral local resolution of 10 µm can be obtained. The experimental setup is shown in Fig. 2(a).

For *in situ* micro-ellipsometry (the experimental setup is shown in Fig 2(b)) an automatic rotating analyser device (AFE Sentech) with a HeNe laser as light source (632.8 nm) is used. For focusing a lens of small numerical aperture (focal length = 13 cm) was been chosen to prevent disturbances of the polarisation state [8,9]. This limits the obtainable resolution. With our multiple angle of incidence ellipsometer a resolution of *ca.* 50 µm is obtained. In [10] an explicit formula for correcting the ellipsometric data deter-

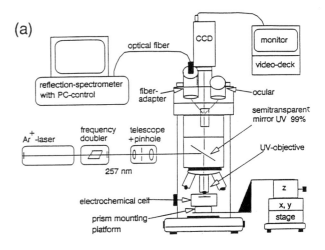

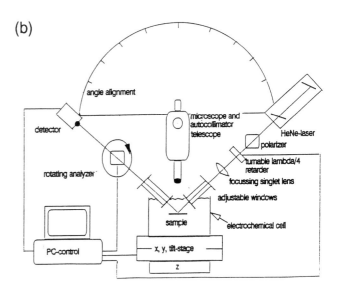

Fig. 2 (a) In situ laser scanning apparatus in combination with reflection spectroscopy and simultaneous microscopic observation via a CCD-camera.
(b) setup for micro-ellipsometry.

mined by micro-ellipsometry is given. Nevertheless focusing errors are neglected in our investigation because their order of magnitude is *ca.* 1%.

The investigations presented here were carried out at an angle of incidence of 70°.

The technique of ellipsometry is based on the measurement of changes in the state of polarisation of a collimated monochromatic light beam caused by the interaction of the beam with the physical system under investigation. In general the light is elliptically polarised after the reflection, the ellipsis is characterised by the two angles delta and psi.

In comparison to other optical techniques ellipsometry has several advantages as for example: high surface sensitivity (below one monolayer), independent measurement of both optical constants n (index of refraction) and k (absorption coefficient) at one wavelength (Kramers-Kronig transform must not be used) with high absolute accuracy (10^{-2}–10^{-4} for n and k) if a two layer model is assumed.

2.2 Electrodes and surface processing

The titanium (99.6 %) plates (10 mm dia.) were first tempered in a vacuum oven (10^{-7} mbar) at 950°C for 165 h to obtain an average grain-size of *ca.* 200 μm. Then the samples were mechanically polished with emery paper 200, 600, 1200 and 2400 and finally electropolished applying a procedure given in [11] to get a smooth surface with open grain boundaries. All presented measurements were carried out with such coarse grained samples. An electrolyte, 0.5 M H_2SO_4 (pH 0.3 at 25°C), was used throughout the paper. Electrochemical potentials are given with respect to the standard hydrogen electrode (SHE).

3. Results and Discussion

3.1 Oxide formation on single Ti-grains

Photoelectrochemical measurements with laser irradiation at high local resolution allow to get information about microscopic inhomogeneities of passive films [12–16]. To get a complete physical picture of ionic and electronic properties, we developed further methods which are presented here.

In Fig. 3 the micro-ellipsometric determined oxide growth during a potentiodynamic sweep is shown for different single Ti grains labelled as grain D, A and C. In contrary to the electrochemical current-potential curve, ellipsometry allows to get locally resolved information for single grains. Significant deviations of the oxide growth (dd/dU), which is given by the slope of the bars, occur. There are grains (grains of class A) which show oxide growth curves that are in good agreement with the electrochemical determined values (coulometry) of *ca.* 2 nm.V^{-1}, but on the other hand

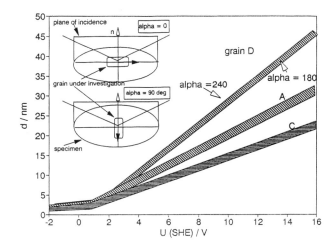

Fig. 3 Micro-ellipsometric determination of the oxide growth rate on various Ti-grains. The width of the bars is a result of the dependence of the delta and psi values on alpha due to the anisotropy. The inset explains the definition of alpha (arbitrary zero).

there are grains which show significant deviations to higher (grain D) as well as to lower (grain C) oxide thicknesses. This result confirms that with electrochemistry only average values are measured. To clear up the mechanisms of for example passive layer formation it is necessary to regard local heterogeneities like single grains.

The micro-ellipsometric data obtained show a strong dependence on the sample rotation around the surface normal due to the anisotropy of Ti. Both the Ti-substratum (hcp-lattice) as well as the oxide-layer (rutile and anatase are tetragonal) are anisotropic. This leads to a sine-like variation of the measured delta and psi values in dependence on the angle alpha which denotes the sample rotation (see inset of Fig. 3). By this reason the absolute layer thickness cannot be determined with high accuracy. Therefore, bars and not thin lines are shown in Fig. 3. This effect is not necessarily a disadvantage, in contrary, it can be used to obtain information on epitaxy and crystal orientation of layers [17] as will be agreed to in section 3.3.

To confirm the ellipsometric result, laser- and reflection spectroscopy scans across the grain structure were carried out as shown in Fig. 4. If the optical constants of the system are known, also reflection spectroscopy allows the determination of layer thicknesses by using the typical interference patterns in the obtained spectra. In Fig. 4 such a thickness determination with a local resolution of 10 μm was carried out, meaning that on each location a whole spectrum was detected and evaluated. A very good correlation between the laser-scanning photocurrent measurement and the reflection spectroscopy scan is obtained. Grains with thicker oxide layers yield lower photocurrents and vice versa. That is due to the E-field effect on the

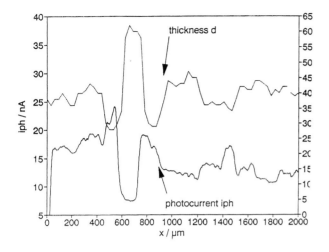

Fig. 4 Laser scanning photocurrent $i_{ph}(x)$ curve across the Ti-grain structure. Potential during the measurement 2 V, laserpower 1 mW, scan rate $1\mu m.s^{-1}$ and determination of the local film thickness by micro-reflection-spectroscopy.

photocurrent, i.e. under potentiostatic conditions thick layered grains yield a lower potential gradient resulting in an inefficient electron/hole pair separation (lower photocurrents). Confirming the ellipsometric result, strong differences in the oxide layer thicknesses occur. The maximum thickness in the reflection spectroscopy scan corresponds for example to a grain of type D, which also exhibited the thickest oxide layer in the ellipsometric measurement.

The strong differences in the oxide layers of single grains is very important for the local corrosion behaviour of such passive materials. Grains with thinner oxide layers may be more susceptible to local corrosion as will be shown in the succeeding sections.

3.2 Laser induced oxide formation on single grains for the simulation of the γ-photoeffect

By UV-laser illumination of the n-type semiconducting Ti oxide layers ($hv > E_{gap}$), e/h pairs are generated. In the presence of an anodic potential, separation of the e/h pairs takes place as well as recombination at recombination centres. The migration of holes to the surface can cause an accumulation of positive charge which yields a change of the potential distribution within the oxide by the photopotential. At potentiostatic conditions the potential drop in the Helmholz layer will be increased. Laser induced oxide growth is a simple anodic ion transfer reaction (ITR) without hole consumption,

$$Ti + 2 H_2O \rightarrow TiO_2 + 4 H^+ + 4 e^-,$$

but it is induced by the mentioned change of the potential distribution [18]. The holes themselves can cause e.g. oxygen formation,

$$4 h^+ + 2H_2O \rightarrow O_2 + 4 H^+,$$

or direct photocorrosion,

$$4h^+ + TiO_2 \rightarrow Ti^{4+} + O_2$$

Figure 5 shows a micrograph of the coarse grain Ti/TiO₂ (20 nm TiO₂) surface. On the three labelled locations laser irradiation with a power-density of *ca.* 1 kW.cm^{-2} was carried out yielding the visible spots of laser induced grown oxide on single grains. The laser transients of Fig. 6 show how the experiment was carried out. First an anodic potential of 2 V was applied to the electrode, than the UV laser light (1 mW) was switched on yielding relative high photocurrents which decrease exponential with time indicating laser induced oxide growth. The shape of these transients shows a large variation for the different grains. The charge between the transient maximum and the line of constant current at illumination is a measure for the amount of formed laser induced oxide, i.e. the susceptibility to laser damage is strongly depending on the grain structure.

The important result is that under the investigated experimental conditions, no local corrosion like pitting was observed, i.e. no γ-radiation induced dissolution of oxide took place. Just oxygen evolution and oxide growth occurred as described above.

3.3 Structure of the laser induced radiation damage

To analyse the structure of the formed laser oxide, anisotropy micro-ellipsometry was carried out. For this, the delta and psi values were measured in dependence on the angle alpha of sample rotation around the surface normal (the ellipsometric plane of inci-

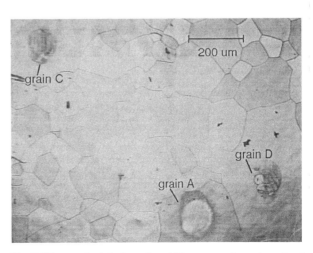

Fig. 5 Micrograph of the investigated Ti coarse grain surface. On the indicated grains laser induced oxide formation (radiation damage) has taken place.

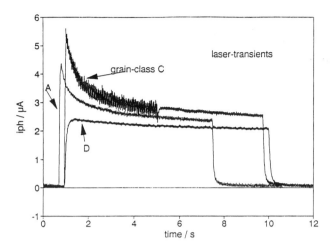

Fig. 6 Laser transient light on light off experiment on single Ti-grains, laserpower 1 mW, potential 2 V.

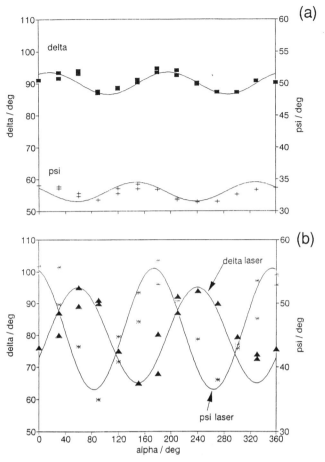

Fig. 7 Anisotropy micro-ellipsometric measurement on the unmodified area of grain A (Fig. 7(a)) and on the laser modified area of that grain (Fig. 7(b)). The sine-like variation of delta and psi in dependence on alpha is due to the anisotropy of the system.

dence resp.). In Fig. 7 the obtained sine-like variation of delta and psi for one grain is shown (grain A in Fig. 5). One measurement was carried out on the unmodified oxide beneath the laser induced oxide spot (Fig. 7(a)) and one on the irradiated spot (7(b)). Three effects are observed:

First the average delta and psi values have changed, that indicates simply an oxide growth. The laser induced oxide has a thickness of *ca.* 100 nm.

The other effects are an increase of the amplitudes and a phase shift of *ca.* 30 deg for the curves of Fig. 7(b) with respect to Fig 7(a). From this the important conclusion that epitaxic growth with a different preferential crystal orientation of the laser induced oxide with regard to the initial oxide takes place, can be drawn. In this context the terminus epitaxy means oxide growth of a preferentially oriented microcrystalline structure in dependence on the structure of the initial substratum orientation and not growth of well oriented single oxide-crystals. The polycrystalline structure of the epitaxic grown initial anodic TiO_2 (unmodified oxide) is known from literature [19–21]. The important result is that the structure of the laser induced oxide is totally different from the initial anodic Ti-oxide.

This conclusion is a result of the following model: The amplitude and the phase of the sine-like variation is determined by the orientation of the optical axis with respect to the ellipsometric plane of incidence. By variation of the angle alpha, the angle between the optical axis and the plane of incidence is varied yielding the observed sine-like variation. If the laser induced oxide grew amorphously, this would have no effect on the anisotropy keeping the orientation of the optical axis constant. If on the other hand simply epitaxic growth with the same orientation as the underlying substratum would take place, again the ori-

entation of the optical axis would not be affected. Only epitaxic growth of optical anisotropic oxide with a different orientation of the optical axis with respect to the substratum orientation yields an effective new optical axis with a changed angle alpha between this axis and the plane of incidence. Solely this change of the angle alpha can cause the observed phase shift in the ellipsometric measurement.

Another hint for the different structure of the laser induced oxide results from micro-reflection spectroscopic measurements shown in Fig. 8. Here reflection spectra on modified and not modified areas of the same grain like in the ellipsometric measurement are shown. The spectrum of the unmodified area shows a significant maximum due to the interference effect. This maximum vanishes on the irradiated area despite of the larger oxide thickness. Moreover the absolute reflectivity decreases. These effects can be due to an arising absorption coefficient k of the laser induced oxide (pure TiO_2 shows no absorption). On the other hand the decreased reflectivity can be due to

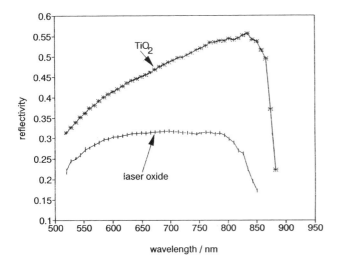

Fig. 8 Micro reflection spectra of the unmodified area of grain A labelled as 'TiO₂' and on the modified area of that grain labelled as 'laser oxide'.

an enhanced diffuse light scattering caused by a changed morphology of the surface. Nevertheless the first interpretation is confirmed by multiple angle of incidence ellipsometry on the irradiated area. From ellipsometry a significant k-value in the range of 0.2 to 0.7 can be deduced. Because of the anisotropy and the inhomogeneity that value cannot be specified with better accuracy.

The arising absorption means that the electronic properties of the oxide are drastically changed by the irradiation. A more metal-like oxide with a higher conductivity due to e.g. a large amount of defect states is generated. Electrochemical measurements show that this oxide can be easily reduced.

The amount of formed laser oxide is a measure for the susceptibility of single grains to photocorrosion. This is a very important result with respect to the local corrosion behaviour of Ti, because the experiments show a strong dependence of the amount of formed oxide on the grain-structure. The presented experiments therefore allow a quantification of the local corrosion resistivity of heterogeneities like single grains.

4. Summary

The effect of γ-irradiation on Ti-passive layers were investigated on single Ti-grains. For this the γ-photoeffect was simulated by *in situ* UV-laser irradiation with high power density at high local resolution. Micro-ellipsometric, micro-reflection spectroscopic and laser scanning examinations of the anodic passive film formation yielded that the layer thickness depends strongly on the grain-structure. On grains with thicker initial oxide layers a smaller radiation damage (smaller amount of laser induced oxide) occurs than on thin

layered grains, i.e. thick layered grains are more resistive to photocorrosion.

Anisotropy ellipsometry and micro-reflection spectroscopy yielded information about the structure of the laser induced oxide. Epitaxic growth with a different crystal orientation of the preferentially oriented microcrystals takes place. A significant absorption-coefficient arises due to enhanced electronic conductivity caused by a large amount of defect states within the bandgap of the formed oxide. Electrochemical investigations show, that the resistivity of this oxide to reduction processes is very low. Nevertheless, the radiation caused no direct dissolution of the passive layer, e.g. no laser induced pitting occurred.

The measurements on single grains proved that with classical standard methods only average values are measured. For the understanding of corrosion mechanisms, especially local corrosion, it is necessary to apply the investigations on hetreogeneities like single grains which affords measurements at high local resolution.

On the basis of the presented experiments, it will be possible to estimate corrosion rates of Ti during γ-irradiation. This knowledge is of great importance for the application of Ti as container material for the disposal of HAW.

5. Acknowledgement

The financial support of the Bundesministerium für Forschung und Technologie is gratefully acknowledged.

References

1. A. V. Bjalobzeskij, in K. Schwabe, Korrosion durch radioaktive Strahlung, Akademie Verlag Berlin, 1971.
2. L. Elfenthal, J. W. Schultze and O. Meyer, Corros. Sci., **29**, 343, 1989.
3. K. Leitner and J. W. Schultze, Ber. Bunsenges. Phys. Chem., **92**, 181, 1992.
4. J. W. Schultze, L. Elfenthal, G. Hansen, Th. Patzelt, B. Siemensmeyer and J. Thietke, Corros. Sci., **31**, 213, 1990.
5. E. Smailos, W. Schwarzkopf, B. Kienzler and R. Koster, Mat. Res. Soc. Symp. Proc. Vol. **257**, 399, 1992.
6. A. Michaelis and J. W. Schultze, Optical *in situ* methods at high local resolution for the investigation of corrosion processes, Ber. Bunsenges. Phys. Chem. (Discussion Meeting 1992) in press.
7. J. W. Schultze, A. Michaelis, O. Karstens and Ch. Buchal, this issue.
8. M. Erman and J. B. Theeten, J. Appl. Phys., **60**, 3, 859, 1986.
9. R. F. Cohn, J. W. Wagner and J. Kruger, Applied Optics, **27**, 22, 4664, 1988.
10. D. O. Barsukov, G. M. Gusakov and A. A. Komarnitskii, Opt. Spectrosc. (USSR) **64**, 782, 1988.

11. Lj. D. Arsov, Electrochim. Acta, **30**, 1645, 1985.

12. A. M. Butler, J. Electrochem. Soc., **130**, 2358, 1983

13. A. M. Butler: J. Electrochem. Soc., **131**, 2185, 1984.

14. M. R. Kozlowski, P. S. Tyler, W. G. Smyrl and R. T. Atanasoski, Surf. Sci., **194**, 505, 1988.

15. M. R. Kozlowski, W. G. Smyrl, Lj. Atanasoski, R. T. Atanasoski, Electrochim. Acta, 34, 1763, 1989.

16. J. W. Schultze and J. Thietke, Electrochim. Acta, **34**, 1769, 1989.

17. A. Michaelis and J. W. Schultze, Thin Solid Films, Proc. ICSE 93, 1993, in press.

18. J. W. Schultze, K. Bade and A. Michaelis, Ber. Bunsenges. Phys. Chem., **95**, 1349, 1991.

19. J. L. Delplancke and R. Winand, Electrochim. Acta, **33**, 1539, 1988.

20. J. L. Delplancke, Ph D-Thesis, University of Bruxelles (1987).

21. Ch. Buchal, J. L. Delplancke, M. Wolff, L. Elfenthal and J. W. Schultze, Radiation Effects and Defects in Solids, **114**, 225, 1990.

45

Passivation of Metals by Plasma-Polymerised Films

W. J. van Ooij, A. Sabata and Ih-Houng Loh*

Armco Research & Technology, Middletown, OH 45044-3999, USA
*Advanced Surface Technology, Inc., Billerica, MA 01821-3902, USA

Abstract

Cold-rolled steel (CRS) substrates were coated with thin films of plasma-polymerised trimethylsilane (TMS) using an RF discharge. Variations in this study were the precleaning of the substrate, the monomer (i.e. either pure TMS or TMS/CO_2 mixtures) and a post treatment of the plasma film in an oxygen plasma. The corrosion properties of the films were determined by Electrochemical Impedance Spectroscopy (EIS) and exposure in a humidity test. Analysis of the film composition was done by AES and TOFSIMS. Best results were obtained with films deposited from plasmas of pure TMS and the substrates pretreated in a hydrogen plasma. Post oxidation of the films in an oxygen plasma deteriorated the passivation behaviour of the films. The results are discussed in light of a simple model in which the crosslinking of the films and their adhesion to the substrate are key parameters.

1. Introduction

Plasma polymerisation is a well-established technique for depositing polymerised films of unusual properties onto semiconductor or polymeric substrates [1,2]. Plasma-polymerised films (PPFs) are usually referred to as highly crosslinked, pinhole free and strongly adherent to any type of substrate. Thus even very thin films should have excellent barrier properties and, therefore, they would be ideally suited for modifying the corrosion behaviour of metals. Remarkably, however, the literature contains only very little information on the use of PPFs for passivation of metals. In our laboratory we have studied the surface and bulk composition and structure of PPFs of simple hydrocarbon and silicon-containing monomers for a number of years [3–6]. In these studies it was observed that PPFs deposited on cold-rolled steels (CRS) or electrogalvanised steel (EGS) sometimes impart unusually good corrosion properties, as compared with conventional organic coatings, such as paints. It was also observed that the properties of PPFs can vary dramatically, apparently depending on the metal pretreatment and the deposition conditions.

In this paper results are presented of a study in which the factors that govern the passivation characteristics of PPFs on CRS were systematically investigated. Only one monomer, trimethylsilane (TMS), was used here. The main variables were the type of metal pretreatment, the monomer (pure TMS or TMS/CO_2 mixtures, with the purpose of varying the amount of oxygen in the films) and the post oxidation of the films by an oxygen plasma. This was done with the objective to vary the surface energy of the PPF. Results obtained with electrogalvanised steel substrates will be published elsewhere.

2. Experimental

2.1 Materials

Cold-rolled steel strip of 0.5 mm thickness of commercial automotive grade was used. It was obtained from Armco Steel Company (Middletown, Ohio). Panels of 10×15 cm^2 were used throughout this study. They were homogeneously coated on both sides with a film of plasma-polymerised trimethylsilane. The TMS gas was of 99% purity and was obtained from Hüls America.

2.2 Plasma Polymerisation

Deposition of the PPFs was done in a laboratory RF discharge reactor which was equipped with a quartz chamber, a mechanical vacuum pump and an optical emission spectrometer. Just prior to deposition, the steel panels were degreased in an organic solvent. The variables in this study were:

(i) the precleaning of the substrates in the plasma reactor, which was done either by an argon plasma or an argon plasma followed by an Ar/H_2 plasma;
(ii) the monomer TMS was mixed with various ratios of CO_2 in some experiments;
(iii) following the PPF deposition, some panels were post-treated in an oxygen plasma. All other parameters, such as pressure, flow rates, deposition times or plasma power were kept constant. These conditions are listed in Table 1.

2.3 Electrochemical Impedance Spectroscopy (EIS)

Samples punched from the coated panels were tested by EIS, the details of which have been published before [5]. Essentially, the impedance of the coated steel samples of 10 mm dia. was determined over the frequency range 0.01 to 100 kHz in aerated 3.5 wt% NaCl solutions at 25°C with the sample held at the open circuit potential. The EIS measurements were made after 10 min of immersion in the NaCl solution. The data were analysed using software developed by Boukamp [7]. A simple equivalent circuit was assumed consisting of the capacitance of the plasma film (C_p), a double layer capacitance (C_{dl}), and two resistances, the pore resistance, R_{po}, and polarisation resistance R_p, respectively. An additional parameter R_1, which is defined as $\log | Z_{100}/Z_{10000} |$, i.e. the ratio of the impedance Z_j at frequencies 10^2 and 10^4 Hz, was also calculated [8]. All of these parameters have been demonstrated to correlate well with the performance of the PPF-coated steels in corrosion tests [5].

2.4 Corrosion testing

10×10 cm panels with sealed edges were exposed in a humidity cabinet at 60°C and 85% relative humidity for up to 96 h. Uncoated CRS shows severe rust formation in this test after exposure for only one day.

2.5 Auger Electron Spectroscopy

AES analysis was done on a Perkin Elmer 590A scanning Auger spectrometer using a 8 kV electron beam of 2000 Å diameter. Depth profiles were recorded by sputtering with a beam of 1.5 keV Ar^+ ions. The sputtering rate was calibrated for a Ta_2O_5 film of known thickness.

2.6 Time-of-Flight SIMS Spectrometry

TOFSIMS spectra were recorded with a Kratos PRISM instrument equipped with a reflectron time-of-flight mass analyser and a 25 kV liquid metal ion source of monoisotopic $^{69}Ga^+$ ions. All spectra were recorded with a total integrated ion dose strictly in the static SIMS regime. The mass resolution of the instrument was *ca.* 3500 (M/ΔM) at mass 27.

3. Results

It is beyond the scope of this paper to discuss the experimental results obtained with all samples. Only four samples, A-D, will be compared, *viz.* those showing the effects on the metal passivation of the major variables, the substrate pretreatment (with or without hydrogen plasma), the post treatment (oxygen plasma), and the presence of CO_2 in the reaction mixture (Table 1). The sample specifications are given in Table 2. All films were investigated by Scanning Electron Microscopy which verified that there were no powder inclusions in the PPFs. Such inclusions are known to have a deleterious effect on the performance of PPFs on metals [1,2].

3.1 AES results

The element depth profiles of the four samples are presented in Fig. 1. All films are between 500 and 1000 Å thick and all contain (apart from hydrogen) Si, C, O and N. The interfacial oxide can clearly been distinguished in all films, even in those with had been pretreated in a H_2 plasma. The effect of the presence of CO_2 in the monomer appears to be an increase of C and a decrease of O in the film. Another effect appeared to be a slight increase in the film thickness. Both effects were reproducible. The oxygen content in the near surface layers has increased in the sample that had been post-treated by oxygen (sample D). Since the concentration of oxygen in the bulk of the film was somewhat variable, it is likely that both oxygen and nitrogen in the bulk of the film stem from the residual air in the plasma reactor. Such reactors are not run

Table 2 Sample identification

Sample	Ar	H₂	Monomer	O₂
A	+	-	TMS	-
B	+	+	TMS	-
C	+	+	TMS+CO₂ (5:1)	-
D	+	+	TMS	+

Table 1 Plasma polymerisation conditions

Step	1. Ar Plasma	2. H₂ Plasma	3. TMS Film	4. O₂ Plasma	5. Quench
Pressure	65 mtorr	80 mtorr	50 mtorr	100 mtorr	10 min after step 3
RF Power	40 W	100 W	120 W	100 W	1 min after step 4
Duration	2 min	20 min	45 min	0.5 min	

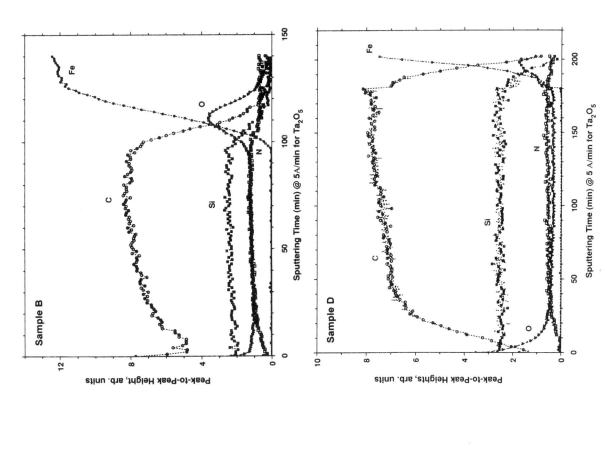

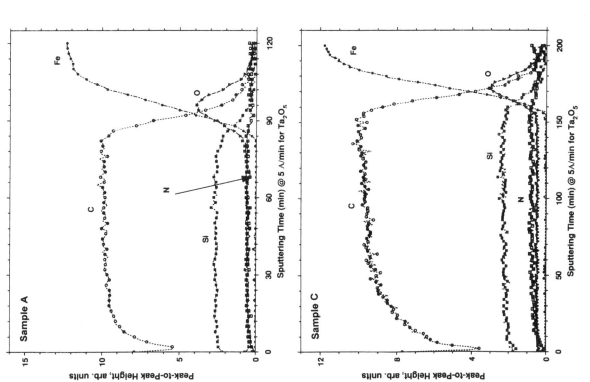

Fig. 1 AES depth profiles of PPF-coated CRS samples of Table 2; sputtering rate 5 Å/min for Ta_2O_5; peak-to peak heights corrected for elemental sensitivities.

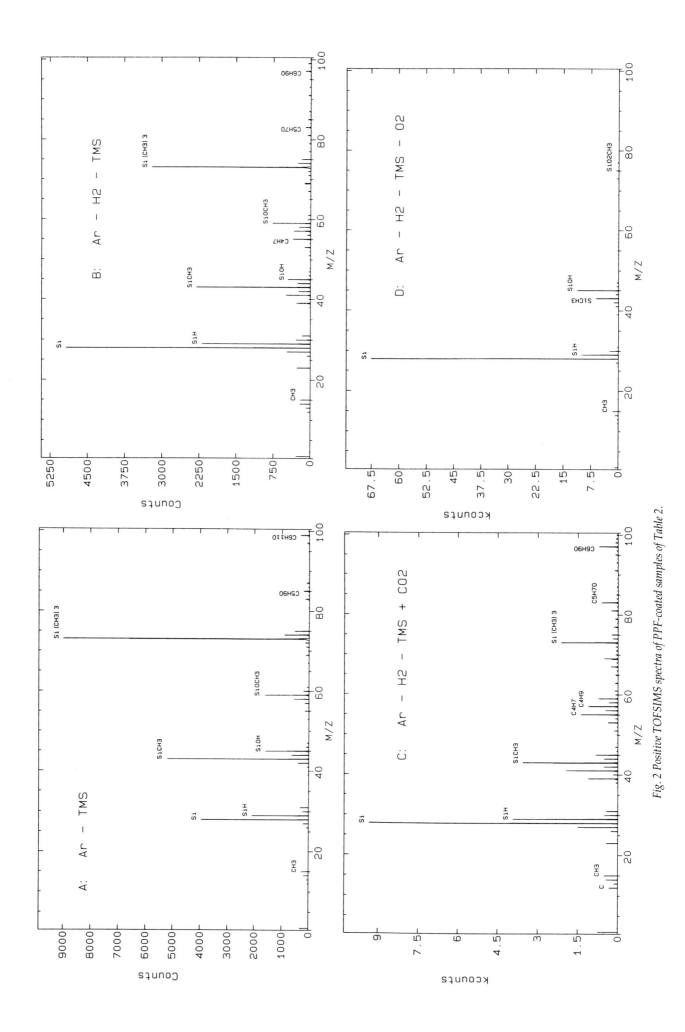

Fig. 2 Positive TOFSIMS spectra of PPF-coated samples of Table 2.

under UHV conditions. The residual atmospheric pressure in plasma reactors is of the order of 1–10 mtorr, which in itself is sufficient to sustain a gas discharge.

3.2 TOFSIMS Results

The positive TOFSIMS spectra of the four films are shown in Fig. 2. Some of the major peaks are labelled in the spectra. The spectrum of film A is typically that of a low-molecular weight polydimethyl siloxane [3]. That of film B is more representative of a high-M_w siloxane, i.e. it seems to contain less oligomeric material [5]. Both spectra show peaks in the 80–100 amu region which are oxidised C_5 and C_6 clusters. Sample B forms the more reduced forms. This is also seen in the spectrum of sample C, which contains more hydrocarbon peaks. The spectrum of the film that had been post-treated (D) resembles that of almost pure SiO_2. Very few hydrocarbon features are detectable in this spectrum. Accordingly, the negative spectrum (not shown) showed a strong increase of the O_2 and OH^- ion intensities. The high oxygen content at the surface can be attributed to post-deposition adsorption effects, related to the presence of long-live radical sites in PPFs [1,2].

3.3 EIS results

The EIS data of the films A–D are presented in Table 3 which also gives the qualitative evaluation and ranking of these films in the humidity exposure test. It is seen that the resistance to rusting in the test decreases in the order: B > C > A > D > Control. The appearance of Film D after exposure was only marginally better than that of the uncoated CRS control sample. Film B showed no rust. The best films were consistently obtained with pure TMS (i.e. without CO_2) and with a hydrogen plasma included in the pretreatment step. The post-treatment in the O_2 plasma had a negative effect on the performance of the films, regardless of the metal pretreatment or whether the PPF was deposited

from pure TMS or TMS/CO_2 mixtures. The AES depth profiles (Fig. 1) indicate that all films had approximately the same thickness. However, the EIS data show the coating capacitance to be different for the four films. This effect may be related to different compositions or crosslink densities in the four materials. The pore resistance is highest for film B which suggests that this film had a better coverage of the steel substrate. The C_{dl} and R_p values are indicative of the corrosion activity of the samples with sample B clearly having the lowest activity.

Comparison of the corrosion performance with the impedance data shows that excellent agreement is obtained between the percentage red rust and all EIS parameters. The films with better corrosion resistance also have the lowest film capacitance, the highest pore resistance (resistance to blistering), the lowest double layer capacitance and the highest polarisation resistance. The parameter R_1 also follows this trend, being highest for the good quality films. Hence the quality of PPFs can be characterised quickly and efficiently by simply measuring this parameter, i.e. the impedance at two frequencies. The value of R_1 is equal to 2.00 only for perfect coatings. For coatings with damage or defects this value is less than 2. Films in the thickness range studied here cannot be considered perfect, but the results show that film B had the lowest number of defects.

4. Discussion

The results described above indicate that excellent protection of CRS against corrosion in an aqueous environment can be obtained by depositing PPFs of TMS and that not more than *ca.* 500 Å thickness is required. It has been reported that the performance drops steeply if the film thickness exceeds 1000 Å [1,2]. Our results also demonstrate that, in addition to the

Table 3 EIS and corrosion data for samples of Table 2

Sample	C_p (F)	R_{po} (Ω)	C_{dl} (F)	R_p (Ω)	R_1	%Rust*
A	$4.3×10^{-6}$	196	$3.0×10^{-5}$	$1.0×10^3$	0.49	60
B	$1.5×10^{-6}$	618	$1.2×10^{-6}$	$9.8×10^4$	1.22	0
C	$1.7×10^{-6}$	182	$1.6×10^{-6}$	$6.5×10^3$	1.02	20
D	$6.0×10^{-6}$	69	$8.6×10^{-5}$	$9.2×10^2$	0.33	80
Control (15 min)	-	-	$5.9×10^{-4}$	$3.6×10^2$	-	100
Control (1 hr)	-	-	$1.6×10^{-3}$	$1.8×10^3$	-	100

* after 96 hours in humidity test; 60°C and 85% r.h.; exposed sample area 5×5 cm² with sealed edges

film thickness, the performance of the films depends strongly on the metal pretreatment and on the deposition conditions. There are no theories or models in the literature with which these effects can be explained satisfactorily. Therefore, we can only give a rather speculative explanation for the observed effects. Clearly, much more work needs to be done before the behaviour of PPFs can be fully understood.

4.1 Effect of H_2 pretreatment

The finding that a H_2 plasma cleaning step improves the performance of PPFs on CRS is in agreement with previous work in which similar films were deposited in a different type of plasma reactor where a DC discharge was used [5]. The EIS data (in particular C_{dl} and Rp) show that film B had the lowest corrosion activity. This effect may be attributable to an improved adhesion of the film if the substrate had been pretreated by a H_2 plasma. The AES depth profiles however do not indicate any difference between the interfaces of systems with or without the H_2 step. The oxide is not removed efficiently by this treatment. A plausible explanation, therefore, is that the effect of the H_2 plasma is crosslinking of the organic residues at the metal oxide surface [9]. Cleaning by sputtering is very unlikely in an H_2 plasma. It has been demonstrated before that precleaning of CRS in an O_2 plasma oxidises the organic residues, but it does not completely remove them [10]. An Ar plasma graphitizes such residues, but does not remove them effectively either [10]. Thus we postulate that PPFs of TMS on CRS have a rather poor adhesion to the metal oxide because a weak boundary layer (WBL) of organic residues (lubricants, etc.) is inherently present between the film and the oxide. Only a hydrogen plasma can remove the effect of this WBL by crosslinking these residues, thus increasing their molecular weight. Ar plasmas also cause some crosslinking but they are less efficient than H_2 plasmas [8].

4.2 Effect of CO_2 in the plasma

The effect of the presence of CO_2, as shown by the AES depth profiles, is to increase the C/Si ratio in the bulk of the PPF. This ratio has previously been observed to vary considerably, depending on the deposition conditions, and a correlation was found between this ratio and the passivation characteristics of the sample [5]. It was postulated that this ratio reflects different degrees of Si–Si crosslinking in the film. This assumption is based on the notion that, as Si–Si crosslinks form, there will be fewer C_xH_y pendant groups in the Si-based network. Thus the film becomes denser and will not swell as much as films with a low crosslink density. Evidence for this effect is the lower film capacitance in those films with lower C/Si ratio which demonstrate

improved performance (Table 3). Thus the negative effect of the addition to CO_2 is explained as a negative effect on the degree of crosslinking of the PPF. The underlying mechanism behind this effect is currently not understood.

4.3 Effect CO_2 Plasma Post Treatment

Here, too, only a speculative explanation can be given. The AES profiles indicate no changes in the film thickness or composition, except for a very thin surface layer. The film surface becomes more SiO_2-like. Therefore, its surface energy and wettability increase strongly [5]. It is thus reasonable to assume that water molecules and hydrated ions responsible for the degradation of the film in humidity and salt exposure, will be adsorbed more readily by O_2-treated PPFs. This explanation implies that a hydrophobic surface of the PPF should be conducive to corrosion performance in an aqueous environment.

5. Conclusions

The results described and discussed above can be summarised as follows:

1. Plasma-polymerised films of TMS of 500–1000 Å thickness on CRS substrates can provide excellent corrosion resistance in an aqueous environment;

2. The best results are obtained by cleaning the substrate *in situ* by an Ar plasma followed by a longer treatment in a H_2 plasma which possibly alleviates Weak Boundary Layer effects caused by organic residues at the oxide surface;

3. Addition of CO_2 to the TMS plasma or a post-deposition oxidation treatment in an O_2 plasma results in poorer films; all variations studied (H_2, O_2 or CO_2) affect the surface composition of the plasma films;

4. Electrochemical Impedance Spectroscopy, EIS, along with analytical methods such as AES and TOFSIMS, are very useful for the characterisation of the structure and the performance of PPFs on metals, and for establishing structure-property relationships.

References

1. H. K. Yasuda, Plasma Polymerization, Academic Press, Orlando, 1985.
2. R. d'Agostino, (ed.), Plasma Deposition, Treatment and Etching of Polymers, Academic Press, Boston, 1990.
3. A. Sabata, W. J. van Ooij and H. K. Yasuda, in Proceedings of the Eighth International Conference on Secondary Ion Mass Spectrometry (SIMS VIII) A. Benninghoven, K. T. F. Janssen, J. Tümpner and H. W. Werner, eds., Amsterdam, September 15–20, 1991, John Wiley & Sons, Chichester, UK, 1992, p. 819.

4. W. J. van Ooij and A. Sabata, Surf. Interface Anal., **19**, 101, 1992.
5. A. Sabata, W. J. van Ooij, H. K. Yasuda and D. Surman, submitted to Prog. Organic Coat.
6. A. Sabata, W. J. van Ooij and H. K. Yasuda, submitted to Surf. Interface Anal.
7. B. A. Boukamp, Proc. 9th Eur. Congr. on Corrosion, Utrecht, The Netherlands, 1989, FU252.
8. F. Mansfeld and C. H. Tsai, Final Report ONR Contract No. N00014-90-J-4123, December 31, 1991.
9. W. J. van Ooij and R. S. Michael, in Metallization of Polymers, ACS Symposium Series No. 440; ACS, Washington, DC, 1990, p. 60.
10. W. J. van Ooij, A. Sabata and B. A. Knueppel, ASTM Symposium on Metal Surface Technology for Adhesive Bonding, St. Louis, Missouri, 7–9 October, 1992, submitted to J. Test. Eval.

IV PASSIVITY BREAKDOWN

46

Fundamental Aspects in the Design of Passive Alloys

D. D. MACDONALD AND MIRNA URQUIDI-MACDONALD

Center for Advanced Materials, The Pennsylvania State University, University Park, PA 16802, USA

Abstract

This paper reviews fundamental aspects in the design of passive alloys as viewed within the framework of the point defect model (PDM) for the growth and breakdown of passive films and the solute–vacancy interaction model (SVIM) for alloying effects. The SVIM attributes alloying effects on passivity breakdown to complexing between mobile cation vacancies and oxidised solute ions in the barrier layer, and predicts that a superior alloying element is one that substitutes preferentially and uniformly into the passive film with an oxidation state that is as high as possible compared with that of the host cation. We also discuss how the PDM and SVIM may be combined with a mechanism-based pit growth model to derive, deterministically, damage functions for pitting attack. In the present treatment, alloying effects are restricted to the nucleation of damage but a more general treatment would include the effects of alloying elements on the growth of damage as well.

1. Introduction

Alloying is the most extensively used method for enhancing the passivity of base metals [1–5]. The art of alloying extends back into antiquity, but it was not until the early part of this century that the highly passive stainless steels were developed [2]. More recently, in response to the need for high strength, corrosion-resistant alloys for gas turbines, various superalloys have been devised that exhibit great resistances to dry oxidation. In this paper, we discuss the role(s) of alloying elements in enhancing passivity in aqueous environments, in terms of the point defect model (PDM) and the solute–vacancy interaction model (SVIM) that we developed a number of years ago [6–14]. These models are used to derive a set of rules for designing new, pitting-resistant binary alloys. The rules have been tested using pit initiation time data for dilute nickel alloys containing Al, Ti, or Mo, and are found to be consistent with many data that exist in the literature. Finally, we illustrate how the PDM and the SVIM are being combined with models for pit growth to yield a deterministic method for predicting the development of damage due to pitting corrosion in alloy systems.

2. Passivity Breakdown and Pit Nucleation

The bulk of the experimental evidence indicates that passive films, which form on metals and alloys in contact with aqueous environments, consist of at least two layers, as depicted in Fig. 1. The inner or 'barrier'

layer forms by movement of the metal/film interface into the metal phase, due to the inward movement of oxygen via the outward movement of oxygen vacancies [1]. On the other hand, the outer layer forms by the hydrolysis of cations ejected from the barrier layer at the film/solution interface. Because the barrier layer forms by a solid-state reaction, it is expected to consist of a disordered oxide containing both anion (oxygen) and cation (metal) vacancies, the relative concentrations of which depend on the thermodynamics of vacancy formation and the kinetics of the vacancy generation and the annihilation reactions (Fig. 2). Assuming that anions from the environment cannot readily enter the barrier layer via the oxygen vacancy structure, one would not expect the barrier layer to incorporate ions from the solution. On the other hand, the outer layer can readily incorporate ions from the environment via coprecipitation, so that the oxide, oxyhydroxide, or hydroxide that comprises the outer layer is expected to contain extraneous species from the environment. This is perhaps best seen in the case of aluminium when anodised in borate buffer solutions, where borate ions are found to be incorporated in the outer layer but are absent from the inner layer [15].

If the passive films on metals, like nickel, and chromium (and alloys thereof) remained completely intact, then the corrosion current flowing across the interface under most industrial conditions would be of the order of $0.01–1.0\ \mu A cm^{-2}$, corresponding to corrosion rates of approximately 1.5×10^{-5} to 1.5×10^{-3} cm

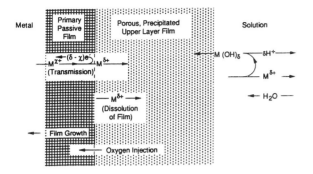

Fig. 1 Schematic of processes that lead to the formation of bilayer passive films on metal surfaces.

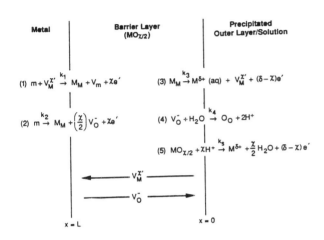

Fig. 2 Schematic of physico-chemical processes that occur within the barrier layer according to the point defect model. m = metal atom, M_M = metal cation in cation site, O_O = oxygen ion in anion site, $V_M^{\chi'}$ = cation vacancy, $V_O^{\cdot\cdot}$ = anion vacancy, V_m = vacancy in metal phase. During film growth, cation vacancies are produced at the film/solution interface, but are consumed at the metal/film interface. Likewise, anion vacancies are formed at the metal/film interface, but are consumed at the film/solution interface. Consequently, the fluxes of cation vacancies and anion vacancies are in the directions indicated. Note that Reactions (1), (3) and (4) are lattice conservative processes, whereas Reactions [2] and [5] are not.

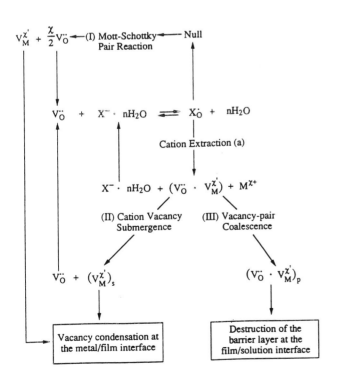

Fig. 3 Summary of proposed reactions leading to passivity breakdown.

per year (i.e. 0.15 to 15 μm/year). For most practical situations, metal loss rates of this order are of no concern, so that our automobiles, bridges, aeroplanes, and industrial systems would last for times extending well beyond the current design lifetimes. Unfortunately, passive films do not remain intact, and corrosion rates many orders of magnitude greater than those indicated above for fully passive substrates are commonly observed, particularly if the attack occurs locally. Passivity breakdown can occur for a variety of reasons, including straining of the substrate metal, the presence of thermal stresses (due to differences in thermal expansivity), fluid flow and cavitation, transpassivity polarisation, and chemically-induced phenomena. In the following discussion, we will deal with only one of the many forms of damage that results from passivity breakdown; pitting corrosion, which is responsible for losses of many billions of dollars every year to the world economy.

2.1 Pitting corrosion

The best known causative agent of 'chemically-induced breakdown' is chloride ion, which shows a remarkable ability to cause pitting on many metals and alloys of industrial interest [16–18]. Assuming that an ion, like chloride, must interact physically with the *barrier* layer to cause passivity breakdown, and hence to nucleate pits, it is of interest to explore, for the moment, how this might happen on an atomistic scale. Accordingly, it is necessary to envisage oneself as a hydrated chloride ion ($Cl^- \cdot nH_2O$, $n \approx 6$) approaching the film/solution interface of the barrier layer (after moving through the precipitated, outer layer). From this vantage point, the barrier layer appears as an undulating surface of charge with positive potentials occurring over cations and negative potentials over anions, with the difference between the peaks depending on the degree of covalent (vs ionic) bonding in the lattice (the greater the extent of covalent bonding the lower the difference between peaks). Occasionally, however, the chloride ion will experience vacancies, with cation vacancies appearing as sites of high negative charge (corresponding to a formal charge of $-\chi e$) and oxygen vacancies appearing as sites of high positive charge (formally $+2e$). Thus, the chloride ion is presented with a number of attractive sites to attack, but which will be favoured? This is a very difficult question to answer unequivocally, because other processes must be considered. For example, a chloride ion could absorb into a surface oxygen vacancy, but this must be done at the expense of considerable dehydration. However, the high co-ordination afforded by neighbouring ions is a positive factor (favouring absorption), although any expansion of the vacancy to accommodate the ion would be energetically costly. On the other hand, the anion could interact electrostatically with a positive centre in the film surface represented by a surface cation; in this case, the interaction might be weaker (because of significant covalent bonding) but, because less dehydration is required in that the ion would not penetrate into the surface, the overall effect might favour absorption at a cation site. These two scenarios could lead to quite different mechanisms for localised attack.

In the first case (anion absorption into a surface oxygen vacancy), the film may respond in a number of different ways, as depicted in Fig. 3. In one way (Case I), the system responds to the loss of oxygen vacancies by generating cation vacancy/oxygen vacancy pairs via a Schottky-pair type of reaction. The oxygen vacancies in turn react with additional anions (e.g. chloride) at the film/solution interface to generate yet more cation vacancies. Importantly, the generation of cation vacancies is autocatalytic, but whether or not the film breaks down depends on the relative rates with which the cation vacancies are transported across the barrier layer and are annihilated by emission of cations from the metal into the film. If this annihilation reaction is incapable of consuming the cation vacancies arriving at the metal/film interface, the excess vacancies will condense and lead to the local detachment of the film from the underlying metal, as depicted in Fig. 4. Consequently, provided the local tensile stresses are sufficiently high and/or the film dissolves locally, the barrier layer will rupture, marking the initiation of a pit. The evidence for this mechanism is discussed elsewhere [1].

This particular case was considered in detail by Lin *et al*. [6], who assumed that the enhanced flux of cation vacancies across the barrier layer could not be accommodated by reaction (1) in Fig. 2, thereby leading to the formation of a cation vacancy condensate. Once the condensate (Fig. 4) has grown to a critical size, dissolution of the film at the film/solution interface and the tensile stresses in the barrier layer induce a mechanical or structural instability, resulting in rupture of the film and hence in rapid localised attack. These ideas were assembled by Lin *et al*. [6] to derive expressions for the critical breakdown voltage and induction time for a single breakdown site as

$$V_c = \frac{4.606RT}{\chi F \alpha} \log \left[\frac{J_m}{J^o u^{-\chi/2}} \right] - \frac{2.303RT}{\alpha F} \log \left(a_{X^-} \right) \quad (1)$$

and

$$t_{ind} = \xi' \left[\exp \left(\frac{\chi F \alpha \Delta V}{2RT} \right) - 1 \right]^{-1} + \tau \quad (2)$$

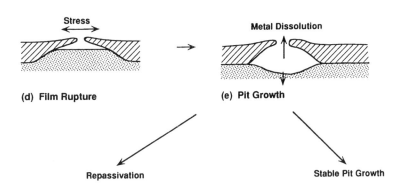

Fig. 4 Schematic outlining various stages of pot nucleation according to the Point Defect Model.

where ΔV is the breakdown overvoltage ($\Delta V = V - V_c$) a_{X^-} is the activity of X^- in the solution,

$$\xi' = \xi / J^o u^{-\chi/2} \left(a_{X^-}\right)^{\chi/2} \exp\left(\frac{\chi\alpha F V_c}{2RT}\right) \quad (3)$$

and ξ is the critical areal concentration of cation vacancies at the metal/film interface. Other parameters are as defined in the original publication [6]. Equations (1) and (2) account for many of the phenomenological characteristics of pitting attack; (i) that the 'pitting potential' (V_c) varies linearly with $\log(a_{X^-})$ with a slope greater than $2.303RT/F$ (i.e. $> 0.05916V$/decade at 25°C) because $\alpha < 1$, (ii) that $\log(t_{ind}) \propto 1/\Delta V$ for sufficiently large overvoltages, and (iii) that t_{ind} is an inverse function of the activity of the aggressive anion.

In deriving eqns (1) and (2), we have assumed that the critical concentration of cation vacancies at breakdown (ξ mol. vacancies/cm^2) is independent of the applied voltage and hence the thickness of the barrier layer. This assumption was made because transmission of cation vacancies through the film can occur only as long as the film is attached to the base metal in order that the vacancies can be annihilated by reaction (1), Fig. 2 (see also Fig. 4). Thus, growth of the condensate normal to the interface cannot occur once the film has separated from the substrate and, since separation occurs by condensation of a single layer of vacancies, our assumption of $\xi \neq f(V, L)$ is justified.

2.2 Distribution functions

On any real surface, a large number of potential breakdown sites exists corresponding to a distribution in the properties of the 'weak spots'. Perhaps the most graphic

illustration of this property is the data of Shibata and co-workers [19, 20], who showed that the 'pitting potential' is near normally distributed and that the induction time follows a distribution that is skewed towards short times. Assuming that the breakdown sites on a surface are normally distributed with respect to the cation vacancy diffusivity, we derived distribution functions for the breakdown voltage and induction time that are of the form [10, 11][†]

$$\frac{dN}{dVc} = -\frac{\gamma D}{\sqrt{2\pi}\sigma_D} e^{-(D-\bar{D})^2/2\sigma_D^2} \quad (4)$$

and

$$\frac{dN}{dt_{ind}} = \left[\frac{\xi u^{\chi/2}}{\sqrt{2\pi}\sigma_D \hat{a}}\right] e^{-(D-\bar{D})^2/2\sigma_D^2} \cdot \frac{e^{-\gamma V}}{a_x^{\chi/2}\left(t_{ind} - \tau\right)^2} \quad (5)$$

where $\gamma = \alpha\chi F/2RT$, σ_D is the standard deviation for the diffusivity for the population of breakdown sites, and the other quantities are as previously defined [10, 11].

For comparison with experiment, we define the cumulative probability in the breakdown voltage and the differential cumulative probability in the induction time as

$$P(V_c) = 100 \int_{-\infty}^{V_c} \left(\frac{dN}{dV_c}\right) dV_c \quad (6)$$

[†]Equations (6) and (7) are given in slightly different form in refs. [10] and [11].

$$\Delta N]_{t_j}^{t_{j+1}} = \int_{t_j}^{t_{j+1}} \left(\frac{dN}{dt_{ind}} \right) dt_{ind} \qquad (7)$$

The latter quantity is defined in this manner so that direct comparison can be made with experimental pitting initiation time data, which are commonly presented as histograms of the number of pits nucleating in successive increments of time.

A fit of $P(V_c)$ to the experimental data of Shibata *et al.* [19, 20] for pitting of Fe–17Cr in 3.5% NaCl solution at 30°C is shown in Fig. 5. This fit was accomplished by adjusting groups of unknown or poorly-known parameters, which affect the location of $P(V_c)$ on the potential axis but not the shape, such that the experimental and calculated distribution functions coincide for a mean diffusivity for cation vacancies of 5×10^{-20} cm^2 s^{-1} (this is approximately the value indicated by electrochemical impedance spectroscopy). Without adjusting any additional parameters, the distribution (histogram) in induction time is found to agree very well with the experimental data for the same system, as shown in Fig. 6. It should be noted that the model described thus far does not consider the 'death' or repassivation of pits however, its omission is appropriate because, in Shibata's analyses, each specimen was taken out of the population once breakdown had occurred. We should also note that similar distributions in Vc and tind are obtained if we assume other distribution functions (e.g. the Student's t and χ^2 distributions) for the breakdown sites with respect to cation vacancy diffusivity.

The analysis outlined above has permitted us to identify factors that make for 'good passivity.' Besides

lowering the total number of potential breakdown sites per unit area of the surface, the parameter that may be manipulated to impact the susceptibility of a passive film to chemically-induced breakdown is the cation vacancy diffusivity. Thus, a decrease in the cation vacancy diffusivity results in an increase in the 'pitting potential' (i.e. $\overline{V}_c$), because a higher voltage is required to produce the same flux of cation vacancies across the barrier layer. Note that metals that have inherently low cation vacancy diffusivities (e.g. Ti, Zr, Ta) are quite resistant to pitting.

3. The Theory of Alloys

The development of a successful theory for the effects of alloying on corrosion resistance would have an enormous impact on how alloys are designed, particularly if the theory is quantitative and, hence, deterministic. Significant progress has been made towards that goal with the development of the Solute Vacancy Interaction Model (SVIM) by Urquidi-Macdonald and Macdonald [1, 13] some years ago. Although this model has now been extended to account for the effects of alloying elements on the distributions in V_c and t_{ind}, it is currently limited to dilute binary alloys. Nevertheless, the SVIM has led to the derivation (to our knowledge) of the first theoretically-inspired set of rules for choosing an alloying element, as discussed later in this paper.

3.1 Segregation of alloying elements

In the analysis that follows, we assume that the alloying element is uniformly distributed throughout the metal phase. Accordingly, the reactions that occur within the metal/solution interphase (as depicted in

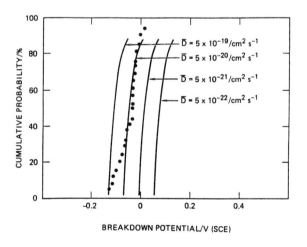

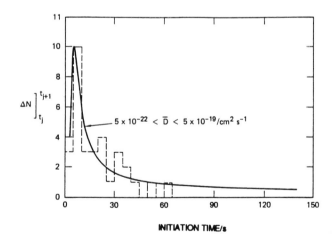

Fig. 5 *Cumulative probabilities for the breakdown voltage as a function of $\overline{D}$ for normal distributions in the diffusivity D.*

σ_D=0.75. *Data for Fe–17Cr in 3.5% NaCl solution at 30°C from Shitbata [19, 20] $\overline{V}_c$ = –0.046V(SCE).*

Fig. 6 *Differential cumulative probabilities for the initiation time as a function of $\overline{D}$ for normal distributions in D.*

σ_D = 0.75. (—)Data for Fe–17Cr in 3.5% NaCl solution at 30°C from Shitbata [19, 20] $\overline{V}_c$ = –0.046V(SCE), V = 0.020 V(SCE), $\bar{t}$ = 7.5 s, t = 0.

Figs 1–3) lead to non-uniform distributions of the elements within the metal (but close to the interface), the barrier layer, and perhaps even in the precipitated outer layer normal to (but not laterally across) the surface. That alloying elements are segregated into the barrier layer (but not necessarily into the upper, precipitated layer) on some alloy systems is shown unequivocally by the SALI (Surface Analysis by Laser Ionisation) data presented in Fig. 7 [21]. These data show that, for a series of Ni–A alloys, with A = Al, Ti, and Mo, the extent of segregation of the alloying element into the barrier layer increases with increasing charge on the solute (i.e. Ni–Al < Ni–Ti < Ni–Mo, for which the solutes may be represented as $Al_{Ni}^{\bullet}$, $Ti_{Ni}^{2\bullet}$ and). Furthermore, it is evident from the existence of the diffusion gradient of the alloying element in the alloy phase that segregation occurs via a solid state reaction at the metal/film interface, as discussed below. The greater segregation of more highly charged solutes into the barrier layer can be explained in terms of the more favourable free energy of these species in the high dielectric film (for example, experimental measurements indicate that the dielectric constant for the passive film on Cr is ≈ 56 whereas that on Type 304SS is 68–107, see citations in Ref. [11]), compared with that for a less highly charged species.

Extensive work on the segregation of chromium into the passive films that form on Fe/Cr alloys in acidic and alkaline solutions has been reported by Strehblow and co-workers [22–24] and by others [25–27]. Using a combination of XPS (X-ray Photoelectron Spectroscopy) and ISS (Ion Scattering Spectroscopy), Strehblow *et. al.* [22–24] found that the inner ('barrier') layers of the passive films formed on Fe–XCr (X = 5–

20%) were substantially enriched in chromium, whereas the outer layers were enriched in Fe(III). For passive films formed under identical conditions (mechanical polishing followed by passivation in 0.5M H_2SO_4 for 3 h at 0.9 V vs SHE) the segregation factor decreases from ~14 for the Fe–5Cr alloy to ~8 for Fe–20Cr, as calculated from the data given in Fig. 3 of Ref. [30]. Furthermore, for passive films formed for 5 min on Fe–15Cr in 1M NaOH, the segregation factor was found to decrease slightly with increasing formation voltage, from ~3 at –0.86V (SHE) to ~2.2 at 0.34V (SHE). A similar trend with formation voltage was found in our work [21] for the Ni–A (A=Al, Ti, Mo) alloys referred to above, although no clear trend in the segregation factor with concentration of A in the alloy could be discerned. A comprehensive theory for the segregation of alloying elements into the barrier layers formed on alloys, on polarisation in aqueous systems, has yet to be developed, so that a quantitative interpretation of the findings discussed above is not possible. In the analysis that follows, we will accept segregation as an experimentally demonstrated phenomenon, with the caveat that it is not yet possible to calculate, on an *a priori* basis, the concentration of alloying element in the barrier layer from the composition of the alloy. Accordingly, all 'compositions' referred to in the remainder of this paper will be those for the barrier layer.

3.2 Solute-vacancy interaction model

Returning now to the role of alloying elements in passivity breakdown, we proposed [12] that the interaction between the substitutionally present (immobile) solute and mobile cation vacancies can be represented as a chemical equilibrium,

$$A_M^{(n-\chi)\bullet} + qV_M^{\chi'} \xleftrightarrow{K_q} \left[A_M \left(V_M^{\chi'} \right)_q \right]^{[n-\chi(1+q)]\bullet} \tag{8}$$

where n is the oxidation charge of the solute (e.g. +6 for Mo^{6+}), and K_q is the equilibrium constant. For example, if M is nickel, A is molybdenum, and q = 1, the association reaction can be written as

$$Mo_{Ni}^{4\bullet} + V_{Ni}^{2'} \Leftrightarrow (Mo_{Ni} \cdot V_{Ni})^{2\bullet} \tag{9}$$

Clearly, further complexing could occur

$$(Mo_{Ni} \cdot V_{Ni})^{2\bullet} + V_{Ni}^{2'} \Leftrightarrow \left[Mo_{Ni} \cdot (V_{Ni})_2 \right] \tag{10}$$

to form a neutral species. Evidently, not only is the strength of the electrostatic interaction between the solute and a mobile cation vacancy greater for more highly charged solutes, but the total number of vacancies that may be complexed is also greater. It is clear, then, that the effect of complexing is to decrease the diffusivity and concentration of mobile cation vacan-

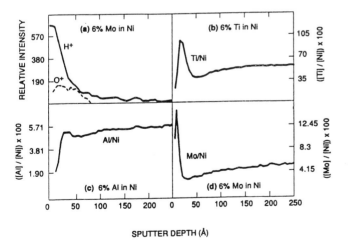

Fig. 7 Concentration profiles of H, O, Al, Ti, and Mo in passive films formed on Ni-6% Al, Ni-6% Ti, and Ni-6% Mo in 0.1N H_3PO_4/NaOH, pH = 12, at 25°C as determined by surface analysis by laser ionisation. V = 0.30V vs SCE, growth time = 12h [21].

cies in the film, all other factors remaining the same. For 1:1 complexes, Urquidi-Macdonald and Macdonald [11] derived the modified vacancy diffusivity as

$$D^* = \frac{D}{2}\left[1 \pm \frac{(\hat{\alpha} - n_A)}{(\hat{\alpha}^2 - n_A n_V)^{1/2}}\right] \qquad (11)$$

and a modified K^* ($= \varepsilon F/RT$) as

$$K^* = 2K\frac{\left[1 - \hat{\alpha}/n_V \pm (\hat{\alpha}^2 - n_A n_V)^{1/2}/n_V\right]}{\left[1 \pm (\hat{\alpha} - n_A)/(\hat{\alpha}^2 - n_A n_V)^{1/2}\right]} \qquad (12)$$

where $\hat{\alpha} = (n_A + n_V + K_1^{-1})/2$, and n_A and n_v are the stoichiometric concentrations of the solute and cation vacancies in the film. By applying ion-pairing theory, as used in solution theory and in solid state physics, we can express the equilibrium constant, K_1, as

$$K_1 = \left[4\pi(t/kT)^3 \int_a^{-t/2kT} e^{\gamma}\gamma^2 d\gamma\right] \qquad (13)$$

where a is the distance of closest approach, $t = z_1 z_2 e^2 / \hat{\varepsilon}\hat{\varepsilon}_o$, z_1 and z_2 are the charges (including signs) on the interacting species, e is the electron charge, $\hat{\varepsilon}$ is the dielectric constant, $\hat{\varepsilon}_o$ is the permitivity of free space, and γ is the variable of integration. The equilibrium constant, K_1, needs to be corrected for screening by the mobile vacancies. This correction is expressed through Debye-Huckel theory as [12]

$$K_1^{corr} = K_1 \ f_A \ f_M \qquad (14)$$

where f_A and f_M are activity coefficients given by

$$f_A = \exp\left\{-\frac{e^2 z_2^2}{2\hat{\varepsilon}\hat{\varepsilon}_o T\ell_D(1+b'/\ell_D)}\right\} \qquad (15)$$

and

$$f_M = \exp\left\{-\frac{e^2 z_1^2}{2\hat{\varepsilon}\hat{\varepsilon}_o kT\ell_D(1+b'/\ell_D)}\right\} \qquad (16)$$

where ℓ_D is the Debye length

$$\ell_D = \left\{\frac{\hat{\varepsilon}\hat{\varepsilon}_o kT}{e^2 4\pi\left[n_V z_1^2 + n_A z_2^2\right]}\right\}^{1/2} \qquad (17)$$

and b' is the distance at which the coulombic interaction energy is equal to kT (the thermal energy)

$$b' = -\frac{z_1 z_2 e^2}{2\hat{\varepsilon}\hat{\varepsilon}_o kT} \qquad (18)$$

Although the SVIM is currently quite crude, in that it does not consider the complexing of more than one cation vacancy per solute and does not employ exponential distributions of vacancies across the film, the model is surprisingly successful in accounting for the effect of molybdenum, for example, on the pitting characteristics of stainless steel. Thus, by combining the SVLM with the distribution functions for passivity breakdown, we calculated P(Vc) as a function of molybdenum concentration in the alloy, assuming that the solute is in the +6 oxidation state (we have also considered the +4 state, but it will not be discussed extensively here), and that preferential segregation of Mo into the barrier layer did not occur (i.e. the concentration of Mo in the barrier layer was assumed to be the same as that in the base alloy). The distribution functions are shown in Fig. 8. In deriving these data, we selected model parameters so that the molybdenum-free case coincided with the experimental data of Shibata [19, 20] for Fe–17Cr in 3.5% NaCl at 30°C; these parameters were then maintained constant for all molybdenum containing alloys. Accordingly, as far as the latter are concerned, there are no arbitrarily adjustable parameters in the model.

The P(Vc) data plotted in Fig. 8 predict that small additions of molybdenum (e.g. 1%) have only a modest impact on the pitting characteristics of stainless steel, but that additions of greater than 2% have a large impact. However, additions of more than 3% have incrementally smaller effects, at least for the parameter values chosen for these calculations. In this regard, it is interesting to note that experience has shown that 2–2.5% Mo is optimal for protecting Type 304SS against pitting in seawater systems with the modified alloy being the well-known Type 316SS. Perhaps a better test of the SVIM is afforded by the data plotted in Fig. 9, where a comparison of (breakdown voltage at the

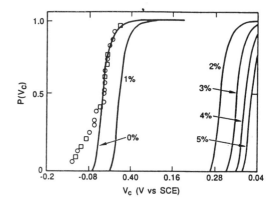

Fig. 8 Effect of solute (Mo) concentration (n_{Mo} wt%) on the cumulative distribution function for V_c for a passive film containing 6^+–3^- complexes. $n_V = 53 10^{20} cm^{-3}$, $K_1 = 1.133 10^{-16} cm^3$.
○, □ Data of Shibata [19, 20] for Fe–17Cr in 3.5% NaCl at 30°C.

50th percentile) is made with experimental data from the literature. Although the experimental data do not display the sigmoid shape predicted by theory (and probably could not because of their limited precision), the agreement between experiment and theory is surprisingly good. Also shown is the prediction of the SVIM assuming that molybdenum is in the +4 state; clearly, the former (Mo6+) provides a better description of the experimental data than does the latter (Mo4+), although the actual oxidation state of molybdenum in the barrier layer on stainless steel has not been established unequivocally. The major impact of shifting $P(V_c)$ in the positive direction is to greatly increase the initiation time for passivity breakdown (i.e. the time required to accumulate a critical concentration of cation vacancies at the metal/film interface). Although we do not show the calculations here, this is precisely what is predicted by the model.

Other data also support the predictions of the model. For example, various studies [3–5] on supersaturated aluminium alloys of the type Al-A (A = 0–8 at.% Mo, Cr, Ta, W) have shown that elements such as molybdenum and, in particular, tungsten, can displace the critical potential for pitting in chloride solution in the positive direction by as much as 2500mV [4]. Both elements form species in the +6 oxidation state, which should complex mobile vacancies, although the extensive segregation of tungsten into the barrier layer has not been demonstrated except in acidic solutions. However, in the presence of a thick outer layer (relative to the barrier layer), even when

using a glancing radiation technique (e.g. EXAFS), as employed in those studies [3–5], it is difficult to establish the extent of segregation into the barrier layer alone, and it is likely that techniques with much higher depth resolution, such as SALI, will be required to fully characterise the composition of the interfacial region.

4. Alloy Design

What makes a good alloy? An answer to this question is of enormous scientific, technological, and economic importance, given that the annual cost of corrosion in any industrial society is *ca.* 4.5% of the GNP (about $230 billion for the US in 1990). The work outlined above provides clear guidance on this question and, recognising that the models are still quite crude, we have formulated a set of principles to aid the alloy designer in devising new systems of superior resistance to passivity breakdown. The rules are as follows:

(i) The alloying element must segregate into the barrier layer, preferably preferentially with respect to the host cation.

(ii) The alloying element must exist in a 'dissolved' (substitutional) state in the barrier layer in as high an oxidation state as possible, and certainly in an oxidation state that is higher than that of the host cation.

(iii) The alloying element should be uniformly distributed throughout the layer or at least should not form a second phase that might introduce heterogeneities into the barrier layer that could act as sites for the nucleation of localised attack.

Other factors also affect the theoretical effectiveness of an active alloying element, according to the SVIM. For example, a decrease in the dielectric constant (Fig. 10) and a decrease in the distance of the closest approach (eqn (13)) will both shift reaction (8) to the right, signifying stronger interaction between the ionised alloying element and the mobile cation vacancies. However, neither of these parameters are readily manipulated, so that their optimisation is not included in the rules outlined above.

We have tested these principles by measuring distribution functions, $P(t_{ind})$, for a series of Ni–xAl, Ni–xTi, and Ni–xMo alloys (x_o= 0–8 at. %) in NaCl/ borate buffer solutions at 25°C [29]. In these experiments, the specimens were held at a constant potential above the 'pitting potential,' and the number of pits nucleated on the surface were counted as a function of time. Although these experiments are complicated by the observation that existing active pits protect the remaining surface at radii that increase with time (and

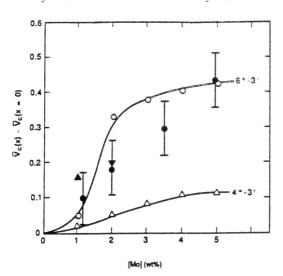

Fig. 9 Effect of molybendum concentration on $V_c(x)$–$V_c(x = 0)$ for 6^+–3^- (○) and 4^+–3^- (△) complexes in the film.
$n_V = 5310^{20} cm^{-3}$, $K_1 = 9.05310^{-21} cm^{-3}$ (D).
● *Lislovs and Bond [28]: Fe–18Cr in 1M NaCl at 25°C.*
▲ *Shibata [19]: Fe–17Cr in 3.5% NaCl at 30°C.*
▼ *Shibata [20]: Fe–18Cr in 3.5% NaCl at 30°C.*

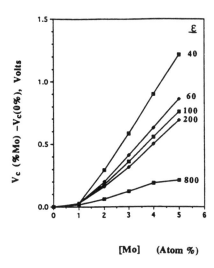

Fig. 10 Calculated shift in the mean critical breakdown potential with molybdenum concentration in the barrier layer as a function of the dielectric constant.

[$\Omega = 30$ cm^3/mol, $\Delta G_{A-1} = -4 \times 10^4$ J/mol, $\phi^o_{f/s} = -0.5$V, $\Delta G^o_s = -4 \times 10^4$J/mol, $\tau = 0$, $D = 5 \times 10^{-20}$cm^2/s, $n_v = 5 \times 10^{20}$cm^{-3}, $\varepsilon = 1.1 \times 10^6$V/cm, T = 303.15K, $\alpha = 0.65$, $\beta = -0.01$, pH = 7, $a_{x^-} = 0.402$, $\xi = 1.0 \times 10^{16}$cm^2, $J_m = 1.587 \times 10^{13}$/cm^2.s, $N_o = 5000$cm^{-2}].

for this reason they have not yet been published), the data are in good qualitative agreement with the predictions of the model, in that the Ni-Mo alloys were the most resistant and the NiAl alloys were the least resistant to passivity breakdown, with the effect being roughly as expected from the charge on the solute (i.e. 4: 2: 1 for).

5. Prediction of Corrosion Damage Functions

A principal goal in corrosion science and engineering is the prediction of damage functions for various forms of corrosion as a function of exposure time. If this goal can be realised, a rational deterministic basis might be established for estimating component lifetimes in industrial systems. Of great importance is the damage function for pitting corrosion, examples of which are shown in Fig. 11 for Type 403SS before and after being buried in the ground for five years [30]. Because this form of attack frequently leads to failure without significant outward signs of corrosion, the scheduling of preventative maintenance in industrial systems prone to this form of damage is an extremely difficult task. The point defect and solute/vacancy interaction models may provide the necessary deterministic basis for predicting localised corrosion damage functions, and

hence for scheduling maintenance, as outlined below.

Consider the cumulative probability function for the initiation time (integral of eqn (5)), as shown schematically in Fig. 12(a). For an increment $\Delta t = t_{j+1} - t_j$, a total of ΔN] pits nucleate and grow for a period u before being observed at t_{obs}. If the rate of growth during this period is r(t), then these pits will grow to a depth of

$$h = \int_o^u r(t)dt \qquad (19)$$

Thus, by carrying out this calculation for all time increments $(t_{j+1} - t_j)$, we are able to construct the damage function for the time of observation, t_{obs} as shown in Fig. 12(b). The observation time is now changed and the procedure outlined above is repeated to generate a family of damage functions corresponding to different times of observation in the future. Note that, in the analysis outlined above, we have not allowed for the repassivation ('death') of localised corrosion events; however, this is easily done by modifying the cumulative probability function to allow for the possibility that a pit may repassivate at some time after nucleation.

We have developed the algorithm outlined above to estimate damage functions for pitting corrosion on condensing heat exchangers in domestic and gas-fired furnaces and for stress corrosion cracking in water cooled nuclear reactors. In each case, we employ deterministic models to estimate pit and crack growth rates, respectively; these models are based on either Beck and Alkire's [31] diffusion model (for pit growth) or on the Coupled Environment Fracture Model (CEFM) for crack growth in sensitised Type 304SS in light water reactor heat transport environments [32]. Determin-

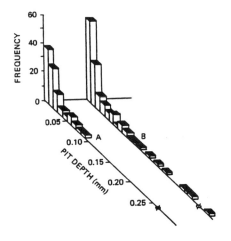

Fig. 11 Pit depth distributions for Type 403 stainless steel (After Ishikawa et al. [30])
A-Initial distribution
B-After five years.

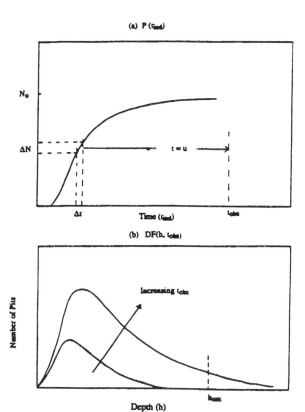

Fig. 12. Schematic of alogrithm for calculating localised corrosion damage function $DF(h, t_{obs})$ from the cumulative probability function for nucleation, $P(t_{ind})$.

ism is assured in the CEFM through charge conservation, by requiring that the net positive current exiting the crack (or pit) be consumed quantitatively on the external surfaces by appropriate cathodic reactions, such as oxygen reduction.

To illustrate how alloying is expected to affect the damage functions for pitting attack, we show the calculations presented in Fig. 13. These damage functions were calculated for three alloying elements of different oxidation states segregated into the barrier layer formed on a M(II) substrate (e.g. on nickel). In all cases, the pit growth rates were estimated using the diffusion model of Beck and Alkire [31], which does not specifically recognise alloying effects. Accordingly, no credence should be given to the depth of attack. Our purpose in presenting these calculations, however, is to illustrate the significant effects that the alloying elements, particularly those that form high oxidation states relative to that of the host cation, are predicted to have on the probability of attack. Thus, for the trivalent alloying element (e.g. Al) the effect of complexing between the solute in the barrier layer and the mobile cation vacancies is expected to be slight, even at solute concentrations in the barrier layer of up to 5 a/o (Fig. 13(a)). However, as the solute charge is increased from 4 (Fig. 13(b)) to 6 (Fig. 13(c)) the effect becomes profound, such that in the latter case essentially no break-

down events are predicted to occur at solute concentrations of 4 a/o and above.

6. Future Development of the Models

As noted earlier in this paper, the SVIM, and to a lesser extent the PDM, are still quite crude, although both have provided good semiquantitative explanations for passivity breakdown and of the role(s) of alloying elements therein. Indeed, the SVIM inspired the design of the Ni-A (A = Al, Ti, Mo) alloys, referred to earlier in this paper, which were found to have pitting resistances that correlate with the oxidation state of the solute as predicted from theory. However, in deriving the SVIM we made a number of simplifying assumptions that need to be addressed in any subsequent work in this area. The most important assumption is that the alloying element and the cation vacancies are uniformly distributed across the barrier layer. Experiment (e.g. Fig. 10) shows that the alloying element is *not* of uniform concentration in the barrier layer, and good theoretical reasons exist for expecting the concentration of cation vacancies to decrease exponentially from the barrier layer/outer layer interface to the alloy/barrier layer interface [8]. Clearly, then, more realistic distribution functions, which are derived from the physics of the problem, need to be employed in describing the interaction between the solute and cation vacancies as a function of distance through the barrier layer.

The second major shortcoming of the present models is that we lack a sound theory for calculating the extent of segregation of an alloying element into a barrier layer. The simple equilibrium model previously proposed [21] has proven to be inadequate, no doubt because the metal/solution interphase on an active metal is not an equilibrium system. Experimentally, we find segregation factors (S) ranging from more than ten to about one, depending on the solute identity, the applied voltage, the concentration of the solute in the alloy phase, and the pH. The observed segregation factors appear to imply the dominance of kinetic effects so that any accurate model for the interphase system should involve kinetic (rather than equilibrium) principles.

At this time, the SVIM has been developed to describe solute-vacancy interactions in the barrier layers on binary alloys. However, many industrially-important alloys are ternary, quaternary, or even more complex systems, so that the model needs to be extended to describe interactions in multicomponent (cation) systems. Accounting for synergistic effects, such as that which exists between chromium and molybdenum in austenitic stainless steels, is of particular importance.

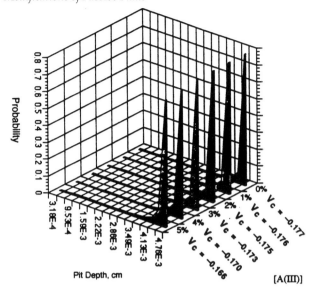

(a) For trivalent alloying element

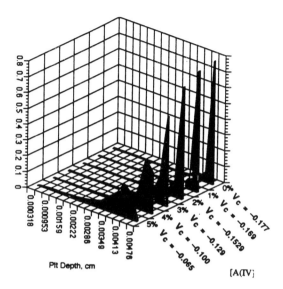

(b) For tetravalent alloying element

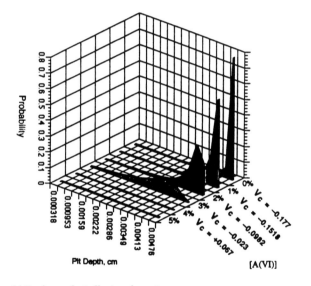

(c) For hexavalent alloying element.

Fig. 13 Calculation damage function for alloys M(II) –xA, where z=concentration (a/o) of the alloying element in the barrier layer, as a function of the oxidation state of A in the film. The parameters used in the calculations are as given in the caption to Fig. 10. Time of observation = 0.63 min. Pit growth rates were calculated using the model of Beck and Alkire [31]. $V_{apl} = 0.15V$.

7. Acknowledgement

The authors gratefully acknowledge the support of this work by the U.S. Department of Energy, Division of Basic Energy Sciences, under Grant No. DE-DG03-84ER45 164.

The method used for estimating the solute/vacancy pairing equilibrium constant is also quite crude, in that it does not take into account vacancy clustering ($q > 1$) around highly charged solutes. While a treatment of this phenomenon is not excluded in the present model, the computational difficulties are formidable. Furthermore, eqn (12) implies that segregation will modify the electric field in the barrier layer. The present treatment attributes this to a solute charge vs pairing phenomenon, rather than by properly incorporating the roles of all charge carriers, in particular highly mobile electrons and holes. Finally, while all of the modifications outlined above are necessary, we regard them to be refinements in nature, and are not expected to change the rules that we have derived for selecting the preferred alloying element.

References

1. D. D. Macdonald, J. Electrochem. Soc., **139**, 3434, 1992.
2. A. J. Sedriks, Corrosion of Stainless Steels, The Electrochemical Soc., Princeton, NJ, 1979.
3. B. A. Shaw, G.D. Davis, T. L. Fritz, B. J. Rees and W. C. Moshier, J. Electrochem. Soc., **138**, 3288, 1991.
4. G. D. Davis, W. C. Moshier, G. G. Long and D. R. Black, J. Electrochem. Soc., **138**, 3194, 1991.
5. P. M. Natishan, E. McCafferty and G. K. Hubler, J. Electrochem. Soc., **133**, 1061 (1986); **135**, 321, 1988.
6. C. -Y. Chao, L. -F. Lin and D. D. Macdonald, J. Electrochem. Soc., **128**, 1187 (1981); **128**, 11 (1981); **129**, 1874, 1982.
7. D. D. Macdonald and M. Urquidi-Macdonald, J. Electrochem. Soc., **137**, 2395 (1990).
8. D. D. Macdonald, S. Biaggio and H. Song, J. Electrochem. Soc., **139**, 170, 1991.
9. D. D. Macdonald and S. I. Smedley, Electrochim. Acta., **35**, 1949, 1990.
10. D. D. Macdonald and M. Urquidi-Macdonald, Electrochim. Acta., **31**, 1070, 1986.

11. M. Urquidi-Macdonald and D. D. Macdonald, J. Electrochem. Soc., **134**, 41, 1987.

12. M. Urquidi-Macdonald and D. D. Macdonald, J. Electrochem. Soc., **136**, 961, 1989.

13. D. D. Macdonald and M. Urquidi, J. Electrochemical Soc., **132**, 555, 1985.

14. S. Lenhart, M. Urquidi-Macdonald and D. D. Macdonald, Electrochim. Acta, **32**, 1739, 1987.

15. G. C. Wood and J. P. O'Sullivan, J. Electrochem. Soc., **116**, 1351, 1969.

16. 'Passivity and Its Breakdown on Iron and Iron-Based Alloys,' R. W. Staehle and H. Okada (Eds), NACE, Houston, TX, 1976.

17. 'Passivity of Metals and Semiconductors,' M. Fromout (Ed), Elsevier, Amsterdam, 1983.

18. 'Passivity of Metals,' R. P. Frankenthal and J. Kruger (Ed), the Electrochemical Society, Princeton, NJ, 1978.

19. T. Shibata, Trans. ISIJ, **23**, 785, 1983.

20. T. Shibata and T. Takeyama, Corrosion, **33**, 243, 1977.

21. D. D. Macdonald, M. Ben-Hain and J. Pallix, J. Electrochem. Soc., **136**, 3269, 1989.

22. C. Calinski and H.-H. Strehblow, J. Electrochem. Soc., **136**, 1328, 1989.

23. S. Haupt and H.-H. Strehblow, Corros. Sci., **29**, 163, 1989.

24. S. Haupt, C. Calinski, U. Collisi, H.W. Hoppe, H.-D. Speckmann and H.-H. Strehblow, Surf. Interf. Anal., **9**, 357, 1986.

25. R. P. Frankenthal and D. L. Malm, J. Electrochem. Soc., **123**, 186, 1976.

26. S. Storp and R. Holm, Surf. Sci., **68**, 10, 1977.

27. K. Asami, K. Hashimoto and S. Shimodaira, Corr. Sci., **16**, 387, 1976.

28. E. A. Lizlovs and A. P. Bond, J. Electrochem. Soc., **122**, 720 1975.

29. D. D. Macdonald, C. English and S. J. Lenhart, unpublished data, 1985.

30. Y. Ishikawa, T. Ozaki, N. Hosaka and O. Nishida, Trans. ISIG, **22**, 977, 1982.

31. T. R. Beck and R. C. Alkire, J. Electrochem. Soc., **126**, 1662 1979.

32. D. D. Macdonald and M. Urquidi-Macdonald, Corros. Sci., **32**, 51, 1991.

47

In Situ AFM Investigation of Pitting

G. GUGLER, J. D. NEUVECELLE, P. METTRAUX, E. ROSSET AND D. LANDOLT

Laboratoire de métallurgie chimique, Departement des matériaux, Ecole polytechnique fédérale de Lausanne (EPFL), MX-C Ecublens, 1015 Lausanne, Switzerland

Abstract

An electrochemical cell has been built for *in situ* studies with the Atomic Force Microscope (AFM). The technique has been applied to the characterisation of the surface of stainless steel and to the study of pit formation.

1. Introduction

Pit nucleation occurs when a passive metal exposed to an aggressive environment such as chloride solution is polarised above the critical pitting potential. The breakdown mechanism of passive films under these conditions has been studied by numerous authors but is not completely understood [1]. The presence of non metallic inclusions, in particular sulphides, favours pitting while other types of inclusions are expected to be less damaging [2]. To study the pit initiation mechanism and the role of different types of inclusions, *in situ* techniques are needed which permit the characterisation of surface topography on a local scale.

Scanning probe techniques derived from the STM [3] are finding increasing applications in all aspects of surface science including electrochemistry [4]. The atomic force microscope (AFM) invented by Binnig, Quate and Gerber [5] permits one to investigate surface topography with submicroscopic or even atomic resolution. The AFM is based on the measurement of a force rather than a current and therefore can be used on conducting as well as non conducting materials. Furthermore, it is particularly well suited for *in situ* studies of electrode surfaces at controlled potential. The present paper describes an electrochemical cell for *in situ* AFM studies of electrodes and it presents preliminary data obtained on stainless steel exposed at constant potential to a chloride electrolyte.

2. Experimental Arrangement

A commercial AFM instrument (Park) equipped with both a $10 \times 10 \mu m$ and $130 \times 130 \mu m$ scanner was used. In this instrument the probe displacement is measured by the deflection of a laser beam (Fig. 1(a)). The probe tip was made of silicon nitride and the opening angle was $90°$. The measurements are made in the repulsive force mode. The scanning frequency is of the order of 1 Hz.

The electrochemical cell built in the authors' laboratory is schematically shown in Fig. 1(b). A 27 mm diameter plexiglas dish of 6 mm internal wall height contains the electrolyte. A metallic holder glued into the dish bottom fits onto the AFM sample stage. A stainless steel sample of typically 3×3 mm size cast in epoxy serves as anode. Contact to the metallic sample holder is made with conducting resin. A platinum–10% rhodium wire counter electrode is placed concentric at the wall of the dish. A glass capillary traverses the wall and is connected to a conventional calomel reference electrode by means of tygon tubing. During the scanning probe experiments the electrolyte is covered flush by an optical glass plate. In this way disturbances in the measurement of the probe displacement by laser deflection can be avoided. The whole cell assembly is placed underneath a metallic cover to avoid disturbance by air drafts.

Two types of stainless steels were used, a titanium stabilised Fe–17Cr ferritic alloy containing a considerable number of titanium nitride and titanium carbide inclusions and a commercial austenitic steel (AISI 303, DIN Nr 1.4305) corresponding to the composition (in wt%): 17-19 Cr, 8–10 Ni, <0.12 C, <0.1 Si, < 2.0 Mn, < 0.060 P, 0.15–0.35 S. Before an experiment the steel samples were mechanically polished in successive steps to a 1 μm diamond powder finish. Unless otherwise specified, a 0.5 M NaCl solution at room temperature and in contact with air was employed. The cell potential was controlled by a potentiostat (Amel 551).

3. Results

Imaging in air and in water is compared in Fig. 2. It shows a TiN inclusion (identified by microprobe analy-

(a) Electrochemical AFM

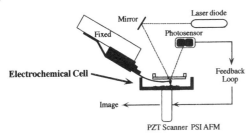

(b) Electrochemical Cell Design

Top View

A-A View

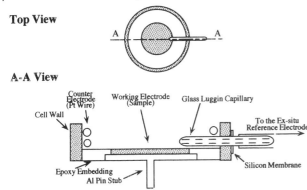

Fig. 1 Schematic view of (a) AFM arrangement and (b) of the electrochemical cell for in situ *studies of pitting.*

sis) on the surface of the Fe17Cr alloy, made visible by a preceding attack in oxalic acid. Picture (a) was obtained in air, picture (b) during immersion in water. The quality of the image of the immersed sample is comparable or even better. Submicron resolution is attainable in both cases.

Figure 3 illustrates pit formation on the Fe17Cr alloy, near a handle shaped inclusion, probably titanium carbide. The picture shown in Fig. 3(a) was made on a surface immersed in the NaCl electrolyte at open circuit (−240 mV(SCE)). The sample was then polarised anodically to 650 mV(sce) for 1 s by application of a potential pulse. Following this the surface was scanned again without removing the sample from the solution. Figure 3(b) shows that a pit has nucleated near the inclusion which itself remains untouched. The pit is too deep for the needle to reach its bottom. Therefore on the photograph one observes just a black spot.

Figure 4 illustrates the appearance of pits represented in 2D or 3D images. It represents a pitted surface of the AISI 303 steel observed *in situ* in 0.5 M NaCl. The pits on 2D images appear as dark spots. This mode of imaging, therefore, is well suited for the study of the evolution of the number of pits with time. The 3D

images on the other hand provide better topographical information. These are most suited for the qualitative appreciation of the shape evolution of surfaces during corrosion. For a more quantitative evaluation of the growth of individual pits, line scans can be used advantageously. This is demonstrated by Fig. 5 for the two pits marked A and B respectively in Fig. 4. The scans shown were obtained from the same data used for construction of 2D and 3D images. In these experiments the surface was exposed to 0.5M NaCl at constant potential for a given time, then the potential was increased and the same procedure was repeated. Every few minutes the surface was scanned *in situ* without interrupting the polarisation (the time needed for a complete surface scan was 5 min typically). The scans shown in the figure illustrate the fact that the two pits monitored increased in size during exposure. In principle such data permit one to study the process of pit growth in a quantitative way. No such attempts was made here because insufficient data are available at this time. It may also be mentioned that the study of pit growth by AFM is feasible only as long as the inclination of the sidewalls of the pits is well below 90∞ in order to permit the needle to image them. Undercutting therefore can not be studied with the AFM technique described here.

4. Conclusions

The present preliminary study shows that *in situ* AFM is an interesting method for the investigation of pit nucleation and growth. The spatial resolution is clearly better than that obtained with the optical microscope, a classical *in situ* technique for the observation of local events. Furthermore, the information obtained with the AFM relying on a scanning probe is easily quantified. Because the acquired probe displacement data are stored in digital form a number of different representations can be used by the researcher depending on the type of problem he wants to investigate.

In the present paper three applications were demonstrated using a ferritic Fe–Cr alloy and an austenitic commercial stainless steel: observation of pit formation at an inclusion using 3 D imaging, observation of the increase of the number of pit nuclei with time using 2D imaging, observation of the growth of individual pits using line scan imaging.

5. Acknowledgement

Financial support by Fonds national suisse is gratefully acknowledged.

(a) In air

(b) In water

Fig. 2 A TiN inclusion on an Fe17Cr alloy surface imaged by (a) AFM in air and (b) in distilled water.

(a) Before Pitting

(b) After Pitting

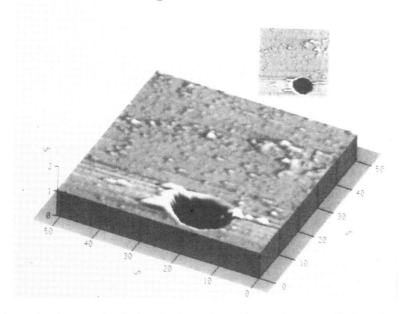

Fig. 3 A Fe17Cr stainless steel surface near a handle shaped inclusion observed by AFM in 0.5M NaCl. Picture (a): surface at the corrosion potential of – 240 mV (SCE). Picture (b): surface after application of a potential pulse of 1 s to 650 mV to initiate pitting, then anodically polarised at 150 mV (SCE). The pitting potential is approx. 350 mV.

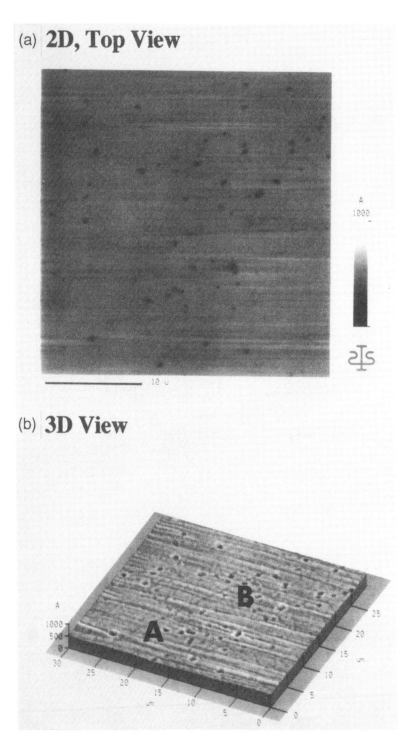

(a) 2D, Top View

(b) 3D View

Fig. 4 2D and 3D representation of a pitted surface of Type 303 stainless steel exposed to 0.5M NaCl under anodic polarisation. The pits labelled A and B are represented in Fig. 5.

Fig.5 (opposite) In situ *study of pit shape evolution as a function of time at constant potential. The figure shows line scans of pits A and B (Fig. 4) after exposure of a Type 303 surface to 0.5M NaCl at the indicated potentials and times.*

References

1. H. H. Strehblow, Surf. Interf. Anal., **12**, 363, 1988.
2. S. Smialowska, Pitting Corrosion of Metals, NACE, Houston, 1986, 486pp.
3. G. Binnig and H. Rohrer, Helv. Phys. Acta, **55**, 726, 1982.
4. H. Siegenthaler, R. Christoph and R. J. Behm Editor, Scanning Tunnelling Microscopy and Related Methods, Kluwer Academic Publishers 1990, p.315.
5. G. Binnig, C. F. Quate and Ch. Gerber, Phys. Rev. Letters,**56**, 930, 1986.

f(Time)

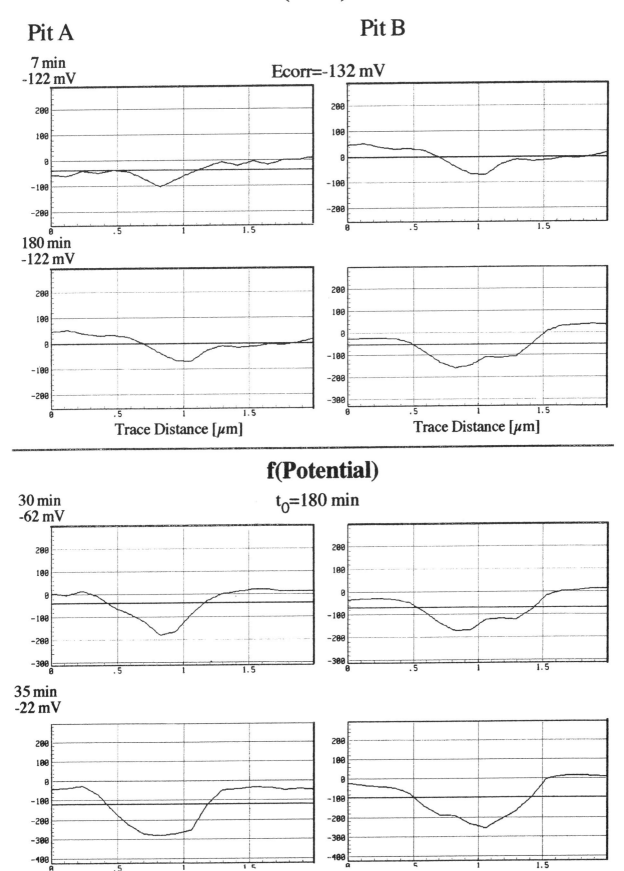

Pit A

7 min
-122 mV

180 min
-122 mV

Trace Distance [μm]

Pit B

Ecorr=-132 mV

Trace Distance [μm]

f(Potential)

t_0=180 min

30 min
-62 mV

35 min
-22 mV

48

Passivity of Non-metallic Inclusions in Stainless Steels and their Effect on Resistance Against Pitting

A. Szummer , M. Janik-Czachor* and S. Hofmann†

Department of Materials Science and Engineering, Warsaw, University of Technology, Warsaw, Poland
*Institute of Physical Chemistry, PAS Warsaw, Poland
†Max-Planck Institut für Metallforschung, Stuttgart, Germany

Abstract

Non-metallic inclusions are known for their detrimental effect on resistance of stainless steels to localised corrosion. In this work the composition of sulphide inclusion in 18Cr–14Ni stainless steel has been modified by varying the level of supplementary alloying elements (Ti,Mo) as well as technological additives and impurities (Mn,S) during the process of steel making. Electrochemical, microscopic (optical, SEM, X-ray electron microprobe) and surface analytical (AES–SAM) investigations revealed a correlation between both the composition of passive film formed at the inclusions and at the adjacent matrix, and the stability of the passive state of this steel in neutral chloride containing solution. In particular SAM revealed that MnS_x inclusions in the material with a common Mn level do not undergo any passivation, thus exposing themselves directly to the aggressive environment and producing large discontinuities within the passivating film at the inclusions themselves. This, in various ways, may facilitate the initiation of pitting.

1. Introduction

Manganese sulphide inclusions in steels are known for their influence on various properties of the steels matrix [1–3] including its resistance to pitting corrosion [4–17]. As a consequence, reducing Mn content in steels has proven to be a successful strategy for increasing pitting resistance of steels [9–11]. In this paper we report the corrosion behaviour of recently developed low Mn steels, with low or none Mo ($\leqslant$Mo), in Cl$^-$ containing solutions. The present results suggest that, at low Mn content, lowering the Mo content in the matrix does not adversely affect stability of its passive state in Cl$^-$ containing solutions.

Microscopic and surface analytical measurements aimed at finding a correlation between the anodic behaviour to the steels, the composition of their non-metallic inclusions as well as the local composition of the passive film at the inclusions, as compared to that at the adjacent metal matrix, are also presented.

2. Experimental Method

2.1 Materials

The metallic materials used for these investigations were:

- three commercial steels alloyed with Ti or Mo (Table 1);
- three modified steels with composition basically similar to that of the commercial ones, but, with a low Mn content (0.2% wt) and with a reduced Mo content, from 0.7wt% up to as indicated in Table 2. Supplementary alloying elements added were: Ti to E-1 and Z-7, Mo to Z-7;
- two laboratory heats of known pitting behaviour [6] with low or high Mn and S content, as shown in Table 3. They were chosen for comparative purposes particularly for surface analytical measurements, since they contained in their structure quite large sulphide inclusions (Cr, Mn)Sx or pure MnS_x, respectively.

2.2 Electrochemical measurements

The electrochemical measurements were performed in deaerated solutions: of 0.1 up to 0.5M NaCl, at various temperatures within the region of 22–70°C. For most cases the measurements were carried out at low scan rate of $\frac{dE}{dt} = 1V/h$ and the c.d. was recorded, both in the active and in the passive region. The anode potential was measured vs 'SCE'.

E_{np}–pit nucleation potential was determined, as described elsewhere [6].

For the surface analytical measurements the specimens were prepassivated in borate buffer of 0.11M H_3BO_3 + 0.027M $Na_2O_4B_7$, pH = 8.4, during 1h, at open circuit potential (aerated electrolyte), or at E = –250 mV_{SCE} in deaerated electrolyte. Both procedures resulted in the same passivation behaviour of the inclusions. After passivation the specimens were removed from the electrolyte, rinsed with pure ethanol, dried and stored in small glass containers. The borate buffer

Table 1 Chemical composition of commercial steels (wt%)

Steel	C	Mn	Si	P	S	Cr	Ni	Mo	Ti
(I)OH17N13M2T	0.040	1.40	0.60	0.028	0.015	16.8	12.3	2.6	0.22
(II) OOH17N14M2	0.020	1.91	0.33	0.022	0.012	17.6	14.7	2.8	–
(III)OOH18N10	0.030	1.65	0.75	0.026	0.012	17.6	11.2	–	–

Table 2 Chemical composition of modified low Mn steels (wt%)

Heat	Chemical composition (wt%)						Additional alloying elements
	C	Mn	P	S	Cr	Ni	
E –1	0.03	0.23	0.006	0.015	18.3	13.9	Ti–0.64
R–4	0.03	0.02	0.008	0.010	18.4	13.9	—
Z–7	0.04	0.31	0.006	0.014	18.7	14.6	Ti–0.65; Mo–0.71

Table 3 Chemical composition of laboratory heats (reference steels, wt%)

Heat	C	Mn	Si	S	P	Ni	Cr	Al	N_2	O_2
A	0.005	0.15	0.3	0.12	0.008	14.1	17.9	0.025	0.031	0.012
B	0.005	2.0	0.3	0.24	0.008	14.1	18.1	0.13	0.026	0.003

was used for the prepassivation because pit nucleation potential for a variety of steels in 0.5M NaCl was equal in the buffered and not buffered solution [6]. Moreover, as it has been shown that chloride ions do not affect passive film composition, but rather its thickness [31], investigations of the film formed in the above nonagressive solution seemed reasonable.

2.3 Microscopic examinations

Microscopic examinations were carried out for all the specimens before and after the electrochemical measurements: optical microscopy, SEM and X-ray microprobe analysis were used. Non-metallic inclusions present within the steels under investigation were characterised. Some typical inclusions were marked by microhardness diamond pyramids for further analysis. The X-ray microprobe analysis was carried out in a similar manner as before [6, 11, 20]. At least 5 specimens for each materials were taken for the analysis. Several areas for each sample were investigated and 5–10 typical sulphide inclusions were carefully analysed with X-ray microprobe. The quantitative 'point' and 'line' analyses of the inclusions have been performed at an accelerating voltage of 15 kV, by using JEOL JXA-3A Microprobe, equipped with two wavelength dispersive spectrometers (WDS). Analyses were performed for Mn, S, Cr, Fe,Ni and additionally for Ti and Mo depending on the type of sulphide inclusions. Pure metals and pyrite, for S, were used as standards in X-ray microanalysis. In the case of quantitative 'point' analysis special correction program for 'small particles' considering size and thickness of the inclusions was developed and applied, thus the volume of material analysed was not the 'critical point' in the procedure of determining the inclusions composition.

2.4 SAM analysis

Specimens with large non-metallic inclusions were selected for AES-SAM. They were removed from the electrolyte after prepassivation, rinsed with pure ethanol, dried, and stored in small glass containers. The SAM analysis was performed with a Perkin-Elmer 600 SAM Spectrometer at 10^{-8} Pa, with 15 nA raster electron beam. The lateral resolution was 200 nm. Depth profiling wa carried out with 3 kV Ar^+ ion beam and a sputtering rate of *ca.* 20 Å/min based on Ta_2O_5 standard measurements. For the AES-SAM line scan the signals for various elements were taken as integrated values of the corresponding characteristic peak areas. For the analysis of the $(Cr,Mn)S_x$ inclusions the major problem was identification and proper semi-quantitative evaluation of the Cr and Mn signals. The 529 eV Cr peak and the 589 Mn peak were considered. However, the 571 eV Cr peak partly over lapped with the 589 Mn peak. At the non-differentiated spectrum for the inclusions the 571 eV Cr peak usually appeared at the side of a large 589 Mn peak (or vice versa) and so the Mn signal could be deconvoluted for a purpose of the semi-quantitative analysis.

3. Results and Discussion

3.1 Electrochemical investigations in neutral chloride solutions

Figure 1 shows a diagram comparing E_{np} for various steels in 0.5M NaCl. The low Mn steels E-1, R-4 and Z-1 exhibit distinctly higher E_{np} than the other steels, thus inferring that lowering the Mn content is an efficient strategy for increasing the resistance of the steels against Cl attack. It should be pointed out that

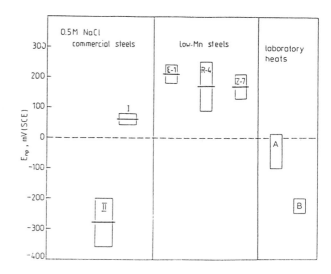

Fig. 1 Diagram summarising pitting resistance (E_{np} in 0.5M NaCl aq) of the commercial steels, novel low Mn modified steel and two special laboratory heats with a common (B) and low Mn content (A).

even the steel R-4, with no Mo or Ti additions, exhibited higher E_{np} than the commercial steels. The special laboratory heats B and A, the latter with low Mn and C content, exhibited a more negative E_{np} than the novel low Mn steels. This points to the role of subtle interplay between various alloying elements in optimising the pitting resistance. Moreover, the effect of the sulphur content and sulphide size (and probably its shape) plays a role here; steel A and B contained a high concentration of S, and rather large inclusions, what apparently increased pitting susceptibility of the material; the effect was pointed out by some other authors already [8,17].

Figure 2 summarises the result obtained by Degerbeck [9] and our results for low Mn steels E-1 and R-4 in terms of E_{np} vs temperature.

In 0.1M NaCl the steel R 442 developed by Degerbeck which contained 0.24% Mn and 2.7% Mo exhibited a much higher E_{np} than AISI 316. Our low Mn steels exhibited lower E_{np} than R 442, still quite close to the values obtained for AISI 316, within a wide temperature range. These results suggest that the low Mn steels are quite resistant.

3.2 Microscopic examinations

Table 4 summarises X-ray microprobe measurements of the compositions of the non-metallic inclusions present in steels under investigation [19, 20]. It should be noted that in the commercial steel (II) Mn rich sulphides were present and in laboratory heat B only MnS_x inclusions occurred. In the low Mn steels, and in heat A mixed $(Cr, Mn)S_x$ or TiS_x inclusions occurred as indicated in Table 4. Large sulphide inclusions occurred in the heats A and B; their diameter, as meas-

Table 4 X-ray electron microprobe characterisation of sulphide inclusions

Steel	Prevailing Inclusions [x]
Commercial (I)	TiS_xN_y (~32%S, ~63%Ti, ~0.4%Mo, ~4.6%N)
Commercial (II)	$(Mn, Cr)S_x$ enriched with Mn
modified E–1	mostly $(Ti)S_xC_yN_z$ (~30%S)
modified R–4	$(Cr, Mn)S_x$(~48%S, ~32%Mn, ~20%Cr)
modified Z–7	$(Ti, Mn) S_xN_y$ (~30%S, ~64%Ti, ~0.4%Mo, ~5, 6%N)
lab. heat A	$(Cr, Mn)S_x$ (~14%Mn, ~38%Cr)
lab. heat B	MnS_x only (~46%S, ~54%Mn)

[x]
The results have been normalised to 100%
The quantitative estimations have been performed by A. Wolowik, M. Sc. (32), see Experimental.

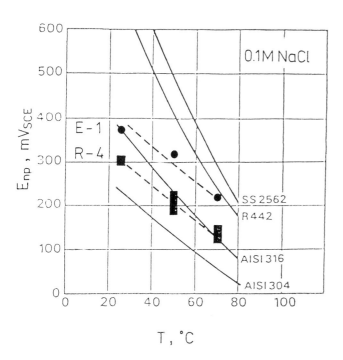

Fig. 2 Effect of temperature on E$_{np}$ for various commercial steels and a low Mn steel R 442 developed by Degerbeck [9], as well as for the novel E–1 and R–4 low Mn steels.

ured with SAM, was 2–5μm. However, some smaller inclusions were also present.

A characteristic feature common to commercial steels was that the Mn content in the matrix adjacent to any sulphide inclusion was relatively high. Figure 3(b) gives a typical example of a line scan across a $(Mn,Cr)S_x$ inclusion in steel (II), demonstrated in Fig. 3(a). The signal for Mn outside the inclusion is relatively high. Moreover, the Cr signal inside the inclusion is rather low as compared to its level outside the inclusion (Fig. 3(b)).

In the low Mn steels R-4 and A some $(CrMn)S_x$ inclusions were typical. Figure 4 shows an example of a line scan across such an inclusion in steel R-4. In this case the signal for Mn outside the inclusion is much lower—by about an order of magnitude—than that observed for the commercial steel (II) discussed above (compare Fig. 3(b)). Moreover, the Cr profile across the $(Cr,Mn)S_x$ inclusion in the steel R-4 is distinctly different from that for the commercial steel (II) in the following aspects:

- the Cr signal inside the inclusion in steel R-4 is much higher than that inside of an analogous inclusion in the commercial steel (II) (Fig. 3(b) and 4). This infers a more stable passivation of these inclusions in the low Mn steel R-4 than in the commercial steel (II);

- for steel R-4 the Cr signal is higher inside the inclusion than outside of it (Fig. 4). This profile is also typical for

283

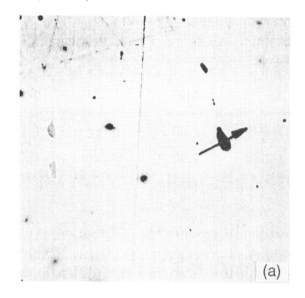

the (Cr,Mn)Sx inclusions in steel A. For the commercial steel (II) the analogous profile was just opposite, as we have seen.

Microscopic examinations, carried out after the corrosion test, suggests that in neutral chloride solutions pits nucleated on sulphide inclusions. No pitting was found away from the sulphide inclusions, at least for all the specimens examined, according to the procedure described in 'Experimental'. Our present observation compares with the previous ones [5, 6, 11, 19, 20, 29, 30], as well as with those reported by other authors [4, 7, 8–10, 12, 14–17]; preferential pit initiation on sulphide inclusions in steels is a rather well known phenomenon.

3.2 Surface analytical measurements

As already mentioned in 'Experimental', rather large inclusions were selected for the more detailed investigations.

Results of surface analytical measurements (AES-SAM) for the samples passivated in borate buffer solution (with no tendency to pitting, see 'Experimental') showed the following:

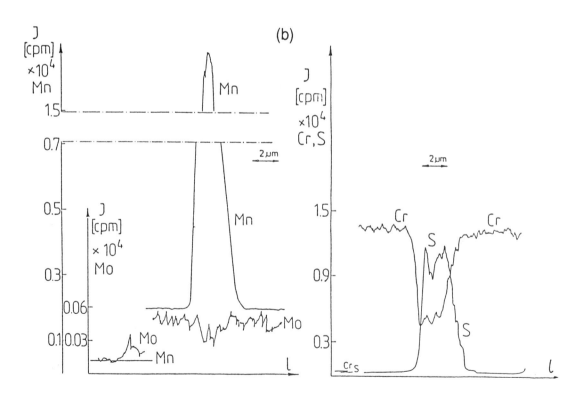

Fig. 3 X-ray microprobe analytical results for a typical $(Mn,Cr)S_x$ inclusion in commercial steel (II); (a) backscattered electron image of the inclusion and the adjacent matrix (Magnification ×1000); (b) line scan across the inclusion, as marked by an arrow, showing distribution of Mn, Mo, Cr and S within the inclusion and outside; cpm - counts per min.

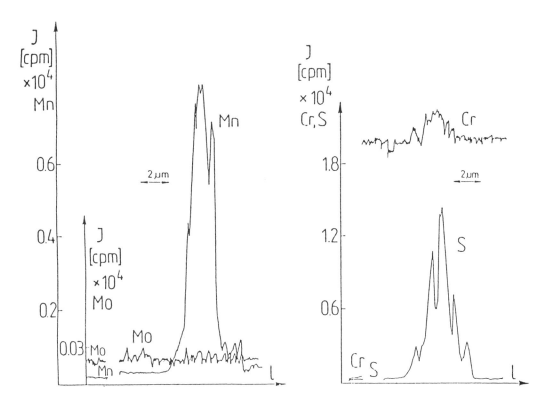

Fig. 4 Line scan across a typical $(Cr,Mn)S_x$ inclusion in the modified steel R-4, showing distribution of Mn, and S within the inclusion and outside. X-ray microprobe analysis; cpm - counts per min.

- The non-metallic inclusions distinctly modified the composition and/or thickness of passivating film, locally.

- A dramatic effect was observed for the steel B, with MnS_x inclusions as compared to steel A, where only mixed $(Cr, Mn)S_x$ inclusions were present (see Table 4).

Figure 5 shows a typical composition profile of the passive film formed at the steel A. The main components are clearly Fe, Cr, O. The adventitious C vanishes with sputtering. Figure 5(b) shows a typical composition profile of the passive film formed at the $(Cr,Mn)S_x$ inclusion, occurring typically within the steel. (The inclusion was localised using secondary electrons imaging). The main components are S, O and Cr; some contaminating C vanishes already after 30 s of sputtering. Clearly the inclusion is passivated with an oxide type film containing Cr and S. Figure 5(a) shows a typical composition profile of the passive film formed on steel B. It is very similar to that in Fig. 5(a), inferring that the film composition of both matrices is quite the same. However, there is a dramatic difference at the sulphide inclusion areas, as compared to that found in the steel with 0.15% Mn, see Figs 5(b) and 6(b).

Figure 6(b) shows the corresponding composition profile at the inclusion side, and infers that oxygen is practically absent there. This suggests the absence of any oxide type passivating film at the inclusion [21]. There are S and Mn signals, whereas O signals (as well as Cr and Fe signals—not shown in the picture for clarity) are below the noise level. In connection with the above findings one should note that the stable existence of MnS in neutral chloride solutions at low anodic potentials is in agreement with thermodynamic predictions; viz. the corresponding Pourbaix diagrams calculated by Eklund [14].

As the composition of various sulphide inclusions in the materials under investigation varied to some extent, the composition (and also probably thickness) of the passive film varied from one inclusion to another. However, it was found in all cases, that in low Mn steels Cr was enriched within the passive film formed at the inclusion, see an example in Fig. 7 (p.287).

In a similarly way it was found that for steels with common Mn level, Cr and O were drastically depleted, or completely absent at the surface of the inclusion, as compared to the level of these elements in the film formed on the matrix, see Fig. 8 (p.287).

Figure 9 (p.288) shows an example of a composition profile of a passivating film formed on the low manganese steel E–1, as well as on the TiS_x inclusion containing also some N and C. Both the matrix and the

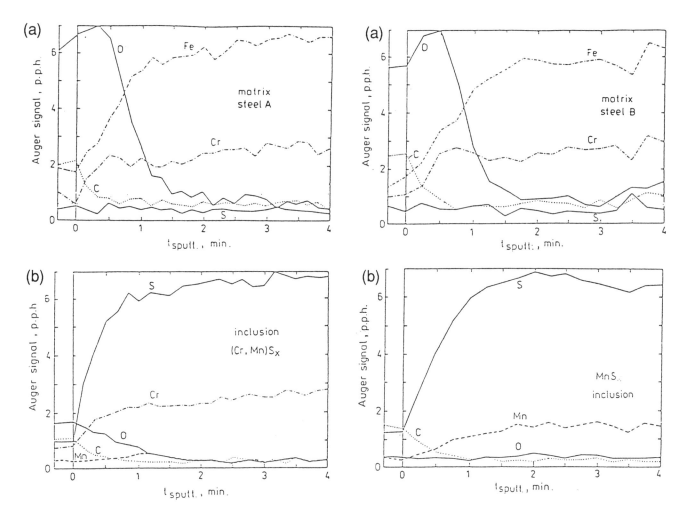

Fig. 5 *Composition profile of the prepassivation surface of steel A with 0.15% Mn (a) metal matrix; (b) sulphide inclusion (Cr, Mn)S.*

Fig. 6 *Composition profile of the prepassivation surface of steel B with 2% Mn (a) metal matrix, (b) sulphide inclusion MnS_x. A lack of oxygen at the inclusion in contrast to its distinct presence at the metal matrix suggest a discontinuity of the passive film at the inclusion.*

inclusion were passivated (see the oxygen profile). Some Cr and Ni containing oxide was present at the matrix, where no Ti was detected.

On the contrary, Ti rich oxide was apparently present at the inclusion.

Summarising the surface analytical results one can say that all the inclusions containing sufficient amount of Cr and/or Ti exhibited an ability to passivate. Only MnS_x inclusions seemed to be unable to passivate, thus producing discontinuities in the film, at low anodic potential in neutral solutions.

3.3 Local modification of the passive film due to the presence of sulphide inclusions, and its role in pit nucleation

In order to understand the corrosion behaviour of the above considered materials the mechanism of pit nucleation in steels should be taken into account. The general opinion is that inclusions containing MnSx or complex oxide/MnS_x are the most susceptible initiation sites.

Our results infer that there may occur a discontinuity of the passivating film at the matrix/MnS_x inclusion interface due to a lack of any passivation for the inclusion itself. This makes it quite clear why these materials are so prone to localised attack in chloride media [6–18]. The above result is in line with the mechanism of pit nucleation on voids within the passive film, suggested by Wood *et al.* [22].

The sulphur species derived from inclusions may additionally contribute to the early stages of pitting on stainless steels [8].

There is an unconfirmed report [23] about production of $S_2O_3^{2-}$ species during anodic oxidation of MnS in near neutral chloride solution. The potency of these species for inducing localised corrosion is well documented [24–26]. However, as shown by Castle [7] MnS_x may be a source of elemental S, which activates the bare metal surface and hinders passivation. The

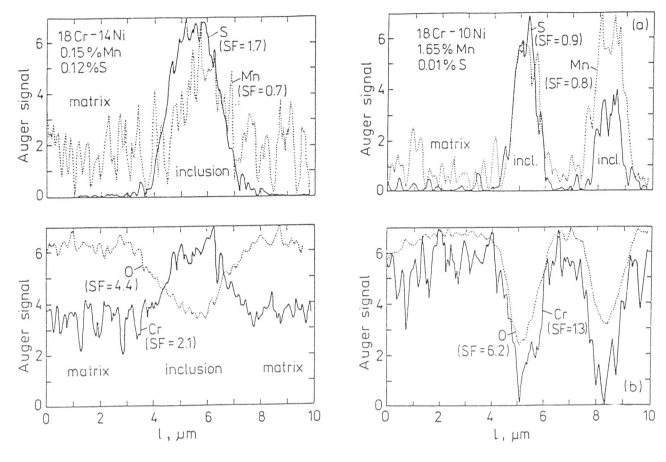

Fig. 7 AES-SAM line scan across an inclusion of $(Cr,Mn)S_x$ and across the adjacent matrix (steel A) passivated electrochemically in a borate buffer solution. The inclusion is covered with a passivating film (O level much above zero) enriched with Cr, Mn-line scan is affected by a partial overlap with Cr signal. SF - scaling factor.

Fig. 8 AES-SAM line scan across two inclusions of MnS_x and across the adjacent matrix (commercial steel II) passivated electrochemically in a borate buffer solution. Signals of both Cr and O are drastically reduced at the inclusions suggesting a very thin or no passivating film at these sides. SF - scaling factor.

latter effect was noted and studied in detail by Marcus [27,28] for Ni contaminated to various degree with S, and may also play a role for Fe alloys. For steels undergoing pitting in nearly neutral solutions a ring of sulphur appeared around certain sulphide inclusions as the inclusions dissolved [7, 29]. An example of such a S-ring revealed by line scan with SAM is given in Fig. 10. The ring was formed after a prolonged (~ 1 day) exposition in aerated borate buffer at corrosion potential. This result suggests that the both anodic and cathodic activity at and around the inclusion, suggested by Castle [7], takes place also in chloride- free solution, leading to the production of elemental S, but not to pitting. Apparently, chloride is required to inhibit repassivation efficiently and to promote pit development. We have seen, moreover, that the ring disappeared after some Ar+-ion sputtering in the AES chamber and the sulphide inclusion itself showed up,

(Fig. 10(b)), thus confirming the observation reported by Castle [7].

In acidic solutions another factor may also contribute to pit nucleation: MnS is not stable in such solutions and when it dissolved it exposes an active surface area. If the remaining hole is shallow and open it can be easily repassivated. On the contrary, it will not be passivated if it forms a narrow crevice. So, not only chemistry of the inclusions, but also their geometry, their adhesion to the matrix, as well as the composition of the adjacent matrix region. are the factors influencing the pitting resistance of steels.

In particular, if the composition of the matrix adjacent to inclusions favours repassivation, pit formation and growth becomes difficult, and so the pitting resistance should improve. This factor has been neglected so far, but this is exactly what is observed in the case of low Mn steels [19]. As shown in Fig. 4 the

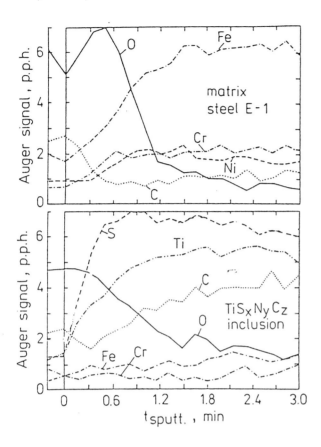

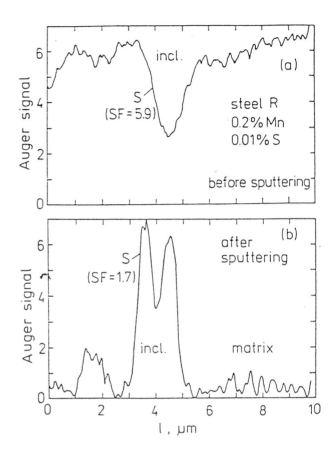

Fig. 9 Composition profile of the prepassivated surface of steel E-1 with low Mn content, supplementary alloyed with Ti (a) metal matrix (b) sulphide inclusion TiS$_x$ containing also some C and N..

Fig. 10 AES-SAM line scan across a sulphide inclusion in steel R-4, passivated, showing a sulphur ring around the inclusion. This suggest that S arises out of the S released from (Mn,Cr)S$_x$ inclusion during corrosion in the electrolyte. SF - scaling factor.

matrix adjacent to the inclusions in the steel contains only small amounts of Mn, but large amount of Cr, which improves repassivation ability. Moreover, the inclusions themselves are chemically more stable, because they are enriched in Cr (or Ti, see ref.[19]).

In the case of the commercial steels the situation is different. The steel (II) contains mostly (Mn,Cr)S$_x$ inclusions. The matrix adjacent to the inclusions contains relatively large amounts of Mn. This is also the case for the Ti bearing steel I [19], where TiS$_x$ sulphides are present.

Mn apparently degrades the ability to passivate of the metal, thus making it more prone to pitting than in the case of the low Mn steels [19].

Coming back to the problem of the chemical stability of individual non-metallic inclusions, as well as their ability to passivate, the following should be taken into account.

In the low Mn steels where (Cr,Mn)S$_x$ or TiS$_x$ inclusions, able to passivate, are occurring the mechanism of pit initiation may be different than at MnS$_x$ and thus requiring higher anodic potentials. The most

probable mechanism seems to be the localised agglomeration of chloride followed by film thinning [31] at the inclusion/matrix interface, as already proposed [30]. The following comment seems adequate to the recent discussion on the pit mechanism [33] in metals:

the passive film at sulphide inclusions in steels in contact with neutral chloride solutions is distinctly different from that at the matrix, and in certain cases even absent, which in various ways, may facilitate the chloride attack there.

4. Conclusions

1. Electrochemical, microscopic, and surface analytical measurements revealed a correlation between the anodic behaviour of the steels, the composition of their non-metallic inclusions as well as the local composition of the passive film at the inclusions as compared to that at the adjacent metal matrix.

2. Electrochemical measurements showed that low Mn, Cr–Ni stainless steels with low (< 1% Mo) or no

Mo exhibit higher pitting resistance in neutral chloride solutions than the analogous commercial steels containing 1.4% Mn and over 2 % Mo.

3. X-ray microprobe analysis revealed that MnS_x inclusions in steels with a common Mn content (~2% Mn) are the most susceptible spots for pit nucleation. In the modified low Mn steels the MnS_x inclusions have been eliminated. Instead some $(Cr,Mn)S_x$ or TiS_x were present. They appeared to be less susceptible to localised attack by chlorides.

4. SAM investigations suggest that there are discontinuities in the passive film due to a lack of passivation of the MnS_x inclusions in steels containing ~2% Mn. Such discontinuities do not exist for low Mn steels, as the $(Cr,Mn)S_x$ inclusions occurring there do passivate. This explains the higher resistance of low Mn steels to localised attack as compared to the commercial ones.

References

1. R. Kiessling, in 'Sulfide Inclusions in Steels', proc. Symp. Port Chester, USA, 1974, Ed. R. Baboian, p.104.
2. H. J. Engell, Stahl und Eisen, **94**,1085, 1974; **93**, 1203,1973.
3. P. Poyet and R. Leveque, Etude des Sulfures dans les Aciers, Influence de divers Elements d'Addition sur leur Composition; Rev. Met., **64**, 1967, 653.
4. L. Tronstad and J. Sejersted, J. Iron Steel Inst. **127**, 1933, 425.
5. M. Smialowski, Z, Smialowska, M. Rychcik and A. Szummer, Corros. Sci., **9**, 1969, 123.
6. M. Janik-Czachor, Bull. Sci. Acad. Polon. ser. sci. chim., **25**, 561, 1977; M. Janik-Czachor and A. Szummer, in 'Passivity of Metals and Semiconductors', Ed. M. Froment, Elsevier 1983, 547.
7. J. E. Castle and E. Ke, Corros. Sci., **30**, 409, 1990.
8. J. Steward and D. E. Williams, Corros. Sci., **33**, 457, 1992.
9. J. Degerbeck and K. Blom, Mater. Perform., 53, July 1983.
10. L. J. Freiman, J. J. Reformatskaya, Ya. M. Kolotyrkin, Yu. P. Konnov and A. E. Volkov, Proc. 10th ICMC, Madras 1987, 3919.
11. Z. Szklarska-Smialowska, A. Szummer and M. Janik-Czachor, Brit. Corros. J., **5**, 159, 1970.
12. M. B. Ives and S. C. Srivastava, Corrosion, **43**, 687, 1987. M. B. Ives and S. C. Srivastava, Proc. Int. Conf. on Localized Corrosion, Orlando 1987.
13. M. Kesten, Corrosion, **32**, 94, 1976.
14. G. S. Eklund, J. Electrochem. Soc., **121**, 467, 1974.
15. P. Forchhammer and H. J. Engell, Werkst. Korros., **20**, 1969, 1.
16. J. Degerbeck and E. Wold, Werkst. Korros. **25**, 1974, 172.
17. J. Scotto, G. Ventura and E. Traverso, Corros. Sci., **19**, 237, 1979.
18. B. Baroux, Poster paper presented at the European Symposium on Modification of Passive Films / Nr. 33, Paris 1993, see also this volume.
19. A. Szummer, K. Lublinska and M. Janik-Czachor, Werkst. Korros., **14**, 1990, 618; JSI International, **31**, 240, 1991.
20. A. Szummer and M. Janik-Czachor, Corros. Sci., in press.
21. A. Szummer, M. Janik-Czachor and S. Hofmann, Materials and Physics, **34**, 181, 1993.
22. Abd. Rabbo, G. C. Wood, J. A. Richardson and C. K. Jackson, Corros. Sci., **14**, 645, 1974; **16**, 677, 1976.
23. S. E. Lott and R. C. Alkire, J. Electrochem. Soc. **136**, 973 and 3256, 1989.
24. C. Newman and E. M. Franz, Corrosion, **40**, 325, 1984.
25. R. C. Newman, M. S. Isaacs and B. Alman, Corrosion, **38**, 261, 1982.
26. R. C. Newman, Corrosion, **41**, 450, 1985.
27. P. Marcus, A. Tessier and J. Oudar, Corros. Sci., **24**, 259, 1984.
28. P. Marcus, I. Olefjord and J. Oudar, Corros. Sci., **24**, 269, 1984.
29. M. Janik-Czachor and A. Szummer, Corr. Reviews, Jerusalem, in press.
30. M. Janik-Czachor, J. Electrochem. Soc., **128**, 513C, 1981.
31. K. E. Heusler, L. Fischer, Werkst. Korros., **27**, 555, 1976.
32. A. Wolowik, Master of Science Thesis, Dept. of Materials Science and Eng. Univ. of Technology, Warsaw 1992.
33. H. J. Engell and R. A. Oriani, Corros. Sci., **29**, 119, 1989.

49

Inclusions in Stainless Steels—Their Role in Pitting Initiation

E. F. M. Jansen, W. G. Sloof and J. H. W. de Wit

Delft University of Technology, Laboratory for Materials Science, Rotterdamseweg 137, 2628 AL Delft, The Netherlands

Abstract

Inclusions in stainless steels are important initiating sites for pitting corrosion. The process of initiation at inclusions is described. Seven stainless steels with a ranging pitting corrosion susceptibility were investigated using potentiodynamic and potentiostatic techniques, Electron Probe X-ray Micro Analysis (EPMA) and Scanning Auger Microscopy (SAM).

It is concluded that the pitting susceptibility is mainly determined by the number and the size of manganese sulphide inclusions. Increasing chloride concentrations enhance the extent of attack, as do low concentrations of thiosulphate. High concentrations of thiosulphate and any acetate concentration reduce the extent of localised attack.

1. Pit Initiation; a Model

Inclusions in commercial stainless steels are regarded as initiation sites for pitting corrosion [1, 2]. Depending on the composition of the inclusion, the electrolyte composition and the potential, certain inclusions will dissolve at defects in the surface film. Especially (manganese) sulphide inclusions are easily dissolved [3–5]. Dissolution of an inclusion can lead to an occluded cell if the surface film is only damaged to a very limited amount just above the inclusion. The damaged surface normally has a much smaller area than the inclusion, but is nevertheless responsible for the dissolution process through the 'cap' over the inclusion. Dissolution products accumulate in the occluded cell and chloride and metal ion concentrations increase. Furthermore, the pH decreases due to the corrosion reactions. All these phenomena depend on the size, composition and shape of the inclusion and on the steel matrix composition and may result in pitting if conditions for repassivation are unfavourable.

Local repassivation can occur when the electrolyte solution in the occluded cell is diluted after the cap is removed. Its stability is thus of importance. A microcrevice or a non-protecting film can impede repassivation.

A stainless steel surface is sketched in Fig. 1. The unstable inclusions (A) lying at the surface will dissolve practically immediately and the fresh steel surface will repassivate. Only when the inclusion is large or the larger part of the inclusion lies underneath the surface, localised dissolution can proceed. The inclusions close to the surface (B) are possible initiation sites for pitting. The passive film, the result of a continuous dissolution and formation process under stationary conditions, above an inclusion will eventually become locally unstable with respect to the surrounding passive film because there is no steel matrix underneath to supply the necessary components for the passive film. Dissolution of the inclusion and the surrounding steel matrix is promoted by chloride ions, a low pH and high anodic potential values.

The dissolution of an inclusion and following repassivation is reflected by a current spike as measured on anodic polarisation of a stainless steel. In Fig. 2 a schematic representation of a current spike is given with the successive stages indicated for a single site. Each current spike can be considered as a non-successful pitting initiation event [6]. The transition to macroscopic pitting is not discussed in this paper.

Some parameters involved in the dissolution and repassivation process were investigated. The results will be discussed in view of the given description of pit initiation.

2. Experimental

The stainless steel specimens, listed in Table 1, were ground with SiC emery paper (final stage P1200 grit) prior to the electrochemical measurements. The specimens used for EPMA and SAM in combination with electrochemical measurements were polished (down to 1 μm diamond paste) and were ultrasonically cleaned in ethanol.

The electrochemical experiments were performed in an 'Avesta' cell [7]. For the potentiodynamic measurements an ECO PGSTAT20 and ECO Autolab were used, for the potentiostatic current–time measurements

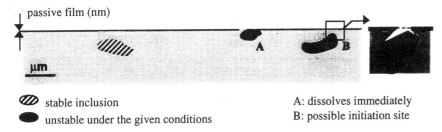

Fig. 1 Schematic representation of a stainless steel surface layer.

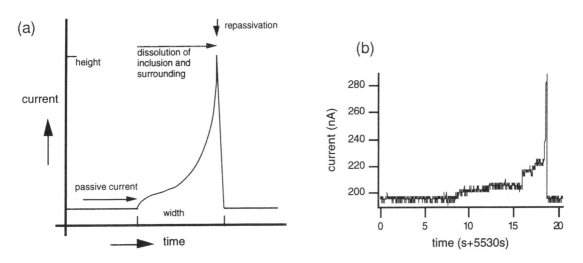

Fig. 2 Schematic representation of a current spike (a) and (b) an example of a measured spike (AISI 304, 250 mV/SCE, 0.1M NaCl solution).

a battery-operated potentiostat Jaissle 1002T-NC was used. (For further details see [8].)

A Jeol JXA 733 electronprobe X-ray microanalyser equipped with four wave-length dispersive spectrometers and an energy dispersive spectrometer, fully automated with Tracor Northern TN5500 and TN5600 systems, was used to determine the size and composition of the inclusions in the stainless steel specimens. The analyses were performed with a 12 keV, 2 nA electron beam. An inverse backscattered electron signal was used to recognise the inclusions. A total area of 1 mm² of each specimen was analysed (magnification × 2000). Additional composition analyses of the inclusions were performed with a PHI 4300 scanning Auger microprobe (Perkin-Elmer).

3. Results and Discussion

Polarisation curves were recorded of all the stainless steels, with a scan rate of 1 mV.s⁻¹ in aerated 0.1M NaCl at room temperature. None of the curves showed the typical current increase due to pitting before the transpassive region, except for AISI 304 and 316L. AISI 304 (Fig. 3) was the only alloy showing an active–passive transition when ground shortly before the

polarisation curve was recorded. It can be seen that the current fluctuations, measured as current spikes in potentiostatic experiments, only appeared after the passive film had formed. Considerable current fluctuations were also observed in the curves of AISI 316, and, to a lesser extent, in the curves of AISI 317L and 34LN, however not in the curves of SAN28 and 254SMO.

The number of spikes in a potentiostatic current-time curve (a typical example is given in Fig. 4) is considered to be a measure for the pitting corrosion susceptibility [8–10]. The number is reduced to only a few if the surface is pickled prior to the measurement (immersion in 20% HNO₃ /5% HF for 4 min at room temperature), see Fig. 4(b). This treatment is known to remove sulphide inclusions in the steel near the surface [9,11,12]. Thus, the observed difference in number of spikes due to pickling suggests that sulphide inclusions are responsible for the spikes. This is confirmed by the relation found between the number of spikes and the number of manganese sulphide inclusions determined with EPMA before and after the potentiostatic current–time measurements.

From the composition analyses the inclusions can be divided in three groups: manganese sulphide, com-

Table 1 The stainless steels and their composition (wt%), tradenames and designations according to UNS(1) or DIN (Werkstoff-Nr)(2)

	304	316L	316	317L	34LN	SAN28	254SMO
	S30400[1]	S31603[1]	S31600[1]	S31600[1]	1.4439[2]	N08028[1]	S31254[1]
C	0.070	0.016	0.033	0.017	0.002	0.014	0.011
Si	0.60	0.49	0.63	0.59	0.48	0.41	0.33
Mn	1.65	1.48	1.41	1.63	1.75	1.76	0.53
Cr	17.5	17.4	17.1	18.3	16.3	27.2	20.0
Ni	8.8	11.1	10.6	14.0	13.5	31.1	17.9
Mo	0.26	2.28	2.53	3.13	4.50	3.47	6.07
P	0.340	0.035	0.029	0.026	0.023	0.020	0.020
S	0.024	0.021	0.025	0.011	0.002	0.003	0.001
N	0.028	0.052	0.062	0.072	0.146	0.062	0.204

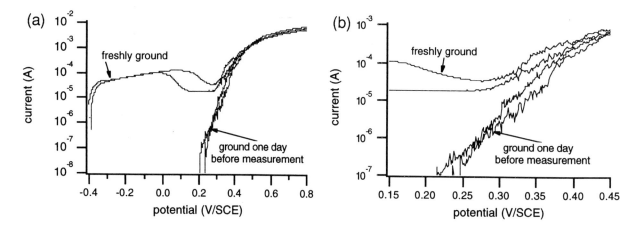

Fig. 3 Polarisation curves of AISI 304 (a), samples freshly ground or ground one day before the measurement, (b) detail of the curves. Scan rate 1 mV/s, aerated 0.1M NaCl solution, room temperature.

plex compounds and iron-chromium oxides. The sulphur found in the investigated AISI 304, 316L, 316 and 317L alloys was present mainly in separate manganese sulphide inclusions. These steels contain relatively much sulphur compared with 34LN, SAN28 and 254SMO. In 34LN sulphur and manganese were found associated with other elements as aluminium, calcium, silicon and titanium. In SAN28 and 254SMO sulphur was only found in combination with magnesium, aluminium, silicon, calcium, titanium and/or vanadium.

The number of manganese sulphide inclusions determined with EPMA was decreased significantly after the potentiostatic current–time measurement (see Table 2), whereas in most cases the number of complex and iron–chromium oxide inclusions remained the

same within the experimental error. This implies that the current spikes are associated with the dissolution of manganese sulphide inclusions. However, the number of spikes is significantly less than the number of manganese sulphide inclusions dissolved. Probably, manganese sulphide inclusions at the surface were dissolved at the beginning of the polarisation step before a reasonably stable passive current was obtained.

The dissolution and repassivation process can be characterised by the width and height of the spikes respectively. The following relations between the size of the manganese sulphide inclusions and the steel matrix composition can be found. Comparing the widths of the current spikes for the steels AISI 316 and 316L, which have practically the same composition

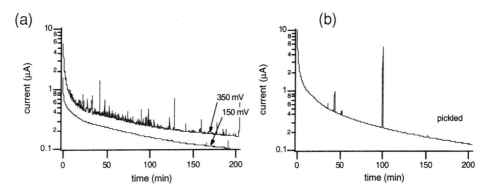

Fig. 4 Typical current-time curve for AISI 316 (a) measured at 350 and 150 mV/SCE and (b) after pickling the surface, 350 mV/SCE. Aerated 0.1M NaCl solutions, room temperature.

Table 2 Comparison of data, S = sulphur content of the alloy, S-incl = sulphur containing inclusion, ΔMnS = difference in number of manganese sulphide inclusions found before and after current-time measurement in 0.1 M NaCl, spikes were counted from 10 to 205 min after applying the anodic potential, 250 mV/SCE (all other steels)

	S	S-incl. number	S-incl area	ΔMnS number	spikes number	average width	average height
	(wt%)	(mm^{-2})	(μm^2)	(mm^{-2})	(mm^{-2})	(s)	(nA)
304	0.024	749	1.3	716	12.6	2.6	181
316L	0.021	319	0.9	291	17.9	4.0	311
316	0.025	679	0.4	497	9.0	1.8	101
317L	0.011	757	0.4	690	14.2	4.5	112
34LN	0.002	75	3.5	0	0.5	3.0	274
SAN28	0.003	6	3.1	0	0.05	0.8	58
254SMO	0.002	19	0.3	0	0	-	-

regarding the main alloying elements, it appears that the dissolution takes longer for the inclusions in AISI 316L than for those in AISI 316. This agrees with the difference in size of the inclusions in these steels (see Table 2). Differences in the heights of the current spikes can be explained by the different area which repassivated and different extent of changes in local solution composition. The size of the manganese sulphide inclusions in AISI 316 and 317L is about the same, but the width of the current spikes is larger, thus the dissolution and repassivation event takes longer for AISI 317L. This is explained if it is assumed that the passive surface layer of this steel forms a more stable passive layer which holds the occluded cell conditions for a longer time. Both the large height and width of the current spikes for 34LN agrees with the large size of the sulphide inclusions and the relatively low chromium content of the alloy. The inclusions in SAN28 and 254SMO seem to be stable under the given condi-

tions. The measurements of 304 must be left out in the comparisons because this steel was polarised to a less anodic potential. A more detailed discussion will be published later [13].

The frequency of the spikes depends on the potential, pH and different anion concentrations. These parameters were investigated for AISI 316. The number of current spikes decreased if the potential was decreased, as shown in Fig. 4, or the pH was increased (not shown). The potential and pH affect the chemical stability of the inclusions. Eklund [3] calculated a pH/potential diagram for the system MnS-H$_2$O-Cl-, showing that manganese sulphide is not stable at low pH and high anodic potentials. The repassivation of the metal matrix is also dependent on pH and potential. This accounts for the observed effects of these variables.

The conditions leading to dissolution of inclusions were further investigated by measuring current spikes at different anion concentrations. The effect of an increasing chloride concentration and of addition of thiosulphate on the frequency of the spikes is shown in Fig. 5. Addition of small amounts of thiosulphate (0.0004M $Na_2S_2O_3$, 0.1M NaCl) had a similar effect on the frequency of spikes as raising the chloride concentration to 0.5M. Initially the number of spikes per 10 min interval increased in both cases, but then this number decreased much faster than for the 0.1M chloride solution. The total number of spikes remained about the same (within the time range of 205 minutes). Higher concentrations of thiosulphate showed a similar effect as additions of acetate: the total number of spikes decreased till zero at 0.1M acetate in a 0.1M NaCl solution for both anions (see Fig. 6).

The effect of acetate depends on the chloride concentration (Fig. 6(b)), indicating that a competitive effect exists. Chloride accumulation at anodic sites is reduced proportional to the concentration of additional anions, so repassivation is less hindered. Thiosulphate however may form complexes with iron or chromium, increasing the rate of attack and acidification and hindering repassivation. At high concentrations however no spikes were found. It seems that dissolution of inclusions is hindered. A critical chloride concentration in relation to other anions is needed for sulphide inclusions to dissolve. In the critical situation, with just not high enough thiosulphate present to prevent all attack, dissolution is strongly enhanced by the still relatively high concentration thiosulphate to which the bare metal is exposed (0.004M and 0.01M in Fig. 6(a)), resulting in macroscopic pitting.

4. Conclusions

The conclusions can be summarised as follows:

- Spikes demonstrate the dissolution of sulphide inclusions.
- The pitting susceptibility of an alloy is mainly determined by the number and size of manganese sulphide inclusions present.
- Chloride promotes localised attack after dissolution of an inclusion, as do low concentrations of thiosulphate. High concentrations of thiosulphate and all concentrations of acetate reduce localised attack.

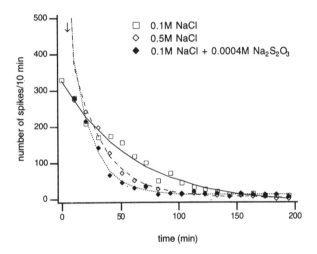

Fig. 5 Frequency of spikes (number of spikes per 10 min interval), AISI 316, 350 mV/SCE, measured in aerated 0.1M NaCl, 0.5M NaCl and 0.1M NaCl + 0.0004M $Na_2S_2O_3$.

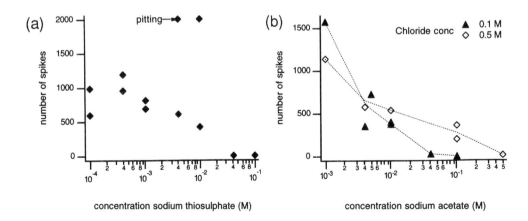

Fig. 6 The effect of (a) sodium thiosulphate added to 0.1M NaCl and (b) sodium acetate added to 0.1M and 0.5M NaCl (all end concentrations) on the total number of spikes measured from 10 to 205 min after polarisation, AISI 316, 350 mV/SCE.

References

1. G. Wranglén, Corros. Sci., **14**, 331–349, 1974.
2. M. B. Ives and S. C. Srivastava, in Advances in Localised Corrosion (NACE-9), NACE, Houston, 1991, pp.295–302.
3. G. S. Eklund, J. Electrochem. Soc., **121**, 467–473, 1974.
4. Z. Szklarska-Smialowska and E. Lunarska, Werkst. Korros., **32**, 478–485, 1981.
5. R. Ke and R. Alkire, J. Electrochem. Soc., **139**, 1547–1580, 1992.
6. H. Bohni and L. Stockert, Werkst. Korros., **40**, 63–70, 1989.
7. R. Qvarfort, Corros. Sci., **28**, 135–140, 1988.
8. E. F. M. Jansen and J. H. W. de Wit, in Proceedings of the 12th Scandinavian Corrosion Congress & Eurocorr '92 **I**, ed. P. J. Tunturi, 1992, pp.91-95.
9. W. M. Carroll and T. G. Walsh, Corros. Sci., **29**,1205-1214, 1989.
10. R. Morach, P. Schmuki and H. Bohni, in Materials Science Forum **111–112**, O. Forsén (ed.), 1992, pp.493–506.
11. G. Hultquist, S. Zakipour and C. Leygraf, in Passivity of Metals and Semiconductors, M. Froment (ed.), 1983, pp.399–404.
12. M. A. Barbosa, A. Garrido, A. Campilho and I. Sutherland, Corros. Sci., **32**, 179–184, 1991.
13. E. F. M. Jansen, thesis, Delft University of Technology, 1993.

50

Influence of Ion Implanted Molybdenum or Molybdate Solution on Pitting Corrosion of 304 Stainless Steel in Chloride Solutions. Statistical Study by Digital Image Analysis of Pit Initiation and Propagation

I. LACOME, P. FIEVET, Y. ROQUES AND F. DABOSI

Laboratoire de Métallurgie Physique, U.R.A. C.N.R.S. 445 118, route de Narbonne 31077, Toulouse Cédex, France

Abstract

The effects of Molybdenun implantation and molybdate ions on pitting corrosion of 304 stainless steel in chloride solution have been studied. Pit initiation time, propagation kinetics and pit surface density were determined by digital image analysis. Addition of molybdate ions slightly increases pit initiation times and decreases pit propagation by a factor of two. Molybdenum ion implantation greatly increases the measured initiation times. Their repartition is no more exponential but complex. The pit surface areas are very small.

The pit surface density is increased both by molybdate addition and by molybdenum implantation.

1. Introduction

The scope of this paper is to study the sensitivity of 304 stainless steel in chloride solutions. Pitting corrosion can be divided in two stages: pit initiation and pit propagation. Pit initiation occurs on defects, inclusions or is linked with the structure of the material. Molybdenum is known to improve the pitting corrosion resistance of stainless steel [1]. The literature shows there is few studies on initiation and propagation of pitting corrosion on ion implanted metals. Molybdenum can be introduced as an alloy element or by surface ion implantation. Positive influence of molybdenum in passive film is shown to result from a bipolar passive film and oxidation of molybdenum as molybdate [2, 3]. In this paper it is proposed to compare the influence of molybdenum in solution (molybdate) and ion implanted on pitting corrosion of stainless steel in chloride media. Pitting initiation times and pit surface areas are determined by an *in situ* image analysis of the corroding surface.

2. Experimental Procedure

2.1 Materials and procedure

The material used in this work was AISI type 304 steel cold rolled with the following composition (mass%):

C 0.053; Mn 1.43; Si 0.485; S 0.0025; P 0.029; Ni 8.28; Cr 17.38; Mo 0.139; Cu 0.145; Ti 0.002; N 0.04.

The specimens were in the form of discs of 12 mm dia. mounted in epoxy resin. They were ground with silicon carbide paper up to 4000 grade then with 1 μm diamond paste and cleaned with deionised water. The surface of some samples was modified by Mo^+ ion implantation (2.5×10^{16} at.cm^{-2}, 100 keV).

Solutions were prepared with deionised water and reagent grade NaCl 0.5 mol.L^{-1} and NaCl 0.5 mol.L^{-1} + Na_2MoO_4 0.3 mol.L^{-1}. Samples corroded in these solutions are referred as 304 and 304 MoO4 respectively.

The pH of the solutions is adjusted to pH = 6 $\pm$ 0.5 to have the same basis of comparison. HCl 0.5 mol.L^{-1} is used for this in order to have the same chloride content. Mo^+ ion implanted samples were corroded in NaCl 0.5 mol.L^{-1} solution and are referred as 304 Mo. Pitting of samples was obtained by a 10 min immersion at open circuit potential followed by a 150 s polarisation at 500 mV/SCE in the same solution without deaeration.

2.2 Apparatus

The experimental apparatus is shown in Fig. 1. It consists in a macroscope (Wild-Leitz), a CCD camera (512$\times$512 pixels), an image analogue digital conver-

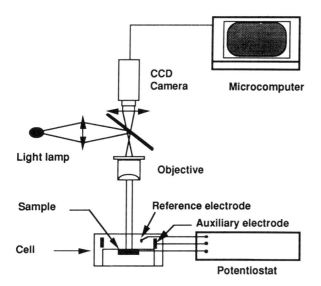

Fig. 1 Experimental setup.

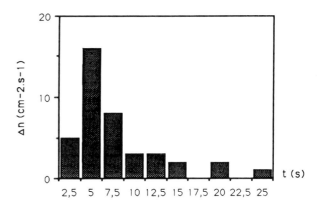

Fig. 2 Histogram of initiation times for 304 sample polarised 150 s at 500 mV/SCE.

sion, a memory card and a microcomputer. The observed field is a 22 mm² area with a resolution of 10 μm. The sample is polarised in a classical three electrodes electrochemical cell. A top glass window allows an *in situ* observation of pitting corrosion of the sample. The software for image processing stores the images at a 3 s time interval. The pits are discriminated from defects: they have a circular shape, and are not present in the first image... thus it is possible to determine the position, the initiation time and the area of each pit in the plane of the surface.

3. Results

3.1 Histogram of pit initiation times

Pits occur at random initiation times. These results can be presented in Fig. 2 by histogram of the pitting frequency Δn (number of pits per unit surface and unit time) versus time intervals. The pitting frequency is very high in the first time intervals then it decreases and tends towards zero. It shows that:

(i) there is a predetermined number of sites leading to pitting;

(ii) the duration of the test (150 s) is large enough to take all the events in consideration. By integration, one can determine the pit density N (total number of pits per unit area). So histograms show the evolution of pitting frequency and pit density with time. Influence of molybdate ions in solution or Mo ion implantation is rather difficult to see on histograms. It is better to consider independently the pit density N and the time influence given by a reduced variable P:

$$P(t) = n(t) / N \qquad (1)$$

n(t) being the number of pits at time t and N the total number of pits at the end of the test. P(t) is the probability of pitting at time t.

So, the experimental results of Fig. 2 can be plot in a −ln(1−P) = f(t) diagram in Fig. 3 curve 304. The linear variation observed shows that:

$$- \ln [1 - P(t)] = \lambda . t \qquad (2)$$

$$P(t) = 1 - \exp [-\lambda . t] \qquad (3)$$

The slope λ is characteristic of pitting frequency. The relation (3) shows that pits initiations are independent. At a given time the pitting probability depends on the number of non pitted sites. The sensitivity of pitting of a site is independent of the story of the sample.

The curve 304 MoO4 in Fig. 3 was obtained in 0.3mol.L⁻¹ MoO4 solution. The variation of n(1−P) with t is linear for shorter times, with slightly lower slope as in chloride solution. The slight shift of curve can be explained by a difference in pit growth rate (as shown below). Pits are not detected at the first step of their lives (atomic level) but they are macroscopic. So a growth stage is included in what is considered as the initiation time. Supposing the same growth kinetics before and after detection, one can correct the initiation time. For times between 20 and 30 s no linear variation is observed but a plateau. This could result from a modification of the passive film or an interaction between pits. For higher times, the number of experimental points is low, thus no significant conclusion can be proposed.

The curve 304 Mo corresponds to pitting initiation

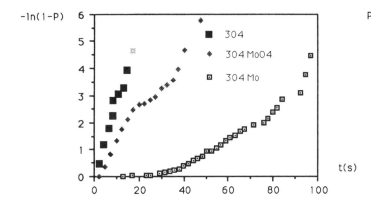

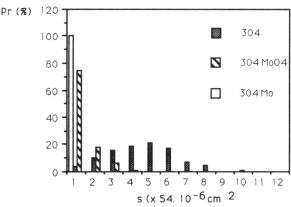

Fig. 3 Influence of molybdate and Mo ionic implantation on pitting probability P for 304, 304 MoO4 and 304 Mo samples polarised at 500 mV/SCE. Diagram –ln(1–P) = f(t).

Fig. 4 Influence of molybdate and Mo ionic implantation on pit area s (cm²) distribution for 304, 304 MoO4 and 304 Mo samples polarised at 500 mV/SCE. Diagram Pr(%) = f(s).

times of an Mo ion implanted sample. No simple analytical relation was found to describe that curve (normal, Weibull...). The complex form of the repartition can be due to an evolution of the sample during the test.

The comparison of curves 304 and 304 Mo shows the initiation times are greatly increased with Mo ion implantation. This can be explained by a passivating effect of Mo dissolving into the solution.The unprotected areas resulting from a rupture of the passive film can be repassivated more easily. The open circuit potential of 304 Mo samples is higher than that of 304 samples.

2.3 Pit surface density

The pitting corrosion sensitivity cannot only be estimated by the initiation times but also by the pit surface density N. The technique used in this paper allows to measure it. The duration of the test is chosen so long as the pitting probability is near null at the end of the test (see histogram in Fig. 2). The pit surface density is a random value so the confidence interval is plotted. In fact, the number of sites nucleating pits obey Poisson process. Comparison of results in Table 1 shows that the $MoO_4^=$ ions and Mo ion implantation increase the number of sites nucleating pits. It is rather paradoxal to see the $MoO_4^=$ ions increase the number of pits because $MoO_4^=$ ions are known to be passivating. As shown below $MoO_4^=$ ions lower the pit growth kinetics.

The comparison between 304 MoO4 and 304 Mo samples can only be qualitative because the concentration of Mo by dissolution of the passive film is not known. The other reason is that the structure of the film may be different according to the sample is ion implanted or not. Ion implantation creates physical defects. Moreover the pits observed in this paper are macroscopic, and have dimensions higher than the implanted layer thickness (50 nm). So not only the properties of the passive film are to be taken into account but also the dissolution rate of the base metal.

2.4 Repartition of pit areas

The area of the pits in the surface of the sample was measured for different images of a test. Thus the kinetics of pit growth was determined [4]. The observed law is mostly a linear variation of surface area with time. That shows the pit growth is under diffusion control [5] For 304 MoO4 samples corrosion products precipitate at the mouth of the pits. So pit surface areas were not measured at each image but only at the end of the test after surface cleaning. The histogram of pit areas is presented in Fig. 4. The frequency of each class is referred to the total number of pits of the same test and expressed in percentage Pr(%). The repartition obtained for 304 sample is rather symmetric with a large standard deviation. In reality, it obeys normal law. The molybdate ions decrease the growth rate of pits areas significantly. The observation of the samples during corrosion shows some pits stop growing. This can be explained by passivation property of molybdate inside pits.

For 304 Mo samples, the pits areas are very small. They belong to the same first class.

Table 1

Sample	304	304 MoO4	304 Mo
Pit density N (cm⁻²)	60 ±20	220±70	800±300

3. Conclusion

Pitting corrosion is a statistical phenomenon. Pits appear at random positions in space and at random initiation times. Two steps are involved: the initiation step and the propagation step. The *in situ* digital image analysis allows to determine the initiation time of each individual pit on a sample, the pit surface density and the pit areas.

The experimental results show:

(i) molybdate ions increase slightly pit initiation times which obey exponential distribution law, increase the pit surface density and decrease the surface area of the pits;

(ii) molybdenum ion implantation greatly increases the measured initiation times. Their repartition is not exponential but complex. The pit surface areas are very small. It seems that the molybdate ions have a small influence on pit initiation times but decrease the pit propagation. The rather high initiation times of molybdenum implanted samples can result from that slow propagation. The pits are detected when they have a given macroscopic size.

The pit surface density is increased in molybdate solution and by molybdenum implantation. There is a correlation between the size of pits and their surface density. A question is to know if it is a result of interaction between pits (diffusion of corrosion products) or an intrinsic phenomenon. Experimental tests are going on.

4. Acknowledgements

This work was supported by the C.N.R.S. (G.S. Corrosion), U.N.I.R.E.C. and Ugine-Savoie which was greatly appreciated.

References

1. K. Sugimoto and Y. Sawada, Corrosion, **32**, 347, 1976.
2. C. R. Clayton and Y. C. Lu, Corros. Sci., **29**, 881, 1976.
3. C. R. Clayton and Y. C. Lu, J. Electrochem. Soc., **133**, 2465, 1986.
4. P. Fievet, Y. Roques and F. Dabosi, Proc. Symp. on 'Critical Factors in Localized Corrosion', J. Electrochem. Soc., **92–9**, 290, 1992.

51

The Effect of pH and Potentiostatic Polarisation on the Pitting Resistance of Stainless Steels: Relation to Non-metallic Inclusions or Passive Film Modifications

B. Baroux* and D. Gorse†

Ugine Research Centre, 73403 Ugine, France
*Also at INP Grenoble, LTPCM, BP75, 38402 Saint Martin d'Hères, France
†CECM-CNRS, 15 rue Georges Urbain, 94407 Vitry/Seine, France

Abstract

For industrial steels, pitting in chloride containing aqueous solutions occurs on Manganese sulphides, when they are present in the steel. Adding a slight amount of Ti as alloying element prevents the MnS formation, since Ti is a stronger sulphide former than Mn. Pits are shown to initiate on Ti nitrides in low chloride containing solutions, but on Ti sulphides for more concentrated ones, indicating that Ti sulphides may dissolve in high chloride containing solutions. One shows that the pitting potential pH dependence is the signature of the type of inclusions which act as pitting sites: as far as sulphides are concerned (MnS or Ti_2S), a strong decrease in pitting potential is observed when the pH becomes smaller than a critical value corresponding to the H_2S formation. Measuring the current fluctuations occurring during a potentiostatic polarisation below the pitting potential shows that such fluctuations occur when pits initiate on sulphur containing inclusions, then repassivate. From this analysis, a critical pH ranging between 4 and 5 is found again. No fluctuations are evidenced when pits initiate on Ti nitrides. Whatever the pits initiation sites, further pitting resistance can be improved by such a prior polarisation in the corrosive medium. This shows the difficulty to separate the roles of the non-metallic inclusions (which act at the scale of some micrometers) and of the passive film itself (whose size scale is of the order of some nanometers) when discussing the pits initiation mechanisms.

1. Introduction

It is well known that, on industrial stainless steels, pitting occurs on non-metallic inclusions[1, 2] and particularly on sulphides [3], the size of which is of the order of some micrometers. From another hand, most of the theories on pitting initiation refer to local passive film breakdown, at a scale of some nanometers, without any reference to these metallurgical defects. Considering this discrepancy, some workers suggest that pitting occurs at the matrix inclusion boundaries, which has been observed in some cases, but is not a general feature. Moreover the chemical effect of the species which dissolve from the non-metallic inclusions are hardly taken into account by the passive film breakdown theories.

It is intended in this work to partially enlighten on these points, by measuring the pitting potential in neutral or acidic chloride media of some specially designed stainless steels, and identifying the corresponding pitting sites. Additionally, the samples were polarised at a constant potential in the corrosive solution but below the pitting potential: the effects of this prepolarisation treatment on the further pitting potentials were recorded and the anodic current fluctuations during the polarisation analysed.

2. Studied Steels and their Non-metallic Inclusions

The steels under investigation are mainly some ferritic FeCr alloys of the AISI 430 type containing also either Nb (steels A and A') or Ti (steels B and B') additions. Their composition is indicated in Table 1. Excepting for steel A, aluminium additions during the melting process are used as deoxidising agent, which leads to Al~ 0.030% and induces the presence of some Al_2O_3 inclusions. Let us note the low sulphur content, which is easily attained with the modern steelmaking techniques and the presence of some stabilising elements (Ti or Nb) which trap the carbon and avoid the formation of chromium carbides. For both steels A and A', Sulphur is trapped under the form of Manganese sulphide, which is considered as a noxious pitting site.

Table 1 Steels composition in wt% or in ppm (brackets)

	Cr	Ni	Si	Mn	Ti	Nb	(S)	(C)	(N)
A	16.4		.4	.45		.5	40	240	n.d.
A'	15.7		.4	.45		.7	50	340	340
B	16.8		.4	.45	.4		30	260	140
B'	17.4		.4	.45	.4		30	250	110
C	17.6	8.2	.4	1.4			20	530	470
D	17.8	9.3	.4	1.3	.4		30	310	170

The difference between steels B and B' is their Cr content; both contain Ti, which also trap Nitrogen under the form of Titanium nitrides and Sulphur under the form of Ti sulphides. This trapping occurs during the steelmaking process before the steel solidification. Additionally some 304 and 321 AISI type steels are considered (resp. C and D), in order to separate the matrix an the inclusion effects. These two steels are austenitic and Ni bearing but the latter contains Ti and is then MnS free, whereas the former contains MnS inclusions. Last it should be noted that the steels are tested in the as annealed condition (800 °C annealing for ferritic steels, 1050 °C solution treatment for austenitic ones). The detailed study of the inclusions and precipitates characteristics is out the scope of this paper. Nevertheless their main features, as far pitting initiation is concerned, are presented on Figs 1(a) and (b) for steel A' and Fig 1(c) for steel B (same for steel B'). A specific preparation procedure for the STEM examinations was used for avoiding the sulphur dissolution, but this is also out of the scope of the paper.

In the case of Steel A', some MnS is found around aluminium oxides. Since these oxides have a poorer ductility than the metallic matrix, the cold rolling process provokes some microdecohesions around the inclusion, where some manganese sulphides are often located, leading to the possible formation of very noxious microcrevices. This phenomenon is not observed on Al free steels, where oxides are mainly malleable silicates. However, the art of the steelmaker consist in avoiding the formation of Cr oxides, which could produce a similar, but worst, effect than Al oxides. In every case, some MnS are also found closely stuck to Nb carbonitrides (Fig. 1b). Since these carbonitrides precipitate at ~ 1200 °C, this shows that

(for the considered sulphur content), the high temperature sulphur solubility is sufficient for the MnS precipitation to occur in the solid steel (lower than 1200 °C): to our knowledge, this result is quite new, since MnS are generally consider to precipitate at the end of the steel solidification. Note that very few isolated MnS precipitates are found, probably because Nb(C,N) acts as precipitation sites. This situation contrasts with the behaviour of steel C (304) for which (i) no carbides may nucleate the MnS precipitation (ii) sulphides can precipitate in austenite at the delta → gamma transformation temperature.

Figure 1(c) shows the location of Ti sulphides on steel B, i.e. around the Titanium nitrides, embedded in a Ti carbide belt. Detailed examination suggests that at high temperature an homogeneous Ti carbosulphide belt precipitates around the Ti nitrides (which formed in the liquid steel). Then, lowering the temperature, Titanium sulphides and Ti carbides separate at some points, producing the facies presented on Fig. 1(c). As an other result, note that Al oxides have been identified in the core of the Ti nitrides, playing probably the role of nuclei for the TiN precipitation. Last, isolated Ti carbides are also present in the steel, but have no relation with pits initiation.

3. pH and Polarisation Effects on the Pitting Potential

The pitting potentials are measured on some mechanically polished samples (SiC grade 1200, under water) using a procedure and an experimental set-up which are described elsewhere [4]. The electrolyte is a deaerated chloride containing aqueous solution (0.02–0.5M NaCl), the pH of which is adjusted by HCl additions from pH 6.6 (near neutral) to pH 3 (which is

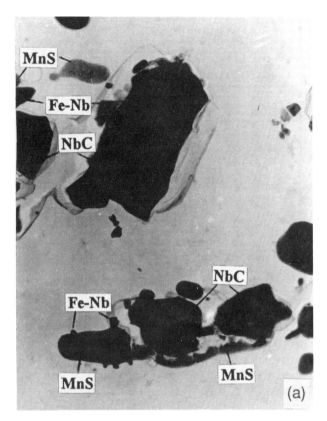

STEM × 18 000

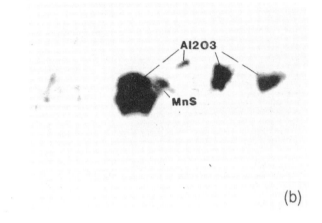

(b)

SEM × 4000

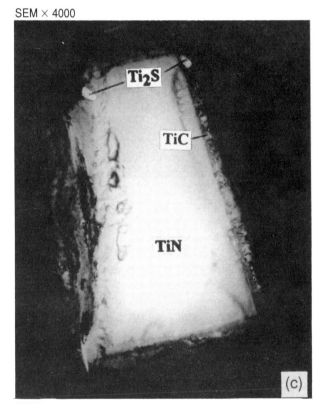

STEM × 18 000

Fig. 1 Non-metallic inclusions in the studied steels.

(a) Steel A': MnS nucleated on an aluminium oxide.

(b) Steel A': MnS nucleated on a Nb carbide (or carbonitride). Intermetallic (Fe,Nb) phases are also observed.

(c) Steel B: Ti nitride surrounded by a Ti carbide, in which some Ti sulphides are embedded.

above the critical pH below which the passive film could suffer general breakdown). The samples were aged for 24 h in air after polishing, then immersed for 15 min in the corrosive solution before beginning the potentiokinetic scan (100 mV/min). A conventional pitting potential V_{pit} was obtained, for which the Elementary Pitting Probability [4] is equal to 0.1 cm^{-2}. Figure 2(a) shows the typical results, obtained for steels A' and B in NaCl (0.02M). No pH dependence is observed for steel B. At the opposite, a sharp pitting potential decrease is evidenced when pH is lowered under a critical value PH_c ranging between 4.5 and 5. Since the main difference between the 2 steels is the

presence or absence of MnS, and that Ti sulphides are known to have a better stability in aqueous electrolytes than MnS, one can assume that this decrease is due to the MnS potential and pH assisted dissolution. Note that the same phenomenon is observed (Fig. 2b) when comparing the two austenitic steels C (MnS containing) and D (MnS free). The effect of the solution pH on the MnS dissolution was studied in details by Eklund [5], as it can produce some harmful sulphur bearing species (from the pits initiation viewpoint). For the studied pH values, the overall reaction to be considered should be: $MnS + 2H^+ \rightarrow Mn^{2+} + H_2S$, which is strongly pH assisted. The existence and the numerical value of the threshold pH will be discussed in a forthcoming paper, together with the potential dependence of the dissolution kinetics.

Figure 2(b) also shows the results obtained for steels A and B', which confirm the herabove findings. The slight difference between steels B and B' may be attributed to the difference in Cr concentration. The strong difference between steels A and A' may be due both to the difference in Cr concentration and sulphur contents, but also, from our other industrial experience, to the noxious effect of Al oxides present in steel A' and around which MnS are found (cf herabove).

The effect of the solution chloride content was also investigated between 0.02 and 0.5M. For neutral pH (6.6), the results are quite consistent with the logarithmic law previously found [4] for this sample preparation procedure. However, things become more complex for MnS containing steels when pH < pH_c. Furthermore, Fig. 2(c) shows that for Ti containing steels, a pitting potential pH dependence is found again for high enough chloride concentrations, suggesting that Ti sulphides are not so stable in such electrolytes.

The final experiment that we performed was to polarise the sample for 1 h at various potential $V_{pol} < V_{pit}$ after the 15 min initial holding at rest potential, then rising the potential up to pitting and measuring the new V_{pit}. Figure 2(d) shows the results obtained in NaCl (0.02M) at pH3 and 6.6 for steels A and B. One can see that the potentiostatic prepolarisation increases the further pitting potential ($dV_{pit}/dV_{pol} \sim 0.5$ in any case, for $V_{pol} = -200$ to $+200$ mV/SCE). For Ti containing steel, the pitting potential does not depend on the pH, but for MnS containing steel, it remains lower at pH3 than at pH6.6, whatever the polarisation potential. Furthermore, polarising steel A 1 h at 150 mV/SCE in acidic solution does not improve but rather decreases the further pitting resistance. Micrographic observation of the polarised sample shows in this case that strong 'prepitting' had occurred during the prepolarisation (see hereunder). The main conclusion of this experiment is that prepolarising a sample in the corrosive solution under conditions where pitting does not occur improve the further corrosion resistance, which is clearly related to a passivity reinforcement. As there is no evidence of some possible passivation of inclusions (pitting sites), one should admit that not only the non-metallic inclusions but also the passive film play a role in the pitting initiation. This is a rather expected result, but apparently difficult to correlate to the major role plaid by the sulphides dissolution. Complementary investigations performed on steel B,

either at pH3 or 6.6, showed that increasing the prepolarisation time up to 1 day provides a further increase in the pitting resistance, motivating fundamental studies on the passive film modifications occurring during this polarisation [6].

4. Pitting Sites

The procedure for evidencing the pitting sites is the following: first the samples are diamond polished in order to obtain a micrographically observable (SEM) surface. Then they are aged for 24 h in air then immersed for 15 min at rest potential in the test solution as hereabove. Last the potential is quickly rose up to pitting, to obtain an anodic current of the order of few microamps. Several limit anodic currents, from 5 to 50µA, were tried. The samples are then examined *ex situ* under the Scanning Electron Microscope, in order to observe the pitting sites. Appropriate samples (i.e. samples for which the pit initiation is observable but the inclusion responsible for the pit initiation is not completely dissolved) are selected and the typical morphologies are recorded.

Figure 3 (p.305) shows the typical pits initiation sites. For steel A', whatever the pH or the chloride content, pits initiate either around Al oxides (Fig. 3a), where some MnS are located as shown above, or on MnS inclusions (Fig. 3b) around Nb(C,N) or (seldom) isolated. This confirms the preceding hypothesis. For Steel B, two different behaviours are observed: (i) in NaCl (0.02M), whatever the pH, pitting occurs at the centre of Titanium nitrides (Fig. 3c), which is consistent with the absence of pH dependence related to a suphide dissolution. (ii) in NaCl(0.5M) solution, pitting occurs often at the TiN boundary (Fig. 3.4), where Ti sulphides are present. Even in this case, some pits still occur on Ti nitrides centre, but the pit iniation at the inclusion boundary becomes much more frequent when the pH decreases below the critical value. This shows clearly that for such chloride concentrations, Ti sulphides act as pitting sites and are then unstable when the potential increases. The basic chemical reactions (a,b,c) discussed above should be rewritten without any reference to the Mn presence, only referring to the potential or pH assisted sulphur species transformations.

5. Prepitting Noise

Recording the anodic current during a polarisation below the 'conventional' pitting potential provides 2 types of information. First the average anodic current decreases with time, corresponding to the onset of an improved passivity. Second, some oscillations of this anodic current ('prepitting noise') are observed (peak amplitude from 0.1 to 1 µA), which, using the same

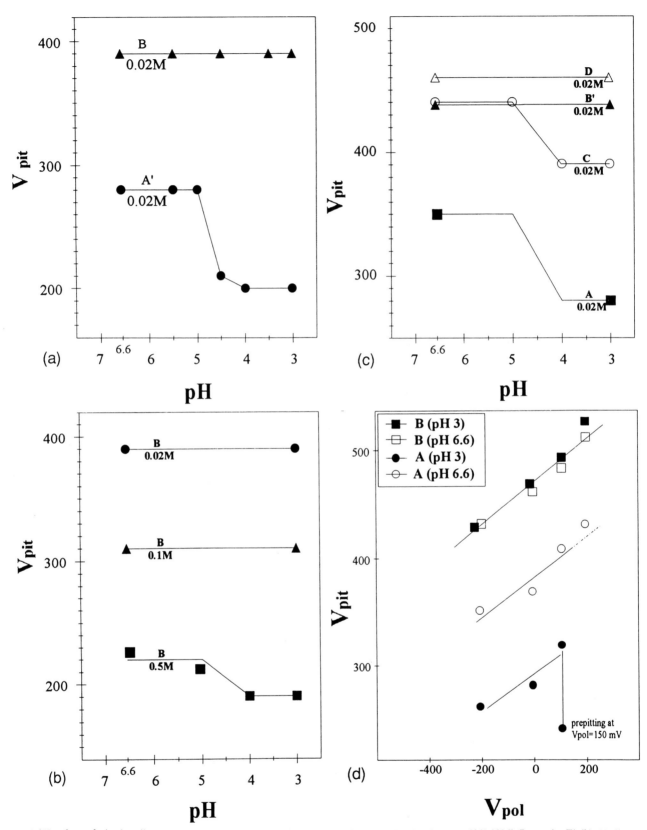

Fig. 2 pH and prepolarisation effects on the pitting potentials (mV/SCE). (a) pH effect on steels A' and B in NaCl (0.02M). B contains Ti. (b) pH effect on steels A, B' (FeCr steels) and C,D (FeCrNi steels in NaCl (0.02M). B' and D contain Ti. A is better than A' since it does not contain Al oxides. (c) Pitting potential pH and chloride content dependence for a Ti containing steel. For low chloride contents (0.02 and 0.1M), there is no pH effects. for 0.5M, lowering the pH decreases the pitting resistance. (d) Effect of a prepolarisation for 1 h at various potentials Vpol = –200 to +200 mV/SCE, in NaCl (0.02M) pH3 or 6.6 on the further pitting potential.

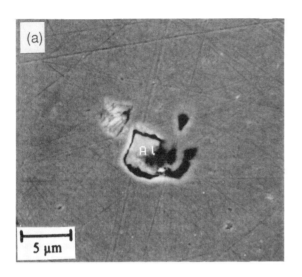

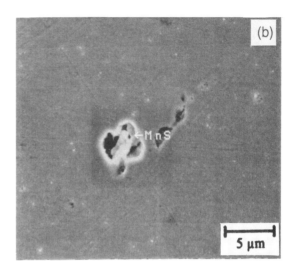

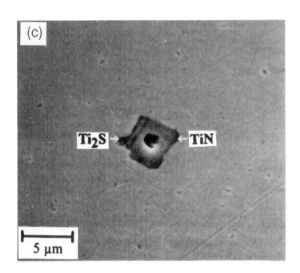

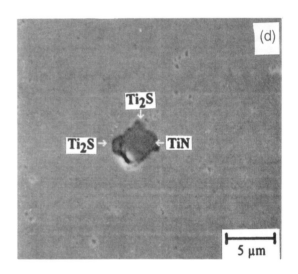

Fig. 3 Pitting sites.

(a) Steel A': Pitting at the boundary of an alumina inclusion.

(b) Steel A': Pitting on a Manganese sulphide.

(c) Steel B: Pitting on a Ti nitride (centre).

(d) Steel B: Pitting at the boundary of a Ti nitride.

(e) Typical view of a semi-developed pit: The inclusion responsible for the pit initiation has disappeared. A thin metallic film is still present and covers a part of the pit. Secondary pits are visible all around the main hole. The pit may either go on or repassivate if the thin 'top' film breaks down.

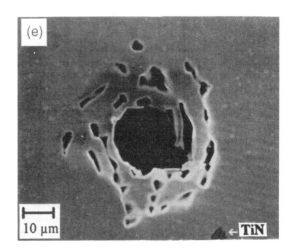

305

SEM observations than above, were shown to correspond to some initiated then repassivated pits. This behaviour is observed only with MnS containing steels (A or A'), the behaviour of steels B or B' being somewhat different (some rare and well separated prepitting events), except in NaCl (0.5M) pH3 where a slight prepitting noise is observed again. It is therefore logical to relate the so called 'prepitting' anodic events to the dissolution and repassivation of sulphides. The problem which arises for analysing the prepitting noise is to separate its signal from the average anodic current decrease which corresponds to the passive film reinforcement. In this purpose we used some well known image analysis techniques [7] whose the description would be out the scope of this paper. The result is to substract from the signal its baseline (corresponding to the average anodic current) without cutting the characteristic fluctuations frequencies. Figure 4(a) shows the typical reduced signal which is obtained; this signal can be analysed in 2 ways. The simpler one is to build up a function N(I) which counts, for each time period (~8 min), the number of events for which the anodic intensity i is larger than I. This function N decreases with I as shown in Fig. 4(b). Two conclusions can be drawn from this analysis: first, the discontinuity of N between pH 4 and 5 is once more in evidence (Fig. 4b). Second, measurements made for different time periods show that the prepitting noise decreases with the polarisation time (Fig. 4c). Further experiments show that the noise increases with the polarisation potential and the chloride content.

A more sophisticated way for analysing the fluctuations signal is to calculate its Fourier transform (FFT algorithm) and, for instance, to derive the Power Spectral Density of this Fourier transform as a function of the signal frequency. The result is consistent with previously published works [8], showing first a plateau at low frequency (cutting frequency = f_c ~ some 10 mHz) followed at higher frequency by a PSD vs frequency –20 dB/decade dependence. From the model presented in [8], one can calculate the pits nucleation (l) and the pits repassivation (μ) rates, which depend on pH in accordance with the herabove sulphides dissolution theory (Fig. 4d), evidencing a critical pH between 4 and 5.

6. Concluding Remarks

It is now well accepted that, for industrial steels, pits initiate on non-metallic inclusions. From another hand, the more sophisticated theories of pits iniation deal with the passive film breakdown mechanisms, with no reference to such metallurgical heterogeneities. The question of the relevance of such theories for industrial

steels is then debatable. On the other hand, manganese sulphides are known as active pitting sites, due to their poor stability in chloride containing media at anodic potentials. Titanium additions increase the pitting resistance due to the better stability of Ti sulphides. The present study shows that these sulphides can also act as pitting sites for high enough chloride contents. Two behaviours are observed:

(i) For Ti containing steels used in 'soft' conditions, Ti sulphides remain stable and Pits initiate on Ti nitrides. No pitting potential pH dependence is observed.

(ii) For MnS bearing, or for Ti containing steels used in more concentrated chloride solutions. Pits initiate on sulphides. A critical pH is then observed below which the pitting potential sharply decreases. This critical pH should be related the reaction kinetics of the sulphur species, irrespectively to the nature of the dissolving sulphides (MnS or Ti2S).

These conclusions motivate further experimental and theoretical investigations on the effect of these kinetics and their incidence on the pit generation limiting process. However, the effect of passive film modifications on pits initiation cannot be so simply ruled out. Ageing at constant potential in the corrosive solution (but below the pitting potential) clearly increases the pitting resistance. The contribution of the passive film should be studied under conditions where sulphides do not act as pitting sites. Several works are currently undergone in our team on this point. It should be noticed that the scale at which the pits 'embryos' are observed, using either electrochemical experiments or electronic microscopy, is much larger that the passive film breakdown typical scale. The latter (few nanometers) is referred to as 'microscopic' and the former (few micrometers) as 'mesoscopic'. Figure 3(e) shows the situation at the near ultimate stage (size scale ~ some 10 μm), just before pit propagates irreversibly: the thin metallic cap above the hollow produced by the pit initiation (with no visible residual inclusion) can still break down and the pit repassivate! Many elementary steps should intervene between the microscopic and the mesoscopic scale, and at each of them repassivation may occur. In a shortcoming paper, we will discuss the deterministic or stochastic character of growth or repassivation mechanisms at each step. The current models in this field are clearly related to the microscopic step, which is never observed experimentally.

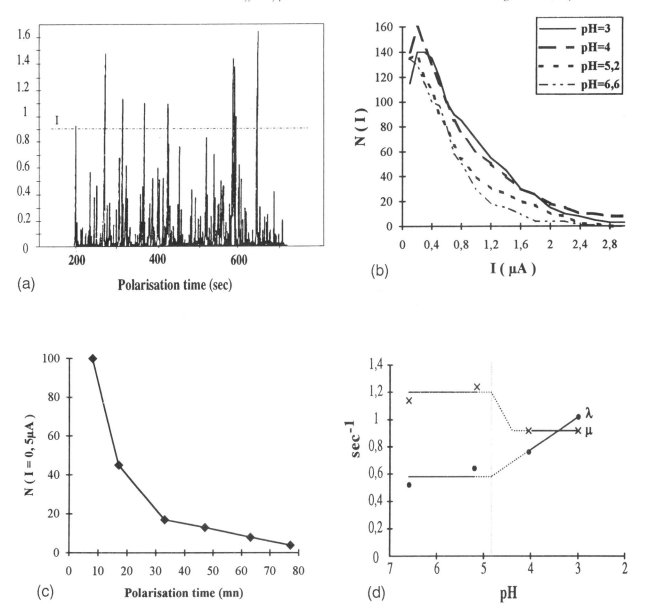

Fig. 4 Prepitting noise for steel A'. (a) Typical prepitting noise at 200 mV/SCE in 0.02M NaCl after subtracting the anodic current baseline. (b) Distribution function N(I) = number of anodic events larger than I. (same conditions as Fig 4(a), but varying the pH). Results at pH 6.6 and 5.2 are similar. A significant noise increase is observed for pH 4 or 3. (c) Effect of the polarisation time: the noise intensily (distribution function) strongly decreases with time (same conditions as Fig 4(a), pH 6.6). (d) pH effects on the nucleation rate (l) and the repassivation rate (m). (same conditions as Fig 4b).

References

1. Z. Szklarska-Smialowska, Pitting Corrosion of metals, NACE publication, 1986.
2. J.-L. Crolet *et al.*, Mem. Sci. Rev. Met. (France), Nov. 1977, p 647.
3. A. Szummer and M. Janik-Czachor, Brit. Corros. J., **9**, 216, 1974.
4. B. Baroux, Corros. Sci., **28**, 969, 1988.
5. G. S. Eklund, J. Electrochem. Soc. **121**, 467, 1974.
6. J. P. Petit, L. Antoni and B. Baroux, this Symposium.
7. D. Lizzarazu, 1992, Diplome d'Etudes approfondies, Institut National Polytechnique de Grenoble (ENSEEG).
8. G. Gabrielli, F. Huet, M. Keddam and R. Oltra, Corrosion (NACE), **46**, 266, 1990.

52

Localised Corrosion of Laser Surface Alloyed 7175-T7351 Aluminium Alloy With Chromium

R. Li, M. G. S. Ferreira, A. Almeida, R. Vilar, K. Watkins* and W. Steen*

Instituto Superior Técnico, 1096 Lisboa Codex, Portugal
*University of Liverpool, Liverpool L69 3BX, UK

Abstract

Laser surface alloying of 7175-T7351 aluminium alloy with chromium using a CO_2 laser modifies the surface composition and the microstructure. In the laser alloyed layer, microcrystalline grains are formed, α-Al phase contains supersaturated Cr element and $CrAl_7$ secondary phase contains supersaturated Al element. The modified composition and microstructure of the alloy enhance the pitting potential of the alloy in deaerated 3%NaCl solution comparatively to their casted and spinned counterparts. The higher the Cr content in the alloyed layer (especially in α-Al phase), the higher the pitting potential is.

1. Introduction

Laser surface modification is a rapidly melting and rapidly resolidifying process which changes both the microstructure and composition of the surface of the substrate or microstructure alone, with foreign elements addition or addition free during the laser processing. Due to these advantages, laser surface modification is a prospective and exciting technique for improving the corrosion resistance of materials. There have been a number of studies on corrosion properties of laser surface melted stainless steel [1, 2], bronze [3], aluminium [4, 5] and zirconium alloy [6], laser chromium alloyed steel [7, 8], laser nickel–chromium alloyed steel [9], laser chromium alloyed and laser molybdenum alloyed aluminium foils [10]. High strength 7000 series aluminium alloys are widely used in aircraft industry as structural materials. These alloys have excellent corrosion resistance to uniform corrosion, but suffer localised corrosion just like other aluminium alloys. The purpose of this study is to improve the localised corrosion resistance of 7175-T7351 aluminium alloy by laser surface alloying with chromium.

2. Experimental Details

2.1 Laser processing

The samples used for laser processing are 7175-T7351 aluminium alloy plates. The laser used in the experiment is a CO_2 laser operated at 2 kW. The alloying powder used in the experiment was a mixture of 75%Al+25%Cr. The powder was blown to the melting pool with delivery gas. To obtain a more uniform composition distribution in the laser melted layer, after laser alloying, the alloyed layer was remelted again by laser beam for one sample. Samples and corresponding laser processing parameters are listed in Table 1. Argon was blown over the melting pool to avoid oxidation of the samples during laser alloying and laser remelting.

2.2 Corrosion tests

All laser alloyed 7175-T7351 samples for corrosion tests were mounted in epoxy resin, abraded with em-

Table 1 Samples and corresponding laser processing parameters

sample	laser alloying				laser remelting		
	power	spot	speed	feed rate	power	spot	speed
Al-1	2KW	1.5mm	10mm/s	0.03g/s	2KW	1.5mm	20mm/s
Al-2	2KW	2mm	10mm/s	0.05g/s			
Al-3	2KW	2mm	10mm/s	0.05g/s			

Al:7175-T7351 aluminum alloy. Al-1 is the sample with laser alloying + laser remelting, Al-2 and Al-3 are samples cut from the same track, but from different places, without remelting by laser. Width of overlapping for all samples is 0.5mm.

ery paper to 1000 grit and polished with 3μm diamond paste. The boundary between the sample and the epoxy resin was sealed with bees wax to avoid crevice corrosion. The surface area of the samples exposed to the solution was 0.2–0.5 cm². Pitting corrosion was studied in 3%NaCl deaerated solution with anodic polarisation. The solution was prepared with analytical grade reagent and distilled water, deaerated with pure nitrogen gas for at least 3 h before each experiment and continuously during the experiment. Before polarisation, the samples were immersed in the solution for 2 h to stabilise the open circuit potential. The potential sweep rate was 30 mV/min. The polarisation tests were performed in a three-electrode cell. The working electrodes were laser alloyed samples, the reference electrode was a saturated calomel electrode and the auxiliary electrode was a platinum foil. A Thompson 251 Ministat, a Thompson DRG16 linear sweep generator and a HP-87 microcomputer connected to the ministat were used to acquire and record the experimental data. After polarisation, the corrosion morphology was observed by optical microscopy and SEM. For comparison, the same kind of procedure was carried out on untreated 7175-T7351 samples under the same experimental conditions.

3. Results and Discussion

Figure 1 presents the anodic polarisation curves for untreated 7175-T7351 and laser surface alloyed 7175-T7351 samples in the deaerated 3%NaCl solution. The average composition of the samples obtained by EDX analysis is shown in Table 2. The pitting potential of untreated 7175-T7351 is –730 mV. The pitting potential of Al-1 with 3.17wt% Cr is –622 mV, the pitting potential of Al-2 with 5.4 wt% Cr is –274 mV and the pitting potential of Al-3 with 4.97 wt% Cr is –378 mV. These

Table 2 Average composition of the alloys and composition of the secondary and matrix phases of the same alloys

	Cr	Cu	Mg	Zn	Al
Al-1 (average)	3.17	1.72	0.65	3.06	91.40
Al-1 (secondary)	10.83	2.31	0.38	2.52	83.96
Al-1 (matrix)	0.97	0.63	0.82	3.89	93.69
Al-2 (average)	5.40	1.56	0.66	3.30	89.07
Al-2 (secondary)	12.04	2.11	0.42	2.67	82.76
Al-2 (matrix)	1.33	0.73	0.33	2.26	95.36
Al-3 (average)	4.97	1.62	0.73	3.41	89.27
Al-3 (secondary)	11.02	1.87	0.35	2.48	84.28
Al-3 (matrix)	1.23	1.36	0.74	3.86	92.81

results show that, with the laser surface alloying parameters shown in Table 1, laser surface alloying of 7175–T7351 with chromium enhances the pitting potential of the alloy. The higher the chromium content in the laser alloyed layer, the higher the pitting potential is.

Figure 2(a) shows the pitting corrosion morphology on Al-2 sample after polarisation. From this figure, it was observed that the interspace between the dendrites is the preferential site for pitting attack. The pits nucleate at those interspace regions leaving the less corroded dendrites forming a dendrite skeleton. As the pits propagate, due to a decrease of the pH inside them, the areas of the attack spread, causing the corrosion of the dendrites and forming larger pits as indicated by the arrows. Figure 2(b) shows the pitting morphology on Al-1 sample which is very similar to that on Al-2. The pits also nucleate at the interspace between the dendrites leaving the less corroded finer dendrites (compared to the dendrites in the Al-2 sample) forming a dendrite skeleton; in the propagation stage, they also spread out forming larger pits. Figure 2(c) shows the crystallographic pitting corrosion morphology, which relates to the tunnel propagation of the pit in the NaCl solution, on 7175-T7351 aluminium alloy as received.

Figure 3 are scanning electron images of the cross sections of the laser surface alloyed layer of samples Al-2 and Al-1 polished with 3μm diamond paste, etched with Keller's reagent. The alloyed layer mainly consists of two phases. One is a matrix phase, the other is a secondary phase appearing as dendrites, dispersed in the matrix. Figure 3(a) shows the microstructure of Al-2. The size of the dendrites ranges from 1 to 10μm. Figure 3(b) shows the microstructure of Al-1. Since the transverse speed of the sample in the laser

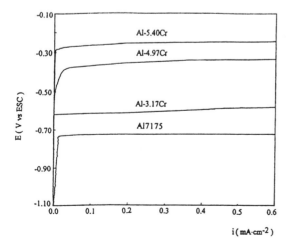

Fig. 1 Anodic polarisation curves of laser alloyed Al7175-T7351 with chromium and untreated one in deaerated 3%NaCl solution.

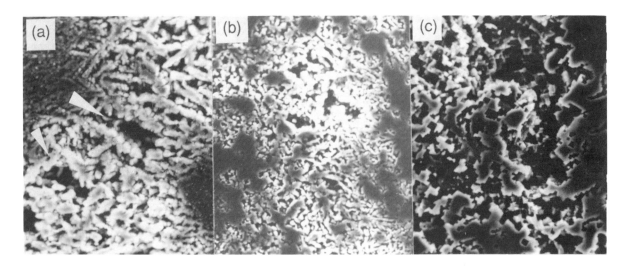

Fig. 2 *Scanning electron micrographs showing the pitting morphologies (× 1000).*
(a) Laser chromium alloyed 7175-T7351 sample (Al-2) after anodic polarisation; (b) laser chromium alloyed + laser remelted 7175–T7351 sample (Al-1) after anodic polarisation; (c) 7175–T7351 sample (as received) after anodic polarisation.

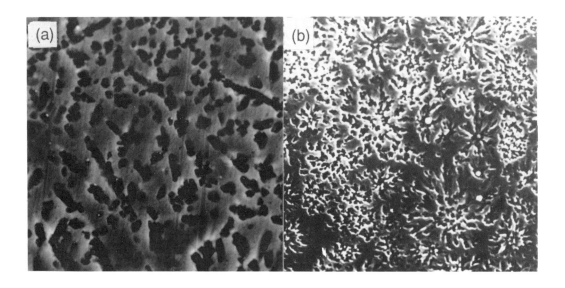

Fig. 3 *Scanning electron images showing the microstructure of the laser surface alloyed layer. (× 2000).*
(a) Laser chromium 7175-T7351 sample(Al-2); (b) laser chromium alloyed + laser remelted 7175-T7351 sample (Al-1).

remelting process is higher than that in the laser alloying process, an higher cooling rate is expected, giving a finer microstructure, compared to that of Al-2. The composition of the phases for each one of the samples obtained by EDX point chemical analysis is also listed in Table 2. From the Al-Cr binary diagram [11] and the EDX analysis, it seems that the laser melted layers of Al-1, Al-2 and Al-3 consist of two phases: α-Al phase in which Cr, Cu, Mg, and Zn are dissolved and $CrAl_7$ phase in which Al, Cu, Mg and Zn are dissolved. The α-Al phase appears as the matrix and the $CrAl_7$ phase appears as the dendrites dispersed in the matrix. From

the Al–Cr binary diagram, at 300° C, the solid solubility of Cr in α-Al phase is *ca.* 0.04wt% and at 400° C, the aluminium content in the $CrAl_7$ phase is *ca.* 78.15wt%. Laser surface modification is a rapidly melting and a rapidly resolidifying process. In the resolidifying process, the solid-liquid interface moves more rapidly and entraps more solute elements than in the equilibrium solidifying process, because there is less time for the diffusion of the solute elements, thus the α-Al phase entraps more chromium, and the $CrAl_7$ phase entraps more aluminium making the α-Al phase supersaturated with chromium and the $CrAl_7$ phase supersaturated with aluminium. The α-Al phase in the laser alloyed layer contains more chromium than that in the conventional cast alloy. The supersaturated chromium in the α-Al phase and the microcrystalline structure in the laser alloyed layer are beneficial to the pitting corrosion resistance of the alloyed layers [12–15].

Figure 4 shows the curve of the pitting potentials of the alloys produced by laser surface alloying vs chromium content. For comparison purpose, the pitting potentials of rapidly solidified aluminium alloys produced by spinning in deaerated 0.5N NaCl (2.9wt%) and of the corresponding conventional cast alloys taken from reference [14] are also included in the figure. The pitting potentials of alloys both produced by laser alloying and by spinning increases with the increase of the chromium content in the alloys, but the pitting potentials of the alloys produced by laser alloying increase much faster with the increase of the chromium content; even the pitting potentials of the alloys produced by laser with 4.97wt% Cr and 5.40wt% Cr are higher than those of the alloys produced by spinning with 7.43wt% Cr and 10.95wt% Cr. As shown in Fig. 2(a) and (b) the α-Al phase was preferentially attacked by pitting corrosion, so the chromium content in the α-Al phase, which is limited by the solid solubility of chromium in the α-Al phase, is more determinative of the pitting corrosion resistance of the alloys. The solid solubility of the solute element is determined by the phase diagram of the alloy and cooling rate of liquid alloy during the solidification process; the higher the cooling rate, the larger the solubility is. It is reasonable to believe that, with the parameters in the Table 1, the cooling rates of the liquid alloys in the laser alloying process are higher than those of the liquid alloys in the spinning process, thus the chromium supersaturated solubility in the α-Al phase in the alloys produced by laser alloying is higher than that in the α-Al phase in the alloys produced by spinning making the alloys produced by the former process more pitting resistant. According to Szklarska-Smialowska's recent work [16] on the effect of alloying elements on the pitting corrosion of aluminium alloy, the Cl⁻ and water can penetrate the passive film through the pre-existing defects forming a pit nuclei in the passive alloy. As the applied potential increases, more Al^{3+} ion is produced by dissolution of the bulk metal and the hydrolysis of the ion reduces the pH value of the solution in the pit nuclei. When the pH is low enough, the passive film in the pit nuclei is dissolved, then stable dissolution of bulk metal occurs, giving place to pitting propagation. When chromium is present, the CrOOH and/or Cr_2O_3 film formed is less soluble in acid solution than aluminium oxide, so a

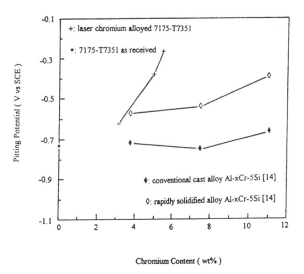

Fig. 4 Curves of the pitting potentials vs chromium content in the alloys.

higher potential is needed to produce enough Al^{3+} ions, that by hydrolysis increase the acidity of the solution in the pit nuclei and dissolve locally the CrOOH film formed on the alloy containing chromium. The pitting potential of the alloy is thus increased by alloying with chromium. The higher chromium content in the laser alloyed layer (especially that dissolved in the α-Al phase), the easier the formation of a CrOOH film, so the pitting potential of Al-2 is higher than that of Al-3 and the pitting potential of Al-3 is higher than that of Al-1.

4. Conclusions

1. Laser surface alloying of 7175-T7351 aluminium alloy with chromium produces a microcrystalline layer on the alloy surface with extended solubility of chromium in α-Al phase and extended solubility of aluminium in the $CrAl_7$ phase.

2. The presence of chromium in the 7175-T7351 laser surface alloy enhances the pitting corrosion resistance of 7175-T7351 greatly by the formation of a CrOOH passive film which retards the propagation of the pitting. That enhancement is higher than the one obtained for their casted and spinned counterparts, due to an increase of the solubility of the chromium in the α-Al phase in the alloy obtained by laser alloying.

3. The higher the chromium content in the laser alloyed layer (especially that dissolved in the α-Al phase), the higher the pitting potential is.

5. Acknowledgement

The authors gratefully acknowledge the support received from the E. C. under contract BREU-CT 91-0494.

References

1. T. R. Anthony and H. E. Cline, J. Appl. Phys., **49**, 1248, 1978.

2. J. Stewart, D. B. Wells, P. M. Scott and A. S. Bransden, Corrosion, **46**, 618, 1990.

3. C. W. Draper, R. E. Woods and L. S. Meyer, Corrosion, **36**, 405, 1980.

4. E. McCafferty, P. G. Moore and G. Peace, J. Electrochem. Soc., **129**, 9, 1982.

5. P. L. Bonora, M. Bassoli and G. Battaglin, Electrochim. Acta, **25**, 1497, 1980.

6. J. Rawers, W. Reitz, S. Bullard and E. K. Roub, J. Electrochem. Soc., **47**, 769, 1991.

7. P. G. Moore and E. McCafferty, J. Electrochem. Soc., **128**, 1391, 1981.

8. S. Chiba, T. Sato and A. Kawashima, Corros. Sci., **26**, 311, 1986.

9. L. Renaud, F. Fouquet, J. P. Millet and J. L. Crolet, Surf. and Coat. Tech., **45**, 449, 1991.

10. P. L. Hagans and R. L. Yates, Laser treatment of thin molybdenum and chromium films on aluminum for enhanced corrosion resistance, in Environmental Degradation of Ion and Laser Beam Treated Surface, (G. S. Was and K. S. Grabowski ed.), The Minerals, Metals and Materials Society, 1989.

11. Kent R. Van Horn, Aluminum, Vol. One: Properties, Physical Metallurgy and Phase Diagrams, American Society for Metals, (1967), pp371-372.

12. W. C. Moshier, G. D. Davis and G. O. Cote, J. Electrochem. Soc., **136**, 356, 1989.

13. G. D. Davis, W. C. Moshier, T. L. Fritz and G. O. Cote, J. Electrochem. Soc., **137**, 422, 1990.

14. H. Yoshioka, S. Yoshida, A. Kawashima, K. Asami and K. Hashimoto, Corros. Sci., **26**, 795, 1986.

15. D. Lui, F. Wang, Ch. Cao, H. Lou, L. Zhang and H. Lin, Corrosion, **46**, 975, 1990.

16. Z. Szklarska-Smialowska, Corros. Sci., **33**, 1193, 1992.

53

The Problem of Convective Diffusion in Laser Pit Initiation

I. O. EFIMOV AND V. A. BENDERSKII

Institute of Chemical Physics. 142432 Chernogolovka, Moscow region, Russia

Abstract

Laminar convective motion of liquid close to an inhomogeneously heated electrode is investigated. It is shown that flow of liquid is directed from the outside of the heated spot to its centre and then near the critical point turns to the bulk of solution. Due to this motion the surface reaction consuming reactant from the bulk of solution is accelerated at the periphery and inhibited at the centre of the heated zone.

1. Introduction

The local rates of electrochemical reactions can be raised by one to three orders of magnitude by laser heating of metal electrode surfaces. It was shown that an acceleration of electrode processes could be due to the Arrhenius dependence of reaction rate of the temperature, to a shift of the equilibrium potential with the temperature [1], to the generation of bare metal after passive film breakdown [2], to plastic deformations of the metal [3] and to convective motion of solution near the heated surface.

The contribution of each of these effects depends on the time and intensity of irradiation. The first two of the above reasons are of major importance when the heating is slight but powerful (the intensity $L < 10^4 \text{W}/\text{cm}^2$, millisecond duration of laser pulse). The third of these reasons is predominant when the heating is powerful ($L > 10^6 \text{W}/\text{cm}^2$) but brief (nanosecond duration of laser pulse).

It was shown that laminar convective flow starts upon heating to the temperatures below the boiling point [1]. Turbulent flow starts when the solution boils up. This agitation of solution gives rise to the increase of the rate of diffusion controlled reactions. In the turbulent mode, reactant concentration becomes uniform near the heated area due to intensive mixing, thus no peculiarities appear in surface reaction rate spatial distribution. However, in laminar mode, on the contrary, the reactant concentration is not averaged out and a spatial dependence of the electrochemical reactions could be expected. This effect has not been investigated so far and the present work fills this gap.

2. Theoretical

The distribution of temperature (T) and liquid motion in the flow near the heated surface is described by continuity equation and equation of convective heat transfer

$$\frac{\partial V_x}{\partial x} + \frac{\partial V_y}{\partial y} = 0 \tag{1}$$

$$V_x \frac{\partial T}{\partial x} + V_y \frac{\partial T}{\partial y} = \chi \left(\frac{\partial^2 T}{\partial x^2} + \frac{\partial^2 T}{\partial y^2} \right) \tag{2}$$

Axis y is directed towards the bulk of solution, axis x - along the surface (Fig. 1), V_x and V_y are the components of velocities in the liquid flow, χ - is the thermal conductivity of solution. We consider a 2D case for the sake of simplicity. The boundary condition for the system (1)-(2) corresponding to the stationary gaussian temperature distribution reads

$$T(x) = T_0 \exp(-x^2/b^2) \tag{3}$$

where b is gaussian parameter of the laser beam, T_0 is the maximum temperature.

Taking into account the symmetry of the system and the fact that the flow contains the turning point ((0.0) in Fig. 1), the solution of (1)–(3) can be readily obtained

$$V_x = -\frac{2\chi}{b^2} x; \qquad V_y = -\frac{2\chi}{b^2} y; \tag{4}$$

$$T(x,y) = T_0 e^{(y^2 - x^2)} \left(1 - \frac{2}{\sqrt{\pi}} \int_0^{y/b} e^{-\xi^2} d\xi \right) \tag{5}$$

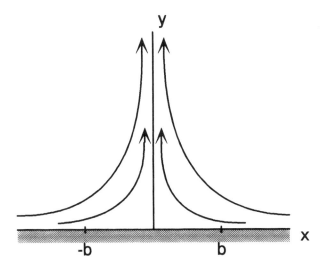

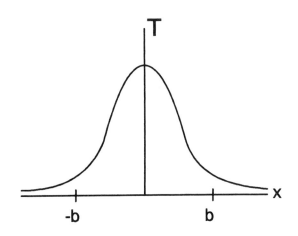

Fig. 1 Motion of liquid near a heated surface segment with the temperature distribution T(x).

Equation (4) describes the flow in the bulk of solution and the first of them should be used as the outer condition U(x) in the Prandtl equation for the boundary layer

$$V_x \frac{\partial V_x}{\partial x} + V_y \frac{\partial V_y}{\partial y} - \nu \frac{\partial^2 V_y}{\partial y^2} = U \frac{dU}{dx} = \frac{4\chi^2}{b^4} x \qquad (6)$$

where ν is the viscosity of liquid. The boundary condition on the surface reads

$$V_x \big|_{y=0} = V_y \big|_{y=0} = 0 \qquad (7)$$

Velocity distribution near the surface (when $y \ll b$) gains the form

$$V_x = -\frac{2\chi^2}{\nu b^4} xy^2; \quad V_y = \frac{2}{3} \frac{\chi^2}{\nu b^4} y^3 \qquad (8)$$

We shall assume now that a substance with the

bulk concentration c_0 is present in the solution. The profile $c(x,y)$ near the heated spot obeys the equation of convective diffusion for the boundary layer

$$V_x \frac{\partial c}{\partial x} + V_y \frac{\partial c}{\partial y} = D \frac{\partial^2 c}{\partial y^c} \qquad (9)$$

The boundary conditions are defined as follows. At $x > b$ the surface is not irradiated and reaction does not proceed.

$$c(|x| > b, \ y = 0) = c_0 \qquad (10)$$

At $x < b$ laser activation is so intensive that reaction occurs under diffusion control, i.e.

$$c (\ |x| > b, y = 0) = c_0 \qquad (11)$$

Problem (9)–(11) with velocity distribution in the form of (8) can be solved analytically. Concentration distribution reads

$$c = c_0 \frac{4}{\Gamma(1/4)} \int_0^\xi e^{-t^2} dt \qquad (12)$$

where

$$\xi = \frac{1}{6^{1/4}} \frac{xy}{b^2} \left[1 - \frac{x^4}{b^4} \right]^{-1/4} \left[\frac{\chi^2}{\nu D} \right]^{1/4} \qquad (13)$$

The lines of constant concentration are determined by equation

$$y = const \frac{(b^4 - x^4)^{1/4}}{x} \qquad (14)$$

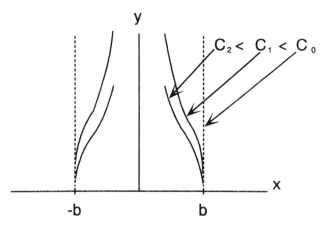

Fig. 2 Constant concentration profiles c (x,y) of substance which reacts at the surface: at |x| < b the reaction is in diffusion regime, at |x| > b the reaction rate is zero. c_0 is bulk concentration.

314

and shown in Fig. 2. Limiting diffusion current density is given by

$$j(x) = \frac{\partial c}{\partial y}\bigg|_{y=0} = \frac{4}{6^{1/4}\,\Gamma(1/4)}\left[\frac{D^3\chi^2}{\nu}\right]^{1/4}\frac{c_0}{b}\frac{x/b}{(1-x^4/b^4)^{1/4}} \qquad (15)$$

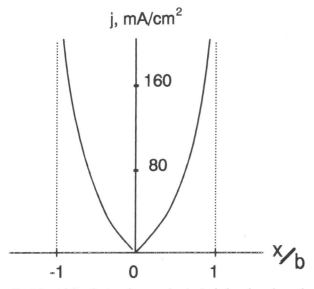

Fig. 3 Spatial distribution of current density in the laser heated spot: for $|x| < b$ calculations were done according to (15) for aqueous solution, $c_0 = 1M$, $D = 10^{-5}\,cm^2/c$, $n = 6.75x = 0.01cm^2/c$, $|x| > b\ j(x) = 0$.

Current density increases from the centre of spot (Fig. 3) and due to idealised boundary conditions tends to infinity at the boundary, but surface integral of current I is finite and does not depend from the heated spot

$$I = 2\int_0^b j(x)dx = (2/3)^{1/4}\frac{16\pi^{3/2}}{5\Gamma^3(1/4)}\left[\frac{D^3\chi^2}{\nu}\right]^{1/4}c_0 \qquad (16)$$

3. Discussion

This model can be applied to analysis of dislocation etching patterns on the (001) plane of a tungsten monocrystal which were obtained in [3] by laser treatment of W-electrode in 2% NaOH. Etched pits are arranged periodically at the outlying area and no etching occurs at the centre. The following polishing and etching show that dislocations fill all the spot. So, determined by dislocation density, electrochemical activity of the surface is uniform. However dislocation etching on W occurs only in the presence of OH⁻ ions. Convective flow acidifies solution in the centre of laser spot and prevents the generation of etching holes.

4. Conclusion

The main result of this treatment is that a surface process consuming reactant from the bulk of solution is inhibited in the centre of the laser spot and asselerated at the boundary. As etching and decoration of dislocations take place only in the presence of anions the described mechanism leads to acceleration of this processes at the outlying area of laser spot.

References

1. J. C. Puipee, R. E. Acosta and R. J. von Gutfeld, J. Electrochem. Soc., **128**, 2539, 1981.
2. R. Oltra, G. M. Indrianjafy and M. Keddam, Materials Science Forum, **44–45**, 259, 1989.
3. V. A. Benderskii, I. O. Efimov and A. G. Krivenko, J. Electroanal. Chem., **315**, 29, 1991.

Fig. 4 Spatially nonuniform dislocation etching induced by laser treatment of (001) plane W-monocrystal in 2% NaOH.

54

Pitting Corrosion of Heat-tinted Stainless Steel

P. C. Pistorius and T. von Moltke*

University of Pretoria, South Africa
*Iscor Ltd.

Abstract

During welding the metal adjacent to the weld becomes covered with a thickened oxide film which is thought to have a deleterious effect on subsequent corrosion behaviour. The thickened oxide film, known as 'heat tint', can thus be regarded as a damaging modification of the passive film. Specimens of AISI type 304 stainless steel were heated in a welding simulator to grow these oxide films and were subsequently examined by Auger electron spectroscopy. The composition of the oxide film showed a clear transition with increasing peak temperature. At intermediate temperatures the film is relatively poor in chromium; these films show the greatest susceptibility to pitting corrosion. The pitting potential measurements are supplemented with electrochemical noise measurements to elucidate the effect of the heat tint film on the nucleation and growth of metastable pits. During the growth and formation of metastable pits, the passive film has two opposing roles—firstly it provides the good corrosion resistance of the metal. but it also forms a cover over the embryonic (metastable) pits, so supporting their initial growth. These heat modified films provide a good way to study the relative importance of these two roles.

1. Introduction

The formation of a stable pit on stainless steel is preceded by metastable pit growth [1]. During metastable growth, the diffusion distance within the pit is insufficient to maintain the pit anolyte at the concentration required for continued growth, and pit growth is sustained by a cover over the pit mouth [2]. This cover consists of the original metal surface. The probability of forming a stable, destructive pit thus depends on both the rate at which metastably growing pits are nucleated, and the probability that each pit will survive long enough to become stable [3]. The surface condition of the steel (i.e. roughness and the nature of the surface film) is expected to affect both pit nucleation and pit survival. Previous work has shown that the poor pitting behaviour of stainless steel cover with heat tint (thermal oxide that forms alongside welds) is associated with the presence of an iron-rich oxide on the metal surface; this oxide forms in the temperature range 400–700°C [4]. The present work is an investigation, by electrochemical noise measurements, into the effect of heat-tint on pit nucleation and pit survival.

2. Experimental

Wire loop specimens (see Fig. 1) of annealed type 304L stainless steel were used (composition in weight percentages 17.76% Cr, 10.98% Ni, 0.72% Mn, 0.21% Mo, 0.015% C, 0.0245% P, 0.0019% S). The diameter of the wire was 0.5 mm. Four specimens each with four different surface finishes were prepared: wet abraded parallel to the wire axis with 600 grit silicon carbide paper; wet abraded and then pickled in 15% HNO_3 at 60°C for 30 min; abraded, pickled and heated in air (with a welding simulator) to two different peak temperatures for 20 s. For temperature control during heating an R type thermocouple was welded by capacitive discharge in the position shown in Fig. 1. The thermocouple position was chosen to keep remnants of the thermocouple wire above the electrolyte during subsequent testing, so avoiding possible galvanic effects. The actual temperature reached by the wire loop was not known exactly, since thermal conduction down the 100 μm dia. thermocouple wires caused the contact point to be colder than the rest of the wire. However, by comparing the film compositions (determined by Auger analysis) with those found in previous work [4], the two peak temperatures were found to be approx. 400 and 800°C respectively.

The loop specimens (exposed area 0.3 cm²) were tested under open-circuit conditions in stagnant 0.1 M $FeCl_3$ at 31 ± 1°C; the solution was not deaerated. No attack was observed at the solution-air interface. Fluctuations in the corrosion potential of the wire was measured relative to a silver-silver chloride reference electrode; the tip of the electrode was placed about 10 mm below the tip of the loop (Fig. 1), and 20 mm to the side of the loop. The corrosion potential of each specimen was measured at a 50 Hz sampling rate for an hour after immersion in the $FeCl_3$ solution. A 16 bit

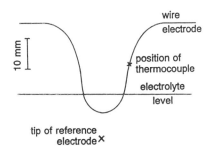

Fig. 1 Shape of wire loop electrode (approximately to scale). The position where the thermocouple was attached during heating is indicated, as well as the immersion depth in the electrolyte.

analogue-to-digital conversion card was used, with a Keithley model 614 electrometer as a high-impedance buffer amplifier.

3. Results and Discussion

Auger electron spectra and depth profiles (measured with a Perkin-Elmer PHI 600 scanning Auger microprobe) are presented in Figs 2 and 3. The spectra in Fig. 2 show the composition of the outer layers of the films (before any sputtering). Comparison of Fig. 2 (a) and (b) shows the familiar increase in chromium content of the passive film which is associated with passivation [5]. The spectra for the thermal oxides show that, in agreement with previous work, no chromium can be detected in the outer layers of the 400°C film, while the outer layers of the 800°C film contain both chromium and manganese in significant amounts, as well as some iron. The depth profiles in Fig. 3, recorded at the same sputtering rate, indicate that the 800°C film is much thicker than the 400°C. This was also clear from the visual appearance: the 400°C specimens had a golden colour, while the 800°C specimens were dark blue. The depth profiles further indicate that the 400°C film has a dual structure, consisting of an iron-rich outer layer over a chromium-rich inner layer, with some chromium depletion of the metal below the oxide. The 800°C film is more homogeneous in composition. Formation of iron-rich oxides on 304 stainless steels at intermediate temperatures has been reported before, and was ascribed to the more rapid kinetics of iron oxide formation, despite the fact that formation of the chromium oxide is thermodynamically favoured [6].

A typical potential transient from the metastable growth and repassivation of a pit is depicted in Fig. 4. The pit grows during the period that the potential falls. Repassivation occurs at the point where the potential starts to rise again. Repassivation is very rapid, being finished in well under 1 s [2]. The slowly rising part of the potential transient does not reflect repassivation,

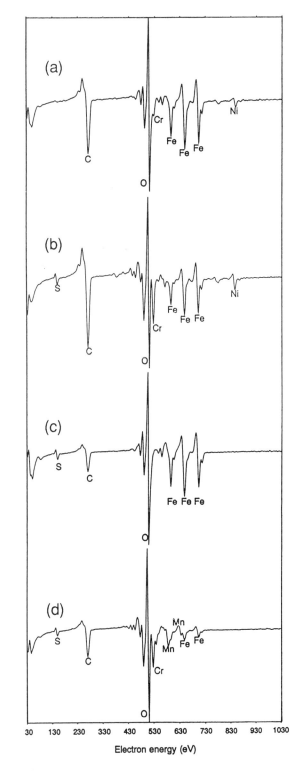

Fig. 2 Auger electron spectra of films on type 304L wires before sputtering. (a) Abraded; (b) Passivated; (c) Heated to ca. 400°C for 20 s; (d) Heated to ca. 800°C for 20 s.

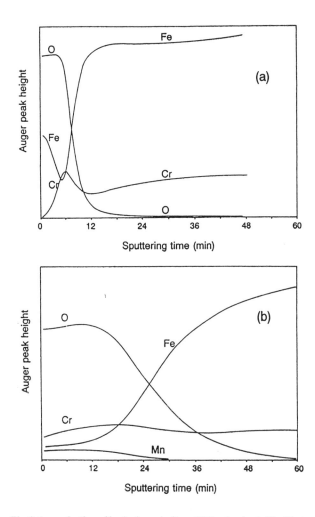

Fig. 3 *Auger depth profiles (redrawn) of type 304L wires heated for 20 s to* (a) 400°C *and* (b) 800°C.

but instead results from recharging of the capacitance of the wire loop electrode [7]. Most of the charge used to support pit growth is drawn from the capacitance, because little current is available from the cathodic reaction on the electrode surface (Fig. 5).

Typical behaviour of passivated and abraded specimens is compared in Fig. 6. Clearly very little pitting activity occurs on the passivated specimen, while many metastable pits form on the abraded specimen. The rate of pit formation decreases with time on the latter specimen, a familiar effect [8]. On average, the potential of the abraded specimen rises with time, indicating that a stable pit has not formed during the exposure time; stable pitting causes a persistently falling potential. No stable pits were observed on the passivated specimens, while two out of four abraded specimens developed stable pits. All the specimens covered with thermal oxides suffered stable pitting.

While few transients were observed on the passivated specimens, the potential variations were large on average. In fact, the largest transient of all (i.e. the largest metastable pit) was observed on a passivated specimen. This is reproduced in Fig. 7, together with typical potential traces from oxidised specimens. These indicate that the 800°C specimen showed few, large fluctuations, while the 400°C exhibited many small transients. This is emphasised by Fig. 8, which shows the detail from corrosion potential recordings of abraded, 400 and 800°C specimens. These traces were taken at the same time intervals after immersion, for specimens with approximately the same average corrosion potential, and when the corrosion potential was on average increasing with time (indicating that the electrode behaviour is not dominated by stable pit-

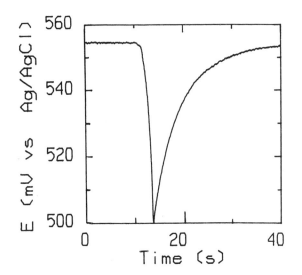

Fig. 4 *Potential transient resulting from metastable pit growth on 304L wire in 0.1 M FeCl₃.*

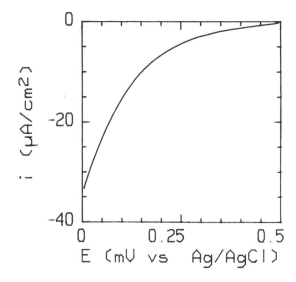

Fig. 5 *Cathodic polarisation diagram of passivated type 304L wire (0.3 cm²) in 0.1 M FeCl₃. Scan rate 5 mV s⁻¹.*

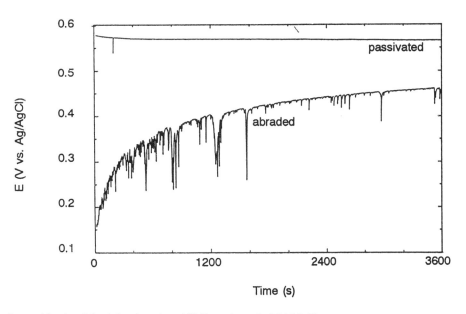

Fig. 6 Comparison of potential noise of abraded and passivated 304L specimens in 0.1 M FeCl₃.

ting). Note that the potential scales in Fig. 8 (a) and (b) are equally sensitive, while that in Fig. 8 (c) is four times as sensitive.

The differences in electrochemical noise are echoed by differences in pit morphology (Fig. 9). After testing, the 400°C specimen was covered by numerous, small pits (which had grown metastably). Far fewer and somewhat larger pits (on average) were present on the abraded specimen. The pits on the 800°C specimen were fewer and larger still, while the passivated specimen had the fewest and largest metastable pits.

These results indicate that the rate of metastable pit nucleation is the highest on the 400°C specimens, lower on the abraded specimens, lower still on the 800°C specimens, and lowest on the passivated specimens. The magnitude of the potential transients reflect the sizes to which the pits grow before repassivating, and can thus be taken as an indication of survival probability. This probability follows the reverse sequence to the nucleation rate: smallest on the 400°C specimens, and largest on the passivated specimens. Thus, while both the 400 and 800°C specimens showed poorer pitting resistance than both the abraded and the passivated specimens, this was for opposite reasons: stable pits form on the 400°C specimen because, while the survival probability of any one pit is low, a very large number of pits are nucleated. In contrast, few pits form on the 800°C specimen, but the survival probability of each is large.

4. Conclusion

These results emphasise that two distinct processes

are involved in the establishment of a stable pit on stainless steel: nucleation, and survival. These two processes may be influenced in different ways by heat tint oxide films. Unfortunately space does not allow discussion of the reasons for these effects—survival is presumably controlled by the mechanical integrity of the film, while nucleation may be affected by the conduction characteristics of the film [9] or by chromium depletion below the film. In the case of passivation the reduction in nucleation rate may be due to sulphide removal, and not to changes in film composition at all [5].

References

1. L. Stokert, Metastabile Lochfrasskorrosion, Dissertation No. 8632, ETH Zurich, 1988.
2. P. C. Pistorius and G. T. Burstein, Phil. Trans. R. Soc. Lond. A, **341**, 531–559, 1992.
3. D. E. Williams, C. Westcott and M. Fleischmann, J. Electrochem. Soc., **132**, 1796–1804, 1985.
4. T. von Moltke, P. C. Pistorius and R. F. Sandenbergh, INFACON 6, Proc. 1st Int. Chromium Steels and Alloys Congress, Cape Town, Vol.2, 185–195, 1992.
5. M. A. Barbosa, A. Garrido, A. Camphilo and I. Sutherland, Corros. Sci., **32**, 179–184, 1991.
6. G. Betz, G.K. Wehner, L. Toth and A. Joshi, J. appl. Phys., **45**, 5312–5316, 1974.
7. H.S. Isaacs and Y. Ishikawa, J. Electrochem. Soc., **132**, 1288–1293, 1985.
8. G. Daufin, J. Pagetti, J.P. Labbe and F. Michel, Corrosion, **41**, 533–539, 1985.
9. G. Bianchi, A. Cerquetti, F. Mazza and S. Torchio, Corros. Sci., **12**, 495-502, 1972.

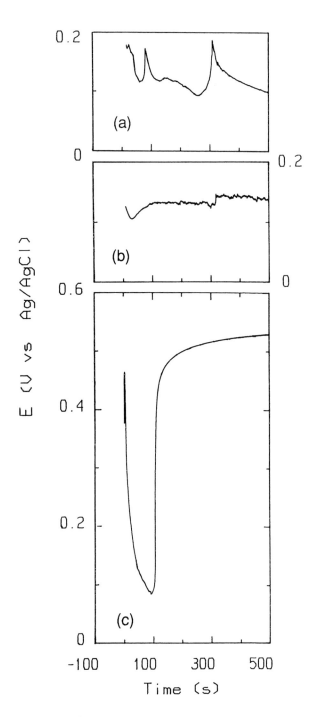

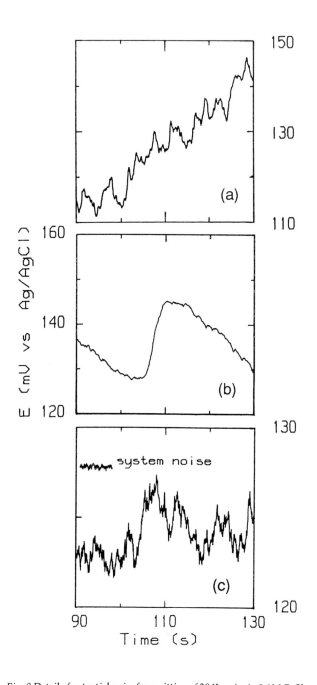

*Fig. 7 Potential noise from pitting of 304L wires in 0.1 M FeCl₃.
(a) Specimen heated to ca. 800°C.
(b) Specimen heated to ca. 400°C. (c) Passivated specimen.*

*Fig. 8 Detail of potential noise from pitting of 304L noise in 0.1M FeCl₃.
(a) Abraded specimen. (b) Specimen heated to ca. 800°C. (c) Specimen
heated to ca. 400°C; inherent noise of measuring system is also shown.*

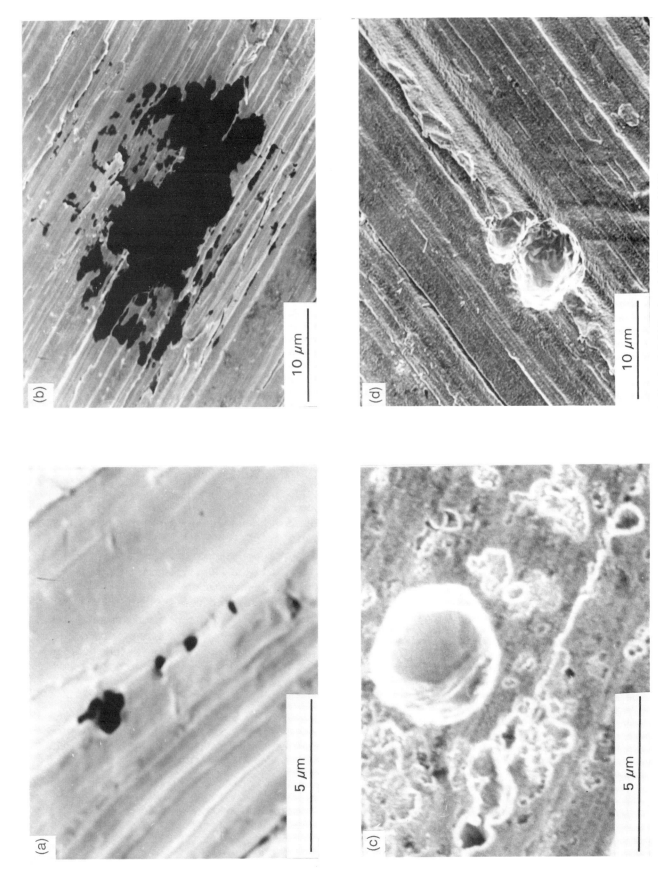

Fig. 9 Scanning electron micrographs of metastable pits on 304L specimens after exposure to 0.1M FeCl₃. (a) Abraded; (b) Passivated; (c) Heated to ca. 400 °C; (d) Heated to ca. 800 °C.

55

Pitting Corrosion Susceptibility of Oxide Layers in the Amorphous Alloy $(Fe_XNi_{1-X})_{0.8}(X, Y)_{0.2}$

M. L. ESCUDERO, A. R. PIERNA*, J. C. GALVÁN AND F. FERNÁNDEZ*

Centro Nacional de Investigaciones Metalúrgicas, Avda. Gregorio del Amo 8, E-28040 Madrid. Spain
*Dpto. Ing. Quim. y Medio Amb., P.O.Box 1379, San Sebastian, Spain

Abstract

$(Fe_XNi_{1-X})_{0.8}(X,Y)_{0.2}$ metallic glasses were obtained by a rapid solidification technique. Reversible oxide layers of variable thickness were generated in a alkaline medium on the surface of these metallic glasses, using an electrochemical technique of potential triangular sweeps. The composition of the metallic matrix as well as the oxide layers electrochemically generated have been studied both by X-ray photoelectron spectroscopy and specular reflactance spectroscopy. Pitting corrosion susceptibility tests were realised in NaCl solution 0.05 M, pH 6.8 ± 0.1 at room temperature on the basis of a study of anodic polarisation curves. The generated layers are not only stable in the medium in which they were generated, but also in other corrosive media with different pH, such as a NaCl neutral solution. The glass with a generation of layers of maximum thickness substantially improve pitting resistance in comparison to control specimens in reception state. The best behaviour corrosion was found in the $Fe_{38.5}Ni_{38.5}Mo_2Si_{13}B_8$ glass, with a typical curve for material in a passive state. Finally, the different peaks observed in the SRS spectra of this last glass insinuate the existence of different chemical species and/or different state oxidation of the elements of these layers when they were generated after 200 or 400 potential cycles.

1. Introduction

Rapid solidification is now used to refer a wide variety of laboratory and industrial techniques processes for cooling molten metals at speeds of over 10^4 Ks^{-1}. The purpose of this new technology is to obtain metal materials with an excellent combination of properties that make them an attractive alternative and often serious competitors to conventional alloys in diverse industrial applications [1].

Metallic glasses stand out amongst these new materials. Their applications are varied: mechanical, chemical, electrical and magnetic [2]. In the field of corrosion, metallic glasses with an Fe base and additions of metalloids and, above all, metallic solutes such as Cr and Mo, are particularly effective in severe corrosive media[3, 4]. It is known from literature that different metalloids show a very different effect on the corrosion behaviour [4–7].

In this work the formation of surfacial layers of oxide in alkaline environments and the corrosion resistance of different metallic glasses are studied.

2. Experimental

The metallic glasses were obtained in the E.U.I.T.I. industrial chemistry laboratory in San Sebastian. Experiments with $Fe_{40}Ni_{40}B_{20}$, widely discussed in literature, were used to upgrade the rapid solidification system. The results also served to model the authors'

solidification process and compare it with models established by other researchers [8].

The metallic glasses used to study electrochemical behaviour had the following nominal composition (in % atoms): $Fe_{40}Ni_{40}B_{20}$, $Fe_{40}Ni_{38}Mo_4B_{18}$ and $Fe_{38.5}Ni_{38.5}Mo_2Si_{13}B_8$. The study was based on the generation and growth of oxide layers in an alkaline medium.

The amorphous ribbons were 20–30 mm thick and 10 mm wide.

The amorphous state of the samples was identified by X-ray diffraction using Co Kα radiation.

The samples were chemically analysed with the help of an atomic absorption spectrophotometer.

Prior to the electrochemical measurements, the samples were degreased in methanol, rinsed in 2N HCl. Reversible oxide layers of variable thickness were generated on the surface of the metallic glasses by means of triangular sweeps of potential, with cathodic and anodic limits of –2200 and + 100 mV respectively, in contract to a reference electrode of saturated calomel (SCE) at a sweep rate of 2 V s^{-1}. The change produced in the surface on the electrode and the thickness of the oxide layer, are calculated by the integration of the I–E curve, which is proportional to the charge related to the oxidation processes (Q_{oxi}).

The composition of the metallic matrix as well as the oxide layers electrochemically generated has been studied both by X-ray photoelectron spectroscopy (XPS) and specular reflectance spectroscopy (SRS). The changes observed in the shape of the SRS spectra are a function both, of the composition of layer analysed and of the oxidation state of the elements [9–10].

Pitting corrosion susceptibility was realised in NaCl solution 0.05 M, pH 6.8 ± 0.1 at room temperature on the basis of a study of anodic polarisation curves. The polarisation rate was 0.16 mVs^{-1}. This test permitted to evaluate the protective capacity of the oxide layer.

3. Results and Discussion

Layers of oxides reversible in solutions with different concentrations of KOH were generated on the surfaces of Fe$_{40}$Ni$_{40}$B$_{20}$, Fe$_{40}$Ni$_{38}$Mo$_4$B$_{18}$ and Fe$_{38.5}$Ni$_{38.5}$Mo$_2$Si$_{13}$B$_8$ glasses. The optimum values of anodic and cathodic limits were + 100 and – 2000 mV respectively, with respect to the SCE at a speed of 2 Vs^{-1}. Using the oxidation load, we calculated the thickness of the oxide layer. Table 1 sets out the cycles applied to the samples, the oxidation charge and the layer thickness formed for the test glasses.

Figure 1 shows an example of a typical voltammogram for the generation of the oxide layers, in this case for the Fe$_{40}$Ni$_{40}$B$_{20}$ glass.

Figure 2 shows the polarisation curves of the three test glasses in as-received state. The metallic glasses are not in the passive state because there is no passivity step, but rather the current density grows with the potential. The corrosion behaviour is independent of the chemical composition of the test glasses. Similar performances were obtained for the oxide layer generated in the Fe$_{40}$Ni$_{40}$B$_{20}$ metallic glass. For the glasses

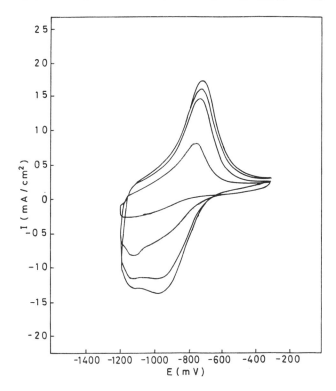

Fig. 1 Typical voltammograms for the generation of the oxide layers in the Fe$_{40}$Ni$_{40}$B$_{20}$ glass.

that changed 2 or 4 atoms of Ni for Mo, Fe$_{40}$Ni$_{38}$Mo$_4$B$_{18}$ and Fe$_{38.5}$Ni$_{38.5}$Mo$_2$Si$_{13}$B$_8$, the generation of oxide layers on the surface made a significant improvement in the pitting resistance, specially the layers generated at a high number of cycles and that hence increased their thickness (curve 3 of Figs 3 and 4).

Table 1 Characteristics of the reversible oxide layers gernerated on the surface of Fe$_{40}$Ni$_{40}$B$_{20}$, Fe$_{40}$Ni$_{38}$Mo$_4$B$_{18}$ and Fe$_{38.5}$Ni$_{38.5}$Mo$_2$Si$_{13}$B$_8$ glasses in KOH solution

SAMPLES	CYCLES	Q$_{oxi}$ (mC.cm^{-2})	THICKNESS (A)
Fe$_{40}$Ni$_{40}$B$_{20}$	200	9.30	170
Fe$_{40}$Ni$_{40}$B$_{20}$	400	28.00	515
Fe$_{40}$Ni$_{38}$Mo$_4$B$_{18}$	200	9.30	170
Fe$_{40}$Ni$_{38}$Mo$_4$B$_{18}$	400	10.90	200
Fe$_{38.5}$Ni$_{38.5}$Mo$_2$Si$_{13}$B$_8$	200	5.10	95
Fe$_{38.5}$Ni$_{38.5}$Mo$_2$Si$_{13}$B$_8$	400	7.25	133

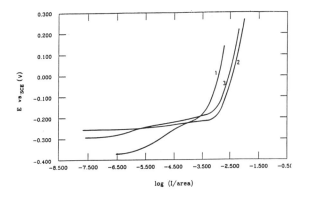

Fig. 2 Anodic polarisation curves for (1) $Fe_{40}Ni_{40}B_{20}$, (2) $Fe_{38.5}Ni_{38.5}Mo_2Si_{13}B_8$ (3) $Fe_{40}Ni_{38}Mo_4B_{18}$ as quenched in 0.05 M NaCl solution.

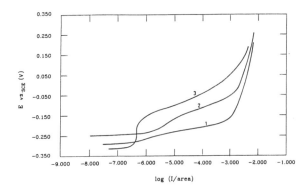

Fig. 5 Anodic polarisation curves for $Fe_{40}Ni_{38}Mo_4B_{18}$, (1) as-quenched, (2) after 200 cycles potential, (3) after 400 cycles in 0.05M NaCl solution.

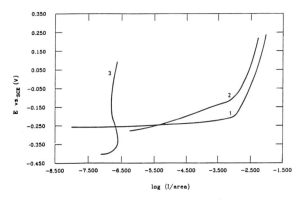

Fig. 3 Anodic polarisation curves for $Fe_{38.5}Ni_{38.5}Mo_2Si_{13}B_8$ (1) as-quenched, (2) after 200 cycles potential (3) after 400 cycles potential in 0.05M NaCl solution.

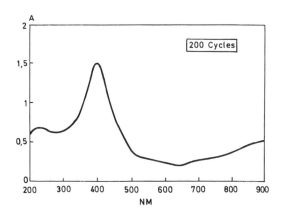

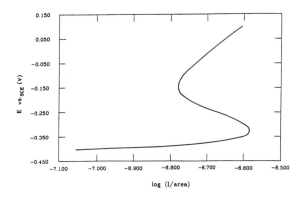

Fig. 4 Curve 3 of Fig. 3 in enlarged scale.

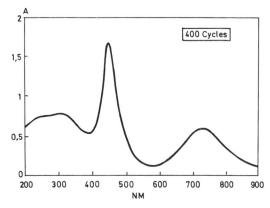

Fig. 6 UV-VIS spectra of the oxide layer of the metallic glass $Fe_{40}Ni_{38}Mo_4B_{18}$.

The layers generated on the surface are not only stable in the medium in which they were generated (KOH) but also in corrosive media such as NaCl with a neutral pH.

The best pitting resistance was found in the $Fe_{38.5}Ni_{38.5}Mo_2Si_{13}B8$, 400 cycles (Fig. 5).

Figure 6 shows the spectra of UV-VIS reflectance spectroscopy obtained on samples with different ox-ide layer thicknesses. Different peaks can be seen to appear, reflecting the differing type of layer formed in function of the applied cycles. The different peaks observed insinuate also the existence of different chemical species and/or different state oxidation of the elements of these layers. An XPS analysis envisaged in subsequent tests, should resolve the composition of these layers.

4. Conclusion

- The conditions for generating layers of oxide in the test glasses were optimised.

- The generated layers are not only stable in the medium in which they were generated, but also in other corrosive media with different pH, such as NaCl.

- The glass with a generation of layers of maximum thickness substantially improve pitting resistance in comparison to reference specimens in as-received state.

- The best corrosion behaviour was found in the $Fe_{38.5}Ni_{38.5}Mo_2Si_{13}B_8$ glass, with a typical curve for a material in a passive state.

- The different peaks observed in the SRS spectra insinuate the existence of different chemical species and/or different state oxidation of the elements of these layers.

- Anyway, necessary additional work is needed in order to confirm the last point. In this way the XPS will be an useful tool to follow the changes in composition of these layers.

References

1. M. Gutierrez, A. M. Irisarri and F. H. Froes, Rev. Metal. Madrid, **25**, 1, 44–53, 1989.
2. F. H. Froes and R. Carbonara, J. of Metals, **40**, 20–27, 1988.
3. M. L. Escudero, J. A. González and A. R. Pierna, Studia Chemical, **15**, 5–12, 1990.
4. M. D. Archer, C. C. Corke and B. H. Harji, Electrochem. Acta 32, **1**, 13–26, 1987.
5. K. Hashimoto, K. Asami and A. Kawashima, International Congress Metallic Corr., **1**, 208–215, 1984.
6. T. P. Moffat, W. F. Flanagan and B. D. Lichter, International Congress Metallic Corr. 3, 454–467, 1984.
7. D. Mukherjee, G. T. Parhiban, C. Srividyarajagopal and N. Karupiah, Bulletin of Electrochemistry, **6**, 578–579, 1990.
8. A. R. Pierna, J. González and A. Lorenzo, Procedings Our Future with new Materials. 1, 1–15, 1991.
9. N. Hara and K. Sugimoto, Proc. Passivity of Metals and Conductor, **1**, 211–218, 1983.
10. C. Gutierrez and B. Beden, Electroanal. Chem., **293**, 1, 253–259, 1990.

56

Influence of the Energetic β⁻ Flux on the Corrosion Behaviour of Zr Alloys

R. SALOT, F. LEFEBVRE AND C. LEMAIGNAN

CEN Grenoble, SECC, 85 X, F 38041 Grenoble Cedex, France

Abstract

Corrosion of Zr alloys is known to be greatly increased under irradiation. Observations of local enhancements of corrosion in front of high energetic β⁻ sources allowed to point out the role of radiolytic species in the phenomenon.

Numerical simulations using the Maksima-Chemist program have been realised to study the evolution of the concentrations of these species in the bulk of water and close to a β⁻ source in a reactor core. They are proved to be increased locally in the last case.

The connection between the radiolysis phenomena and the development of a porosity in the oxide layer is made. An explanation of the general acceleration based on the radiolysis of the water in the pores of zirconia is then presented.

Finally, a recently developed electrochemical cell under irradiation is described.

Zr alloys are mainly used as structural and fuel cladding materials in nuclear reactors. In water cooled power reactors, the most usual ones are the Zircaloys 2 and 4 which present a good resistance to oxidation in these high temperature water conditions [1–3]. However, the corrosion rate obtained under irradiation is observed to be higher compared to the rate obtained in autoclave testing (Fig.1). Moreover, local enhancements of the corrosion happen to be observed. In most cases, that can be correlated to the presence of strong energetic β⁻ sources in the vicinity [4].

The acceleration of the cladding oxidation in reactor conditions can then be analysed in terms of irradiation induced water radiolysis enhanced by energetic, β⁻ emission.

In the reactor, various means of production exist for high energy , β⁻ electrons. They can result from the decay of activated materials such as manganese in stainless steel, copper or platinum. They can also be created by the conversion of high-energy γ rays in a pair of β⁻, β⁺ particles. These high energy β⁻ electrons increase locally the dose rate. For example, close to a Gd_2U_3-UO_2 pellet, the dose rate due to neutrons, γ photons and the electron irradiation reaches $4.E+21$ $eV.l^{-1}.s^{-1}$ while its average value in the core, due to neutron and γ photon irradiations, is $5.E+20$ $eV.l^{-1}.s^{-1}$.

This energy deposition in the water induces radiolysis phenomena which produce oxidising species such as:

$$O_2, HO_2, O_2^- HO_2^-, O, H_2, H_2O_2, H, OH \text{ [5–7]}$$

In the case of pressurised water, the three types of irradiation (neutrons, γ photons and electrons) have the same efficiency to produce radiolytic species [8]. Since experimental measurement of their concentration is impossible in reactor, numerical simulations proved to be useful. Some forty radiolysis reactions are generally taken into account with time constant ranging from E–8 to E+10 s⁻¹. Specific computer programs are then necessary [9].

We have performed numerical simulations using the Maksima-Chemist program in order to evaluate the radiolytic species concentrations in nominal PWR conditions (Fig. 2). After a short period (#1s), steady

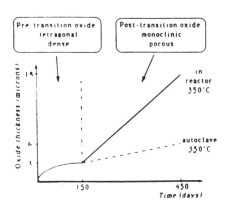

Fig. 1 Evolution of the oxide layer with time.

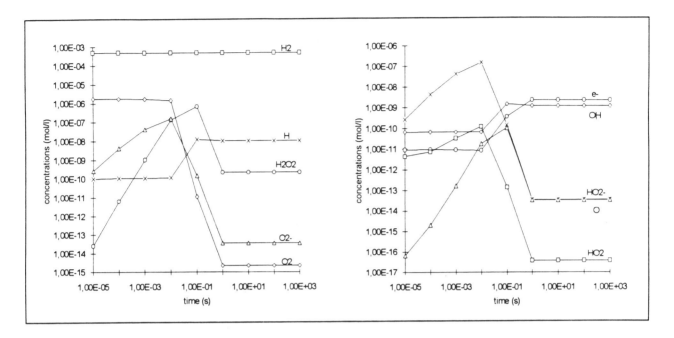

Fig. 2 Numerical simulation of the concentrations of radiolytic species. T = 300°C, Dose Rate = 5. 10²⁰ eV/l.s.

state concentrations are reached. The evolution of these concentrations has also been calculated under an increment of dose rate due to a , β^- source [10] (Fig. 3): during 5.10^{-3} s the dose rate is multiplied by ten inducing a significant increase of the radiolytic species concentrations. As soon as the dose rate is back to its initial value the concentrations decrease rapidly to the steady state values.

Thus the concentration of most species appears to be significantly increased just in front of a β^- source.

In Pressurised Water Reactors, the cladding corrosion process can be divided in two steps:

• First, the oxide layer grows by a diffusion controlled process of the O_2^- ions and the produced zirconia is dense with a tetragonal structure [11].

•Then, after this stage, a phase transition from the tetragonal to a monoclinic form occurs which leads to the creation of an external porous oxide [12].

Since the acceleration of the corrosion is reported only under irradiation and after that transition, as shown in Fig. 1, i.e. when radiolytic species are created and large Surface/Volume pores are present in the oxide, it may signify that interface reactions are the controlling step of the oxidation process.

Indeed, the created radiolytic species can undergo two competitive types of reactions. Either they can recombine in the bulk or they can be involved in

reactions at the oxide/water interface as illustrated in Fig. 4. Then, bulk recombination processes would be easier in the pre-transition conditions and radiolysis phenomenon would have little impact. On the contrary, after the transition, the existence of pores would increase the role of surface phenomena compared to bulk recombinations. Consequently, greater quantities of radiolytic oxidising species would be allowed to react on the walls of the pores increasing the oxidising boundary conditions, and then the corrosion rate.

This analysis is in good agreement with the local acceleration of the corrosion near an energetic, β^- source; indeed in this case, we have simultaneously a local increase of oxidising species such as O and the presence of pores in the oxide. It can also be proposed for the general acceleration of corrosion under irradiation since an average dose rate of 5.E+20 eV.l^{-1}. s^{-1} is high enough to create significant concentration of radiolytic species as shown on Fig. 2.

•To study the electrochemical behaviour of Zr alloys under these particular conditions, an experimental system has been set up under a high energy , β^- flux. The β^- source is made of a pre-irradiated Iridium pellet the decay of which produces 0.6 MeV, β^- electrons. Since this decay also produces γ photons, the electrochemical cell and the source have been placed in a leaded cell. After a 24 hours pre-irradiation in the Grenoble experimental nuclear reactor, the produced dose rate (10²⁰ eV.l^{-1}.s^{-1}) is almost the same as in

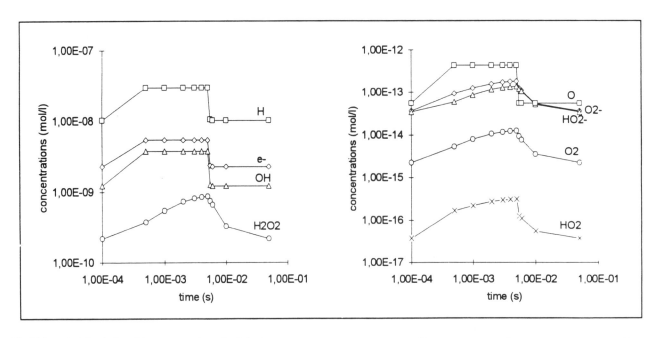

Fig. 3 Numerical simulation of the concentrations of radiolytic species. T = 300 °C, Dose Rate = 4.10²⁰eV/l.s. during 5.10⁻³s and then Dose Rate = 5.10²⁰eV/l.s.

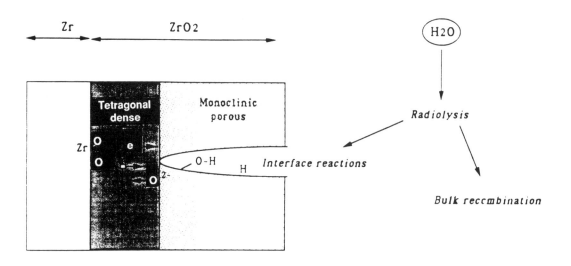

Fig. 4 Two competitive types of reaction for radiolytic species.

nuclear reactors. Maksima-Chemist has been used to evaluate the concentrations of the induced radiolytic species in front of that Ir source in cold water (Fig. 5). In this case, γ photons and electrons have the same efficiency towards radiolysis reaction, different from that of neutrons. A comparison with the results found in PWR conditions (Fig. 2) shows that the concentrations obtained in the electrochemical cell are higher (excepted for HO_2^- and e^-) than in a reactor core. That is mainly due to the difference between the initial

concentrations and to the absence of pressure in our experiment. Nevertheless, the results give the concentrations to be encountered in the volume delimited by the electrons stopping range. Beyond it a gradient of concentrations is expected due to diffusion phenomena. That allows the Zircaloy sample to be exposed to different concentrations of radiolytic species. Moreover, these concentrations can be modified by changing either the dose rate or the initial concentrations. Since the aim of the experiment is not to reproduce exactly

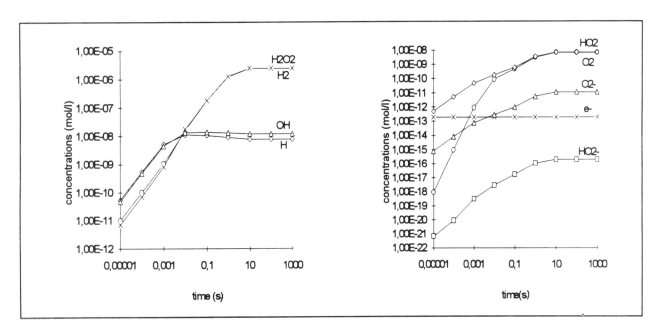

Fig. 5 Numerical simulation of the concentrations of radiolytic species under β⁻ flux produced by a pre-irradiated Ir sample. T = 25°C, Dose Rate = 10^{20} eV/l.s.

the conditions of a reactor core but to create radiolytic species this experimental device is a tool relevant to study the influence of irradiation induced radiolysis on the Zr alloys corrosion.

The experimental device is sketched in Fig. 6. The working electrode is a pellet of Zircaloy located in front of the , β⁻ source. The distance between them can be modified allowing us to expose the Zircaloy sample to different local chemical conditions.

References

1. C. Lemaignan and A. T. Motta, in Nuclear Materials (Vol.10), Materials Science and Technology Series, eds R. W. Cahn, P. Haaser, E. J. Kramer, VCH, 1993.
2. E. Hillner, ASTM STP, **633**, 211, 1977.
3. F. Gazarolli and R. Holzer, Nucl. Energy, **31**, 1, 65, 1992.
4. C. Lemaignan, J. Nucl. Mater., **187**, 122, 1992.
5. K. Ishigure, B. Johnson, B. Cox, C. Lemaignan and N. Nichaev, An assessment of Irradiation Corrosion Mechanisms for Zr alloys in high temperature water, Technology Review, AIEA, in press.
6. W. G. Burns and P. B. Moore, Radiation Effects, **30**, 233, 1976.
7. E. Roth, Chimie nucleaire appliquée, Masson et Cie Ed, 1968.
8. S.R. Lukac, Rad. Phys. Chem., **33**, 3, 223, 1989.
9. M.B. Carver, D.V. Hanley and K.R. Chaplin, AECL 6413, 1979.
10. R. Salot, Rapport DEA, CEN Grenoble, 1992.
11. B. Cox and J. P. Pemsler, J. Nucl. Mat., **28**, 73, 1968.
12. B. Cox, J. Nucl. Mater., **29**, 50, 1969.

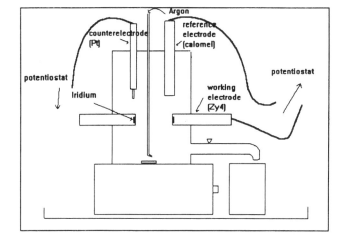

Fig. 6 Experimental device.

57

Influence of Ion Implantation on Corrosion and Stress Corrosion Cracking of 304L Stainless Steel

*R. Karray, C. Sarrazin, D. Desjardins and B. Darriet**

Laboratoire de Mécanique Physique - URA 867 du CNRS, Université Bordeaux I, 351 cours de la libération,
33405 Talence cédex, France
*Laboratoire de Chimie du Solide du CNRS

Abstract

The effects of ion implantation on electrochemical and stress corrosion cracking behaviours of an austenitic stainless steel (AISI 304L) in chloride media has been investigated. Nitrogen, molybdenum, neon and krypton ions have been implanted at room temperature with ion fluences ranging from 2.5×10^{16} to 2.10^{17} ions cm^{-2}. The implanted alloys have been characterised by transmission electron microscopy and Auger electron spectroscopy. The modifications of the different treated steels have been studied by electrochemical tests and classical tests used in stress corrosion cracking.

Ion implantation of reactive species such nitrogen and molybdenum modifies the electrochemical and stress corrosion cracking behaviours of the alloy. On another hand, implantation of inert gas ions (krypton, neon) has given little changes.

1. Introduction

Ion implantation is a process for injecting atomic species in the form of an ion beam into a solid material. The purpose of this technique is to alter the surface properties of the base material by changing its composition.

Ion implantation of various elements into iron-based alloys has been found to enhance surface corrosion properties [1–3].

It is known that the addition of molybdenum to austenitic stainless steel increases their pitting corrosion resistance in chloride solution [4]. The enhance is attributed to the presence of molybdenum in the protective passive film [5, 6].

Numerous works studying the effects of implantation of nitrogen ions on the resistance of steel to wear and friction have demonstrated the efficiency of this surface treatment in improving tribological properties [7–10]. Furthermore, several studies have indicated that the addition of nitrogen leads to improvements in passivation and pitting resistance [11–13].

Implantation of neon and krypton were done to provide a distinction between the chemical effects produced by nitrogen and molybdenum respectively and the physical effects produced by ion implantation.

The aim of this work is to study the effect of ion implantation on electrochemical and stress corrosion cracking behaviours of an austenitic stainless steel in chloride medium.

2. Experimental Procedure

The specimens used for the different tests are discs of 12.5mm diameter and cylindrical 1.5mm dia. wires.

The chemical composition of these samples (AISI 304L) is reported in Table 1. Samples (before implantation and electrochemical testing) were first mechanically polished on abrasive discs (grade 600–1200) then with diamond pastes (3–1μm). They were finally degreased in acetone, ethanol and rinsed with distilled water then dried. The cylindrical wires used for stress corrosion cracking tests were annealed at 1150°C for 30 min, water quenched and then electropolished in perchloric acid followed by an acid etching.

Ion implantation experiments were performed with the 'multi-ion' implanter of the UNIREC research centre (France). Implantation parameters for the specimens are presented in Table 2.

Electrochemical behaviour in chloride solution is studied using potentiodynamic polarisation test and immersion tests. The electrolyte used in both tests was $MgCl_2$ (30wt%) with a boiling temperature of 117°C. The electrochemical tests were carried in a double-walled glass cell.

Stress corrosion cracking behaviour is characterised by classical tests like constant load tests. The load used in constant load tests is 150 MPa which is just over the yield stress of the studied steel.

Various analytical techniques were used to characterise the specimens. Electron diffraction patterns and microstructures were recorded using a TEM operated at 200 keV. Surface analysis of samples before and after immersion tests were performed using Auger Electron Spectroscopy (AES) in conjunction with sputter etching with 2 keV argon ions. The sputerring rate was obtained by ion etching of a Ta sample covered by a Ta_2O_5 layer of known thickness.

Table 1 Chemical composition (wt%) of steels

Steel	C	Mn	Si	S	P	Ni
304L(disc)	0.053	1.427	0.485	0.0025	0.029	8.28
304L(wire)	0.064	1.76	0.4	0.011	0.031	8.69

Cr	Mo	Cu	Ti	Al	N	B
17,38	0.139	0.145	0.002	0.002	0.040	0.0004
18,59	-	-	-	-	-	-

Table 2 Implantation parameters

Ion beam	Fluence (ions/cm^2)	Energy (keV)
Mo $^+$	2.5×10^{16}	100
N $^+$	2.5×10^{16}	100
N $^+$	2×10^{17}	100
Kr $^+$	2.5×10^{16}	100
Ne $^+$	2.5×10^{16}	50

3. Experimental Results and Discussion

3.1 TEM analysis

Microstructural observations of unimplanted 304L stainless steel (disc) reveal an austenitic f.c.c matrix composed of grains *ca.* 20μm in size and having a lattice parameter of 3.58Å. Few defects (dislocations and stacking faults) were observed. No precipitation is seen in grain boundaries (Fig. 1). The observations of implanted steels are discussed in the following section.

3.2 Surface analyses results before electrochemical tests

Figures 2(a)–(d) give the AES profiles obtained with unimplanted and implanted alloys (discs). All implanted samples present a surface contamination (carbon). Ions that penetrate in the surface form a gaussian concentration. The depth penetration depends on the type of ion, the number of implanted ions, and the beam energy. Nitrogen the mass of which is lower than molybdenum penetrates deeper in the substrate. The nitrogen implantation presents a flat concentration profile compared with that for molybdnemum.

3.3 Electrochemical tests

3.3.1 Polarisation tests

The samples were initially cathodically polarised for a few minutes (300–350 mV below open circuit potential E_{ocp}) to yield a cathodic current in order to reduce the air formed oxide films. When the current density was constant, the specimen were subjected to potentiodynamic anodic polarisation with a scanning rate of 30 mV/min. All potentials were measured with respect to a satured calomel electrode (SCE). The results obtained for cylindrical wires and discs samples are comparable.

A schematic representation of potentiodynamic curve is given Fig. 3 and corresponding parameters obtained for cylindrical wires are given in Table 3.

The effect of molybdenum implantation leads to a more anodic potential which is indicative of an increased passivation tendency. The difference between E_{ocp} and E_p (pitting potential) is more important for Mo-implanted steel than for the unimplanted one. An electrochemical study about Mo-implanted austenitic stainless steel [14] revealed that molybdenum implantation at sufficient energy and at doses sufficient to

Fig. 1 Diffraction pattern and image of unimplanted 304L (disc).

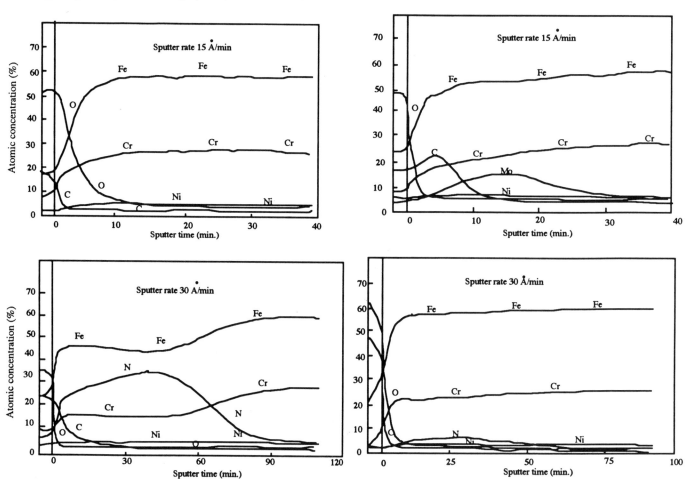

Fig. 2 Auger Profiles of unimplanted and implanted steels. (a) Auger Depth Profile for unimplanted 304L (disc).
(b) Auger Depth Profile for a 2.5×10^{16} Mo$^+$.cm^{-2} implanted 304L (disc). (c) Auger Depth Profile for a 2×10^{17} N$^+$ cm^{-2} implanted 304L (disc) . (d) Auger Depth Profile for a 2.5×10^{16} N$^+$cm^{-2} implanted 304L (disc).

Table 3 Electrochemical parameters obtained for cylindrical wires

Steel	Potentials(mV/ECS)		
	E_{ocp}	E_p	ΔE
304	-370	-290	80
304-Mo	-370	-270	100
304-Kr	-350	-300	50
304-N ($2,5.10^{16}$ N$^+$/cm^2)	-360	-260	100
304-Ne	-350	-290	60

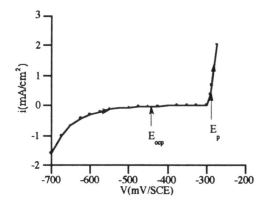

Fig. 3 Potentiodynamic current—potential curve in MgCl₂ 30wt%, 117°C. Scanning rate: 30 mV min⁻¹.

produce a near surface concentration greater than 10at.% can significantly reduce the pitting corrosion susceptibility of stainless steel under cathodic polarisation control in marine like environments. Implantation of krypton, which does not interact chemically with the substrate, does not bring any improvement in the corrosion resistance of the stainless steel. Indeed, the potential corrosion being near the pitting potential, the localised corrosion risk is considerably enhanced.

There is an improvement in the corrosion behaviour with respect to the pitting for nitrogen implanted steel when compared with either neon implanted steel or untreated steel. This result agrees with some studies which suggested that nitrogen increases pitting resistance by inhibiting the anodic dissolution of stainless steels at sites where the passive film break down [13].

3.3.2 Immersion tests

Figures 4(a)–(c) show the AES profiles of unimplanted and implanted steels (discs) after immersion in MgCl₂ (30wt%) at 117°C for *ca.* 20 h.

By considering that the thickness of the oxide layer is determined as the etch depth [15] at which the oxygen signal has dropped to half of the value between the highest and the lowest intensities, it seems that the

molybdenum is present in the protective layer (*ca.* 5at.%) whose thickness given by the sputter profiles is *ca.* 30Å. This remark agrees with several studies suggesting the beneficial role of molybdenum which is attributed to its enrichment in the passive film [5, 6] and consequently retards the pitting of stainless steel in chloride solutions.

From the profile of nitrogen implanted steel (high dose), it seems that nitrogen is enriched at the surface and is incorporated in the passive film. Indeed, some authors suggested that nitrogen is segregated to the alloy surface where it formed a relatively stable interstitial nitride phase enriched in Cr, Ni [11]. The mechanism of surface enrichment of beneficial element is suggested due to the formation of thermodynamically stable interstitial chromium nitrides which are uniformly distributed on anodic polarised surface [13].

3.3.3 Constant load tests

Experiments were carried out under open circuit conditions. After potential stabilisation which lasts for *ca.* 2 h, the load (150 MPa) which is over the yield strength was applied to the sample. The change of the elongation vs time during creep is recorded in a semi-logarithmic scale indicated in Fig. 5.

Stress corrosion cracking process may be schematically divided into two stages which are crack initiation and crack propagation. These two periods depend on medium, steel composition, applied stress, temperature. Indeed, crack initiation is a parameter which depends strongly on surface and so on ion implantation. It was largely shown [16] and presented in Fig. 5 how the elongation curve vs time enables a separation between the initiation time and the propagation time to be made: during the initiation period, the elongation follows the law obtained in inert environments at the same temperature which is:

$$l(t) = lo + \alpha \log t$$

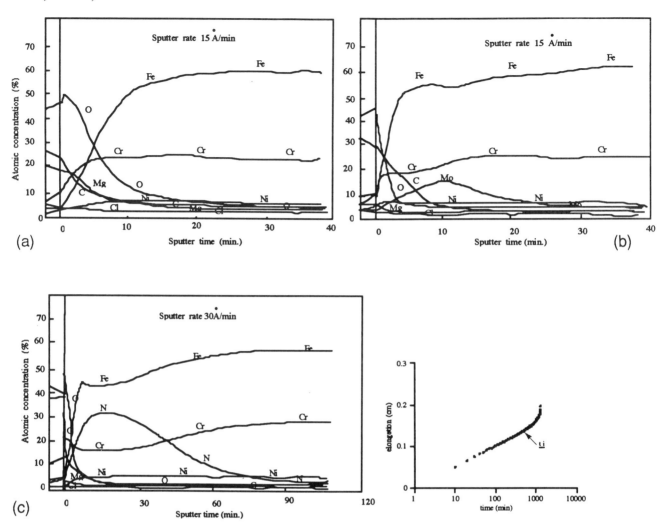

Fig. 4 *Auger Profiles after immersion tests. (a) Auger Depth Profile for unimplanted 304L (disc). (b) Auger Depth Profile for a 2.5 × 10¹⁶ Mo⁺.cm⁻² implanted 304L (disc). (c) Auger Depth Profile for a 2 × 10¹⁷ N⁺.cm⁻² implanted 304L (disc).*

Fig. 5 *Constant load test in MgCl₂ 30wt%, 117°C. Stress: 150MPa Sample elongation vs time (semi-logarithmic scale).*

Table 4 Initiation times of different treated steels

Steel	304	304-N	304-Mo	304-Ne	304-Kr
t_i	400	3000	700	200	200

where t is the time, lo the elongation during the first minutes after loading, $\alpha = \alpha(\sigma, T)$ depends on stress and temperature. In the propagation period, the elongation law diverges from this law until failure. This period is not affected by surface treatments. Table 4 reveals initiation times obtained for the different treated steels which correspond to an average of three tests.

These results agree with potentiodynamic polarisation observations. Mo-implanted 304 stainless steel improve initiation time in comparison with untreated steel by a factor of two but the most beneficial effect is the initiation time of nitrogen implanted steel. This result confirms again the beneficial effect of nitrogen implantation in situation of stress corrosion cracking.

In the case of inert ions such as krypton and neon, the initiation times detected on implanted steels decrease by a factor of two in comparison with unimplanted steel. This result confirms the detrimental effects of inert implants and suggests the important role of chemical effects associated with ion implantation.

4. Conclusion

An austenitic stainless steel (AISI 304L) has been implanted with nitrogen, molybdenum, neon and krypton ions at room temperature with fluences ranging from 2.5×10^{16} to 2×10^{17}ions cm^{-2}. Nitrogen and molybdenum ions implantations result into a clear modification of behaviour in comparison with unimplanted steel.

The electrochemical behaviour of the implanted alloys has been characterised by polarisation tests and immersion tests. The results show that nitrogen and molybdenum implanted ions enhance pitting resistance of 304L stainless steel in chloride solution. However, inert species like neon and krypton do not affect the electrochemical behaviour of 304L stainless steel.

The stress corrosion cracking behaviour of the implanted steels has been investigated by means of constant load tests. The crack initiation times given by these tests are considerably improved by active ions implantations. While for alloys implanted with inert gas ions (neon, krypton), the initiation times decrease in comparison with unimplanted steel.

From these results, it can be concluded that the improvement of electrochemical and stress corrosion cracking behaviours of 304L stainless steel is attributed to chemical effects produced by implanted species.

5. Acknowledgements

This research is financially supported by the CNRS, Ungine ACG, Unirecusinor Salicor France in a GS Corrosion.

References

1. V. Ashworth, R. P. M. Procter and W. A. Grant, Thin Solid Films, **73** , 179–188, 1980.
2. V. Ashworth, W. A. Grant and R. P. M. Procter, Corros. Sci., 16, 661–675, 1976.
3. E. Mc Cafferty, P. G. Moore, J. D. Ayers and G. K. Hubler, in 'Corrosion of metals processed by direct energy beams', Louisville, Kentucky, October 1981, Ed. Metallurgical Society, 1982, 1–21.
4. A. P. Bond and E. A. Lizlovs, J. Electrochem. Soc., **115**, 1130–1135, 1968.
5. C. R. Clayton and Y. C. Lu, J. Electrochem. Soc., **133**, 2465–2473, 1986.
6. K. Sugimoto and Y. Sawada, Corrosion Science, **17**, 425–445, 1977.
7. S. Fayeulle, Thesis, Lyon, 1987.
8. S. Fayeulle, D. Treheux and C. Esnouf, Applied Surface Science, **25**, 288–304, 1986.
9. M. Iawaki, T. Fujihana and K. Okitaka, Mater. Sci. Engng, **69**, 211–217, 1985.
10. C. Chabrol and R. Leveque, Mem. Et. Sci. Rev. Met., **1**, 43–54, 1988.
11. R. Willenbruch, C. R. Clayton, M. Oversluizen, D. Kim and Y. Lu, Corros. Sci., **31**, 179–190, 1990.
12. Sh. Song, W. Song and Zh. Fang, Corros. Sci., **31**, 395–400, 1990.
13. Y. C. Lu and M. B. Ives, Bulletin du Cercle d'Etudes des Metaux, **16**, 1–6, 1987.
14. M. B. Ives, U. G. Akano, Y. C. Lu, Guo Ruijin and S. C. Srivastava, Corros. Sci., **31**, 367–376, 1990.
15. P. Marcus and I. Olefjord, Corros. Sci., **28**, 589-602, 1988.
16. M. C. Petit, D. Desjardins and B. Darrieutort, Mem. Et. Sci. Rev. Met., 241, 1976.

58

Development of a Neural Network Model for Predicting Damage Functions for Pitting Corrosion in Condensing Heat Exchangers

M. Urquidi-Macdonald, M. N. Eiden and D. D. Macdonald*

Department of Engineering Science and Mechanics, The Pennsylvania State University, University Park, PA 16802, USA
*Center for Advance Materials, The Pennsylvania State University, University Park, PA 16802, USA

Abstract

Pitting attack occurs in *ca*. 10% of domestic and industrial gas fired heat exchangers, and generally appears during the first five years of operation. The causes of pitting corrosion are several, including the use of chlorinated solvents in the ambient environment, the quality of the gas burned, and the material used to fabricate the heat exchanger.

Several attempts have been made to develop predictive models based upon observed pitting data, but they are limited in their predictive capabilities. Recently, we have initiated a program to develop a deterministic model to predict the damage resulting from pitting corrosion. However, the problem is complicated, and several restrictive assumptions have had to be made to render the problem tractable. An alternative approach, which is developed here, is to assume that we have no intrinsic information concerning the physico–chemical mechanisms involved in the nucleation and growth of pits, but that we are able to discern relationships between the observed damage and various input parameters which may be used to extrapolate the damage to future times. Probably the most efficient method of establishing these relationships is to use artificial intelligence techniques. Accordingly, we describe here an Artificial Neural Network (ANN) for predicting pitting damage functions for condensing heat exchangers. When the net is trained with reliable data and knowledge, we are able to predict accurately damage functions under significantly different conditions.

1. Introduction

An outstanding problem in corrosion science and engineering is to predict, *a priori*, damage functions for localised corrosion[1–3]. This problem is important because localised attack, such as pitting corrosion, normally occurs on highly passive alloys with small weight loss, and often is not detected until component failure occurs.

The problem of estimating damage functions (number of pits vs. pit depth) for pitting corrosion is made all the more difficult by the observation that the parameters characterising pitting damage (e.g. pit depth and nucleation time) are distributed. Accordingly, it is necessary to consider the probability of nucleation, in addition to the kinetics of localised attack and pit repassivation, when deriving models from which damage functions can be estimated.

We outline in this paper a methodology that may enhance our ability to predict damage functions for one of the most important forms of localised attack, pitting corrosion, with particular emphasis on lifetime prediction and reliability analyses for domestic and commercial gas-fired condensing heat exchangers.

2. Pitting in Condensing Heat Exchangers

On the basis of laboratory studies [4], and through the analysis of field data collected over the past decade by Battelle Columbus Laboratory [5], several factors have been identified as contributing to pitting in gas fired heat exchangers in domestic and industrial service:

(i) The type of alloy used for fabricating the heat exchanger
(ii) Chloride concentration in the gas and condensate
(iii) Temperature
(iv) Exposure time
(v) Ambient vs indoor air
(vi) pH
(vii) Electrochemical potential.

Unfortunately, few of these factors are simply related to the damage functions or to one another. Accordingly, it is seldom possible to establish a simple empirical equation for predicting pitting damage as a function of these variables. Therefore a new tool is needed to find an 'equation' to relate various environmental parameters to the damage due to pitting corrosion. The required tool is found in Artificial Neural Net (ANN) methodology.

3. The Neural Network Model

An Artificial Neural Network is a highly interconnected system inspired by the brain and formed by simulated 'neurons' represented by a transfer function, and 'weights' associated to the connections of the 'neurons'. The back propagation training algorithm allows experimental acquisition of input/output mapping knowledge within multilayer networks. Because we have experimental data on the numbers of pits vs pit depth (knowledge) as a function of the time of observation and applied potential, we decided to use an ANN backward propagation technique for a supervised learning system. During training of the ANN, the number of pits and the pit depth were used as 'outputs' and the applied potential and time of observation as 'inputs'. We explored several topologies to obtain the best compromise between learning and computing time for an ANN with 2 hidden layers. The maximum training time was set to 12 h on a Macintosh II microcomputer with a minimum threshold of 0.1% of the normalised input values.

The ANN has the following features:

(i) Heteroassociative memory, for which the patterns on recall from the memory are purposely different from the input pattern, because the inputs and outputs are different and belongs to different classes of information.
(ii) Delta rule type of learning, where the neuron weights are modified to reduce the difference between the desired output and the actual output of the processed element. The weights are changed in proportion to the error calculated. This rule also limits the learning, if the error at the output of the network is lower than a given threshold. The learning rates of those layers close to the output are set lower than the learning rates of the other layers.
(iii) A momentum term which is used to smooth out the changes.
(iv) A sigmoid transfer function, which is a monotonicaly continuous mapping function.

4. Data Sets

In this study, we employ laboratory data [4] for pitting damage functions initially to train the ANN. These data were measured as the pit density (number/unit area) versus pit depth on AL29-4C (Fe $\approx$ 67%, Cr = 28.75%, Mo = 3.78%) measured at 35°C in synthetic condensate (200 ppm Cl^-, 40 ppm F^-, 20 ppm SO_4^{2-}) at a constant pH, under controlled electrochemical potential conditions (0.9, 0.85, 0.8 and 0.74 V Ag/AgCl). Experiment details of these measurements are given elsewhere [4], but is important to note that the data were obtained under well defined experimental conditions and hence are well suited for training of the net.

Note also that our data set employs only two independent variables; potential and time. Accordingly, the ANN could not be trained with respect to any dependence of the damage function on pH, chloride concentration, temperature, or metallurgical properties of the substrate, all of which are expected to be important in any comprehensive analysis of pitting damage in condensating heat exchangers. In the present case, it was observed that the damage function was strongly dependent on both the applied potential and the time of exposure.

The damage functions employed in this study were obtained at applied potentials of 0.90 V, 0.85 V, 0.80 V, and 0.74 V (Ag/AgCl); and exposures times ranging from 2 to 2804 h. The pits were counted at each applied potential and at each time of exposure. The depth of the pits were also measured in micron increments from 1 to 102 μm. Thus, the inputs to the ANN were applied potential and time of observation. The pit depth and pit number data were used as 'output' during training of the ANN.

The square root of the sum of the difference between the ANN prediction and the target was used to 'measure' the speed of learning of the net. The input and output data were normalised between [–1,1], and the error threshold was set at 0.001. Some of the data set that were used to train the net as well as data sets that were used to test the performance of the net are summarised in Table 2.

5. Results and Discussion

Figure 1 shows a set of results on training the ANN, on experimental data for a voltage of 0.9 V and a time of observation of 44 h while Fig. 2 compares the ANN predictions (tests) with the measured data for all available data (see Table 1). As noted in Table 1, training of the net involved four sets of experimental data, with potentials ranging from 0.74 to 0.9V for observation times from 763 h to 44 h. Because we do not have a way to measure the error introduced in counting pits and in measuring their depth, a full error analysis of these predictions is very difficult; however, we consider it reasonable to associate an uncertainly of ± 10% error with the laboratory data. The ANN predictions of the total number of pits agrees with the measured data to within a 10% error margin, but the level of agreement for each category is less satisfactory (up 20%, see Fig. 2). Nevertheless, given the estimated uncertainty in the measured data, we regard the training activities to have been highly successful.

We developed 'synthetic' data for other potentials (0.95, 0.825 and 0.7 V (Ag/AgCl)) and for intermediate and extrapolated observation times (10, 50,100, 250, 375, 500, 1000, 2000 and 4000 h). We then used the net

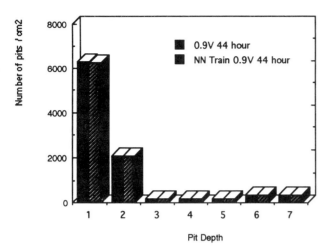

Fig. 1 Measured and NN predicted number of pits vs pit depth Al29–4C 0.9V, 35 °C, 44 h (1) 0–5 µm; (2) 6–10 µm; (3) 11–15 µm; (4) 16–20 µm; (5) 21–30 µm; (7) 46–80 µm.

to predict number of pits and pit depth. In these calculations, we arbitrarily selected a critical dimension of 102 µm, which might correspond (for example) to the thickness of a condenser wall. Accordingly, our calculations estimate the time at which the deepest pit is predicted to exceed the critical dimension.

Figure 3(a) shows the evolution of the deepest pit with time for 3 applied potentials. At low applied potential (0.70 V vs Ag/AgCl), the rate of pit growth is very slow, so that the pit depth is predicted to only double after 4000 h of exposure. However, the pit growth rate is found to be a very sensitive function of potential. Thus, for a potential of 0.825 V, the pit reaches the maximum depth (failure = 102 µm) in 100

h; whereas at 0.95 V the critical dimension is achieved in 50 h. Figure 3(b) shows that the number of pits grows very slowly at a potential of 0.7 V, but at higher potentials (0.825 and 0.95 V), the number of breakdown sites reaches a maximum after 1000 and 500 h, respectively. These results indicate that the number of available breakdown sites is potential dependent, and that at the two highest potentials all the available sites are activated within the first thousand hours. However, at the lowest applied voltage, sites continue to activate after 4000 h.

Figures 4(a) and 4(b) on p.340 show calculated damage functions corresponding to potentials employed in the laboratory (0.74 and 0.90 V (Ag/AgCl)), but for times interpolated or extrapolated by the ANN (100, 500, 1000, and 4000 h). It is interesting that (1) the total number of pits grows slowly with time; and (2) the pit distribution do not grow in a uniform manner; i. e. at 500 h the ANN predicts that there are 8000 pits with size 0–1 µm, but at later time (4000 hours) those 8,000 pits do not all achieve the same depth indicating that the rate of growth is distributed. Accordingly, for this case at least, any empirical or deterministic model must recognise this distribution in growth rate if accurate damage functions are to be predicted.

Figure 4(b) shows that the pit distribution at the highest potential (0.90 V (Ag/AgCl)) reaches a maximum number of breakdown sites sooner than does the sample at 0.74 V. It is also interesting to note that the pit distribution (evolution with time) appears to change little with time after the initial period of 100 h.

Similar results to those obtained for the experimental potentials and synthetic times were obtained

Table 1 Data sets

Potential (V)	Time (Hours)	Comments
0.74	763	Training
	1464	Testing
	2804	Testing
0.8	310	Training
	476	Testing
	598	Testing
0.85	207	Training
	310	Testing
	452	Testing
0.9	22	Testing
	44	Training
	96	Testing

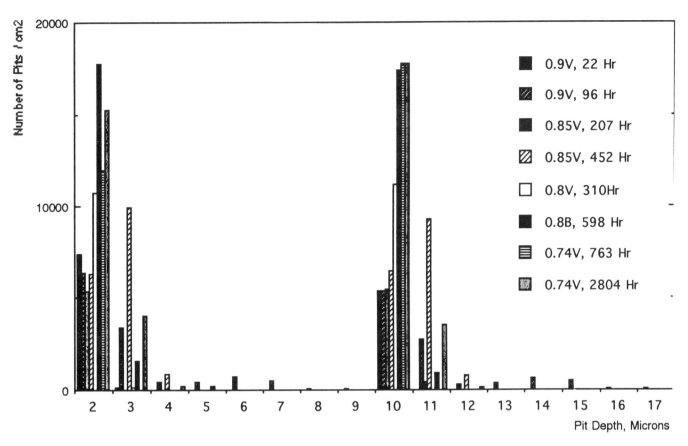

Fig. 2 Number of pits vs pit depth measured (Categories 2–9) and ANN calculated (Categories 10–17). Categories: (2) and (10) 0–5μm; (3) and (11) 6–10 μm; (4) and (12) 11–15 μm; (5) and (13) 16–20 μm; (6) and (14) 21–30 μm; (7) and (15) 31–45 μm; (8) and (16) 46–80 μm; (9) and (17) 81–102 μm.

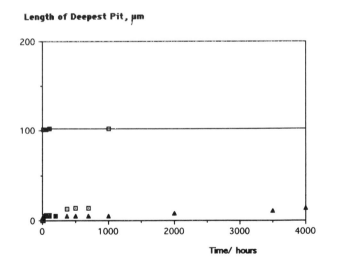

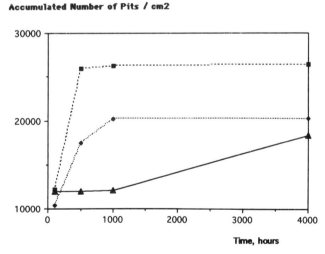

Fig. 3(a) Length of deepest pit vs observational time at 3 applied potentials (▲: 0.7 V; ⊡: 0.825 V; ■: 0.9 V (Ag/AgCl)). Alloy AL29-4C, T = 35°C, 200 ppm Cl.

Fig. 3(b) Accumulated number of pits /cm² vs observational time at 3 applied potentials (—: 0.7 V;....: 0.825 V; – – –: 0.9 V(Ag/AgCl)). Alloy AL29-4C, T = 35°C, 200 ppm Cl.

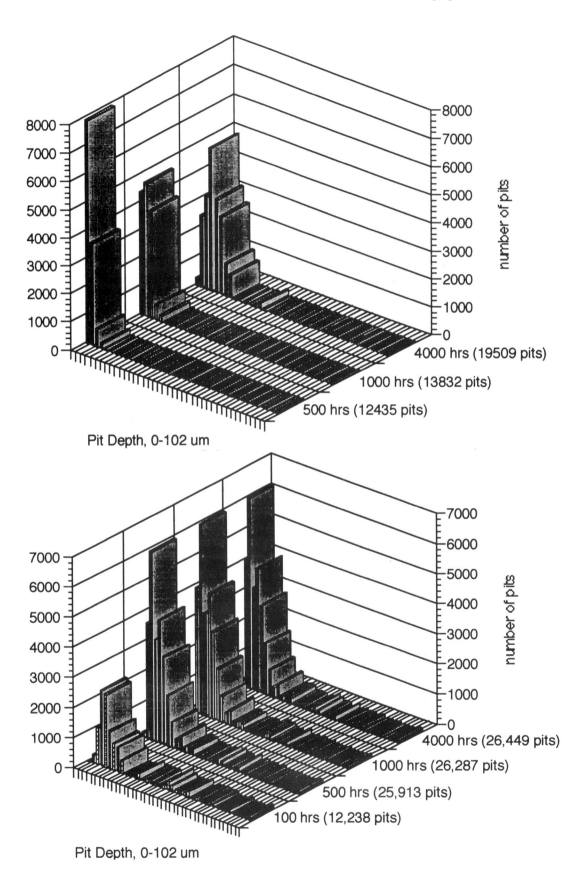

Fig. 4(a) Pit distribution for three observation times as indicated. Applied potential = 0.74 V (Ag/AgCl).

number of pits

8000
7000
6000
5000
4000
3000
2000
1000
0

4000 hrs (19509 pits)

1000 hrs (13832 pits)

500 hrs (12435 pits)

Pit Depth, 0-102 um

number of pits

7000
6000
5000
4000
3000
2000
1000
0

4000 hrs (26,449 pits)

1000 hrs (26,287 pits)

500 hrs (25,913 pits)

100 hrs (12,238 pits)

Pit Depth, 0-102 um

Fig. 4(b) Pit distribution for four observational times as indicated. Applied potential = 0.9 V (Ag/AgCl).

for synthetic potentials and times (0.70 and 0.95 V; 100, 500, 1000, and 4000 h). Figures 5(a) and 5(b) on p.342 show predicted damage functions for voltages of 0.74 and 0.90 V, respectively. Again, the pit growth rate appears to be distributed at the lower potentials but not at the higher voltage.

The most apparent explanation of the finding of this work is that the damage occurs at inclusions that intersect the surface, as has been found in stainless steels[6]. We depict the sequence of events that are envisaged to occur in Fig. 6 (p.343), in which we differentiate between initiation and growth. In this model pits are envisaged to nucleate at the intersection between the passive film and the emergent inclusions [7], possibly because those are regions of high cation vacancy diffusivity, as proposed in the Point Defect Model (PDM) for passivity breakdown, and hence are the 'weak spots' on the surface. Growth of the pits then proceeds by the dissolution of the inclusions while new pits nucleate, and this progressive nucleation and growth process continues until the active inclusions have been fully dissolved. At this point, the progressive development of damage becomes time-independent (or only weakly dependent on time) as shown in Figs 4(b) and 5(b). If the inclusion dissolution process is assumed to be an electrochemical reaction (e.g. MnS $\longrightarrow Mn^{2+} + S + 2e^-$), then the rate of dissolution will be potential dependent. Therefore, the final damage profile (Fig. 6(c)), defined by the size distribution of the inclusions, will be attained more rapidly at higher potentials. According to this model, the most probable inclusion size is of the order of 2–10 μm (corresponding to the maximum in the damage function), which is typical for MnS inclusions in stainless steel, with a few inclusions (possible stringers) extending to 102 μm. However, the deepest pits may in fact also reflect dissolution of the alloy matrix within the crevices formed by distribution of the inclusions. Furthermore, the experimental data suggest that the number of inclusions that activate is potential dependent, and that more sites activate as the potential is made more positive. It is interesting to note that Williams *et al.* [6] have found that inclusions below a minimum size do not initiate stable pits on Type 304SS, because the resulting embryonic crevices cannot sustain the conditions that are necessary for active dissolution and hence growth. Possible, over much longer exposure periods, growth of the crevices into the alloy matrix become the principal mechanism for development of damage, but the present work suggests that this is a much slower process than the initial dissolution of the inclusions.

We end this discussion by noting that our principal purpose in this study was to explore the usefulness of Artificial Neural Networks for interpreting and extending experimental damage functions for pitting corrosion. This is desirable because of the difficulties inherent in measuring damage functions directly, and because of the large number of variables that may affect the development of damage. We have shown here that ANNs may be adequately trained on experimental data sets, and then be used to interpolate and extrapolate the data sets even though no analytic equation exists for the damage function. Clearly, the reliability of the extrapolation, in particular, depends critically on the quality of the data and on the design of the ANN [8], and, sensibly, variables that are not included in the training can not be extrapolated. However, one of the strengths of this technique is that ANNs provides a convenient means of assessing the sensitivity of the damage functions to the input parameters, even when the input data are sparse or 'noisy'. A further advantage is that the ANN can be made to progressively learn as additional data are added to the input. This is a powerful feature that should find application in the management of large and evolving data bases.

6. Conclusion

The conclusions of this work are as follows:

1. An Artificial Neural Network can be trained to represent pitting damage functions for alloy AL29–4C in synthetic condensate as function of applied potential and observation time.
2. Damage functions interpolated and extrapolated from the training data sets to other voltages and times are found to be in reasonable agreement with experimental data.
3. The rate at which pitting damage (number of pits and pit depth) develops in AL29–4C in synthetic condensate is a strong function of the applied voltage, with the damage accumulating most rapidly at higher voltages.
4. At low applied potentials, the pit growth rate appears to be distributed, but higher potentials the damage does not develop further to any significant extent for times exceeding 500 h for 0.90 V and 100 h for 0.85 V.
5. The findings of this work are consistent with rapid, potential dependent attack at inclusions intersecting the alloy/environment interface, as it is commonly observed in stainless steels.

7. Acknowledgements

The authors are grateful to the Gas Research Institute contract No. 5090-260-1969, and to Dr. Kevin Krist for his support.

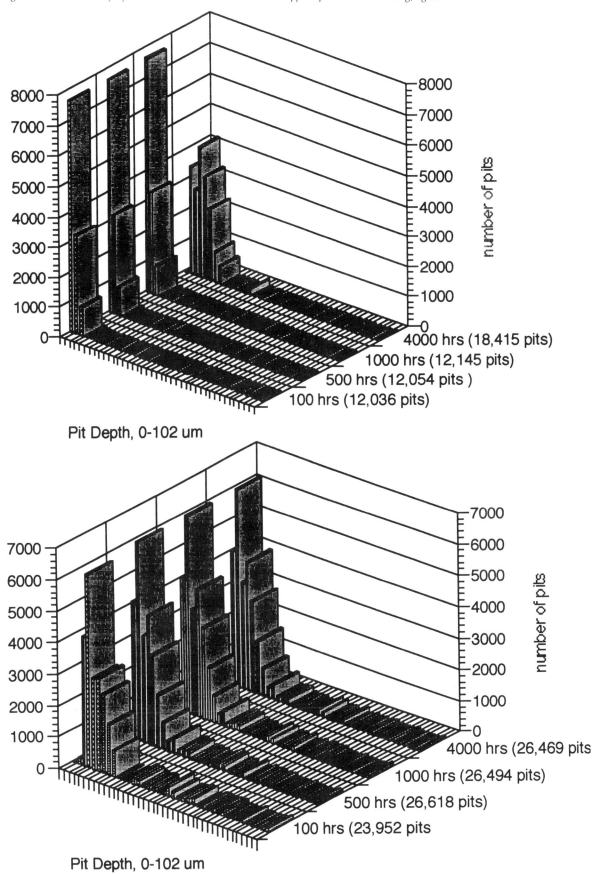

Fig. 5(a) Pit distribution for four observational times as indicated. Applied potential = 0.7 V (Ag/AgCl).

Pit Depth, 0-102 um

4000 hrs (18,415 pits)
1000 hrs (12,145 pits)
500 hrs (12,054 pits)
100 hrs (12,036 pits)

number of pits

Pit Depth, 0-102 um

4000 hrs (26,469 pits
1000 hrs (26,494 pits)
500 hrs (26,618 pits)
100 hrs (23,952 pits

number of pits

Fig. 5(b) Pit distribution for four observation al times as indicated. Applied potential = 0.9 V (Ag/AgCl).

342

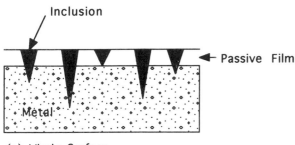

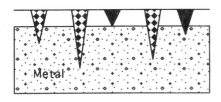

(a) Virgin Surface

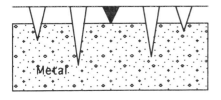

(b) Progressive nucleation
(passivity breakdown and
growth)

(c) Final damage

Fig. 6 Cartoon outlining the development of pitting damage in AL29–4C synthetic condensate.

References

1. J. Kruger, in 'Passivity and Its Breakdown on Iron and Iron-Base Alloys,' Ed. R. W. Staehle and H. Oakda, NACE, Houston, Texas, pp. 91, 1976.
2. M. Janik-Czachor, J. Electrochem. Soc., **129**, 513C, 1981.
3. N. Sato and G. Okamoto, in 'Comprehensive Treatise in Electrochemistry,' Ed. J. O'M. Bockris, B. E. Conway, E. Yeager and R. E. White, Plenum Press, N.Y., Vol. 4, pp. 193–245, 1981.
4. SRI Final Report. June 1989-July 1991. Material Research Laboratory; SRI International; Menlo Park, CA.
5. R. Razgaitis, J. H. Payer, S. G. Talbert, B. Hindin, E. L. White, D. W. Locklin, R. A. Cudnik and G. H. Stikford, 'Condensing Heat Exchanger for Residential/Commercial Furnaces Boilers,' Phase IV, Battelle Report to DOE/BNL, BNL Report No. 51943, October 1985.
6. A. R. J. Kucernak, R. Peat and D. E. Williams, J. Electrochem. Soc., **138**, 2337–2340, 1992.
7. J. E. Castle and J. H. Qiu, 'The application of ICP-MS and XPS to study of ions selectivity during passivation of stainless steel', J. Electrochem. Soc., **137**, 7, 2931–2038, 1990.
8. M. Eiden and M, Urquidi-Macdonald, Internal report, ESM. The Pennsylvania State University, University Park, PA December 1991.